Nanostructure Science and Technology

Series editor

David J. Lockwood, FRSC
National Research Council of Canada
Ottawa, Ontario, Canada

More information about this series at http://www.springer.com/series/6331

Fan Zhang

Photon Upconversion Nanomaterials

Springer

Fan Zhang
Department of Chemistry
Fudan University
Shanghai
China

ISSN 1571-5744 ISSN 2197-7976 (electronic)
Nanostructure Science and Technology
ISBN 978-3-662-45596-8 ISBN 978-3-662-45597-5 (eBook)
DOI 10.1007/978-3-662-45597-5

Library of Congress Control Number: 2014955327

Springer Heidelberg New York Dordrecht London

Printed on acid-free paper

Springer-Verlag GmbH Berlin Heidelberg is part of Springer Science+Business Media
(www.springer.com)

Preface

Starting from the late 1990s, the field of photon upconversion nanomaterials has undergone significant expansion and has become one of the most active research areas within the nanoscience community. Exciting developments have been made at a very fast pace by many research groups. The vast number of papers published on upconversion nanomaterials over the past two decades clearly witness it, which has increased exponentially, with most of the activity and development happening in the last 10 years. These kinds of materials can emit ultraviolet/visible/near-infrared light under near-infrared excitation (anti-Stokes emission). This unique optical property precludes background fluorescence and light scattering from biological systems. The emission of multiple and narrow emission lines is an additional hallmark of upconversion nanoparticles that opens up new avenues for optical applications. In the past decades, the related theories, methods and techniques have been explored. As a consequence, novel photon upconversion materials are increasingly emerging, and their applications extend from traditional fields, for example, optical communication amplifiers and solid-state lasers to high-tech fields, including biotechnology, sensors, solar cell and photocatalysis, etc. Researchers can therefore give a deep insight into the synthesis strategies, pathways, and phenomena for photon upconversion nanomaterials, and in particular, establish the relationship for structure—function—synthesis.

This book contains 12 chapters. The Introduction (Chap. 1) involves the general introduction of upconversion luminescence materials, including the energy transfer mechanism, category, and chemical composition of such materials. From the viewpoint of the "wet" chemical synthesis for photon upconversion nanomaterials, Chap. 2 summarizes the most widely used synthetic pathways and the formation mechanism of crystallization (including nucleation and growth) for monodisperse nanocrystals, as well as the corresponding characterizations for the optical, chemical, and structural upconversion of nanoparticles. Provided that these factors and mechanisms can be fully grasped, researchers even beginners can easily obtain high-quality upconversion nanomaterials. In Chap. 3, the optical properties of upconversion lanthanide ions doped nanocrystals including the upconversion emission color tunability, enhancement of luminescence efficiency, core@shell structure-based luminescence engineering, and the optical properties comparison between upconversion nanocrystals and quantum dots and dyes are summarized. In recent years, the development of multifunctional nanomaterials with fantastic physical, chemical, and biological properties has become an attractive research topic. As some of the most important luminescence nanomaterials, upconversion nanomaterials were also used to fabricate multifunctional nanocomposites. In Chap. 4, we summarize recent advanced upconversion nanoparticles-based nanocomposites, including upconversion-mesoporous SiO_2, upconversion-magnetic nanoparticles, upconversion-metal and upconversion-semiconductor nanocomposites, etc. Chapter 5 is devoted to the surface modification and bioconjugation of upconversion nanoparticles. Surface modification of the UCNPs not only improves the photostability of the nanoparticles with desirable interfacial properties, but also provides a potential platform for attaching biological macromolecules for various biomedical applications. In Chaps. 6–11, we deal with the applications of photon upconversion nanomaterials including biomedical imaging, bioassay, biosensor, thermal sensing, light activated therapy, solar cell and photocatalysis, etc. It is apparent that the photon upconversion nanomaterials field is eager for more and more researchers from other fields to explore attractive applications. Finally, the latest progress in photon upconversion nanomaterials are reviewed, and the next stages are put into outlook.

Photon upconversion nanomaterials have been experiencing rapid development in the past decade. A comprehensive review is thus necessary, and is the main purpose of this book. Besides a review, this book also includes an understanding, induction, and summary from the authors. The book is organized along the following guidelines: (1) following the forefront of current research, and striving to reflect the latest progress and developments; (2) comprehensive review with focus on basic fundamental research; and (3) practical research experience in methodology, experiment skills, and data analysis. Particularly, we spent huge efforts on the basic knowledge of photon upconversion nanomaterials. Therefore, this book is especially readable for beginners and graduate students who have just entered this field. We hope that they can, through reading this book, fully understand the chemistry of photon upconversion nanomaterials, grasp synthesis skills, obtain high-quality materials, and therefore, deeply explore the chemical and physical properties of the materials and their applications. Based on the guidelines, most of the chapters were written by Professor Fan Zhang at Fudan University and his students. Dr. Xiaomin Li helped to draft several chapters and dedicated to editing them for publication (Chaps. 1–4), Ph.D. candidates Rui Wang (Chaps. 1, 6, 10), Chi Yao (Chap. 5), Lei Zhou (Chap. 7), Chengli Wang (Chap. 10), and Master candidate Lei Chen (Chap. 8) and Zhenzhen Guo (Chap. 9) also dedicated to editing several chapters. Master candidates Jichuang Mao and Exchange students Vishaal Varahamurthy and Daniel Yur from University of California, Santa Barbara, dedicated to editing the book for publication. Here I express my heartfelt appreciation to them. The authors also thank the financial support from NSFC.

This book is a distillation of the authors' intelligence and great efforts. We wish that this book will help and inspire researchers working in the fields of chemistry and materials science, especially on luminescent materials. We also hope it can provide a reference source or serve as a textbook for undergraduate and graduate students who major in chemistry, chemical engineering, physics, materials and biology, as well as for those readers interested in upconversion luminescent materials. As this book covers relatively wide areas and numerous contents connected to complex scientific issues, errors and omissions may be unavoidable due to the limited knowledge and competence of the authors, therefore we sincerely appreciate the criticism and comments from the readers.

August 2014 Fan Zhang

Contents

Abbreviations

AA	Ascorbic acid
ABMD	9,10-Anthracenediylbis(methylene)dimalonic acid
ADPA	9,10-Anthracenedipropionic acid
AEP	2-Aminoethyl dihydrogen phosphate
AFM	Atomic force microscopy
AFP	A-Fetoprotein
ALD	Atomic layer deposition
APBA	Aminophenylboronic acid
APTES	(3-Aminopropyl)triethoxysilane
ATP	Adenosinetriphosphate
AuNPs	Au nanoparticles
BE	Binding energy
BHMT	Bis(hexamethylene)triamine
BPD-MA	Benzoporphyrin derivative monoacid ring A
BSA	Bovine serum albumin
BTB	Bromothymol blue
C18TMS	Octadecyltrimethoxysilane
Cat D	Cathepsin D
CB	Conduction band
CCD	Charge-coupled device
CCK-8	Cell counting kit-8
CEA	Carcinoembryonic antigen
CLSM/LSCM	Confocal laser scanning microscopy
CNPs	Carbon nanoparticles
CR	Cross relaxation
CSM	Concentric shell model
CSU	Cooperative sensitization upconversion
CT	Computed tomography
CTAB	Cetyltrimethylammonium bromide
CW	Continuous-wave
D2PA	Disk-coupled dots-on-pillar antenna array

DA	Dopamine
DAPI	4′-6-Diamidino-2-phenylindole
DC	Downconversion
DCL	Down luminescence
DDA	1,10-Decanedicarboxylic acid
DEG	Diethylene glycol
DHMA	1,8-Dihydroxy-3-methylanthraquinone
DMF	N,N-Dimethylformamide
DMMA	2,3-Dimethylmaleic anhydride
DMSO	Dimethylsulfoxide
DOPC	1,2-Dioleoyl-sn-glycero-3-phosphocholine
DOX	Doxorubicin
DPBF	1,3-Diphenylisobenzofuran
DSSCs	Dye-sensitized solar cells
EDS/EDX	Energy dispersive x-ray spectroscopy
EDTA	Ethylenediaminetetraacetic acid
EELS	Electron energy-loss spectroscopy
ELO	Epitaxial lift-off
EMCCD	Electron-multiplying charge-coupled device
EMU	Energy migration-mediated upconversion
EPR	Enhanced permeability and retention
EQE	External quantum efficiency
ESA	Excited-state absorption
ESA/GSA	Excited-state absorption
ETU	Energy transfer upconversion
FA	Folic acid
FBS	Fetal bovine serum
FDA	Food and drug administration
FDOT	Fluorescence diffuse optical tomography
FIR	Fluorescence intensity ratio
FITC	Fluorescein isothio-cyanate
FRET	Förster resonance energy transfer
FTO	Fluorine-doped tin oxide
FWHM	Full width at half maximum
GFP	Green fluorescent protein
GNCs	Gold nanoclusters
GNPs	Gold nanoparticles
GO	Graphene oxide
GSA	Ground state absorption
HA	Hyaluronic acid
HAADF	High angle annular dark field imaging
HAC_{60}	Hyaluronated fullerene
hCG	Human chorionic gonadotropin
HEC	Heterogeneous doping core
HET	High energy transfer

HOC	Homogeneous doping core
HP	Hematoporphyrin
HPD	Hematoporphyrin derivative
HPPH	2-(1-Hexyloxyethyl)-2-devinyl pyropheophorbide-a
HRTEM	High-resolution transmission electron microscopy
ICP-MS	Inductively coupled plasma mass spectrometry
IO	Iron oxide
IONPs	Iron oxide nanoparticles
IPET	Interparticle energy transfer
IR	Infrared light
LAT	Light activated therapy
LBL	Layer-by-layer
LCST	Lower critical solution temperature
LFE	Local field enhancement
LMP	Low-melting point
LOD	Limit of detection
LRET	Luminescence resonance energy transfer
LSB	Lauryl sulfonate betaine
LSS	Liquid−solid-solution
LSUCLM	Laser-scanning upconversion luminescence microscopy
LUMO	Lowest unoccupied molecular orbital
M^3D	Maskless mesoscale material deposition
MB	Methylene blue
MC540	Merocyanine 540
MD	Molecular dynamics
MES	2-(N-Morpholino)ethanesulfonic acid
ML	Monolayer
MMP-2	Matrix metalloproteinase-2
MNPs	Magnetic nanoparticles
MPS	Mononuclear phagocyte system
MR	Magnetic resonance
MRI	Magnetic resonance imaging
mTHPC	Meta-tetrahydroxyphenylchlorin
MTS	3-(4,5-Dimethylthiazol-2-yl)-5-(3-carboxymethoxyphenyl)-2-(4-sulfophenyl)-2H-tetrazolium, inner salt
MTT	3-(4,5-Dimethylthiazol-2-yl)-2,5-diphenyltetrazolium bromide
MUA	11-Mercaptoundecanoic acid
NGO	Nanographene oxide
NIR	Near-infrared
NMR	Nuclear magnetic resonance
NPs	Nanoparticles
NVP	*N*-Vinylpyrrolidone
OA	Oleic acid
OAM/OM	Oleylamine
OCMC	*O*-Carboxymethyl chitosan

ODE	1-Octadecene
OM	Oleylamine
OQLCS	Octadecyl-quaternized lysine-modified chitosan
OTA	Ochratoxin A
PA	Photon avalanche
PAA	Poly(acrylicacid)
PAH	Poly(allylamine hydrochloride)
PAMAM G1	Polyamidoamine generation 1
PBS	Phosphate-buffered saline
PCR	Polymerase chain reaction
PDT	Photodynamic therapy
PEG	Polyethylene glycol
PEGDA	Poly(ethylene glycol) diacrylate
PEI	Polyethyleneimine
PET	Positron emission tomography
PL	Photoluminescence
PLH	Poly-L-histidine
PMAO	Poly(maleic anhydride-alt-1-octadecene)
PMMA	Poly(methyl methacrylate)
PMPD	Poly-m-phenylenediamine
PNIPAM	Poly(N-isopropylacrylamide)
PS	Photosensitizer
PSA	Polysialic acid
PSPI	Trans-4-[p-(pyrrolidinyl)styryl]-N-methylpyridinium iodide
PSS	Poly(sodium styrene sulfonate)
PTA	Photothermal ablation
PTT	Photothermal therapy
PVP	Polyvinylpyrrolidone
QDs	Quantum dots
QDSCs	Quantum dots solar cell
QRs	Quantum rods
QY	Quantum yield
REDC	Rare earth-based downconversion
RES	Reticuloendothelial system
RET	Resonance energy transfer
RIE	Reactive ion etching
ROS	Reactive oxygen species
RT	Radiotherapy
SA	Succinic anhydride
SAED	Selected area electron diffraction
SBR	Signal-to-background ratio
SDBS	Sodium dodecyl benzenesulfonate
SGM	Spherical grating monochromator
SHG	Second harmonic generation
SLBL	Successive layer-by-layer

SNPs	Sacrificial nanoparticles
SNR	Signal-to-noise ratio
SPCD	Silicon phthalocyanine dihydroxide
SPECT	Single-photon emission computed tomography
SPION	Superparamagnetic iron oxide nanoparticles
SPM	Superparamagnetic
SPR	Surface plasmon resonance
SQ	Squaraine
STEM	Scanning transmission electron microscopy
STPA	Simultaneous two-photon absorption
SWIR	Short-wavelength infrared
SWNTs	Single-walled carbon nanotubes
TBAs	Thrombin aptamers
TEM	Transmission electron microscopy
TEOS	Tetraethyl orthosilicate
TESPIC	(3-Isocyanatopropyl) triethoxysilane
TGA	Thioglycolic acid
TM	Transition metal
TOP	Trioctylphosphine
TOPO	Trioctylphosphine oxide
TPA	Two-photon absorption
TPP	Tetraphenyl-porphyrin
UC	Upconversion
UCL	Upconversion luminescence
UCNC-B	Blue-emissive upconversion nanocapsule
UCNC-G	Green-emissive upconversion nanocapsule
UCNC-Y	Yellow-emissive upconversion nanocapsule
UCNPs	Upconversion nanoparticles
uPAR	Urokinase plasminogen activator receptor
UV	Ultraviolet
VB	Valence band
XPS	X-ray photoelectron spectroscopy
XRD	X-ray diffraction
ZnPc	Zinc(II) phthalocyanine

Chapter 1
General Introduction to Upconversion Luminescence Materials

Abstract Upconversion nanoparticles can emit ultraviolet/visible/near-infrared light under near-infrared excitation (anti-Stokes emission). This unique optical property precludes background fluorescence and light scattering from biological materials. The emission of multiple and narrow emission lines is an additional hallmark of upconversion nanoparticles that opens up new avenues for optical applications. The unique optical properties make the upconversion nanoparticles ideal for bio-imaging applications with attractive advantages such as no auto-fluorescence from bio-tissues and a large penetration depth. In this chapter, we give a general introduction to the upconversion luminescence materials from the aspects of energy transfer mechanism, category, chemical composition, nanosized upconversion materials, and so on.

Keywords Upconversion nanoparticles · Lanthanide luminescence · Stokes luminescent materials · Anti-Stokes emission · Energy transfer mechanism

Luminescent materials play an important role in many technologies, including display screens, optical communication amplifiers, lamps, and solid-state lasers. To this end, a tremendous volume of fundamental research has been devoted to the design and characterization of new luminescent materials displaying novel photophysical effects. In accordance with the different relative positions between the excitation band and emission band, the well-developed luminescence materials can be divided into two groups, Stokes luminescence materials and anti-Stokes luminescence materials.

For most of luminescent materials (e.g, organic dyes and quantum dots), the emission spectrum always locates at the longer wavelength site compared with the excitation spectrum. That is to say, the energy of the excitation photon is higher than the emission photon. This rule is known as Stokes' law, and this energy loss was called Stokes' shift. In most circumstances, the photoluminescent materials follow the Stokes' law. Most of the organic dyes and quantum dots are typical Stokes luminescence materials. For organic dyes, the photoluminescence mechanism can be

Rui Wang, Xiaomin Li, and Fan Zhang contributed together to this chapter.

F. Zhang, *Photon Upconversion Nanomaterials*,
Nanostructure Science and Technology, DOI 10.1007/978-3-662-45597-5_1

explained by the Jablonski energy diagram (Fig. 1.1a) [1, 2]. The multi-color emission of organic dyes can be accomplished by controlling organic fluorophores. Conventional dye molecules impose stringent requirements on optical systems used to make luminescence measurements [3]. Their narrow excitation spectra make simultaneous excitation difficult in most cases. Furthermore, the broad emission spectrum with a long tail at red wavelengths introduces spectral cross-talk between different detection channels, making quantitation of the relative amounts of different probes difficult. The photoluminescence of quantum dots also abide by the Stokes' law [4]. The difference with the organic dyes is that the width of band gap is fixed, which is determined by its identity. However, the situation changes in the case of nanoscale semiconductor particles with small sizes (normally less than ~10 nm) because of the regime of quantum confinement (Fig. 1.1b) [5, 6]. It means that electronic excitations "feel" the presence of the particle boundaries and respond to changes in the particle size by adjusting their energy spectra. This phenomenon is known as the quantum size effect, and the nanoscale particles that exhibit the effect are often referred as quantum dots. As the size of quantum dots decreases, the spacing between the valence band and conduction bands (i.e., the band gap) increases, leading

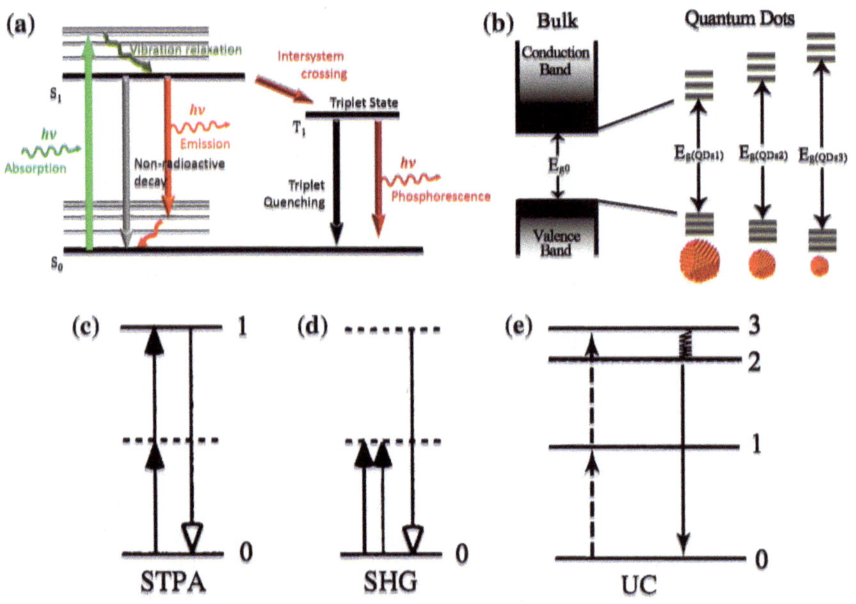

Fig. 1.1 **a** Jablonski energy diagram, showing the excitation and energy dissipation way of the organic dye. **b** A bulk-state semiconductor with continuous conduction and valence energy bands separated by a fixed energy gap, E_{g0} (*left*), while a QD is characterized by discrete atomic-like states with energies that are determined by the radius (*right*). **c–e** Schematic representation of three common nonlinear processes that convert low-energy photon pump sources into higher energy output: *SHG* Second-harmonic generation, *STPA* Simultaneous two-photon absorption, *UC* Up-conversion. **c–e** Reprinted with the permission from Ref. [8]. Copyright 2000 American Chemical Society

to a blueshift of the emission wavelength. The quantum dots can be excited efficiently at any wavelength shorter than the emission peak, and the excitation wavelength has no influence on the characteristic narrow, symmetric spectrum properties of the emission [7].

In some special cases, the emitted photon has a higher energy (shorter wavelength) than the absorbed one, and this is called anti-Stokes shift emission. It has been shown that the anti-Stokes process often occurs in three ways: simultaneous two-photon absorption (STPA), second-harmonic generation (SHG), and upconversion [8, 9]. STPA could be referred to as the process in which two long-wavelength photons are absorbed simultaneously to excite two-photon absorption molecules to the excited states. SHG could happen with highly hyperpolarizable materials under coherent radiation, in which the emission has half the wavelength and twice the frequency of the incident light. In 1961, SHG was first discovered by Franken et al. [10], who observed second-harmonic light with half the wavelength when ruby laser light was getting through a quartz crystal. Similar to STPA and SHG, upconversion is also an anti-Stokes process, but it is more efficient with no requirement of coherent or high-intensity radiation. There are several metastable energy states participating in a typical upconversion process, such as metastable states 1, 2, and 3 in Fig. 1.1e. Upon absorption of a low-energy photon, the ground state is populated to metastable state 1, which is then promoted to a higher energy state, metastable state 3, by subsequently absorbing another low-energy photon. Following some internal relaxation processes, metastable state 3 relaxes to a slightly lower excited energy state, metastable state 2, and then jumps to the ground state by releasing a photon radiation. This upconversion process exhibits higher efficiency than STPA, in which virtual intermediate energy levels are used for accommodating the absorbed photons.

In this book, we discuss several specific topics of current interest in the area of lanthanide-based upconversion materials. Solid-state lanthanide luminescent materials have the combined advantages of typical lanthanide luminescence and photophysical stability. In these materials, lanthanide ions are trapped into the rigid crystal host lattice, avoiding adverse quenching by environmental oscillators. Two types of luminescence modes, downconversion and upconversion, are known for lanthanide luminescence (Fig. 1.2). Upconversion relates to a nonlinear optical process where two or more near-infrared photons are absorbed sequentially to several real intermediate energy states and luminescence is emitted with a wavelength shorter than the absorbed light. Since 1960s, when Auzel [11], Ovsyankin, and Feofilov [12] independently discovered and studied this phenomenon for the first time, the upconversion process has aroused enormous research interests. Nowadays, upconversion materials are widely used in optical devices such as temperature sensors, infrared quantum counters, and compact solid-state lasers [13]. However, bulk glass or crystalline upconversion materials have remained the major research focus, and no biological applications have been reported during the first 30 years of research on upconversion. Stimulated by the prevalence of nanotechnology since the late 1990s, upconversion nanoparticles have been manufactured and have attracted enormous attention in biological sciences [7, 13–18]. In this chapter, we will give a general introduction to

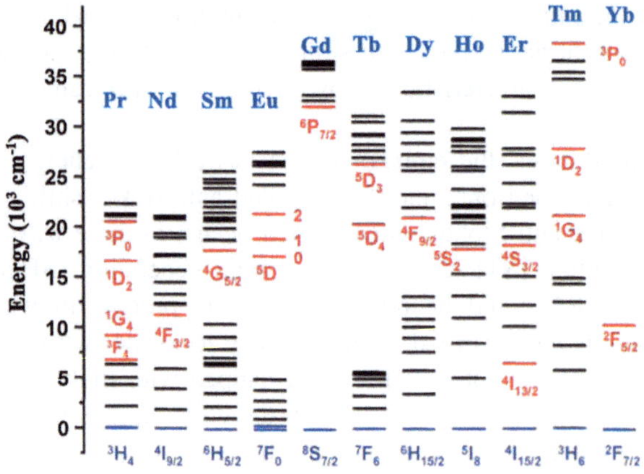

Fig. 1.2 Partial $4f^n$ energy diagrams for lanthanide ions. The main luminescent levels are drawn in *red*, while the ground states are indicated in *blue*. Reprinted with the permission from Ref. [56, 57]. Copyright 2005 and 2013 Royal Society of Chemistry

the upconversion luminescence materials from the aspects of energy transfer mechanism, category, chemical composition, nanosized upconversion materials, and so on.

1.1 Energy Transfer Mechanism of the Upconversion Materials

The upconversion processes are in reality more complicated than that illustrated in Fig. 1.1c. Several mechanisms are involved, such as excited-state absorption (ESA), energy transfer upconversion (ETU), cooperative sensitization upconversion (CSU), cross-relaxation (CR), and photon avalanche (PA), and these are illustrated in Fig. 1.3 [19].

Excited-state Absorption. In the case of ESA, excitation takes the form of successive absorption of pump photons by a single ground-state ion. The general energy diagram of the ESA process is shown in Fig. 1.3a for a simple three-level system [19]. When an ion is excited from the ground state to the E1 level, a second pump photon that promotes the ion from E1 to higher-lying state E2 results in upconversion emission, before it decays to the ground state. Consequently, upconverted emission will occur from the E2 level. To achieve highly efficient ESA, a ladder-like arrangement of the energy states of ions is required. Only a few lanthanide ions such as Er^{3+}, Ho^{3+}, Tm^{3+}, and Nd^{3+} have such energy level structures [11], which also find a good excitation wavelength match with the output of commercially available diode lasers (at ~975 and/or 808 nm) and are currently the most common emitters (activators) in ESA-based upconversion phosphors.

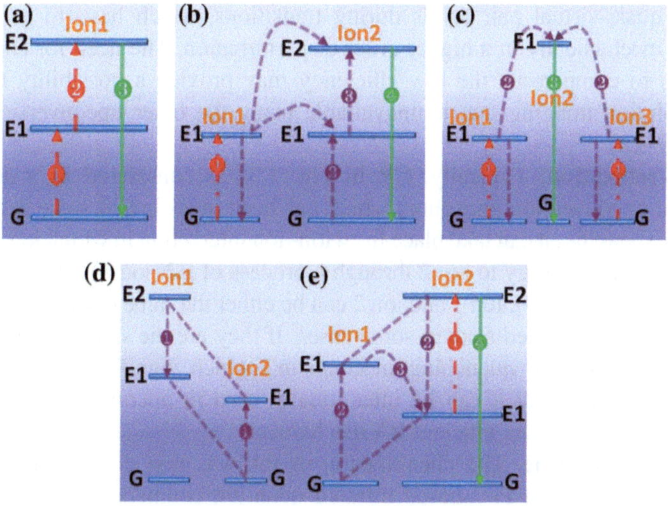

Fig. 1.3 Principal upconversion processes for lanthanide-doped upconversion nanoparticles: **a** excited-state absorption (ESA), **b** energy transfer upconversion (ETU), **c** cooperative sensitization upconversion (CSU), **d** cross-relaxation (CR), and **e** photon avalanche (PA). The *red*, *violet*, and *green lines* represent photon excitation, energy transfer, and emission processes, respectively. **a–e** Reprinted with the permission from Ref. [19]. Copyright 2014 American Chemical Society

Energy Transfer Upconversion. ETU is similar to ESA in that both upconversion processes utilize sequential absorption of two photons to populate the metastable level. The essential difference between ETU and ESA is that ESA is operated within a single lanthanide ion, while ETU involves two neighboring ions [19]. In an ETU process, an ion 1 known as the sensitizer is first excited from the ground state to its metastable level E1 after absorbing a pump photon; then ion 1 transfers its harvested energy to the ground state G and the excited state E1 of ion 2, known as the activator, exciting ion 2 to its upper emitting state E2, while sensitizer ion 1 relaxes back to ground state G (Fig. 1.3b). The dopant concentration that determines the average distance between the neighboring dopant ions has a strong influence on the UC efficiency of an ETU upconversion process. In contrast, in an ESA process, the upconversion efficiency is independent of the dopant concentration due to its single ion characteristic.

Cooperative Sensitization Upconversion. CSU is a process involving the interaction among three ion centers, and ion 1 and ion 3 generally can be considered as sensitizers in ETU process (Fig. 1.3c) [19]. After irradiation with excitation photons, both ion 1 and ion 3 can be excited to the excited state E1, respectively. Then ion 1 and ion 3 will interact with ion 2 (activator in ETU process) simultaneously and excite ion 2 to a higher state E1. The excited ion 2 can relax back to its ground state along with emitting an upconverted photon. The efficiency of CSU is generally orders of magnitude lower than that of ESA or ETU process, because it involves quasi-virtual pair levels during transitions, which have to be described quantum mechanically in a higher order of perturbation. The need for confining an excitation to compensate the low efficiency may provide a possibility to achieve high-resolution imaging that is unavailable using the other upconversion mechanisms [19–22].

Cross-relaxation. Typically, CR in Fig. 1.3d is considered as a detrimental process in upconversion phosphors which has been investigated since decades ago. Generally speaking, it can take place from ion–ion interaction in which ion 1 transfers part of its excited energy to ion 2 through a process of E2 (ion 1) + G (ion 2) → E1 (ion 1) + E1 (ion 2) [19]. Ion 1 and ion 2 can be either the same or different, and ion 2 can also be in its excited state in some cases. If they are the same, CR process will arouse "concentration quenching mechanism" which significantly quenches the emission intensity. But it can be intentionally used in upconversion color-tuning materials or constructing efficient PA mechanism.

Photon Avalanche. The phenomenon of PA was first discovered by Chivian and coworkers [23] in Pr^{3+}-based infrared quantum counters. PA-induced upconversion features an unusual pump mechanism that requires a pump intensity above a certain threshold value (Fig. 1.3e) [19]. When the pump intensity is below the threshold, very little upconverted luminescence is produced, while the photoluminescence intensity increases by orders of magnitude when it is above the pump threshold. Actually, the PA is a looping process that involves processes of ESA for excitation light and an efficient CR that produces feedback. The PA process is initiated with population of level E1 of ion 2 by non-resonant weak ground-state absorption, followed by resonant ESA process to populate upper visible-emitting level E2. Then, an efficient CR process of E2 (ion 2) + G (ion 1) → E1 (ion 2) + E1 (ion 1) between ion 1 and ion 2 occurs, resulting in both ions occupying the intermediate level E1. The two ions readily populate level E2 to further initiate cross-relaxation and exponentially increase level E2 population by ESA, producing strong UC emission as an avalanche process. In addition to high pump intensity, PA-induced upconversion also requires long response time to excitation source (up to several seconds) [19, 23].

1.2 The Category of the Upconversion Materials

A typical upconversion material is composed by two components, an inorganic host matrix and doping ions (emission centers). The principle strategy for obtaining new upconversion material involves two steps: variation of the host lattice and variation of the dopant ions in the lattice. Changing the host lattice may dramatically influence the radiative and non-radiative (multi-phonon relaxation as well as energy transfer) properties, leading to entirely different upconversion luminescence behavior. For example, one can reduce the efficiency of multi-phonon relaxation processes by changing from a lattice having high phonon energies (oxide and fluoride) to one having low phonon energies (chloride, bromide, and iodide), with the drawback that most of these compounds are hygroscopic. Choosing a host with specific optical and/or magnetic properties may also influence the upconversion emission properties of a dopant ion through sensitization or perturbation by exchange interactions. A change in the metal dopant ions has a dramatic effect on the emission properties of a compound. The most obvious effect is a change in the color of the emitted light. The freedom of choice for the host lattice, dopant, and (possibly) codopant covers a very broad spectrum of possibilities in the development of compounds with new and unprecedented upconversion properties, leaving much to the imagination and creativity of the researcher. This section will focus on rare earth upconverters, transition metal (TM) upconverters, mixed rare earth/ transition metal systems, and the upconversion nanoparticles, respectively [24].

1.2.1 Lanthanide-based Upconverters

Most of the well-known upconverting phosphors contain trivalent rare earth ions as the active components, typically Pr^{3+}, Nd^{3+}, Er^{3+}, Tm^{3+}, or Yb^{3+}, though others occur too [15]. The reason why the vast majority of all the work in the field has been devoted to this relatively small collection of ions lies in the fact that for these ions, their inner 4*f* shell contains the spectroscopically relevant electronic states. As their outer 5*s* and 6*p* shells are involved with the bonding, the f states are rather insensitive to the precise nature of the surrounding host lattice (i.e., the exact crystal field and, to a lesser extent, the local site symmetry). As a result, a significant number of the excited *f* states have lifetimes that are long enough to allow for upconversion processes to play a role [7]. Not surprisingly, rare earth ions were also the first ions in which upconversion was observed, discovered by Auzel [11] and independently by Ovsyankin and Feofilov [12] in the mid-1960s.

Inorganic crystals in most cases do not exhibit upconversion luminescence at room temperature, typically, the luminescent center is provided by the dopants in it. Therefore, a good lanthanide-based upconverter has two especially important factors: (1) highly stable host material as protector and (2) efficient dopant ion as luminescent center. In the case of the sensitized luminescence, the dopant ion radiates upon its excitation to a higher energetic state obtained from the non-radiative transfer of the energy from another dopant ion. The ion that emits the radiation is called an activator, while the donator of the energy is the sensitizer. Although upconversion luminescence can be observed in many lanthanide activator–sensitizer combination systems, and most of crystalline materials can be explored as host materials, efficient upconversion only occurs by using a small number of well-selected dopant–host combinations [17].

1.2.1.1 Activators

The upconversion process requires multiple metastable levels which can be provided by lanthanide ions. The lanthanide ions, which are associated with the filling of the 4*f* shell, commence with the lanthanum (La) and end with the lutetium (Lu). The completed filled $5s^2$ and $5p^6$ subshells can shield the 4*f* electrons of Ln^{3+}, resulting in weak electron–phonon coupling which is responsible for its important optical phenomena such as sharp and narrow *f*–*f* transition bands. In addition, the *f*–*f* transitions are Laporte-forbidden, resulting in low transition probabilities and substantially long-lived (up to 0.1 s) excited states. With the exception of La^{3+}, Ce^{3+}, Yb^{3+}, and Lu^{3+}, the lanthanide ions commonly have more than one excited 4*f* energy level [11, 17].

Owing to the above-mentioned advantages, lanthanide ions hold a great promise in upconversion luminescence. Indeed, upconversion emission can be theoretically expected for most lanthanide ions [17]. However, to generate practically useful upconversion emission, the energy difference between each excited level and its lower-lying intermediate level (ground level) should be close enough to facilitate photon absorption and energy transfer steps involved in upconversion processes. Therefore, many lanthanide ions are not suitable for this type of excitation because the lack of appropriate energy level structure. Er^{3+}, Tm^{3+}, and Ho^{3+} are currently the most common emitters (activators) in upconversion phosphors due to their ladder-like arranged energy levels (Fig. 1.4). The upconversion efficiency of Er^{3+} is particularly high due to the similar energy gap in different energy levels. For example, the energy difference in Er^{3+} ($\sim 10{,}350\ cm^{-1}$) between the $^4I_{11/2}$ and $^4I_{15/2}$ levels is similar to that ($\sim 10{,}370\ cm^{-1}$) between the $^4F_{7/2}$ and $^4I_{11/2}$ levels. Thus, the energy levels of $^4I_{15/2}$, $^4I_{11/2}$, and $^4F_{7/2}$ can be used to generate upconversion emission using $\sim$980-nm excitation. In addition, the energy difference between $^4F_{9/2}$ and $^4I_{13/2}$ states is in the same region, and hence, at least three different transitions in Er^{3+} ions are induced by $\sim$980-nm excitation, thus leading to emission of green and red light after the sequential absorption of two photons [11, 13].

Since 4*f*–4*f* transitions are Laporte-forbidden, the upconversion luminescence efficiency suffers from inefficient absorption of the exciting light, especially in the case of thin samples of lanthanide-doped materials. Generally speaking, the absorption can be improved by increasing the doping concentration of the lanthanide ions in the material. But non-radiative multi-phonon relaxation can occur, and the process of cross-relaxation severely limits the range of useful dopant concentrations. Non-radiative multi-phonon relaxation rate between energy levels is another important factor that dictates the population of intermediate and emitting levels and subsequently determines the efficiency of the upconversion process. The multi-phonon relaxation rate constant k_{nr} for 4*f* levels of lanthanide ions is described as [17, 25]

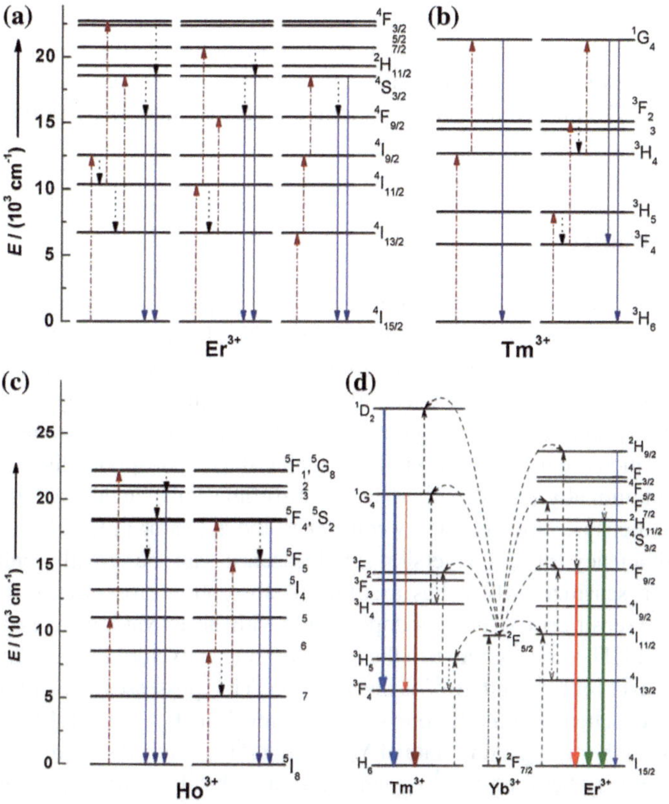

Fig. 1.4 Schematic energy level diagrams showing typical upconversion processes for Er^{3+} (**a**), Tm^{3+} (**b**), and Ho^{3+} (**c**). The *dashed-dotted*, *dotted*, and *full arrows* represent excitation, multi-phonon relaxation, and emission processes, respectively. The excitation originates from either direct photo excitation or energy transfer. Since energy transfer can occur with the assistance of phonons, the energy differences between each key excited level and its key lower-lying level can be a little inconsistent. **d** Proposed energy transfer mechanisms showing the upconversion processes in Er^{3+}, Tm^{3+}, and Yb^{3+}-doped crystals under 980-nm diode laser excitation. The *dashed-dotted, dashed, dotted,* and *full arrows* represent photon excitation, energy transfer, multi-phonon relaxation, and emission processes, respectively. Only visible and NIR emissions are shown here. **a–d** Reprinted with the permission from Ref. [17]. Copyright 2009 Royal Society of Chemistry

$$k_{\mathrm{nr}} \propto \exp\left(-\beta \frac{\Delta E}{\hbar\omega_{\max}}\right) \tag{1.1}$$

where β is an empirical constant of the host, ΔE is the energy gap between the populated level and the next lower-lying energy level of a lanthanide ion, and $\hbar\omega_{\max}$ is the highest energy vibrational mode of the host lattice. The energy gap law implies that the multi-phonon relaxation rate constant decreases exponentially with increasing energy gap. As shown in Fig. 1.4a–c, Er^{3+} and Tm^{3+} have relatively large energy gaps and thus low probabilities of non-radiative transitions among various excited levels of the ions. In agreement with the energy gap law, the most efficient UCNPs known to date are obtained with Er^{3+} and Tm^{3+} as the activators [17, 26–39].

1.2.1.2 Sensitizer

In the case of doped UCNPs, the range of useful dopant concentrations is severely limited due to the cross-relaxation. High doping levels can lead to the quenching of excitation energy. Thus, the concentration of activator ions should be kept low and precisely adjusted to avoid the quenching effect. But, the low doping levels can lead to low absorption of the pump light, resulting in low emission luminescence efficiency. To increase the absorption of lanthanide-doped phosphors, the host material is often additionally doped with strongly absorbing ions called sensitizers, which should also ensure efficient energy transfer to the activator [13, 17].

The efficiency of upconversion luminescence can be greatly enhanced by codoping a sensitizer which has a sufficient absorption cross-section in the NIR region. The sensitizer is usually codoped along with the activator to take advantage of the efficient ETU process between the sensitizer and activator [17]. The most widely used sensitizer for Er^{3+} and Tm^{3+} is the Yb^{3+} ion. Yb^{3+} possesses an extremely simple energy level scheme with only one excited 4*f* level of $^2F_{5/2}$ (Fig. 1.4d). The energy separation of the $^2F_{7/2}$ ground state of Yb^{3+} and its $^2F_{5/2}$ excited state matches well the transition energy between the $^4I_{11/2}$ and $^4I_{15/2}$ and the $^4F_{7/2}$ and $^4I_{11/2}$ states of Er^{3+}, thus allowing for efficient energy transfer between the two ions. The similar principle can be used in Tm^{3+} and Ho^{3+}. These optical characteristics make Yb^{3+} particularly suitable for use as an upconversion sensitizer. Usually, Yb^{3+} is codoped into the lattice in high concentrations (18–20 mol%), while the activator concentration is relatively low (<2 mol%), minimizing cross-relaxation energy loss [13, 17].

1.2.1.3 Host Materials

The properties of the host lattice and its interaction with the dopant ions have a strong influence on the upconversion process, because the lattice of host materials determines the distance between the dopant ions, their relative spatial position, coordination numbers, and the type of anions surrounding the dopant. Therefore, selection of appropriate host materials is also essential in the design of UCNPs with favorable optical properties such as high efficiency and tunable emission color. There are two important factors which are generally required for the host materials: low phonon energy and small lattice mismatch to the dopant ions [13, 17].

Ideal host materials should have low lattice phonon energies, which have a great influence on the upconversion efficiency. In order to minimize non-radiative loss and maximize the radiative emission [17], the selection of host material composition is discussed in many reported reviews [13, 17]. Typically, heavy halides such as chlorides, bromides, and iodides generally exhibit low phonon energies of less than 300 cm^{-1}. However, they are hygroscopic and are of limited use. Oxides exhibit high chemical stability, but their phonon energies are relatively high and generally larger than 500 cm^{-1} due to the stretching vibration of the host lattice. In comparison, fluorides usually exhibit low phonon energies ($\sim$350 cm^{-1}) and high chemical stability and thus are often used as the host materials for upconversion [13, 17]. Besides above-mentioned host materials, semiconductor nanocrystals such as ZnS have also been proposed as upconversion host materials. However, it is still debatable whether lanthanide ions in these nanocrystals are incorporated homogeneously in the host lattice or located on the outermost layer of the nanocrystals due to largely mismatched ionic size between the host and dopant ions [17, 40].

Trivalent lanthanide ions-based inorganic compounds are ideal host materials for upconversion phosphors because all trivalent lanthanide ions exhibit similar ionic size and chemical properties. In addition, host lattices based on cations such as Na^+, Ca^{2+}, Sr^{2+}, and Ba^{2+} with ionic radii close to those of the lanthanide dopant ions prevent the formation of crystal defects and lattice stress, and therefore, generally Na^+ and Ca^{2+} fluorides are superior host materials for upconversion phosphors. To sum up, the best host materials for upconversion are Na^+-, Ca^{2+}-, and Y^{3+}-based fluorides. For instance, the upconversion efficiency of $NaYF_4$:Yb^{3+},Er^{3+} is 20 times higher than that of La_2O_3:Yb^{3+},Er^{3+} and six times higher than that of $La_2(MoO_4)_3$: Yb^{3+},Er^{3+} [13].

It is also known that the optical properties of upconversion materials can be significantly influenced by the crystal structure of the host materials. To date, hexagonal $NaYF_4$ (β-$NaYF_4$) is the most efficient host material for green and blue upconversion phosphors among the fluoride hosts. The upconversion efficiency of hexagonal-phase $NaYF_4$:Yb/Er bulk materials is approximately 10 times stronger than that in cubic phase counterparts [41]. The phase-dependent optical property can be ascribed directly to the different crystal fields around trivalent lanthanide ions in matrices of various symmetries. Low-symmetry hosts typically exert a crystal field containing more uneven components around the dopant ions compared to high-symmetry counterparts. The uneven components enhance the electronic coupling between 4*f* energy levels and higher electronic configuration and subsequently increase *f*–*f* transition probabilities of the dopant ions. In addition, the upconversion efficiency can also be increased by the decrease in cation size (or unit-cell volume) of the host which leads to the increase in the crystal field strength around the dopant ions. For example, the bulk material of $NaYF_4$:Yb/Er exhibits a upconversion luminescence two times stronger than that of $NaLaF_4$:Yb/Er [17].

1.2.1.4 The Nonlinear Nature and Efficiency of Upconversion Luminescence

As mentioned above, the upconversion luminescence arises from the *4f–4f* orbital electronic transitions with concomitant wave functions localized within a single lanthanide ion. These intra *4f* electronic transitions of lanthanide ions are electric dipoles forbidden by quantum mechanical selection rules, which, however, are relaxed due to local crystal field-induced intermixing of the f states with higher electronic configurations [42, 43]. The very long lifetimes (up to tens of millisecond) for energy levels of lanthanide ions can be attributed to the forbidden nature of the *4f–4f* transition, thus favoring the sequential excitations in the excited states of a single lanthanide ion as well as permitting energy transfers between two or more lanthanide ions. These resulting features of lanthanide dopants determine the basic upconversion mechanisms (see Sect. 1.1). The upconversion photoluminescence intensity generally has a nonlinear dependence on the excitation light density [19, 44–46]:

$$I_{\mathrm{UC}} = KP^{n} \tag{1.2}$$

I_{UC} is the photoluminescence intensity, P is the power of pump laser, K is material-related coefficient, and n is the number of the excitation photons required to produce the upconversion photoluminescence. The log–log relation of Eq. 1.2 is very useful to determine the "*n*" values for the upconversion photoluminescence peaks at very low excitation density [44–46]. However, it should be noted that "*n*" values can be affected by the competition process between "the decay rate" and "the upconverted rate" at the intermediate states, leading to smaller "*n*" values (not integral numbers) at high excitation density. This phenomenon is termed the "saturation effect," which needs to be carefully avoided when determining the photon processes of upconversion photoluminescence peaks [19]. The upconversion quantum yield (UCQY) or the efficiency of a upconversion process is strongly dependent on the excitation density due to the nonlinear nature of upconversion process. UCQY is defined as the ratio of the number of the emitted upconverted photons to the number of the absorbed NIR photons, according to the following formula [19]:

$$\begin{aligned}\mathrm{UCQY} &= \frac{\text{photons emitted}}{\text{photons absorbed}} \\ &\propto \frac{\text{emitted UC light}}{\text{absorbed excitation light}} \\ &= \frac{I_{\mathrm{UC}}}{\alpha P}\end{aligned} \tag{1.3}$$

where α is the absorption coefficient of the host material at the excitation wavelength. By inserting Eq. 1.2 into Eq. 1.3, it can be easily obtained that:

$$\mathrm{UCQY} \propto P^{n-1} \tag{1.4}$$

According to Eq. (1.4), we can conclude that UCQY will significantly depend on the excitation power density. Therefore, the UCQY for a specific upconversion material at one particular excitation density can be easily measured because of the dependence between UCQY and excitation power density [19].

1.2.2 Transition Metal Upconverters

Since the spectroscopically active *d* orbitals in transition metals are the valence orbitals, any environmental perturbation is expected to be more influential in these ions than in the more commonly studied trivalent rare earth metal ions [24]. As discussed in the previous section, the spectroscopically active *f* orbitals are strongly shielded from their environment, resulting in nearly host lattice-insensitive energy level positions for the spectroscopic states of the rare earth ions. However, it is the accessibility of the frontier *d* orbitals that make TM ions extremely attractive in the search for new upconversion materials. The reason for the attraction lies in the fact that the environmental sensitivity of TM ions suggests mechanisms by which their photophysical properties may be modified or even controlled using external chemical, redox, or magnetic perturbations. Furthermore, such *d* metals have spectroscopically active states that are strongly influenced by the local crystal field, allowing for chemical variation of the photophysical characteristics such as the energy levels, excited-state dynamics and lifetimes, multi-phonon relaxation rates, and electron–phonon interaction strengths [24].

Recent developments indicate that several TM ions are also suitable for multi-photon excitation, and sometimes show properties different from those of the lanthanide ions, including the increased likelihood of encountering broad ligand-field-dependent excited-state features and strict spin selection rules [8, 47]. $Ni^{2+}(3d^8)$, long studied for its laser potential in a variety of host lattices, was observed early on to show upper excited-state emission due to ESA processes. This ESA is detrimental to the objective of laser action in such crystals [48]. Ti^{2+} $(3d^2)$ upconversion was also observed several years ago. More recently, a systematic effort to discover new transition metal upconversion systems, as well as to explore the properties of these two old systems, has been initiated. The three examples of Os^{4+} $(5d^4)$ [49–51], Mo^{3+} $(4d^3)$ [52, 53], and Re^{4+} $(5d^3)$ [53, 54] have now been reported. In all of these systems, the same general building blocks as in rare earth-doped systems are observed. The mechanisms of each are summarized in Table 1.1, along with key physical parameters that relate to their photophysical properties.

Table 1.1 Transition metal ions in doped halide lattices for which upconversion luminescence has been demonstrated, including relevant mechanistic and electronic structural information

Ions	Electrons	Host	UC mechanism	Phonon energy (cm^{-1})
Ti^{2+}	$3d^2$	$MgCl_2$	GSA/ESA	100
Ni^{2+}	$3d^8$	$CsCdCl_3$	GSA/ESA, avalanche	620
Mo^{3+}	$4d^3$	Cs_2NaYCl_6	GSA/ESA	600
Re^{4+}	$5d^3$	Cs_2ZrCl_6	ETU	2,300
Os^{4+}	$5d^4$	Cs_2ZrBr_6	GSA/ETU, GSA/ESA, avalanche	2,600

Reprinted with the permission from Ref. [8]. Copyright 2000 American Chemical Society

1.2.3 Mixed Rare Earth/Transition Metal Upconverters

The combination of trivalent rare earth (Ln^{3+}) and TM ions in the same host lattice introduces new cooperative upconversion mechanisms involving both ions [24]. Mixed Ln^{3+}/TM compounds are particularly interesting, as the absorption and luminescence properties of TM ions are susceptible to changes in the ligand environment and the coordination geometry. In contrast, the Ln^{3+} *f–f* transitions are almost independent of the local environment. Thus, the choice of the host lattice allows one (to some extent) to selectively optimize the energetic positions of the absorption and luminescence of the TM ion. Furthermore, it paves the way toward upconversion materials where sharp line excitation in the NIR leads to broadband emission in the visible spectral region and vice versa. Such features are potentially interesting for new visible laser sources as well as for lighting and display devices. Two fundamentally different upconversion schemes in mixed Ln^{3+}/TM compounds are possible: lanthanide-sensitized transition metal upconversion luminescence or transition metal-sensitized lanthanide upconversion luminescence.

1.2.4 Organic Upconverters

In contrast to the inorganic and ionic phosphors that have been discussed so far, there is not a large variety of mechanisms available when one wishes to investigate the upconversion properties of organic molecules. Because almost all excited state to ground state transitions are considerably more allowed in organic phosphors than in inorganic ones, the associated lifetimes of the excited states in organic phosphors are extremely short (typically in the nanosecond range). This very fast decay time means that an organic upconverter cannot be long enough in its excited state to allow for ETU [24]. As a result, the few organic upconversion phosphors that have been identified in the literatures all require the excited-state absorption mechanism to populate the emitting state.

One of the best documented organic upconversion phosphors is the organic dye, trans-4-[p-(pyrrolidinyl)styryl]-N-methylpyridinium iodide (abbreviated as PSPI) [55]. This dye is capable of normal (downconversion) laser action and has a significant two-photon absorption cross-section (under 930-nm excitation). The absorption spectrum shows a strong absorption band with a peak wavelength of 482 nm, corresponding to the first excited singlet state S1. Another absorption band at wavelengths shorter than 350 nm corresponds to the second excited singlet state S2. As the lifetime of the S1 state is $\sim$100 ps, there can be enough population in this state for excited-state absorption via the allowed S1 to S2 transition, after which emission from the S2 state is observed to the ground state. It remains doubtful if one can achieve upconversion laser action on the S2 to S0 transition, but in principle, this may be possible.

1.2.5 Lanthanide-doped Upconversion Nanomaterials

The research in the field of photon upconversion has been devoted almost exclusively to bulk materials in the solid state. Therefore, the idea of inducing upconversion in a nanocrystalline material (perhaps even while suspended in the fluid state) is interesting by itself. If it is possible to prepare sufficiently small and efficient upconversion units, one could use them as new luminescent markers for the detection of bio-molecules or as security labels. Compared to conventional luminescent markers, which are excited in the ultraviolet or blue spectral region, excitation in the near-infrared obviously has the great advantage that the emission in which one is interested is spectrally well separated from the normal downconversion of the material. As a result, the commonly observed auto-fluorescence that presents a large background in biological systems is not present when using nanosized upconversion marker. In addition, photodegradation is strongly reduced, and many materials such as biological tissue are much more transparent to near-infrared light than to ultraviolet or blue wavelengths.

As reviewed in several papers and books [19], lanthanide-doped upconversion nanoparticles are guest–host systems where trivalent lanthanide ions are dispersed as a guest in an appropriate dielectric host lattice with a dimension of less than 100 nm. The dopants of lanthanide are optically active centers, which produce emission when excited. Through judicious selection of lanthanide dopants, upconversion nanoparticles can display wavelength (color)-selective upconversion, such as NIR to shorter NIR, visible (blue, green, and red), or UV. Upconversion nanoparticles have optical properties similar to those of the corresponding bulk forms. They both generally produce the same upconversion luminescence peaks due to the well-shielded 4*f*–4*f* orbital electronic transitions by the outer complete 5*s* and 5*p* shells. However, the efficiency and the relative intensity between different luminescence peaks of upconversion nanoparticles are quite different from their bulk counterpart due to nanosize-induced surface effects. Because of the high surface-to-volume ratio of the nanoparticle, most of the lanthanide dopants are exposed to surface deactivations (due to surface defects, as well as to ligands and solvents that possess high phonon energy). Surface-related deactivations encompass two ways: (i) Dopants located on or around the nanocrystal surface can be deactivated directly by neighboring surface quenching centers and (ii) the energy contained in dopants located in the center of upconversion nanoparticles can randomly migrate and travel a long distance to the dopant on/around surface or directly to the surface quenching sites. In addition to the influence of phonons from the host lattice, the optical properties of upconversion nanoparticles are significantly affected by the surface properties, which are not so pronounced in bulk crystals. Indeed, upconversion nanoparticles are found to exhibit much lower efficiency than their bulk form due to surface-induced deactivations. To produce high efficiency upconversion nanoparticles, appropriate nanochemistry for engineering the surface of these nanocrystals plays a pivotal role (see Chap. 3). Upconversion nanoparticles have intrinsic attributes such as an absence of blinking and photobleaching, high chemical stability, sharp emission bands, NIR light excitation, and large anti-Stokes emissions, making them promising for theranostic applications.

The upconversion nanoparticles have to meet many key technical requirements for bio-applications [19]: (i) The first is high efficiency and multi-color emissions. High efficiency is required for all upconversion nanoparticles to accomplish improved limit of detection in sensing, high signal–background ratio bio-imaging, and increased therapy effect. Single-wavelength-excited multi-color emissions are of critical importance for multiplexing (simultaneously detecting multiple analytes/targets in a sample) capabilities in imaging or the throughput increase of bio-assays. (ii) Next is monodispersed small size, uniform shape, and stoichiometric composition. The monodispersed small size and uniform shape are required to have identical optical properties as well as cellular uptake and biological effects for intracellular theranostics, while a precise stoichiometric composition is necessary to allow control over the concentration of lanthanide dopants to manipulate the optical attributes. (iii) The third is a nanochemically engineered surface for phase transfer and for coupling to targeting ligands. The synthesized raw upconversion nanoparticles are generally hydrophobic due to their capping by long-chain hydrophobic ligands. Engineering the surface of upconversion nanoparticles to allow their dispersion in aqueous phase for biological applications is required. In addition, upconversion nanoparticles lack the targeting moieties to identify regions of interest in the body such as angiogenic tumor area. Appropriate coupling chemistry is required to incorporate a range of antibodies, peptides, biotin, avidin, and proteins onto the nanocrystal surface. (iv) The final requirement is bio-compatibility. Upconversion nanoparticles should be non-toxic and bio-compatible to cells and to the body.

1.3 Summary

In this chapter, the energy transfer mechanism, category, and chemical composition of the upconversion materials were briefly summarized. Until the late 1990s, nanoscience and nanotechnology have experienced fast momentum. In the nanomaterials family, upconversion nanomaterial is a new shining star. So, at the end of this chapter, we also give a general introduction to the lanthanide-doped upconversion nanomaterials. With nearly 20 years' development, upconversion nanomaterials became more and more mature. A lot of outstanding achievements have been obtained in the fields of chemical syntheses, optical properties, multi-functional upconversion nanomaterials, bio-imaging, bio-sensing, and bio-medicine, etc. Although there are a few books about the upconversion materials, they devote almost exclusively to bulk materials in the solid state. For the combination between the upconversion materials and the nanotechnology, a systematic summary remains absent and needs to be addressed. So in the following chapters, we focus on the recent progress in upconversion nanomaterials, including excitation and emission energy levels guided design of upconversion nanostructure (Chap. 3); the techniques to fabricate the nanostructure with optimum energy level structure (Chap. 2); the fabrication of upconversion nanoparticles-based nanocomposites to extend the working range of upconversion nanoparticles by introducing the additional functional components (Chap. 4); the surface modification of the high-quality upconversion nanoparticles (Chap. 5); the various applications based on the high-quality upconversion nanoparticles (Chaps. 6–11).

References

1. Jablonski, A.: Efficiency of anti-stokes fluorescence in dyes. Nature **131**, 839–840 (1933)
2. Jablonski, A.: About the mechanism of photo-luminescence of dye phosphors. Z. Angew. Phys. **94**, 38–46 (1935)
3. Tan, X., Luo, S.L., Wang, D.C., Su, Y.P., Cheng, T.M., Shi, C.M.: A NIR heptamethine dye with intrinsic cancer targeting, imaging and photosensitizing properties. Biomaterials **33**, 2230–2239 (2012)
4. Hong, G.S., Robinson, J.T., Zhang, Y.J., Diao, S., Antaris, A.L., Wang, Q.B., Dai, H.J.: In vivo fluorescence imaging with Ag_2S quantum dots in the second near-infrared region. Angew. Chem. Int. Ed. **51**, 9818–9821 (2012)
5. Bawendi, M.G., Steigerwald, M.L., Brus, L.E.: The quantum mechanics of larger semiconductor clusters ("quantum dots"). Annu. Rev. Phys. Chem. **41**, 477–496 (1990)
6. Wang, Y., Herron, N.: Nanometer-sized semiconductor clusters: materials synthesis, quantum size effects, and photophysical properties. J. Phys. Chem. **95**, 525–532 (1991)
7. Li, X.M., Zhang, F., Zhao, D.Y.: Highly efficient lanthanide upconverting nanomaterials: progresses and challenges. Nano Today **8**, 643–676 (2013)
8. Gamelin, D.R., Gudel, H.U.: Design of luminescent inorganic materials: new photophysical processes studied by optical spectroscopy. Accounts. Chem. Res. **33**, 235–242 (2000)
9. Jin, J., Wong, W.-T.: Encyclopedia of Inorganic and Bioinorganic Chemistry. Wiley, London (2011)

10. Franken, P.A., Weinreich, G., Peters, C.W., Hill, A.E.: Generation of optical harmonics. Phys. Rev. Lett. **7**, 118–119 (1961)
11. Auzel, F.: Upconversion and anti-stokes processes with f and d ions in solids. Chem. Rev. **104**, 139–173 (2004)
12. Ovsyanki, V.V., Feofilov, P.P.: Mechanism of summation of electronic excitations in activated crystals. Sov. J. Exp. Theor. Phys. Lett. **3**, 322–325 (1966)
13. Haase, M., Schafer, H.: Upconverting nanoparticles. Angew. Chem. Int. Ed. **50**, 5808–5829 (2011)
14. Gu, Z.J., Yan, L., Tian, G., Li, S.J., Chai, Z.F., Zhao, Y.L.: Recent advances in design and fabrication of upconversion nanoparticles and their safe theranostic applications. Adv. Mater. **25**, 3758–3779 (2013)
15. Wang, R., Zhang, F.: NIR luminescent nanomaterials for biomedical imaging. J. Mater. Chem. B **2**, 2422–2443 (2014)
16. Li, X., Wang, R., Zhang, F., Zhou, L., Shen, D., Yao, C., Zhao, D.: Nd^{3+} sensitized up/down converting dual-mode nanomaterials for efficient in-vitro and in-vivo bioimaging excited at 800 nm. Sci. Rep. **3**, 3536 (2013)
17. Wang, F., Liu, X.G.: Recent advances in the chemistry of lanthanide-doped upconversion nanocrystals. Chem. Soc. Rev. **38**, 976–989 (2009)
18. Li, X., Zhang, F., Zhao, D.: Lab on upconversion nanoparticles: optical properties and applications engineering via designed nanostructure. Chem. Soc. Rev. doi:10.1039/C4CS00163J (2014)
19. Chen, G.Y., Qju, H.L., Prasad, P.N., Chen, X.Y.: Upconversion nanoparticles: design, nanochemistry, and applications in theranostics. Chem. Rev. **114**, 5161–5214 (2014)
20. Liang, H.J., Chen, G.Y., Li, L., Liu, Y., Qin, F., Zhang, Z.G.: Upconversion luminescence in Yb^{3+}/Tb^{3+}-codoped monodisperse $NaYF_4$ nanocrystals. Opt. Commun. **282**, 3028–3031 (2009)
21. Dwivedi, Y., Thakur, S.N., Rai, S.B.: Study of frequency upconversion in Yb^{3+}/Eu^{3+} by cooperative energy transfer in oxyfluoroborate glass matrix. Appl. Phys. B-lasers. O **89**, 45–51 (2007)
22. Pushkar, A.A., Uvarova, T.V., Kiiko, V.V.: Up-conversion multiwave (white) luminescence in the visible spectral range under excitation by IR laser diodes in the active $BaY_2F_8:Yb^{3+}$, Pr^{3+} medium. Opt. Spectrosc. **111**, 273–276 (2011)
23. Joubert, M.F.: Photon avalanche upconversion in rare earth laser materials. Opt. Mater. **11**, 181–203 (1999)
24. Ronda, C.: Luminescence: From Theory to Applications. WILEY-VCH Verlag GmbH & Co. KGaA, Weinheim ISBN: 978-3-527-31402-7 (2008)
25. Vandijk, J.M.F., Schuurmans, M.F.H.: On the nonradiative and radiative decay rates and a modified exponential energy gap law for 4f–4f transitions in rare-earth ions. J. Chem. Phys. **78**, 5317–5323 (1983)
26. Chen, G.Y., Ohulchanskyy, T.Y., Kumar, R., Agren, H., Prasad, P.N.: Ultrasmall monodisperse $NaYF_4:Yb^{3+}/Tm^{3+}$ nanocrystals with enhanced near-infrared to near-infrared upconversion photoluminescence. ACS Nano **4**, 3163–3168 (2010)
27. Sun, Y.J., Chen, Y., Tian, L.J., Yu, Y., Kong, X.G., Zhao, J.W., Zhang, H.: Controlled synthesis and morphology dependent upconversion luminescence of $NaYF_4$:Yb, Er nanocrystals. Nanotechnology **18**, 275609 (2007)
28. Wang, L.Y., Li, Y.D.: Green upconversion nanocrystals for DNA detection. Chem. Commun. **42**, 2557–2559 (2006)
29. Kramer, K.W., Biner, D., Frei, G., Gudel, H.U., Hehlen, M.P., Luthi, S.R.: Hexagonal sodium yttrium fluoride based green and blue emitting upconversion phosphors. Chem. Mater. **16**, 1244–1251 (2004)
30. Wang, L.Y., Li, Y.D.: $Na(Y_{1.5}Na_{0.5})F_6$ single-crystal nanorods as multicolor luminescent materials. Nano Lett. **6**, 1645–1649 (2006)
31. Suyver, J.F., Aebischer, A., Biner, D., Gerner, P., Grimm, J., Heer, S., Kramer, K.W., Reinhard, C., Gudel, H.U.: Novel materials doped with trivalent lanthanides and transition

metal ions showing near-infrared to visible photon upconversion. Opt. Mater. **27**, 1111–1130 (2005)

32. De la Rosa, E., Salas, P., Desirena, H., Angeles, C., Rodriguez, R.A.: Strong green upconversion emission in ZrO_2:Yb^{3+}–Ho^{3+} nanocrystals. Appl. Phys. Lett. **87**, 241912 (2005)
33. Wang, G.F., Peng, Q., Li, Y.D.: Upconversion luminescence of monodisperse CaF_2:Yb^{3+}/Er^{3+} nanocrystals. J. Am. Chem. Soc. **131**, 14200–14201 (2009)
34. Wang, F., Liu, X.G.: Upconversion multicolor fine-tuning: visible to near-infrared emission from lanthanide-doped $NaYF_4$ nanoparticles. J. Am. Chem. Soc. **130**, 5642–5643 (2008)
35. Chen, G.Y., Ohulchanskyy, T.Y., Kachynski, A., Agren, H., Prasad, P.N.: Intense visible and near-infrared upconversion photoluminescence in colloidal $LiYF_4$:Er^{3+} nanocrystals under excitation at 1490 nm. ACS Nano **5**, 4981–4986 (2011)
36. Chen, D.Q., Lei, L., Yang, A.P., Wang, Z.X., Wang, Y.S.: Ultra-broadband near-infrared excitable upconversion core/shell nanocrystals. Chem. Commun. **48**, 5898–5900 (2012)
37. Chen, G.Y., Liu, H.C., Somesfalean, G., Liang, H.J., Zhang, Z.G.: Upconversion emission tuning from green to red in Yb^{3+}/Ho^{3+}-codoped $NaYF_4$ nanocrystals by tridoping with Ce^{3+} ions. Nanotechnology **20**, 275609 (2009)
38. Rakov, N., Maciel, G.S., Sundheimer, M.L., Menezes, S.L.D., Gomes, A.S.L., Messaddeq, Y., Cassanjes, F.C., Poirier, G., Ribeiro, S.J.L.: Blue upconversion enhancement by a factor of 200 in Tm^{3+}-doped tellurite glass by codoping with Nd^{3+} ions. J. Appl. Phys. **92**, 6337–6339 (2002)
39. Wang, L.L., Qin, W.P., Liu, Z.Y., Zhao, D., Qin, G.S., Di, W.H., He, C.F.: Improved 800 nm emission of Tm^{3+} sensitized by Yb^{3+} and Ho^{3+} in beta-$NaYF_4$ nanocrystals under 980 nm excitation. Opt. Express **20**, 7602–7607 (2012)
40. Bol, A.A., van Beek, R., Meijerink, A.: On the incorporation of trivalent rare earth ions in II-VI semiconductor nanocrystals. Chem. Mater. **14**, 1121–1126 (2002)
41. Heer, S., Kompe, K., Gudel, H.U., Haase, M.: Highly efficient multicolour upconversion emission in transparent colloids of lanthanide-doped $NaYF_4$ nanocrystals. Adv. Mater. **16**, 2102–2105 (2004)
42. Judd, B.R.: Optical absorption intensities of rare-earth ions. Phys. Rev. **127**, 750–761 (1962)
43. Ofelt, G.S.: Intensities of crystal spectra of rare-earth ions. J. Chem. Phys. **37**, 511–520 (1962)
44. Chen, G.Y., Somesfalean, G., Liu, Y., Zhang, Z.G., Sun, Q., Wang, F.P.: Upconversion mechanism for two-color emission in rare-earth-ion-doped ZrO_2 nanocrystals. Phys. Rev. B **75**, 195204 (2007)
45. Suyver, J.F., Aebischer, A., Garcia-Revilla, S., Gerner, P., Gudel, H.U.: Anomalous power dependence of sensitized upconversion luminescence. Phys. Rev. B **71**, 195204 (2005)
46. Pollnau, M., Gamelin, D.R., Luthi, S.R., Gudel, H.U., Hehlen, M.P.: Power dependence of upconversion luminescence in lanthanide and transition-metal-ion systems. Phys. Rev. B **61**, 3337–3346 (2000)
47. Gamelin, D.R., Gudel, H.U.: Upconversion processes in transition metal and rare earth metal systems. Transit. Met. Rare Earth Compd. Excited States, Transitions, Interact. Ii **214**, 1–56 (2001)
48. Moncorge, R., Benyattou, T.: Excited-state absorption of Ni^{2+} in MgF_2 and MgO. Phys. Rev. B **37**, 9186–9196 (1988)
49. Wermuth, M., Gudel, H.U.: Photon avalanche in Cs_2ZrBr_6:Os^{4+}. J. Am. Chem. Soc. **121**, 10102–10111 (1999)
50. Wermuth, M., Gudel, H.U.: Photon avalanche in Cs_2ZrCl_6:Os^{4+}. J. Chem. Phys. **114**, 1393–1404 (2001)
51. Wermuth, M., Güdel, H.U.: Upconversion luminescence in a 5d transition-metal ion system: Cs_2ZrCl_6: Os^{4+}. Chem. Phys. Lett. **281**, 81–85 (1997)
52. Gamelin, D.R., Gudel, H.U.: Excited-state dynamics and sequential two-photon upconversion excitation of Mo^{3+}-doped chloro- and bromo-elpasolites. J. Phys. Chem. B **104**, 10222–10234 (2000)

53. Gamelin, D.R., Güdel, H.U.: Two-photon spectroscopy of d3 transition metals: near-IR-to-visible upconversion luminescence by Re^{4+} and Mo^{3+}. J. Am. Chem. Soc. **120**, 12143–12144 (1998)
54. Gamelin, D.R., Gudel, H.U.: Spectroscopy and dynamics of Re^{4+} near-IR-to-visible luminescence upconversion. Inorg. Chem. **38**, 5154–5164 (1999)
55. Zhou, G., Wang, D., Wang, X., Shao, Z., Jiang, M.: Two-photon absorption and excited state absorption properties of an organic dye PSPI. Opt. Commun. **241**, 215–219 (2004)
56. Liu, Y., Tu, D., Zhu, H., Chen, X.: Lanthanide-doped luminescent nanoprobes: controlled synthesis, optical spectroscopy, and bioapplications. Chem. Soc. Rev. **42**, 6924–6958 (2013)
57. Bunzli, J.C.G., Piguet, C.: Taking advantage of luminescent lanthanide ions. Chem. Soc. Rev. **34**, 1048–1077 (2005)

Chapter 2
"Wet" Chemical Synthesis and Manipulation of Upconversion Nanoparticles

Abstract Development of facile synthesis strategies for high-quality lanthanide-doped upconversion nanoparticles with controlled composition, crystalline phase, shape, and size is crucial in tuning their chemical and optical properties and exploring their potential applications in diverse fields. This chapter focuses primarily on the formation mechanism of the crystallization (including nucleation and growth) for monodisperse nanocrystals and various synthetic procedures for the upconversion nanoparticles. Finally, various optical, chemical, and structural characterizations of upconversion nanoparticles are also summarized.

Keywords Wet chemical synthesis · Upconversion nanoparticles · Upconversion luminescence · Top-down approach · Bottom-up approach · Monodisperse nanocrystals · Optical characterization · Transmission electron microscopy · X-ray photoelectron spectroscopy · X-ray diffraction

2.1 Introduction

Since 1990s, the synthesis of nanocrystals with the sizes ranging from 1 to 100 nm have been intensively pursued, not only for their fundamental scientific interest, but also for their many technological applications. Nanocrystals exhibit very interesting size-dependent electrical, optical, magnetic, chemical, and biological properties that cannot be achieved by their bulk counterparts [1–5]. For many future applications, the synthesis of uniform-sized nanocrystals is of key importance, because the electrical, optical, magnetic, and biological properties of these nanocrystals are strongly dependent on their dimensions. The most popular demonstration of the size-dependent characteristics of nanocrystals are the continuous fluorescent emission from semiconductor nanocrystals (also known as quantum dots) which covers the entire visible spectrum as a function of their size [6, 7]. Monodispersed

Xiaomin Li and Fan Zhang contributed together to this chapter.

F. Zhang, *Photon Upconversion Nanomaterials*,
Nanostructure Science and Technology, DOI 10.1007/978-3-662-45597-5_2

magnetic nanocrystals are essential for the next-generation multi-terabit (Tbit/in^2) magnetic storage media [8–10]. For the upconversion nanoparticles, the size has a very big influence on its optical and magnetic properties, which makes the upconversion nanoparticles have wide application prospects in optical and MRI bio-imaging [11–13].

There are two different approaches to fabricate nanocrystals: the "top-down" approach, which utilizes physical methods, and the "bottom-up" approach, which employs solution-phase colloidal chemistry (wet chemistry) [10]. The advantage of the physical methods are the production of a large quantity of nanocrystals, whereas the synthesis of uniform-sized nanocrystals and their size control is very difficult to achieve by using the top-down approach. In contrast, wet chemical synthetic methods can be used to synthesize uniform nanocrystals with controlled particle size, although generally only subgram quantities are produced. Furthermore, various-shaped nanocrystals, including nanosphere, nanorods, nanoflowers, nanodisks, etc., can be synthesized by varying the reaction conditions.

Development of facile synthesis strategies for high-quality lanthanide-doped upconversion nanoparticles with controlled composition, crystalline phase, shape, and size is crucial in tuning their chemical and optical properties and exploring their potential applications in diverse fields. With the development of nanotechnology, a variety of methods have been established to synthesize different kinds of upconversion nanoparticles with controlled crystalline phases, sizes, and shapes, (Fig. 2.1) including coprecipitation [14–22], thermal decomposition [23–27], hydro (solvo)thermal synthesis [28–37], sol-gel process [38–41], and combustion synthesis [42], which have been reviewed in many papers [43–54]. This chapter focuses primarily on the formation mechanism of the crystallization (including nucleation and growth) for monodisperse nanocrystals and the various synthetic procedures for the upconversion nanoparticles. Finally, various characterizations for the optical, chemical, and structural properties of upconversion nanoparticles are also summarized.

2.2 Mechanism for the Synthesis of Monodisperse Nanocrystals

The formation processes of monodisperse nanocrystals mainly experience two stages, nucleation and growth. Both of the stage needs to be controlled very carefully. The formation of the monodisperse nanocrystals has been investigated from theory to practice and summarized in some review papers [10]. According to the literature [10], we elaborate the mechanism for the synthesis of monodisperse nanocrystals from three aspects in this section: nucleation, growth, and synthetic techniques for the separation of nucleation and growth.

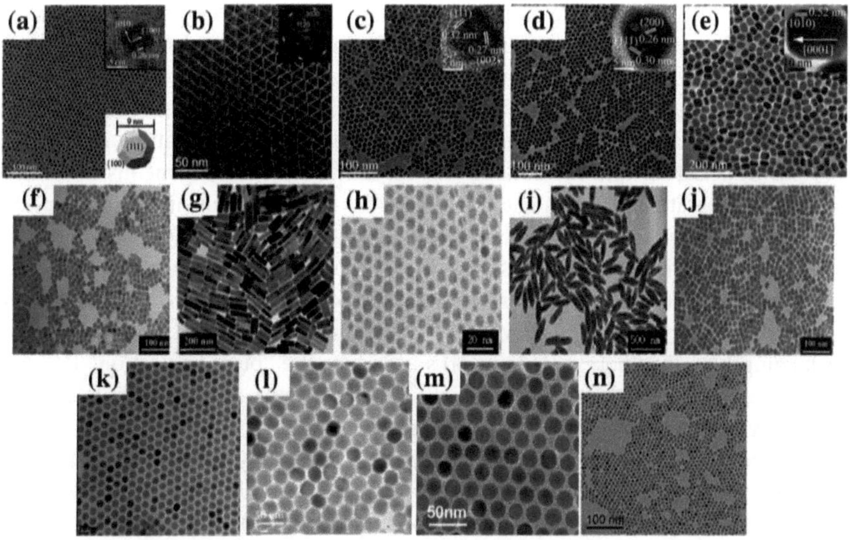

Fig. 2.1 Typical TEM images of lanthanide-doped upconversion nanoparticles. **a** LuOF, **b** LaF_3, **c** α-$NaYF_4$, **d** $NaYbF_4$, **e** β-$NaEuF_4$ synthesized by the thermolysis method, **f** α-$NaYF_4$:Yb^{3+},Er^{3+}, **g** β-$NaYF_4$:Yb^{3+},Er^{3+}, **h** LaF_3, **i** YF_3, **j** α-$NaYF_4$, synthesized by the hydro(solvo)-thermal method, **k–n** β-$NaYF_4$ or CaF_2 nanoparticles. **a** Reprinted with the permission from Ref. [55]. Copyright 2007 Wiley-VCH Verlag GmbH & Co. KGaA. **b** Reprinted with the permission from Ref. [26]. Copyright 2005 American Chemical Society. **c–e** Reprinted with the permission from Ref. [24]. Copyright 2006 American Chemical Society. **f**, **g** Reprinted with the permission from Ref. [45]. Copyright 2011 American Chemical Society. **h–j** Reprinted with the permission from Ref. [37]. Copyright 2005 Nature Publishing Group. **k** Reprinted with the permission from Ref. [56]. Copyright 2012 American Chemical Society. **l** Reprinted with the permission from Ref. [57]. Copyright 2008 Wiley-VCH Verlag GmbH & Co. KGaA. **m** Reprinted with the permission from Ref. [58]. Copyright 2008 American Chemical Society. **n** Reprinted with the permission from Ref. [59]. Copyright 2013 Wiley-VCH Verlag GmbH & Co. KGaA

2.2.1 Nucleation

It is much more difficult to elucidate the mechanism for nanometer-sized particles because of their high surface-to-volume ratios. Although the nucleation and growth processes are strongly correlated, we discuss these two processes separately for the sake of simplicity. The research on the preparation of uniform colloidal particles dates back to the 1940s. LaMer and his colleagues pioneered the concept of "burst nucleation" during the preparation of various oil aerosols and sulfur hydrosols [10, 60, 61]. In this process, many nuclei are generated at the same time, and then these nuclei start to grow without additional nucleation simultaneously. This is the essence of "burst-nucleation" process which makes it possible to control size distribution of particles during growth. Otherwise, if nucleation process occurred throughout the whole particle formation process, the growth history of the particles would differ largely from one another, and consequently, the control of the size distribution would be very difficult. "Burst nucleation" has been adopted as an important concept in the synthesis of monodispersed nanocrystals. It is well known that to prepare highly uniform nanocrystals, it is necessary to induce a single nucleation event and to prevent additional nucleation during the subsequent growth process. As a synthetic strategy, this method is often referred to as "the separation of nucleation and growth" [10, 60, 61]. LaMer and his colleagues first utilized the homogeneous nucleation process to separate nucleation and growth. In the homogeneous nucleation process, nuclei appear in a homogeneous solution without any seed for heterogeneous nucleation. In this homogeneous nucleation process, there exists a high-energy barrier to nucleation, because the system spontaneously changes from homogeneous to heterogeneous phase. The LaMer plot, as shown in Fig. 2.2a, is very useful for visualizing how the energy barrier works to induce the "burst nucleation" [62, 63]. The concentration of "monomer," which is the minimum subunit of bulk crystals, constantly increases with time. Note that precipitation does not occur in stage I even under supersaturated conditions ($S > 1$), because the energy barrier for spontaneous homogeneous nucleation is extremely high. In stage II, during the nucleation occurs, the degree of supersaturation is high enough to overcome the energy barrier for nucleation, thus resulting in the formation and accumulation of stable nuclei. Since the rate of monomer consumption resulted from the nucleation and growth processes exceeds the rate of monomer supply, the monomer concentration decreases until it reaches the level at which the net nucleation rate (the number of nuclei formed per unit time) is zero. Below this level, the system enters the growth stage (stage III), in which nucleation is effectively stopped and the particles keep growing as long as the solution is supersaturated. Once monomer concentration is sufficiently depleted, growth can proceed by Ostwald ripening (stage IV). Here, sacrificial dissolution of smaller (higher-surface-energy) particles results in growth of larger particles and, thereby, fewer particles in the system.

According to literature [10, 61, 64–67], the energy barrier to the homogeneous nucleation is interpreted thermodynamically as follows: The Gibbs free energy of

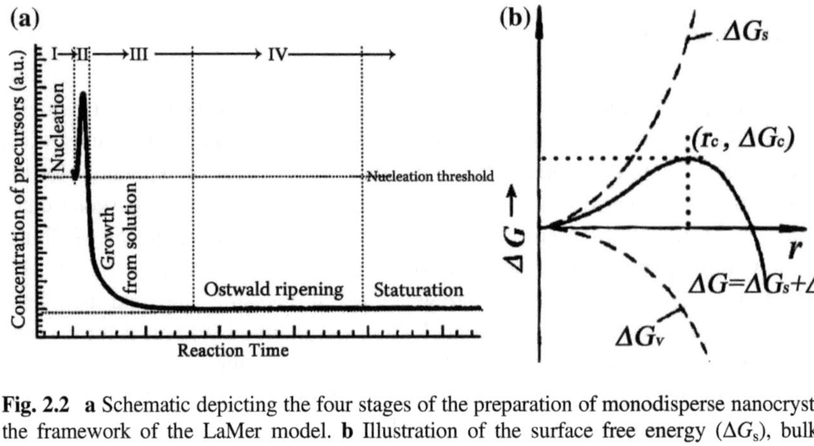

Fig. 2.2 **a** Schematic depicting the four stages of the preparation of monodisperse nanocrystals in the framework of the LaMer model. **b** Illustration of the surface free energy (ΔG_s), bulk free energy (ΔG_v), and overall free energy (ΔG) as a function of the growth particle size r

formation of spherical crystals with radius r from the solution with supersaturation S is given in Eq. (2.1), in which σ is the surface free energy per unit area and ΔG_v is the free energy change between the monomers in the solution and unit volume of bulk crystal, V_m is the molar volume of bulk crystal.

$$\Delta G = 4\pi r^2 \sigma + \frac{4}{3}\pi r^3 \Delta G_v = 4\pi r^2 \sigma + \frac{4}{3}\pi r^3 (-RT \ln S)/V_m \qquad (2.1)$$

σ is always positive, and ΔG_v is negative as long as the solution is supersaturated. Consequently, a plot of ΔG versus r has a maximum. The value of r at which ΔG is maximum is called the critical radius r_c; this is the minimum radius of a nucleus that can grow spontaneously in the supersaturated solution. Setting $d\Delta G/dr = 0$ allows determination of r_c.

$$r_c = \frac{-2\sigma}{\Delta G_v} = \frac{2\sigma V_m}{RT \ln S} \qquad (2.2)$$

Equation (2.2) imposes the first necessary condition for supersaturation with homogeneous nucleation. Because r_c is the minimum radius that will persist and not dissolve away in solution, S should be sufficiently high for r_c to be smaller than the size of the crystal embryos that form the nuclei for the homogeneous nucleation process [64, 65]. Although little is known about the identity of the crystal embryos, their sizes might be less than 1 nm, which is comparable to the size of inorganic molecular clusters. Substituting Eq. (2.2) into Eq. (2.1) gives the critical free energy ΔG_c (Eq. 2.3), which is the free energy necessary to form a stable nucleus.

$$\Delta G_c = \frac{16\pi\sigma^3}{3(\Delta G_v)^2} = \frac{16\pi\sigma^3 V_m^2}{3(RT \ln S)^2} \qquad (2.3)$$

If the rate of increase of the number of particles N is defined as the rate of nucleation (k_B is the Boltzmann constant), it can be written in the Arrhenius form in terms of ΔG_c (Eq. 2.4).

$$\frac{dN}{dt} = A \exp\left[-\frac{\Delta G_c}{k_B T}\right] = A \exp\left[\frac{16\pi\sigma^3 V_m^2}{3k_B^3 T^3 N_A^2 (\ln S)^2}\right] \qquad (2.4)$$

At this point, it should be noted that, in contrast to the simple LaMer plot, it is hard to define exactly the critical supersaturation level at which nucleation begins, because nucleation and redissolution can happen at any concentration, as a result of the energy fluctuation in the solution. In fact, the nucleus can still form even in unsaturated solution, and the particles formed could re-dissolve unless they are stable enough to resist the free energy fluctuation of their surroundings. However, from the practical point of view, it is reasonable to establish the critical supersaturation level (S_c) at which stable nuclei form in an appreciable number per unit time and start to accumulate. To see how this condition affects S, we rewrite Eq. (2.4) so as to express S in terms of N' (=dN/dt) (Eq. 2.5).

$$\ln S = \sqrt{\frac{16\pi\sigma^3 V_{\mathrm{m}}^2}{3k^3T^3\ln(A/N')}} \tag{2.5}$$

This equation shows another necessary condition pertaining to the degree of supersaturation: To start the accumulation and the growth of the nuclei, the nucleation rate should be high enough to equilibrate or to surpass the redissolution rate of the particles. Taken together, S_c is the point at which the nucleation rate is so high that the number of nuclei increases even while smaller nuclei dissolve away.

However, the thermodynamic model discussed so far has some limitations with respect to nanocrystals. Whereas it is generally assumed that σ and ΔG_v are constant, these two values are strongly size-dependent for nanometer-sized particles [66]. As the particle size decreases, the ratio of surface atoms to the bulk atoms dramatically increases. As a result, there is a strong driving force, especially for nanocrystals with a size of few nanometers, to minimize the surface free energy by reconstructing the surface structure or changing the crystal structure (e.g., through phase transitions [67] or lattice contraction [61, 68]).

2.2.2 *Synthetic Techniques for the Separation of Nucleation and Growth*

Both homogeneous and heterogeneous nucleation processes have been utilized to synthesize monodisperse nanocrystals by separating nucleation and growth. The seed-mediated growth method is the most apparent case for the separation of nucleation and growth, wherein nucleation is physically separated from growth by using preformed nanocrystals as seed nuclei. This method utilizes heterogeneous nucleation to suppress the formation of additional nuclei by homogeneous nucleation [69–72]. In this method, preformed nuclei are introduced into the reaction solution and then the monomers are supplied to precipitate on the surface of the existing nuclei. The monomer concentration is kept low during growth to suppress homogeneous nucleation. Seed-mediated growth is further divided into two categories: the synthesis of homogeneous particles [69, 72] and the production of heterogeneous structures, such as core/shell structures [70, 71]. There have been several reports on the fine size control of nanocrystals by separating nucleation and growth by the seed-mediated growth process [71, 72]. However, the seed particles need to be uniform to produce monodisperse nanocrystals.

There are two techniques that utilize homogeneous nucleation to synthesize monodisperse nanocrystals in the organic solutions: "hot-injection" [73–75] and "heating-up" methods [76–79]. The "hot-injection" technique was introduced by Bawendi and coworkers in their report on the synthesis of cadmium chalcogenide nanocrystals [68]. This technique produces high degree of supersaturation by the rapid injection of excess precursor into a hot surfactant solution, resulting in burst nucleation by relieving the excess free energy of the supersaturation. During the nucleation process, the monomer concentration in the solution sharply decreases and thus nucleation rate slows down. This "hot-injection" method has been widely used to synthesize nanocrystals of metal chalcogenides [68, 73], transition metals [74], noble metals [75], and upconversion nanoparticles [80]. The heating-up method is a batch process in which the precursors, reagents, and solvent are mixed at a low temperature and heated up to a certain temperature to initiate the crystallization reaction. The heating-up method is particularly advantageous for large-scale production, because of its simplicity. Although this synthetic procedure is very simple, the size uniformity of the nanoparticles yielded by the heating-up method is often comparable to the best results obtained from the "hot-injection" method [76, 77]. Characterizing the burst-nucleation process is challenge. Characteristic for this process is that there is a point of time at which the number of particles stops increasing and the particle concentration reaches a maximum. After this point, the reaction system enters the growth stage and number of particles either remains constant or decrease. According to LaMer's model, the end of the nucleation stage is closely related to the decrease of the monomer concentration. Consequently, to prove the "burst-nucleation" model experimentally, both the particle concentration and the monomer concentration should be traced simultaneously.

2.2.3 Growth

In the previous section, we mentioned that the separation of nucleation and growth makes it possible to control the size distribution of an ensemble of particles. In this section, the significance of growth without additional nucleation as a necessary condition for a narrow size distribution of the ensemble of particles will be shown [10].

The first theoretical study on the narrowing of the size distribution during the growth process was performed by Reiss [81]. In his model, known as the "growth by diffusion" model, the growth rate of spherical particles depends solely on the flux of the monomers supplied to the particles (J). In this case, the relationship between the monomer flux and the growth rate dr/dt is given by Eq. (2.6).

$$J = \frac{4\pi r^2}{V_m}\frac{dr}{dt} \tag{2.6}$$

If the average distance between the particles is large enough, then the diffusion layer formed at the periphery of each particle is undisturbed. Consequently, it is possible to treat each growing particle independently. For a single spherical particle in a homogeneous medium, there is concentration gradient around a particle with spherical symmetry. Fick's law (Eq. 2.7) gives the flux J of monomers diffusing through the surface of a sphere enclosing the particle (D is the diffusion coefficient, C is the concentration, and x ($\geq r$) is the distance from the center of the particle).

$$J = 4\pi x^2 D \frac{\mathrm{d}C}{\mathrm{d}x} \tag{2.7}$$

If J is assumed to be constant for x, the integration of $C(x)$ from r to $r + \delta$ with respect to x gives Eq. (2.8).

$$J = 4\pi D \frac{r(r+\delta)}{\delta}\left[C_{(r+\delta)} - C_{\mathrm{s}}\right] \tag{2.8}$$

C_{s} ($=C_{\mathrm{r}}$) is the concentration at the surface of the particle. For sufficiently large values of δ ($r \ll \delta$), Eq. (2.8) is reduced to Eq. (2.9), in which C_{bulk} is the concentration of the bulk solution.

$$J = 4\pi r D (C_{\mathrm{bulk}} - C_{\mathrm{s}}) \tag{2.9}$$

Equation (2.10) follows from Eqs. (2.6) and (2.9).

$$\frac{\mathrm{d}r}{\mathrm{d}t} = \frac{V_{\mathrm{m}} D}{r}(C_{\mathrm{bulk}} - C_{\mathrm{s}}) \tag{2.10}$$

If C_{s} and C_{bulk} are constant for all particles, the growth rate of a particle is inversely proportional to its radius. This result can be understood intuitively as follows: The number of monomers diffused onto the surface of a particle increases in proportion to the square of its radius, whereas the volume of a particle consisting of the monomers increases in proportion to the third power of its radius. Thus, the growth rate of a particle is decreased as the radius increases. With this result, it can be shown that for an ensemble of spherical particles, the variation of the radius distribution σ^2 is decreased during growth. From Eq. (2.10), the value of σ^2 can be obtained as Eq. (2.11), in which r and $(1/r)$ are the mean values of r and $1/r$, respectively.

$$\frac{\mathrm{d}\sigma^2}{\mathrm{d}t} = 2V_{\mathrm{m}} D (C_{\mathrm{bulk}} - C_{\mathrm{s}})\left[1 - \bar{r}\overline{\left(\frac{1}{r}\right)}\right] \tag{2.11}$$

Because the arithmetic mean differs from the harmonic mean, $(1/r)$ is always greater than $1/r$. Thus, for $C_{\mathrm{bulk}} > C_{\mathrm{s}}$, the right-hand side of Eq. (2.11) is always negative. In other words, the variance of the size distribution of an ensemble of particles always decreases regardless of the initial size distribution as long as all of the particles are growing and no additional nucleation occurs. This is the self-regulating mechanism of the size distribution during the growth process and is often referred to as the "focusing" effect [63].

However, the model described by Reiss is an oversimplification because it does not consider the reaction kinetics of crystal growth and its dependence on the particle size. As a result, the strong counter effect for the "focusing" mechanism is missing. During the growth process, there are two reactions acting in opposition to each other, namely, precipitation and dissolution (Eq. 2.12).

$$nM^{s} \underset{k_d}{\overset{k_p}{\rightleftharpoons}} M_n^{c} \qquad (2.12)$$

M^s and M^c refer to monomers in solution and in the crystal, k_p and k_d are the reaction rate constants for precipitation and dissolution, respectively. It is assumed that the precipitation is the first-order reaction with respect to C_s and that the dissolution rate is independent of C_s. Then, at equilibrium $k_p\, C_{s,eq} = k_d$, which can be rewritten to give the surface concentration $C_{s,eq}$ (Eq. 2.13)

$$C_{s,eq} = \frac{k_d}{k_p} \qquad (2.13)$$

The change in chemical potential $\mu(r)$ of a spherical crystal with radius r with respect to that μ^o of the bulk crystal arises from the surface free energy of area (A) (Eq. 2.14).

$$\Delta\mu = \mu(r) - \mu^o = \gamma \frac{dA}{dn} \qquad (2.14)$$

Because $dA = 8\pi r dr$ and $dn = 4\pi r^2 dr/V_m$, Eq. (2.14) can be rewritten as the Gibbs–Thomson relation Eq. (2.15).

$$\Delta\mu = \frac{2\gamma V_m}{r} \qquad (2.15)$$

The activated complex theory is adopted to assess the effect of the chemical potential change of a crystal on the precipitation and dissolution reactions. The variation of k_p and k_d with $\Delta\mu$ is given as Eqs. (2.16) and (2.17).

$$k_p = k_p^o \exp\left[-\alpha \frac{\Delta\mu}{RT}\right] = k_p^o \exp\left[-\alpha \frac{2\gamma V_m}{rRT}\right] \qquad (2.16)$$

$$k_d = k_d^o \exp\left[(1-\alpha)\frac{\Delta\mu}{RT}\right] = k_d^o \exp[\left[(1-\alpha)\frac{2\gamma V_m}{rRT}\right] \qquad (2.17)$$

In these equations, α is the transfer coefficient and k^o is the rate constant for the bulk crystal ($r = \infty$) [82]. Qualitatively, Eqs. (2.16) and (2.17) reveal that the smaller the radius of a particle is, the harder it is to grow but the easier it is to dissolve, because of its higher chemical potential. This is the effect in contrast to the "focusing" mechanism, in which smaller crystals grow faster. To combine this effect with the model of Reiss, the assumption that C_s is constant for all particles should be modified. The fluxes of the monomers toward the surface of a particle by precipitation and dissolution (J_p and J_d, respectively) for a particle with radius r are given by Eqs. (2.18) and (2.19)

$$J_p = 4\pi r^2 k_p^o C_s \exp\left[-\alpha\frac{2\gamma V_m}{rRT}\right] \tag{2.18}$$

$$J_d = -4\pi r^2 k_d^o \exp\left[(1-\alpha)\frac{2\gamma V_p}{rRT}\right] \tag{2.19}$$

The net flux J, then, is given by Eq. (2.20), and the equation for C_s (Eq. 2.21) is obtained by equating the expressions for J in Eqs. (2.9) and (2.20)

$$J = J_p + J_d = 4\pi r^2 k_p^o C_s \exp\left[-\alpha\frac{2\gamma V_m}{rRT}\right] - 4\pi r^2 k_d^0 \exp\left[(1-\alpha)\frac{2\gamma V_m}{rRT}\right] \tag{2.20}$$

$$C_s = \frac{k_d^o r \exp\left[(1-\alpha)\frac{2\gamma V_m}{rRT}\right] + DC_{bulk}}{k_p^o \exp\left[-\alpha\frac{2\gamma V_m}{rRT}\right] + D} \tag{2.21}$$

Substituting this result into Eq. (2.10) and using Eq. (2.13) leads to Eq. (2.22). $C_{s,eq}^o$ is the equilibrium surface concentration of the bulk crystal ($r = \infty$), and S is the degree of supersaturation, which is defined as $S = C_{bulk}/C_{s,eq}$. This result can be rewritten in simplified form as Eq. (2.23) [82].

$$\frac{dr}{dt} = V_m DC_{s,eq}^o \left[\frac{S - \exp\left[\frac{2\gamma V_m}{rRT}\right]}{r + \frac{D}{k_p^o}\exp\left[\alpha\frac{2\gamma V_m}{rRT}\right]}\right] \tag{2.22}$$

$$\frac{dr^*}{d\tau} = \frac{S - \exp(1/r^*)}{r^* + K\exp\left(\frac{\alpha}{r^*}\right)} \tag{2.23}$$

The variables and parameters in Eq. (2.23) can be normalized to dimensionless forms (Eqs. 2.24–2.26)

$$r^* = \frac{RT}{2\gamma V_m} r \tag{2.24}$$

$$\tau = \frac{R^2T^2DC_{s,eq}^o}{4\gamma^2 V_m} t \tag{2.25}$$

$$K = \frac{RT}{2\gamma V_m}\frac{D}{k_p^o} \tag{2.26}$$

Equation (2.23) is a modified version of Eq. (2.10) after both the mass transport and the reaction kinetics are considered. The growth rate of a single particle for various values of K and S are calculated from Eq. (2.23) and plotted in Fig. 2.3a, b.

In Reiss' simple model, the "focusing" effect is derived by considering only mass-transport processes. However, the opposite effect comes from a kinetic process related to the Gibbs–Thomson effect. Figure 2.3a, b shows how these two effects compete with each other for the growth of an individual particle. A typical plot of growth rate for small values of K ($\ll 1$) and large values of S ($\gg 1$) has a maximum at $r^* = r^*_{max}$. For $r^* > r^*_{max}$, the size dependence of the crystal chemical

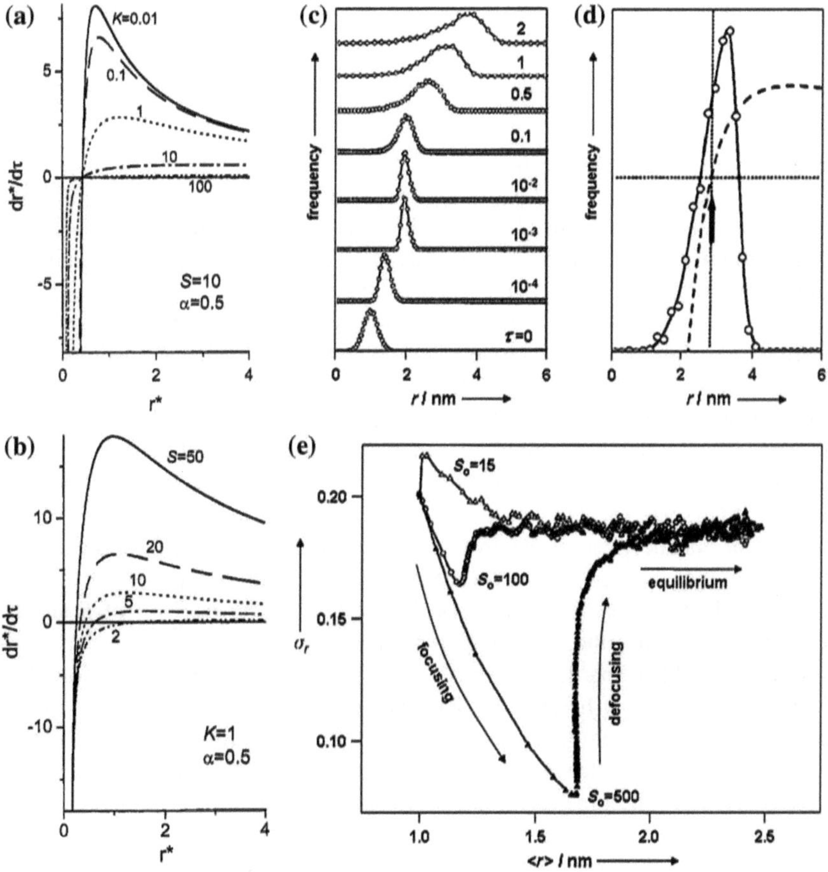

Fig. 2.3 Growth rate as a function of radius calculated from Eq. (2.23) for various values of K (**a**) and S (**b**). $\alpha = 0.5$, $S = 10$ in (**a**), and $K = 1$ in (**b**). **c** Time evolution of the size distribution of the ensemble of particles. **d** The size distribution of particles (*open circles* and *solid line*) and the growth rate as a function of radius (*dashed line*) in the later period of the reaction. The *arrow* indicates the position of r_{zero}. **e** Relative standard deviation versus mean radius for different initial levels of supersaturation. **a–c**, **e** Reprinted with the permission from Ref. [82]. Copyright 2001 American Chemical Society. **d** Reprinted with the permission from Ref. [83]. Copyright 2002 American Chemical Society

potential is relatively small such that the variation of the growth rate with r^* mainly depends on mass-transport effects rather than on kinetic effects. Consequently, the slope of the graph is negative and a narrowing of the size distribution occurs in this region (the "focusing" region). In contrast, for $0 < r^* < r^*_{\text{max}}$, the situation is reversed. In this region, the crystal chemical potential is highly sensitive to the particle size. As r^* decreases, a particle becomes more unstable and the dissolution rate increases so fast that it dominates the net growth rate. As a result, larger particles have a higher growth rate and the slope of the graph is positive (the "defocusing" region). At $r^* = r^*_{\text{zero}}$, the growth rate is zero, and the rates of precipitation and dissolution are balanced. The value of r_{zero} can be obtained from Eqs. (2.23) and (2.24) (Eq. 2.27).

$$r_{\text{zero}} = \frac{2\gamma V_{\text{m}}}{RTr^*_{\text{zero}}} = \frac{2\gamma V_{\text{m}}}{RT \ln S} \tag{2.27}$$

It is notable that this value is equal to r_{c}, which is evaluated by Eq. (2.2) from the nucleation model. From Fig. 2.3a, b, the factors affecting the evolution of the size distribution can also be deduced. In Eq. (2.27), K represents the ratio of the rate of diffusion to the rate of the precipitation reaction. If the value of K is very small, the overall growth reaction rate is controlled by the diffusion rate, that is, the rate of mass transfer. This condition is called diffusion-controlled growth and is similar to the growth condition in Reiss' model. Consequently, the smaller the value of K is, the more effective the narrowing of the size distribution becomes. However, if the value of K is very large, the growth rate is mainly determined by the reaction rate. This condition is called reaction-controlled growth and the "focusing" effect is weakened under this condition. This trend is depicted in Fig. 2.3a, which shows a steep negative slope for small values of K. An increase in the value of S always results in the enhancement of the growth rate according to Eq. (2.23). However, the increment of the growth rate with S is decreased by a factor of $1/[r^* + K\ \exp(\alpha/r^*)]$ and thus is larger for smaller particles. In short, both the increase of S and the decrease of K enhance the "focusing" mechanism.

For an ensemble of particles, it is very difficult to trace the time evolution of the size distribution, mainly because C_{bulk} is not a constant but rather a function of the size of all the particles in the ensemble. Furthermore, the growth rates also depend on C_{bulk}. This mutual dependence makes it very difficult, if not impossible, to derive analytically the time evolution of the size distribution of the particles from Eq. (2.23).

Talapin et al. [82] presented another approach to solve this problem. They performed a numerical simulation of the time evolution of the particle size in an ensemble by using the Monte Carlo method (Fig. 2.3c–e). They used Eq. (2.23) to calculate the growth rate for an individual particle. The initial size distributions of the ensembles of particles were set to normal distributions with various relative standard deviations and a mean value of $r_0 = 1$ nm. With these given ensembles of particles, simulations of the growth process were started with an initial supersaturation S_0. Figure 2.3c shows the time evolution of the size distribution of the particle ensemble. The initial reaction solution is highly supersaturated and the growth reaction operates under the diffusion-controlled condition. Two periods in the growth process can be distinguished in this figure. Initially ($0 < \tau < 10^{-2}$), the mean radius increases rapidly and the size distribution becomes narrower. In this period, the supersaturation is so high that r^*_{zero} is far below the mean radius $\overline{r^*}$, and consequently, all of the particles are in the "focusing" region. In the second stage ($10^{-2} < \tau$), the growth rate declines sharply and the size distribution broadens. In this period, the supersaturation is low because of the rapid consumption of the monomer during the early period. As a result, the value of r^*_{zero} becomes comparable to that of $\overline{r^*}$ and many of the particles in the ensemble fall into the "defocusing" region (Fig. 2.3d). Figure 2.3e depicts the relationship between the mean radius and the relative standard deviation of the size distribution for different initial supersaturations: A high initial supersaturation causes the "focusing" period to be maintained for a large mean radius, resulting in a low relative standard deviation at the end of the "focusing" period. Figure 2.3e also shows that the "defocusing" period leads to a similar equilibrium relative standard deviation regardless of the initial supersaturation. In the "defocusing" and equilibrium periods in which r_{zero} lies near $\overline{r^*}$, Ostwald ripening occurs. In this process, smaller particles dissolve and larger particles grow by receiving the monomers from the dissolving particles (This process was described in the previous section as an explanation for the decrease of the particle concentration during the growth stage). When the Ostwald-ripening process is under pseudoequilibrium state, the dissolving rate and reprecipitation rate of the monomers are balanced and the degree of supersaturation declines very slowly. Generally, the Ostwald ripening broadens the standard deviation of the particle size distribution. At the same time, the mean size of the particle ensemble is also increased. As a result, the relative standard deviation (the standard deviation divided by the mean value) converges to a certain value as the reaction system enters the pseudoequilibrium state. According to the simulation results by Talapin et al. [82], the relative standard deviation in the equilibrium period is almost independent of the initial size distribution but is lowered when the surface free energy is high.

The theoretical works discussed so far explain the behavior of the size distribution during the growth process when there is no additional nucleation. The theoretical studies and simulations of Talapin et al. [81–83] reveal two underlying mechanisms for the control of the size distribution: (1) The "focusing" effect is a kinetically driven process that actively reduces the variance of the particle size distribution during the growth process. It works when the growth process is diffusion controlled and the degree of supersaturation is high. (2) Ostwald ripening occurs when the supersaturation is low. Experimental evidence for the "focusing effect" was provided by Alivisatos and coworkers, who reported that the size distribution of the semiconductor nanocrystals is strongly correlated with the degree of supersaturation [63]. They synthesized CdSe nanoparticles by the hot-injection method and traced the time evolution of the size and the size distribution of the nanocrystals by using photoluminescence (PL) spectroscopy. They observed that a fast increase of the mean particle size and a narrowing of the size distribution occurred simultaneously during the initial period of the growth process. In the later period, the growth rate decreased and the size distribution gradually broadened. The injection of additional monomers during this period resulted in the same effect as that observed during the initial period. This result confirms the relationship between the supersaturation level and the size "focusing" by growth and, consequently, supports the "focusing" mechanism.

2.3 Synthesis Routes for the Upconversion Nanoparticles

Based on the above-mentioned formation mechanism of the monodisperse nanocrystals, a range of synthetic approaches such as thermal decomposition, hydro (solvo)thermal synthesis, Ostwald-ripening method, sol-gel processing, coprecipitation method, as well as ionic liquid-based synthesis have been investigated to synthesize monodisperse high-quality lanthanide-doped upconversion nanoparticles. This section mainly focuses on thermolysis, hydro(solvo)thermal synthesis, etc., which are the most widely used methods, as they can offer precise control over the phase, shape, size, and stoichiometric composition of the core-only and/or the core/shell upconversion nanoparticles. Table 2.1 is a summary of upconversion nanoparticles of varied host materials prepared by these methods.

2.3.1 Thermolysis Strategy

The thermolysis strategy generally employs organometallic compounds as precursors, which decompose in a high boiling point organic solvent with the assistance of surfactants at an elevated temperature. The commonly used precursors are metallic trifluoacetate salts or oleatum; the solvent can be 1-octadecene (ODE), paraffin oil, etc.; the surfactants can be oleic acid (OA), oleylamine (OM), or trioctylphosphine oxide (TOPO), which typically contain a functional capping group to coordinate the metallic elements and the long hydrocarbon chain to prevent nanoparticle aggregation. It is important to note that the rapid decomposition of metallic trifluoroacetate creates a burst of nucleation, which is essential for producing monodispersed nanoparticles. By carefully tailoring the experimental variables, such as the nature of solvents, concentration of metal precursors, reaction temperature, and time, high-quality upconversion nanoparticles with a narrow size distribution, good crystallinity, and exceptional optical properties can be readily obtained from the thermolysis process. The drawback of the thermolysis strategy is that the toxic by-products of fluorinated and oxyfluorinated carbon species mandate a carefully exercised synthetic procedure. Due to the hydrophobic properties of the ligands, the other disadvantage of the thermolysis method is the indispensably post-surface hydrophilic modification, especially for the bio-applications.

Table 2.1 Typical upconversion host materials and their synthetic strategies

Synthetic method	Solvent	Hosts	Morphology	Size (nm)	References
Thermolysis	OA, ODE	$NaYbF_4$	Sphere, polyhedron, sphere, polyhedron	15–30	[24, 84, 85]
		$NaGdF_4$		6–15	[24, 86–88]
		$NaYF_4$	Hexahedron, sphere, nanorod, nanocubic, nanoplate, nanoprism	10–100	[23, 24, 89–91]
		$LiYF_4$	Nanospindle, hexahedron, sphere, nanorod, nanocubic, sphere, nanocubic	20–100	[92–94]
		$NaLnF_4$ (Ln = Nd, Sm, Eu, Tb, Dy, Ho, Er, Tm, Lu)		10–150	[24]
		CeO_2		5–80	[95, 96]
		LaF_3	Nanozigzag, nanoplate	10–50	[26, 55]
		YF_3	Sphere, parallelogram	3–10	[55, 97]
		MF_2 (M = Mg, Ca, Sr)	Nanoneedle, nanocubic	30–100	[98]
		LnF_3 (Ln = Ce, Pr, Nd, Sm, Eu, Gd, Tb, Dy, Ho, Tm, Yb, Lu); LnOF (Ln = Er, Tm, Yb, Lu, Tb, Dy, Ho, Y)	Nanoplate, sphere, nanozigzag, truncatedoctahedron, nanorods, nanopolyhedra	5–20	[55]
		$NaScF_4$	Nanocubic, nanohexahedron	20–30	[99]
	OAM	$NaYF_4$	Sphere	10–50	[100]
		LaOF	Sphere	3–10	[55, 101]
		GdOF	Sphere, nanorods	2–10	[101]

(continued)

Table 2.1 (continued)

Synthetic method	Solvent	Hosts	Morphology	Size (nm)	References
	OAM, ODE	LnOCl	Nanorod, nanoplate	3–200	[102]
	OA, OAM	$NaYF_4$	Sphere	5–20	[103, 104]
		EuOF	Nanowire, polyhedron	3–5	[55]
		GdOF	Nanorod, polyhedron	3–5	[55, 101]
		LnF_3 (Ln = Dy, Ho, Er)	Triangular, nanoplate, nanozigzag,	2–20	[55]
		LnOF (Ln = Ce, Pr, Nd, Sm, Tb, Dy, Ho, Y)	Nanoquadrilateral, nanopolygonal, Nanorod	2–20	[55]
		Ln_2O_3 (Ln = La, Pr, Nd, Sm, Eu, Y)	Nanodisk, nanoplate, nanocubic	5–20	[105]
		$LnPO_4$ (Ln = Ln, Eu, Tb, Y, Ho)	Nanopolyhedron, nanowire, nanoplate, nanorod, nanowormlike wire, nanoquasirod	3–200	[106]
	OA, OAM, ODE	Eu_2O_2S	Nanoplate, nanorod	10–15	[107]
		La_2O_2S	Nanoplate	25–30	[108]
		Gd_2O_3	Nanoplate, nanodisk	8–10	[105, 106]
		Ln_2O_3 (Ln = Tb, Dy, Ho, Er, Tm, Yb, Lu, Y)	Nanoplate, nanodisk	5–20	[105]
		$NaLnS_2$ (Ln = La, Sm, Tb, Ce)	Nanoplate, nanocubic	10–200	[109]
		$NaLnF_4$ (Ln = Pr, Nd, Sm, Eu, Tb, Dy, Ho, Er, Tm, Yb, Lu, Gd)	Nanopolyhedron, nanorod, hexahedron	5–500	[24]
		$NaYF_4$	Nanopolyhedron, nanorod, hexahedron, nanocage	10–200	[24, 110]
		$LiYF_4$	Sphere	10–12	[111]
	OA, TOPO	Y_2O_3	Nanodisk	4–6	[112]
		$NaYF_4$	Nanorod, sphere, nanoplate, nanocubic	9–150	[113–115]

(continued)

Table 2.1 (continued)

Synthetic method	Solvent	Hosts	Morphology	Size (nm)	References
Hydrothermal		$NaYF_4$	Sphere, nanoplate, nanocubic	5–500	[30, 36, 37, 58, 116–119]
		Er_2O_3	Sphere, nanorod	3–3,000	[120]
		$BaYF_5$	Sphere, microcubic	10–1,000	[121]
		CeO_2	Sphere, microcubic	5–1,5	[122]
		SrF_2	Sphere	5–6	[123]
		CaF_2	Nanocubic	5–8	[31]
		LaF_3	Sphere, nanoplate	5–65	[116, 117, 124]
		GdF_3	Sphere	10–20	[116]
		$NaYbF_4$	Sphere, polyhedron	5–25	[30, 116]
		CeF_3	Sphere, nanocubic	5–8	[125, 126]
		$NaCeF_4$	Nanowire, nanorod	100–1,000	[125]
		Na_3MF_7 (M = Zr, Hf)	Nanocubic	6–8	[127]
		BaY_2F_8	Nanobelt	75 × 1,000	[128]
		$YbPO_4$, $LuPO_4$	Sphere	5–10	[21]
		$NaLnF_4$ (Ln = Pr, Nd, Sm, Eu, Gd, Dy, Ho, Er, Tm, Tb)	Polyhedron, nanotube, nanodisk, nanorod	3–800	[30]
		KYF_4	Sphere	20–50	[129]
		$NaLuF_4$	Nanoplate	15–150	[30, 130]

The thermolysis strategy can be divided into three groups according to the solvents (i.e., surface ligands) used in the synthesis.

Thermolysis in oleic acid-based mixed solvents: This thermolysis method was first developed by the Yan group to synthesize highly monodispersed LaF_3 nanoparticles [26]. The approach was later extended as a common route to synthesize high-quality cubic lanthanide-doped $NaYF_4$ upconversion nanoparticles [23–25]. For example, Capobianco and coworkers reported the synthesis of cubic NaYF4 nanoparticles codoped with Yb^{3+}/Er^{3+} or Yb^{3+}/Tm^{3+} via thermal decomposition of metal trifluoroacetate precursors in the presence of OA and ODE [25]. In their studies, the non-coordinating ODE was used as the primary solvent due to its high boiling point. OA was chosen not only as a solvent but also as a passivating ligand that prevents the nanoparticles from agglomeration. The same approach was further refined by Capobianco et al. to synthesize cubic upconversion $NaYF_4$ nanoparticles with a remarkably narrow size distribution without the need for size-selective fractionation [23]. The approach was based on temporal separation of nucleation and crystal growth by slow addition of the precursors to the solution and subsequent manipulation of the temperature. Another outstanding demonstration was reported by Murray and coworkers who prepared highly uniform hexagonal $NaYF_4{:}Yb^{3+}/Er^{3+}$ nanoparticles with controlled sizes and morphologies (spheres, nanorods, and hexagonal prisms) by using the same synthetic strategy [90]. These as-prepared nanoparticles can be easily assembled into large-area super-lattices under appropriate conditions. The decomposition method has also been extended to synthesize other RE fluorides, oxides, and oxyfluorides such as LiREF4 [92–94], KRE_3F_{10} [131], $NaScF_4$ [99], YF_3 [55, 97], MF_2 (M = Mg, Ca, and Sr) [98], $BaREF_5$ [132], CeO_2 [95, 96], and LnOF (Ln = Er, Tm, Yb, Lu, Tb, Dy, Ho, Y) [55].

Thermolysis in oleylamine (OAM)-based mixed solvents: Yan and coworker also presented systematic and general synthesis of high-quality Ln_2O_3 (Ln = Tb, Dy, Ho, Er, Tm, Yb, Lu, Y) [105] and LnOF (Ln = Ce, Pr, Nd, Sm, Tb, Dy, Ho, Y) nanoparticles with diverse shapes in OA/OAM, as well as $NaYF_4$ and $NaLnF_4$ (Ln = Pr, Nd, Sm, Eu, Tb, Dy, Ho, Er, Tm, Yb, Lu, Gd) nanoparticles in OA/OAM/ODE solvents [12, 24, 55, 98, 101, 105, 133]. They refined this technique to the general synthesis of a whole range of rare earth fluorides, rare earth oxide, and rare earth oxyfluoride nanoparticles with multiform crystal phases and morphologies by manipulating the ratio of precursors, solvent composition, reaction temperature, and time. For the synthesis of fluoride nanoparticles, it was found that the combined use of coordinating ligands of the OA and the OAM could produce sub-10-nm cubic lanthanide-doped $NaYF_4$ nanoparticles, which was generally much smaller than the ones produced using OA as the ligand and the ODE as a high-boiling solvent [104]. It is interesting to note that $\sim$10 nm monodisperse hexagonal phase $NaYF_4$: Yb^{3+}/Er^{3+} (or Tm^{3+}) nanoparticles can be synthesized through thermal decomposition of sodium and lanthanide trifluoroacetates by using the single solvent of OM, which served both as a reaction medium and as a capping ligand [100]. However, the resulting hexagonal phase $NaYF_4{:}Yb^{3+}/Er^{3+}$ (or Tm^{3+}) upconversion nanoparticles had relatively poor shape. Subsequently, there was a series of excellent reports on the thermolysis approach in using OA/OAM, OA/OAM/ODE, or OAM solvents to prepare monodispersed lanthanide-doped Eu_2O_2S [107], La_2O_2S [108], Gd_2O_3 [105, 106], LnOCl [102], $LiYF_4$ [111], and $NaLnS_2$ (Ln = La, Sm, Tb, Ce) [109] nanoparticles. The advantage of utilizing OAM for the preparation of upconversion nanoparticles lied in its ability to produce ultrasmall upconversion nanoparticles that are attractive for bio-imaging due to their more efficient clearance from the body. In addition, the weak coordination bond between the amine group and the lanthanide ions on the surface of upconversion nanoparticles facilitated the utilization of a popular ligand exchange procedure for phase transfer.

Thermolysis in TOPO-based mixed solvents: Besides the most frequently used organic capping agents OA, OAM, and ODE, Shan et al. [113–115] first reported the use of TOPO as a coordinating ligand for the synthesis of ultrasmall, monodispersed, and pure hexagonal phase $NaYF_4$ upconversion nanoparticles. As compared to other available coordination solvents such as OA and OAM, the free-energy barrier for the cubic-to-hexagonal phase transition was significantly reduced in TOPO. Unfortunately, the solubility of the as-obtained TOPO-capped nanoparticles was low in common organic solvents such as cyclohexane or chloroform. This problem was solved by combined use of trioctylphosphine (TOP)/OA to replace TOPO to synthesize $NaYF_4$ nanoparticles. It was found that the cooperative actions of OA and TOP also could decrease the energy barrier of α/β phase transition, allowing the preparation of hexagonal $NaYF_4$ at lowered temperatures [113–115].

2.3.2 Hydro(Solvo)Thermal Strategy

The hydrothermal technique is becoming one of the most important tools for advanced materials processing, particularly owing to its advantages in the processing of nanostructural materials for a wide variety of technological applications such as electronics, optoelectronics, catalysis, ceramics, magnetic data storage, biomedical, and bio-photonics. The hydrothermal technique not only helps in processing monodispersed and highly homogeneous nanoparticles, but also acts as one of the most attractive techniques for processing nanohybrid and nanocomposite materials. The term "hydrothermal" is purely of geological origin. It was first used by the British geologist, Roderick Murchison (1792–1871), to describe the action of water at elevated temperature and pressure, in bringing about changes in the earth's crust leading to the formation of various rocks and minerals.

Hydrothermal processing can be defined as any heterogeneous reaction in the presence of aqueous solvents or mineralizers under high pressure and temperature conditions to dissolve and recrystallize materials that are relatively insoluble under ordinary conditions. Definition for the word hydrothermal has undergone several changes from the original Greek meaning of the words "hydros" meaning water and "thermos" meaning heat. Recently, Byrappa and Yoshimura define hydrothermal as any heterogeneous chemical reaction in the presence of a solvent (whether aqueous or non-aqueous) above the room temperature and at pressure greater than 1 atm in a closed system [134]. Possible advantages of this method over other types of synthetic methods include the ability to create highly crystalline phases at lower temperatures and to implement a set of reactions at the same time. Disadvantages of the method include the need for specialized reaction vessels known as autoclaves and the impossibility of observing the nanocrystal as it grows. The autoclaves usually consist of thick-walled steel cylinders with a hermetic seal for carrying out chemical reactions under pressure and high temperatures for prolonged periods of time. Furthermore, protective contact-type inserts typically made of Teflon and titanium materials are generally required to prevent solvent corrosion of the internal cavity of the autoclave.

In a typical procedure of hydro(solvo)thermal synthesis, appropriate reaction precursors, solvents, and surfactants with functional groups are mixed and then heated in a autoclaves. Surfactants such as polyethylenimine (PEI) [14], ethylenediaminetetraacetic acid (EDTA) [121], cetyltrimethylammonium bromide (CTAB), and OA provide chelating ability with cationic ions to regulate their reacting concentration [28], which are essential for the control of the crystalline phase, size, and morphology as well as the surface functional groups of resulting upconversion nanoparticles. The most representative example of hydro(solvo) thermal synthesis is provided by the work of Li and coworkers who reported on a general "liquid–solid-solution (LSS)" strategy for the synthesis of monodisperse (semiconductor, metallic, and dielectric) nanoparticles [37], whereby the reaction, phase transfer, and separation take place at the interfaces. Adapted from the LSS method, a variety of lanthanide-doped upconversion nanoparticles with well-controlled crystal phase, size, and shape have been synthesized, such as $NaYF_4$ [135–138], $NaLaF_4$ [28], $NaLuF_4$ [30, 130], $BaGdF_5$ [139], $KMnF_4$ [140], YF_3 [141–143], LaF_3 [28, 116, 117], LaOF [144], GdF_3 [116], CaF_2 [31], and SrF_2 [123]. Another interesting development was reported by Zhao and coworkers, who utilized an oleic acid-mediated hydrothermal method for the synthesis of uniform $NaYF_4$ nanorods, nanotubes, and flower-patterned nanodisks [34]. Very recently, Liu et al. developed a novel Gd^{3+}-doping approach to give simultaneous control over the crystal phase, size, and optical properties of the resulting $NaYF_4$ upconversion nanoparticles during the hydro(solvo)thermal synthesis [145]. It was discovered that Gd^{3+} doping not only resulted in a rapid cubic-to-hexagonal phase transformation within 2 h, but also lead to the reduction of size, producing ultrasmall hexagonal $NaYF_4$ upconversion nanoparticles at a substantially decreased reaction temperature. The phase transformation and size and morphology optimization induced by impurity doping were further verified by some other groups. Li and coworkers reported the synthesis of $NaLuF_4$:Gd^{3+}/Yb^{3+}/Tm^{3+} nanoparticles with controllable size and phase by doping different amounts of Gd^{3+} ions [146]. The addition of Gd^{3+} ions promoted the phase transformation from cubic to hexagonal and reduced the particle size from large microtubes to small cubes. Wang and coworkers extended this impurity doping strategy for the controlled synthesis of MF_2 (M = Ca, Sr, and Ba) and LnF_3 (Ln = La, Ce, and Pr) nanoparticles [123].

In addition to the aforementioned fluoride nanoparticles, monodispersed Ln^{3+}-doped inorganic oxide nanoparticles could also be readily obtained by using the hydro(solvo)thermal synthesis, as exemplified by the synthesis of Er_2O_3 [120], $BaYF_5$ [121], CeO_2 [122], $NaYbF_4$ [30, 116, 147], $NaCeF_4$ [125], Na_3MF_7 (M = Zr or Hf) [127], $YbPO_4$ [21], $LuPO_4$ [21], KYF_4 [129], and $NaLnF_4$ (Ln = Pr, Nd, Sm, Eu, Gd, Dy, Ho, Er, Tm, Tb) [30].

2.3.3 Other Methods for the Synthesis of the Lanthanide-Doped Luminescent Nanoparticles

Coprecipitation is perhaps one of the most convenient techniques for synthesizing ultrasmall lanthanide-doped nanocrystals with narrow size distribution. Compared to other techniques, there is no need for costly equipment, stringent reaction conditions, and complex procedures, resulting in less time consumption. In some rare instances, crystalline nanoparticles were formed directly by coprecipitation, eliminating the need for a calcination step or post-annealing process. One of the earliest examples of this technique was demonstrated by van Veggel and coworkers [22], who made down-conversion LaF_3 nanocrystals doped with Ln^{3+} (Ln = Eu, Er, Nd, and Ho). The approach was expanded and refined by Chow and coworkers [18], who synthesized LaF_3 nanocrystals with smaller particle size ($\sim$5 nm) and narrower size distribution from simple water soluble inorganic precursors (Fig. 4a, b). In their studies, synthetic ammonium di-*n*-octadecyldithiophosphate was used as a capping ligand to control particle growth and to stabilize the nanocrystals against aggregation. These sub-10-nm crystals can be redispersed in solutions, offering promising applications as luminescent probes for bio-molecules with dimensions from several nanometres to tens of nanometres. In addition to LaF_3 nanocrystals, $NaYF_4$:Yb/Er (Tm), $LuPO_4$:Yb/Tm, and $YbPO_4$:Er nanocrystals were also synthesized via the coprecipitation approach coupled with heat treatment (or post-annealing process) for enhanced upconversion emission by the groups of Haase, Gudel, Chen, and Li [17, 19, 20, 21]. Commercially available ligands such as polyvinylpyrrolidone (PVP) and polyethylenimine (PEI) are also widely used to control particle growth and endow the nanocrystals with solubility and surface functionality [15, 16]. Particularly, the PEI-coated nanoparticles provide a platform for direct surface functionalization of bio-molecules by bio-conjugate chemistry [14].

Sol-gel processing is a typical wet-chemical technique for the fabrication of upconversion nanocrystals for applications as thin film coating and glass materials. The sol-gel process is characterized by the hydrolysis and polycondensation of metal alkoxide (or halide)-based precursors. To improve the crystallinity that is directly associated with luminescence efficiency of the nanocrystals, calcination at high temperatures is often needed. Prasad et al. [41] developed an interesting variation of the sol-gel method that produces Er^{3+}-doped ZrO_2 nanocrystals.

The method involved a sol-emulsion-gel technique that utilized reverse micelles formed in emulsions as the reactors for growing nanocrystals. The sol-gel process was also developed for a variety of upconversion nanocrystals with metal oxides as host materials including TiO_2:Er, $BaTiO_3$:Er, $Lu_3Ga_5O_{12}$:Er, and YVO_4:Yb/Er [38–40]. Despite extensive research efforts in this area, the sol-gel-derived nanocrystals were not particularly suitable as luminescent probes for biological assays due to lack of particle size control and considerable aggregation of the particles when dispersed in aqueous solutions.

In stark contrast to sol-gel and hydro(solvo)thermal methods that often require heating for a prolonged time period up to several days at high temperatures to complete a synthesis, controlled explosions in reactions known as combustion synthesis can give reaction products in minutes. Once initiated by a heat source, highly exothermic reactions with temperatures ranging typically from 500 to 3,000 °C occur in the form of a combustion wave that propagates through the reaction materials in a self-sustained manner without requiring additional heat. The energy saving method represents one of the attractive techniques to synthesize a wide variety of oxide and oxysulfide upconversion nanocrystals (Y_2O_3, La_2O_2S, and Gd_2O_3) reported by the groups of Capobianco, Luo, and Zhang [152–154].

2.3.4 Ultrasmall Upconversion Nanoparticles with Bright Upconversion Luminescence

Although upconversion nanoparticles show many advantages in the non-blinking property and high signal-to-noise ratio, however, compared with QDs and dyes, one of the biggest shortcomings is too large in size ($\sim$20 nm). The optimal size for bio-imaging probes is reported to be less than 10 nm, so that the nanoparticles can be efficiently cleared from the body. A few papers have reported the synthesis of the ultrasmall upconversion nanoparticles (Fig. 2.4). However, the low efficiency is often accompanied by the decrease in size. For example, it has been demonstrated that when the size was reduced to $\sim$5 nm, the upconversion efficiency sharply declined to below 0.01 %, which is much lower than that of the particles with larger size. In regard to this, much effort has been input to obtain small (<10 nm) nanoparticles with intense upconversion emission and many positive achievements have been obtained [104, 148, 149, 155–157]. For example, Prasad's group reported the synthesis of monodispersed $NaYF_4$:Yb^{3+}/Tm^{3+} nanocrystals with ultrasmall size (7–10 nm) and high upconversion efficiency by thermolysis trifluoacetate salts in the OA/OAM mixed solvents [104]. The intensity of NIR emission could be increased up to 43 times along with an increase in the relative content of Yb^{3+} ions from 20 to 100 %. The resultant ultrasmall $NaYbF_4$:2 %Tm^{3+} nanocrystals manifest NIR photoluminescence emission, which was 3.6 times more

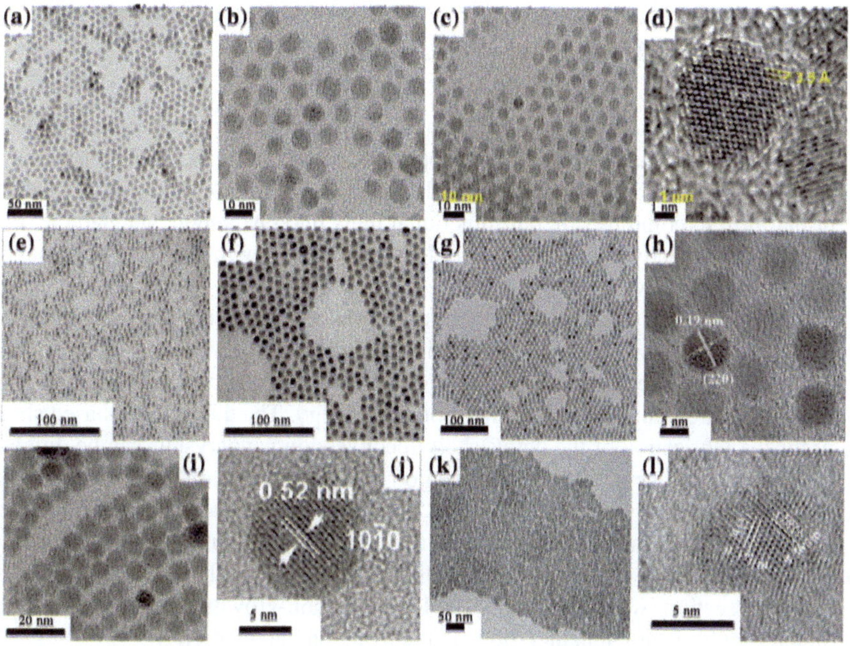

Fig. 2.4 TEM images of the ultrasmall (<10 nm) lanthanide-doped upconversion nanoparticles. **a**, **b** Cubic phase $NaYF_4$:Yb,Tm; **c**, **d** hexagonal phase $NaYF_4$:Yb,Er; **e**, **f** hexagonal phase $NaGdF_4$; **g**, **h** cubic phase CaF_2:Yb,Er@CaF_2; **i**, **j** $NaLu_xGd_{1-x}F_4$:Yb,Er; **k**, **l** Orthorhombic YF_3: Yb,Er. **a**, **b** Reprinted with the permission from Ref. [104]. Copyright 2010 American Chemical Society. **c**, **d** Reprinted with the permission from Ref. [148]. Copyright 2012 American Chemical Society. **e**, **f** Reprinted with the permission from Ref. [149]. Copyright 2011 American Chemical Society. **g**, **h** Reprinted with the permission from Ref. [150]. Copyright 2013 Wiley-VCH Verlag GmbH & Co. KGaA. **i**, **j** Reprinted with the permission from Ref. [146]. Copyright 2011 American Chemical Society. **k**, **l** Reprinted with the permission from Ref. [151]. Copyright 2012 The Royal Society of Chemistry

intense than that from $NaYF_4$:20%Yb^{3+}/2%Tm^{3+} nanocrystals with a size of 25–30 nm reported previously. Cohen's group also described the controlled synthesis of β-phase $NaYF_4$:20%Yb^{3+},2%Er^{3+} nanocrystals with a protein size from 4.5 to 15 nm by using the thermolysis strategy in the OA/OAM/ODE solvents [148]. The core/shell β-phase $NaYF_4$:Yb^{3+},Er^{3+}/$NaYF_4$ nanoparticles less than 10 nm exhibited higher luminescence efficiency (0.18 ± 0.01 %) than the core nanoparticles with 37 nm in size (0.14 ± 0.01 %). Most recently, Liu et al. [146] also synthesized the sub-10-nm hexagonal $NaLuF_4$, which displayed bright luminescence with a QY of 0.47 ± 0.06 %. Mahalingam et al. also demonstrated the upconversion process via interparticle energy transfer in Ln^{3+}-doped $BaLuF_5$ (Ln = Tm and Yb) nanocrystals with a sub-5 nm size for the first time [156].

2.4 Synthesis Routes for the Fabrication of the Core@Shell Structured Upconversion Nanoparticles

In addition to Ln^{3+}-doped core-only nanoparticles, much progress has also been made in the synthesis of composite nanoparticles with core@shell structures in the past decade. Such composite core@shell nanoparticles not only provide an excellent platform to combine different functional units into one single nanoparticle, but also offer an effective solution to problems associated with the core-only nanoparticles such as low PL efficiency.

It has been demonstrated that the thermolysis strategy can offer precise control over the phase, shape, size, and stoichiometric composition of the upconversion nanocrystals. So the routes for the fabrication of the core@shell-structured upconversion nanocrystals are mainly based on thermolysis strategy. The thermolysis-based strategy for the fabrication of the core@shell upconversion nanocrystals can be divided into three routes: one-pot heating-up method for the synthesis of both core only and core@shell upconversion nanoparticles [158–161]; one-pot successive layer-by-layer (SLBL) [162–164] and Ostwald-ripening methods [56]. It should be noted that all the three routes are developed under the direction of the LaMer nucleation-growth model. So, in this section, we focus on the LaMer nucleation-growth model directed controllable synthesis of lanthanide-doped upconversion nanoparticles.

2.4.1 *One-Pot Heating-Up Method*

For one-pot heating-up method, the nucleation and growth match with the LaMer model for the synthesis of monodisperse upconversion nanoparticles. For the synthesis of the core upconversion nanoparticles, the precursors are mixed at a low temperature and then brought to the temperature at which precursor reaction or decomposition occurs sufficiently quickly to result in supersaturation (stage I). Supersaturation is again relieved by a "burst nucleation" (stage II), after which temperature is controlled to avoid additional nucleation events, allowing monomer addition to existing nuclei to occur more rapidly than new monomer formation (stage III).

The seed-mediated growth method for the synthesis of core@shell nanocrystals is the most apparent case for the separation of nucleation and growth, wherein nucleation is physically separated from growth by using preformed nanocrystals as seed nuclei. This method utilizes heterogeneous nucleation to suppress the formation of additional nuclei by homogeneous nucleation. In this method, the reaction can enter into stage III directly. Preformed nuclei are introduced into the reaction solution, and then the monomers are supplied to precipitate on the surface of the existing nuclei. The lower monomer concentration should be kept during growth to suppress homogeneous nucleation. So, the one-pot heating-up method can be used not only in the synthesis of core-only nanoparticles but also make a great success in the fabrication of core@shell upconversion nanoparticles (Fig. 2.5a). The whole experiment procedure is nearly the same as the synthesis of the core nanoparticles except the presence of the seed nanoparticles. The first step is to synthesize the core nanoparticles which are then used as seeds in the second step to induce further epitaxial growth. To ensure an epitaxial growth and to create a homogeneous interface between the core and the outer shell, the host material of the shell should exhibit a low lattice mismatch with the core material. Then, the epitaxial growth can be implemented in the growth solution. The monomer of shells in the growth solution, produced by thermal decomposition of corresponding chemical precursors in the solvents of OA and ODE, is able to gradually and stably deposit on the seeding core nanoparticles and finally form the shell layer (stage III in LaMer plot).

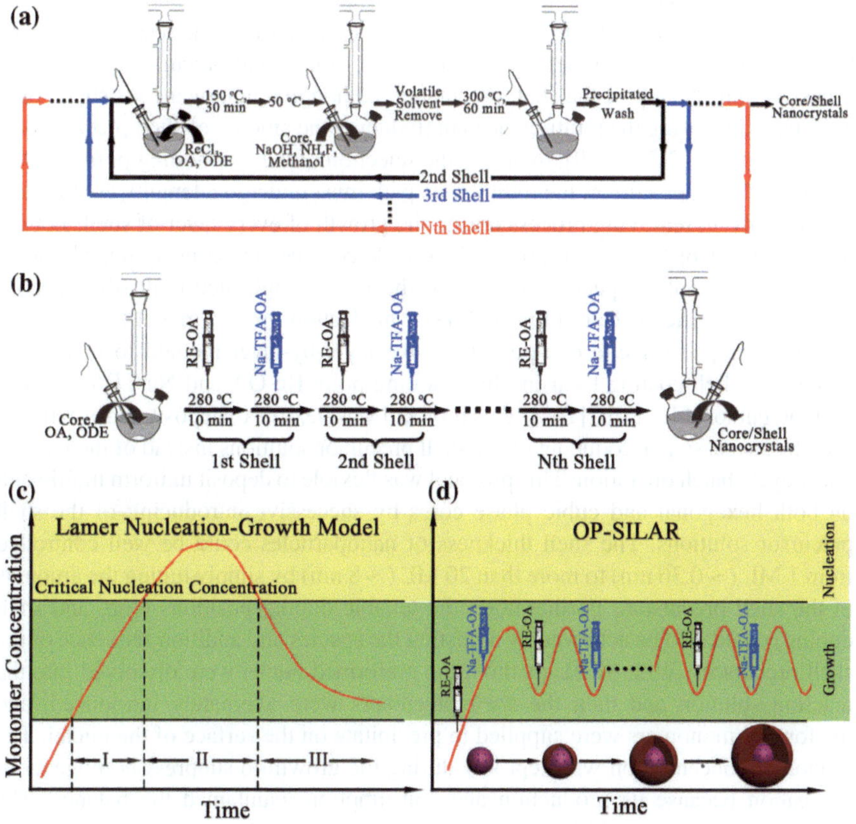

Fig. 2.5 **a**, **b** Schematic depiction of conventional heating-up multi-cycle method (**a**) and the SLBL method (**b**) for the synthesis of core@shell upconversion nanoparticles. **c**, **d** Schematic illustration of the LaMer plot and the SLBL method for the controlled synthesis of core@shell upconversion nanoparticles by using the SLBL method. **a**–**d** Reprinted with the permission from Ref. [164]. Copyright 2013 American Chemical Society

2.4.2 Successive Layer-by-Layer Strategy for the Fabrication of Multi-shell-Structured Upconversion Nanoparticles

Although one-pot heating-up method has been considered to the most commonly used method for the fabrication of core@shell upconversion nanoparticles, a few shortcomings should still be improved. For example, recent work by van Veggel et al. showed that the core@shell nanoparticles synthesized by the one-pot heating-up method did not have a uniform shell (the core is only partially covered by shell), indicating that the core nanoparticles could not be passivated efficiently to decrease the core surface defects and insulated from the environmental effect [161, 165]. On the other hand, it was found that the size distribution of the obtained core@shell nanoparticles was larger than that of the original core. Furthermore, bimodal size distribution is always observed when a high volume of shell precursors is introduced [158]. It is assumed that the non-uniform shell and broader or bimodal size distribution of the obtained core@shell upconversion nanoparticles is resulted from the fast deposition speed or independent nucleation of the shell precursors induced by excessive introducing of the shell precursors. Although a homogeneous shell could be obtained by using the multi-cycle path (reducing the amount of shell precursors of each cycle) (Fig. 2.5a) [159], owing to the selection of the low boiling point NaOH-NH_4F-methanol precursor, the multi-cycle path must undergo a lengthy cooling and volatile solvent removing process during the growth of every layer of shell. For the reason of batch operation, this process is considered to be time-consuming, laborious, and user unfriendly, especially for the synthesis of complicated multi-shell upconversion nanoparticles. Based on the LaMer nucleation-growth model [10, 60], our group developed a facile one-pot successive layer-by-layer method to realize the successive shell coating by using high boiling-point Re-OA and Na-TFA-OA as a shell precursor (Fig. 2.5b) [164]. This protocol was flexible to deposit uniform multi-shell by successive introducing of the shell precursor solutions instead of the lengthy multi-cycle batch operation. This protocol was flexible to deposit uniform multi-shell on both hexagonal and cubic phase cores by successive introducing of the shell precursor solutions. The shell thickness of nanoparticles could be well controlled from 1 ML ($\sim$0.36 nm) to more than 20 ML ($\sim$8 nm) by simply tuning the amounts of the shell precursors. Furthermore, the tunable doping positions (core and shell doping) could also be achieved by adjusting the species and addition sequence of the shell precursors. With SLBL method, the preformed nuclei were dissolved into the reaction solution and then the shell precursors were alternately introduced, and the formed monomers were supplied to precipitate on the surface of the nuclei. The monomer concentration was kept low during the growth to suppress homogeneous nucleation because its production and consumption maintained the balance. The released monomers could be continuously consumed to deposit as a shell without homogeneous nucleation (Fig. 2.5c). The most distinguished feature of core@shell upconversion nanoparticles synthesized from SLBL method was that the shell was very uniform. It means that the cores were completely covered by shell, which is quite different from the non-uniform shell nanoparticles synthesized by commonly used one-pot heating-up method (the core was only partially covered by shell). As a result of the high-quality uniform shells, the optical properties of the obtained core@shell nanoparticles could be improved in upconversion luminescence efficiency and stability. Most recently, our group reported for first time that the homogeneous doping approach based on the SLBL method could greatly improve the efficiency of the upconversion nanoparticles [166]. The quantum yield as high as 0.89 ± 0.05 % was realized for the homogeneous doping $NaGdF_4$:Yb,Er/$NaYF_4$ upconversion nanoparticles, which was nearly 2 times higher than that of the heterogeneous doping $NaGdF_4$:Yb,Er/$NaYF_4$ upconversion nanoparticles (0.47 ± 0.05 %). The influences of spatial distributions and local relative concentrations of the dopants on the optical properties of upconversion nanoparticles were investigated in the single particle level. It was found that heterogeneous doping indeed existed during the spontaneous growth process of the nanoparticles. The heterogeneous doping property can further induce many negative effects on the optical properties of upconversion nanoparticles, especially the luminescent efficiency. The spatial distributions and local relative concentrations of the dopants can be well controlled by the SLBL homogeneous doping method on the monolayer level and homogeneously distributed in the single particle level. With the similar SLBL method, the highly efficient $LiLuF_4:Ln^{3+}$ @$LiLuF_4$ core@shell upconversion nanoparticles and dual-mode up/down converting $NaGdF_4$:Nd@$NaYF_4$@$NaGdF_4$:Nd,Yb,Er@$NaYF_4$ core@shell1@shell2@shell3 has been well established [162, 163].

As mentioned above, the SLBL method can realize the monolayer control of the shell thickness. The spherical concentric shell model (CSM) can be employed to calculate the amount of shell precursor necessary for the growth of each monolayer (ML) [159, 167]. This model has been used extensively for the hexagonal CdS and ZnS shell deposition during the core–shell nanocrystal synthesis process [167]. Because of the highly symmetric structure of the CdS, the ML was inferred to a thickness equal to half the length of the c-lattice parameter. Although the $NaGdF_4$ shell adopted here has also hexagonal structure ($P6_3/m$, $a = 6.02$ Å, $c = 3.60$ Å), the cation sites are of three types. A onefold site occupied by RE^{3+}, another site occupied randomly by $1/2Na^+$ and $1/2RE^{3+}$, and a twofold site occupied randomly by Na^+ and vacancies [145, 159]. Therefore, referral to a $NaGdF_4$, the ML could be taken to mean a thickness equal to the *c*-lattice parameter of the bulk material, 0.36 nm in the case of the hexagonal $NaGdF_4$. The required Gd-OA precursor amount for every layer of one particle can be calculated by the following equations:

$$\begin{aligned} m_{\text{layer}(n)} &= m_{\text{particle}(n)} - m_{\text{particle}(n-1)} = \rho\left(V_{\text{particle}(n)} - V_{\text{particle}(n-1)}\right) \\ &= \rho\frac{4}{3}\pi\left(r_{\text{particle}(n)} - r_{\text{particle}(n-1)}\right) \end{aligned} \tag{2.28}$$

Density of the $NaGdF_4$ materials (ρ) is calculated as follows:

$$\rho = \frac{m}{V} = \frac{M}{\left(N_A * c * \frac{\sqrt{3}}{2} a^2\right) / N'} \tag{2.29}$$

Here M represents relative molecular mass of the material of $NaGdF_4$, N' means the number of $NaGdF_4$ units that one crystal cell contains and according to crystal structure of the hexagonal $NaGdF_4$, $N' = 1.5$. So, for $NaGdF_4$, the cell parameter is $a = 6.02$ Å, $c = 3.60$ Å, molecular weight (M) = 256.3, and $\rho = 5.65$ g/cm^3. For $NaYF_4$, cell parameter is $a = 5.96$ Å, $c = 3.53$ Å, molecular weight $M = 187.9$, and $\rho = 4.31$ g/cm^3.

Particle number of one molar (N) the $NaGdF_4$ (N) is calculated as follows:

$$N = \frac{V_{\text{per molar}}}{V_{\text{per particle}}} = \frac{\frac{M_0}{\rho_0}}{\frac{4}{3}\pi r_0^3} \tag{2.30}$$

Because the particle number is constant during the particle growth from smaller to bigger, the particle number of the obtained initial seeds is nearly the same as N.

2.4.3 Ostwald Ripening Strategy for the Fabrication of Core@Shell Upconversion Nanoparticles

The Ostwald-ripening process is referred to the last stage in LaMer plot, in the process larger particles with smaller surface-to-volume ratios are favored over energetically less stable smaller particles, resulting in the growth of larger particles at the expense of smaller ones. The fabrication of core@shell upconversion nanoparticles by using the common physical phenomenon of Ostwald ripening (i.e., self-focusing by Ostwald ripening) was developed by van Veggel and coworkers [56]. In their study, small α-$NaYF_4$ sacrificial nanoparticles (SNPs) were synthesized first and then used as shell precursors. The subsequent injection of SNPs into a hot solution of $NaYF_4$ core nanoparticles (defocusing) can induce the rapid dissolution of SNPs and their deposition onto the larger core nanoparticles (self-focusing), thereby resulting in the formation of genuine core@shell-structured upconversion nanoparticles with a well-defined shape, narrow size distribution, and tunable shell growth. More importantly, the shell thickness and multiple layers of such core@shell upconversion nanoparticles can be precisely controlled by manipulating the number of SNPs injected and/or repeating the above defocusing and self-focusing cycle. Furthermore, this method is flexible and can be easily extended to prepare other high-quality core@shell nanoparticles with different compositions.

2.4.4 *Cation Exchange Strategy for the Fabrication of Core@Shell Upconversion Nanoparticles*

Besides the self-focusing by Ostwald-ripening strategy, another facile way to prepare monodisperse Ln^{3+}-doped core@shell nanoparticles is through a cation exchange reaction. One of the earliest examples of this technique was demonstrated by van Veggel and Dong [168], who prepared GdF_3@LnF_3 core@shell nanoparticles by exposing GdF_3 core nanoparticles to an aqueous solution containing excess Ln^{3+} ions, where the Gd^{3+} ions in GdF_3 nanoparticles were partially replaced by the Ln^{3+} in the solution. By refining this synthetic route, they synthesized multifunctional $NaYF_4$:Yb/Tm@$NaGdF_4$ core@shell nanoparticles with tunable shell thickness, enhanced upconversion luminescence and outstanding paramagnetic performance via cation exchange of $NaYF_4$:Yb/Tm nanoparticles with Gd^{3+} ions, making these nanoparticles promising for bimodal imaging, namely, optical bioimaging and magnetic resonance imaging (MRI) [165]. Similarly, Li and coworkers developed a novel approach based on rare earth cation-assisted ligand assembly to synthesize Gd^{3+}-modified $NaYF_4$:Yb/Er core@shell nanoparticles [169]. In their study, $NaYF_4$:Yb/Er nanoparticles were first synthesized via a modified cothermolysis process in oleylamine, then Gd^{3+} ions were introduced onto the surface of the $NaYF_4$:Yb/Er nanoparticles by cation-assisted ligand assembly, which thereby greatly increased the positive charge on the surface of the $NaYF_4$:Yb/Er nanoparticles so that the functional molecules (i.e., folic acid and aminocaproic acid) can be attached.

2.5 Characterizations

2.5.1 *Optical Characterization*

The most direct and immediate probes for upconversion nanoparticles are the fluorescence spectra. In a typical fluorescence (emission) measurement (Fig. 2.6a), the excitation wavelength is fixed and the detection wavelength varies. Two general types of instruments exist: (1) Filter fluorometers use filters to isolate the incident light and fluorescent light. (2) Spectrofluorometers use diffraction grating monochromators to isolate the incident light and fluorescent light. Both types use the following scheme: The light from an excitation source passes through a filter or monochromator and strikes the sample. A proportion of the incident light is absorbed by the sample and some of the molecules in the sample fluoresce. The fluorescent light is emitted in all directions. Some of this fluorescent light passes through a second filter or monochromator and reaches a detector, which is usually placed at 90° to the incident light beam to minimize the noise created from radiant power fluctuations and the risk of transmitted or reflected incident light reaching the detector.

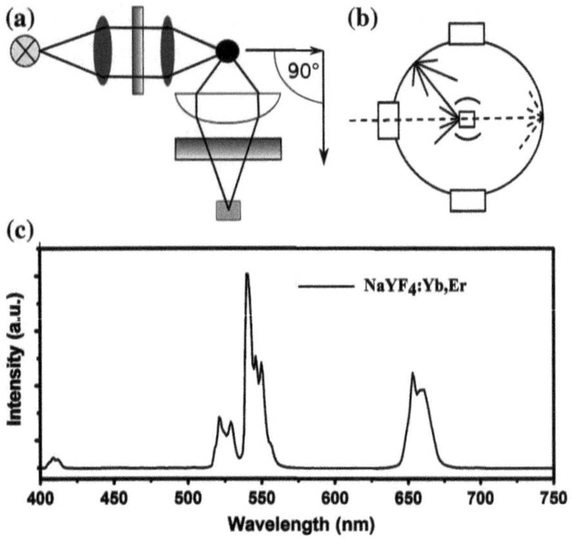

Fig. 2.6 **a** Scheme illustration of the fluorescence spectroscopy. **b** Diagram of the integrating sphere setup for luminescence measurements. *Dashed line* excitation light; *solid line* sample emission. **c** Typical upconversion emission spectra of the $NaYF_4$:Yb,Er upconversion nanocrystals

As mentioned before, the fluorescence is most often measured at a 90° angle relative to the excitation light. This geometry is used instead of placing the sensor at the line of the excitation light at a 180° angle in order to avoid interference of the transmitted excitation light. No monochromator is perfect and it will transmit some stray light, that is, light with other wavelengths rather than the targeted. An ideal monochromator would only transmit light in the specified range and have a high wavelength-independent transmission. When measuring at a 90° angle, only the light scattered by the sample causes stray light. This results in a better signal-to-noise ratio, and lowers the detection limit by approximately a factor of 10,000, when compared to the 180° geometry. Furthermore, the fluorescence can also be measured from the front, which is often done for turbid or opaque samples. The detector can either be single-channeled or multi-channeled. The single-channeled detector can only detect the intensity of one wavelength at a time, while the multi-channeled detects the intensity of all wavelengths simultaneously, making the emission monochromator or filter unnecessary. The different types of detectors have both advantages and disadvantages.

In traditional commercial available fluorescence spectroscopy, xenon arcs and mercury-vapor lamps light sources are always used as excitation sources. A mercury-vapor lamp is a line lamp, meaning it emits light near peak wavelengths. By contrast, a xenon arc has a continuous emission spectrum with nearly constant intensity in the range from 300 to 800 nm and a sufficient irradiance for measurements down to just above 200 nm. Filters and/or monochromators may be used to get the excitations with narrow wavelength interval. A monochromator transmits light of an adjustable wavelength with an adjustable tolerance. The most common type of monochromator utilizes a diffraction grating, that is, collimated light illuminates a grating and exits with a different angle depending on the wavelength.

The monochromator can then be adjusted to select which wavelengths to transmit. For allowing anisotropy measurements, the addition of two polarization filters is necessary: One after the excitation monochromator or filter, and one before the emission monochromator or filter. For the fluorescence spectroscopy used in the measurement of upconversion nanomaterials, the excitation sources should be replaced by a NIR laser (~980, 800 nm etc.) which can only emits light of high irradiance at a very narrow wavelength interval. Because (1) the xenon arc has a continuous emission spectrum range from 300 to 800 nm, which cannot be used as the excitation source of upconversion nanomaterials; (2) the power density of the narrow wavelength excitation, which is obtained from the xenon arc by using filters and/or monochromators, is too low to excite the upconversion luminescence. The excitation monochromator or filter is unnecessary when a laser is used as the excitation source. Figure 2.6c shows the typical upconversion emission spectra of the $NaYF_4$:Yb,Er upconversion nanocrystals.

Unlike in UV/visible spectroscopy, "standard", device-independent spectra are not easily attained. Several factors influence and distort the spectra, and corrections are necessary to attain "true", i.e., machine-independent, spectra. The different types of distortions will here be classified as being either instrument- or sample-related. Firstly, the distortion arising from the instrument is discussed. As a start, the light source intensity and wavelength characteristics varies over time during each experiment and between each experiment. Furthermore, no lamp has a constant intensity at all wavelengths. To correct this, a beam splitter can be applied after the excitation monochromator or filter to direct a portion of the light to a reference detector. Additionally, the transmission efficiency of monochromators and filters must be taken into account. These may also change over time. The transmission efficiency of the monochromator also varies depending on wavelength. This is the reason that an optional reference detector should be placed after the excitation monochromator or filter. The percentage of the fluorescence picked up by the detector is also dependent upon the system. Furthermore, the detector quantum efficiency, that is, the percentage of photons detected, varies between different detectors, with wavelength and with time, as the detector inevitably deteriorates.

Two other topics that must be considered including the optics used to direct the radiation and the means of holding or containing the sample material (called a cuvette or cell). For most UV, visible, and NIR measurements, the use of precision quartz cuvettes is necessary. In both cases, it is important to select materials that have relatively little absorption in the wavelength range of interest. Quartz is ideal because it transmits from 200 to 2,500 nm; higher grade quartz can even transmit up to 3,500 nm, whereas the absorption properties of other materials can mask the fluorescence from the sample. Correction of all these instrumental factors for getting a "standard" spectrum is a tedious process, which is only applied in practice when it is strictly necessary. This is the case when measuring the quantum yield or when finding the wavelength with the highest emission intensity for instance.

As mentioned earlier, distortions arise from the sample as well. Therefore, some aspects of the sample must be taken into account too. Firstly, photodecomposition may decrease the intensity of fluorescence over time. Scattering of light must also be taken into account. The most significant types of scattering in this context are Rayleigh and Raman scattering. Light scattered by Rayleigh scattering has the same wavelength as the incident light, whereas in Raman scattering the scattered light changes wavelength usually to longer wavelengths. Raman scattering is the result of a virtual electronic state induced by the excitation light.

Frequently, the upconversion emission spectra were used in the qualitative analysis of the upconversion nanocrystals. It is very hard to get the quantitative information from the upconversion emission spectra, especially for the spectra obtained from different equipments or samples. For comparison, the integrating sphere is always used to quantify the intensity of fluorescence. An integrating sphere (also known as an Ulbricht sphere) is an optical component consisting of a hollow spherical cavity with its interior covered with a diffuse white reflective coating, with small holes for entrance and exit ports (Fig. 2.6b). Its relevant property is a uniform scattering or diffusing effect. Light rays incident on any point on the inner surface are, by multiple scattering reflections, distributed equally to all other points. The effects of the original direction of light are minimized. An integrating sphere may be thought of as a diffuser which preserves power but destroys spatial information. It is typically used with some light source and a detector for optical power measurement. It has become a standard instrument in photometry and radiometry. It has the advantage over a goniophotometer for measuring the light produced by a source that total power can be obtained in a single measurement.

Until now, the low upconversion efficiencies of upconversion nanoparticles are still the most serious limitation for commercialization, resulting that in most of applications having to be operated at high excitation densities. The efficiency of an upconverter is determined by many variables which influence the radiative, non-radiative, and energy transfer rates, e.g., phonon energies, spectral overlap between sensitizer emission and activator absorption, temperature, and defects that quench the emission. Very importantly, as upconversion is a nonlinear process, the efficiency strongly depends on excitation power. This makes it difficult to define the upconversion efficiency detected under different conditions. There is no generally excepted definition on upconversion efficiency. When the efficiency is determined, it is still not clear how to interpret the values reported. For example, for one emitted photon, at least two photons are absorbed. This means that the maximum internal quantum efficiency is 50 %, analogous to the fact that downconversion can have an efficiency of 200 %. Still, sometimes this maximum efficiency is reported as 100 %.

In 2010, van Veggel et al. reported the measurement of absolute efficiencies of upconversion nanoparticles by using the integrating sphere [155]. They employed a barium sulfate-coated integrating sphere (150 mm in diameter) from Edinburgh instruments. The integrating sphere was mounted on the fluorimeter with the entry and output ports of the sphere located in 90° geometry from each other in the plane of the spectrometer. The powder or colloidal samples were held in a quartz cuvette located in the center of the integrating sphere. Samples were excited with a ~980 nm laser diode coupled to a 105-μm fiber. The emission from the fiber tip was collimated to a beam diameter of 1 mm and directed on the samples using a fiber coupler. Baffles were employed on both sides of the sample holder to ensure that no scattered excitation light or emissions would be collected before scattering off the inside of the sphere. All the spectroscopic data collected were corrected for the spectral response of both the fluorimeter and the integrating sphere. The response of the detection systems in photon flux (sphere, monochromators, and detectors) was determined using a calibrated tungsten lamp. These normalization curves were then applied to all measured spectra.

The QY is defined as:

$$QY = \frac{\text{Photons emitted}}{\text{Photons absorbed}} = \frac{L_{[\text{sample}]}}{E_{[\text{reference}]} - E_{[\text{sample}]}} \tag{2.31}$$

where QY is the quantum yield, L_{sample} is the emission intensity, $E_{\text{reference}}$ and E_{sample} are the intensities of the excitation light not absorbed by the sample and the reference sample, respectively. The above equation expresses the number of photons emitted (numerator) divided by the number of photons absorbed (denominator), so it runs from 0 to 100 % for a one-photon processes and from 0 to 50 % for a two-photon processes.

According to the above method reported by Boyer and van Veggle, an increasing number of papers began to involve data of the efficiency (Table 2.2). For example, Liu et al. [146] reported that hexagonal $NaLuF_4$ nanocrystals codoped with Gd^{3+}, Yb^{3+}, and Er^{3+} with sub-10 nm size displayed bright luminescence with a QY of 0.47 ± 0.06 % under the excitation laser at 980 nm. Cohen et al. reported on the controlled synthesis of protein-sized (4.5–15 nm) β-$NaYF_4$:Yb^{3+},Er^{3+} nanocrystals with the efficiency as high as 0.18 ± 0.01 % [148]. As a result of the high-quality uniform multi-shell epitaxial growth by using the one-pot successive layer-by-layer (SLBL) strategy developed by our group, the optical properties of the obtained core–shell nanocrystals could be highly improved in upconversion luminescence efficiency (up to 0.51 ± 0.08 %) [164]. Prasad's group realized the efficiency as high as ~1.2 ± 0.1 % under the excitation of 1,490 nm for the green photoluminescence of $NaYF_4$:Yb^{3+},Er^{3+} nanocrystals [94]. It is almost four times higher than the photoluminescence efficiency reported to date for hexagonal $NaYF_4$:Yb^{3+},Er^{3+} with a size of ~100 nm using excitation at ~980 nm, but it is still lower than the bulk materials. In 2012, Xu et al. [170] reported that the $NaYF_4$:Yb,Tm@$NaYF_4$ upconversion nanoparticles showed the QY of 3.5 % for 800 nm emission under excitation intensity of 78 W/cm^2. At a low excitation intensity of 21.7 mW/cm^2, the QY was determined to be 3.8×10^{-4}. Although this is the highest efficiency reported to date in literatures for the upconversion nanoparticles, QY measurement method and the detected emission wavelength are different from those reported by van Veggel et al. [155]. In this work, the standard fluorophores were employed to measure the QY with the integrating sphere-based system. The resulting QY of the upconversion nanoparticles was calculated by an indirect comparison. So the upconversion efficiency results should be interpreted with caution and further studies of the efficiency of upconversion nanoparticles are certainly needed in the future work.

Table 2.2 The efficiency of the upconversion nanoparticles with different crystal structure, size, and excitation power reported in the literatures

Upconversion nanoparticles	Crystal structure	Average size (nm)	Power density (W/cm^2)	Efficiency (%)	References
$NaYF_4$:20%Yb^{3+}, 2%Er^{3+}	Hexagonal	Bulk	20	3.0 ± 0.3	[155]
$NaYF_4$:20%Yb^{3+}, 2%Er^{3+}	Hexagonal	100	150	0.30 ± 0.10	[155]
$NaYF_4$:20%Yb^{3+}, 2%Er^{3+}	Hexagonal	30	150	0.10 ± 0.05	[155]
$NaYF_4$:20% Yb^{3+},2%Er^{3+}	Hexagonal	8–10	150	0.005 ± 0.005	[155]
$NaYF_4$:20%Yb^{3+}, 2%Er^{3+}@$NaYF_4$	Hexagonal	30	150	0.30 ± 0.10	[155]
$NaLuF_4$:24%Gd^{3+}, 20%Yb^{3+},1% Tm^{3+}	Hexagonal	7.8	17.5	0.47 ± 0.06	[146]
$NaYF_4$:20%Yb, 2%Er	Hexagonal	5.4	1,000	0.0022 ± 0.0001	[148]
$NaYF_4$:20%Yb, 2%Er@$NaYF_4$	Hexagonal	9	1,000	0.18 ± 0.01	[148]
$NaYF_4$:20%Yb, 2%Er	Hexagonal	37	1,000	0.14 ± 0.01	[148]
$NaGdF_4$:20%Yb, 2%Er	Hexagonal	5	100	0.016 ± 0.008	[164]
$NaGdF_4$:20%Yb, 2%Er @$NaYF_4$	Hexagonal	17	100	0.51 ± 0.01	[164]
$LiYF_4$:Er^{3+}	Tetragonal	85	–	1.2 ± 0.1	[94]
$NaYF_4$:Yb, Tm@$NaYF_4$	Hexagonal	42	78	3.5	[170]

2.5.2 Chemical and Structural Characterization

2.5.2.1 Transmission Electron Microscopy

One of the typical characters of the upconversion nanomaterials is the small particle size. Although some structural features can be revealed by X-ray and neutron diffraction, direct imaging of nanoparticles is only possible using transmission electron microscopy (TEM) and scanning probe microscopy. TEM is unique because it can provide a real space image of the atomic distribution in the nanocrystals and on their surface. Today's TEM is a versatile tool that provides not only atomic-resolution lattice images, but also chemical information at spatial resolution of 1 nm or better, allowing direct identification the chemistry of a single nanocrystal.

A TEM is composed of an illumination system, a specimen stage, an objective lens system, a magnification system, a data recording system, and a chemical analysis system. In addition to the morphology information, TEM can also provide us information about the crystal structure by the selected area electron diffraction (SAED), the chemical composition and distribution of the nanoparticles by the chemical analysis system. The chemical analysis system is making use of the energy-dispersive X-ray spectroscopy (EDS) and the electron energy-loss spectroscopy (EELS), both can be used complimentary to quantify the chemical composition of the specimen. EELS can also provide information about the electronic structure of the specimen.

SAED is a crystallographic experimental technique that can be performed inside a TEM. In a TEM, a thin crystalline specimen is subjected to a parallel beam of high-energy electrons. As TEM specimens are typically ~100 nm thick, and the electrons typically are accelerated with the energy of 100–400 kV, allowing them to pass through the electrons pass through the sample easily. In this case, electrons are treated as wave-like, rather than particle-like (wave-particle duality). Because the wavelength of high-energy electrons is a few thousandths of a nanometer, and the spacing between atoms in a solid is about a hundred times larger, the atoms act as a diffraction grating to the electrons, which are diffracted. That is, some fraction of them will be scattered to particular angles, determined by the crystal structure of the sample, while others continue to pass through the sample without deflection. As a result, the image on the screen of the TEM will be a series of spots, i.e., the selected area diffraction pattern, each spot corresponding to a satisfied diffraction condition of the sample's crystal structure. If the sample is tilted, the same crystal will stay under illumination, but different diffraction conditions will be activated, and different diffraction spots will appear or disappear. Figure 2.7a shows the typical SAED of the hexagonal and cubic phase $NaYF_4$ nanoparticles [164]. It can be seen that the two SAED show quite different diffraction rings. Figure 2.7a1 shows four distinct rings and the calculated d-spacings matched well with interplanar distances of (111), (200), (220), and (311) planes of cubic phase $NaYF_4$ (JCPDS, 77-2042). In contrast, the *d*-spaces of the lattice planes measure from Fig. 2.7a2 are

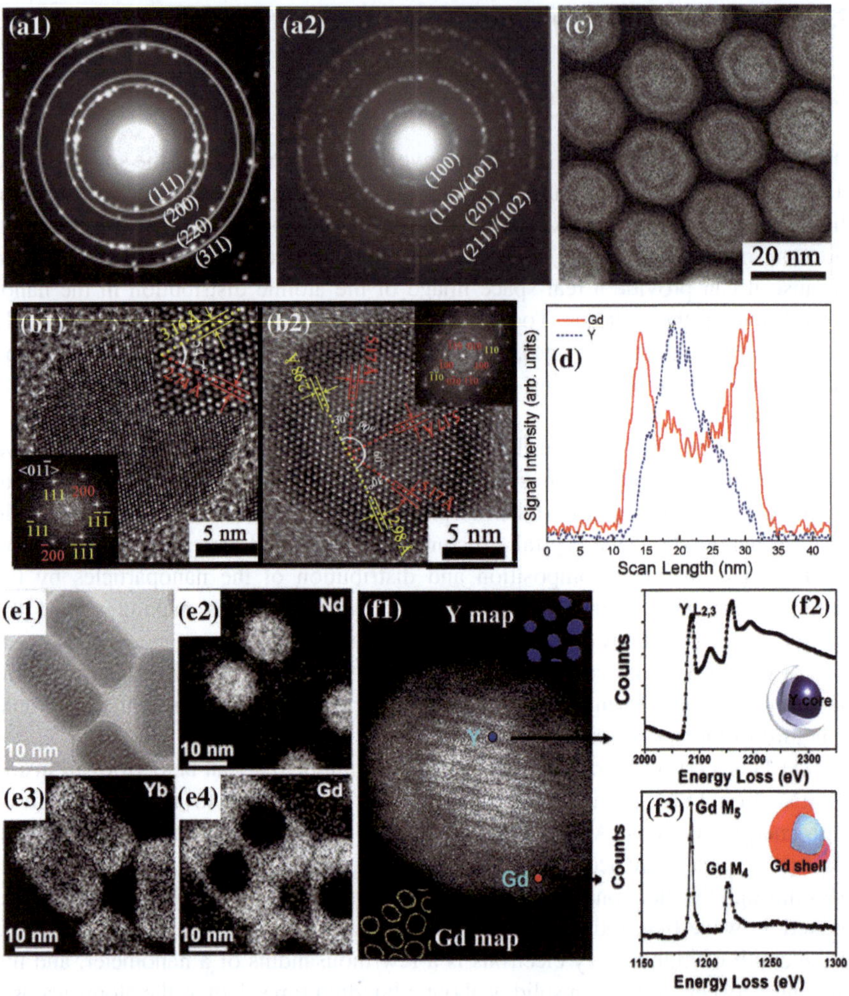

Fig. 2.7 **a** Typical selected area electron diffraction of cubic phase (**a1**) and hexagonal phase (**a2**) $NaYF_4$. **b** Typical high-resolution transmission electron microscopy images of cubic phase (**b1**) and hexagonal phase (**b2**) $NaYF_4$. **c** Typical HAADF-STEM image of the core@shell1@shell2@shell3-structured $NaGdF_4/NaYF_4/NaGdF_4/NaYF_4$ nanocrystals. **d** Energy-dispersive X-ray spectroscopy line scan across a single $NaYF_4/NaGdF_4$ core@shell NC showing *Y* in the core of the particle and Gd located in the shell. **e** Element maps of Nd, Yb, and Gd in the $NaYbF_4$:Nd@Na(Yb, Gd)F_4:Er@$NaGdF_4$ nanoparticles. **f** HADDF image of a single $NaYF_4$:Yb, Er@$NaGdF_4$ nanocrystal (**f1**) and the corresponding EELS spectra of yttrium $L_{2,3}$ (**f2**) and gadolinium $M_{4,5}$ (**f3**) edges taken from the probe location on the inferred $NaYF_4$:Yb,Er core and $NaGdF_4$ shell. **a**, **b** Reprinted with the permission from Ref. [164]. Copyright 2013 American Chemical Society. **c** Reprinted with the permission from Ref. [162]. Copyright 2013 Nature Publishing Group. **d** Reprinted with the permission from Ref. [161]. Copyright 2011 American Chemical Society. **e** Reprinted with the permission from Ref. [171]. Copyright 2013 Wiley-VCH Verlag GmbH & Co. KGaA. **f** Reprinted with the permission from Ref. [159]. Copyright 2012 American Chemical Society

corresponding to the lattice parameters of the (100), (110)/(101), (201), and (211)/(102) planes of β phase $NaYF_4$ (JCPDS, 16-0334).

High-resolution transmission electron microscopy (HRTEM) is an imaging mode of the TEM that allows for direct imaging of the atomic structure of the sample. HRTEM is a powerful tool to study properties of materials on the atomic scale. Figure 2.7b shows the typical HRTEM images of the hexagonal and cubic phase $NaGdF_4$ nanoparticles [164]. It can be seen that the two particles show quite different crystal lattices. The *d*-spaces of the lattice planes measure from Fig. 2.7b1 are ~2.7 and ~3.2 Å, with an angle of the two planes ~55°, corresponding to the lattice parameters of the (111) and (200) planes of α phase $NaGdF_4$. The *d*-spaces of the lattice planes measure from Fig. 2.7b2 are ~5.1 and ~3.0 Å, with an angle of the two planes ~30°, corresponding to the lattice parameters of the (100) and (110) planes of β phase $NaGdF_4$. The results are consistent with the above-mentioned SAED.

As mentioned above, the core@shell structure is one of the most important structures of the upconversion nanoparticles. However, the core–shell structure is commonly inferred from optical measurements, but problems with this arise because luminescence measurements give no insight into the mechanism of shell growth or coverage. Standard TEM is unable to differentiate between core and shell materials in lanthanide-based NCs because of similar material lattice constants and very little contrast difference between core and shell materials. Scanning transmission electron microscopy (STEM), coupled to techniques such as EELS and EDS, is well suited to the characterization of materials with complex compositions on the nanoscale, providing both structural and chemical information. A STEM is a type of TEM. As with any transmission illumination scheme, the electrons pass through a sufficiently thin specimen. However, STEM is distinguished from conventional TEM by focusing the electron beam into a narrow spot which is scanned over the sample in a raster. The rastering of the beam across the sample makes these microscopes suitable for analysis techniques such as mapping by energy-dispersive EDX spectroscopy, EELS, and high-angle annular dark-field imaging (HAADF). These signals can be obtained simultaneously, allowing direct correlation of image and quantitative data. By using HAADF-STEM, it is possible to form atomic-resolution images where the contrast is directly related to the atomic number (*Z*-contrast image). The directly interpretable *Z*-contrast image makes STEM imaging with a high-angle detector appealing, especially for the determination of the core@shell structure. Figure 2.7c shows a typical HAADF-STEM image of the core@shell1@shell2@shell3-structured $NaGdF_4$/$NaYF_4$/$NaGdF_4$/$NaYF_4$ nanocrystals [162]. It can be seen that the nanocrystals show clearly multilayer sandwich structure (bright-dark-bright-dark) in which the brighter parts correspond to the heavier Gd elements ($Z = 64$) and the darker parts correspond to the lighter *Y* elements ($Z = 39$).

As the aforementioned, we can only judge the formation of the core@shell structure from the contrast difference of the HAADF-STEM images. More characterizations are needed to obtain the chemical compositions of the core@shell nanoparticles. Energy-dispersive X-ray spectroscopy (EDS, EDX, or XEDS) is an analytical technique used for the elemental analysis or chemical characterization of a sample. It relies on an interaction of X-ray excitation and a sample. Its characterization capabilities are due in large part to the fundamental principle that each element has a unique atomic structure allowing unique set of peaks on its X-ray spectrum. To stimulate the emission of characteristic X-rays from a specimen, a high-energy beam of charged particles such as electrons or protons, or a beam of X-rays, is focused into the sample being studied. At rest, an atom within the sample contains ground state (or unexcited) electrons in discrete energy levels or electron shells bound to the nucleus. The incident beam may excite an electron in an inner shell, ejecting it from the shell while creating an electron hole where the electron was. An electron from an outer, higher-energy shell then fills the hole, and the difference in energy between the higher-energy shell and the lower energy shell may be released in the form of an X-ray. The number and energy of the X-rays emitted from a specimen can be measured by an energy-dispersive spectrometer. As the energy of the X-rays is characteristic of the difference in energy between the two shells, and of the atomic structure of the element from which they were emitted, this allows the elemental composition of the specimen to be measured. Figure 2.7d, e shows the two work modes of EDS, line scanning, and elemental mapping. Figure 2.7d shows the EDS line scan across a single $NaYF_4$@$NaGdF_4$ core@shell nanoparticle [161]. It can be seen that *Y* is indeed located at the center of the particle and the Gd is distributed around *Y*. Figure 2.7e shows the EDS elemental mapping of the $NaYbF_4$:Nd@Na(Yb, Gd)F_4:Er@$NaGdF_4$ nanoparticles [171]. It illustrates that the elemental distributions of the nanoparticles are very consistent with the designed compositions, confirming the core@shell@shell structure of the nanoparticles.

Although the EDS can be used to analyze the chemical composition and distribution of the specimen, beam damage to nanomaterials, caused by the long scan times required for EDS, is still a challenge in the field and a limitation in some cases. Most recently, our group reported for the first time the direct imaging of the $NaYF_4$:Yb, Er@$NaGdF_4$ nanocrystal core@shell structure at the subnanometer on the basis of the cryo-transmission electron microscopy, rigorous EELS, and high-angle annular dark-field investigations on the core–shell structure under a low operation temperature (96 K) (Fig. 2.7f) [159].

2.5.2.2 X-ray Photoelectron Spectroscopy

X-ray photoelectron spectroscopy (XPS) is well known to be one of the most direct and effective methodologies for the determination of surface molecular and electronic structure of a material providing a good deal of information about features at the surfaces and interfaces that dominate properties of nanostructured materials. XPS easily allows a qualitative and semiquantitative analysis of the elements by measuring the binding energy (BE) of core electrons throughout the evaluation of the kinetic energy of the photoelectrons, emitted by appropriate excitation of a sample. Different oxidation states, geometries and in general chemical environments, induce BE shifts.

These features have been studied to assess the size of nanoparticles, their local structure, coatings, and interactions. XPS provides information about the actual composition, chemical state of surfaces and interfaces, and allows the evaluation of a semiquantitative concentration of the elements in the material under investigation. In fact, by evaluating the intensities and relative areas of core level signals of the elements, photoionization probability factors and photoelectrons mean free path semiquantitative data can be obtained. The sampling depth is one of the most characteristic aspects of XPS: By the imprinting X radiation with sampling depth of many hundreds Angstroms, the escape depth of the excited photoelectrons in the continuum is limited to about three times the photoelectrons mean free path, comprised between 10 and 100 Å. Therefore, this technique yields information about the outmost surface layers of the investigated materials; this is an important peculiarity when the surface plays a fundamental role, as in the case of nanostructured materials. Relative signal intensities can be useful in obtaining information about layered structures and in some cases particle sizes by using angle resolved XPS. In fact, by measuring peak intensities as a function of emission angle, it is possible to achieve information on layering, elemental enrichment, or depletion and have a detailed surface depth profile and a 3D map of the material. For example, XPS spectroscopy has been applied for the investigation of the core@shell structure of $NaYF_4$@$NaGdF_4$ nanoparticles. In the study by van Veggel et al. the XPS spectra of the core@shell NCs were measured at varying photon energies (from 250 to 1,050 eV) to determine the internal composition of the particles [89]. Experiments were performed at the Canadian Light Source, Inc. using the spherical grating monochromotor (SGM) undulator beamline, 11ID.1, in Saskatoon, Saskatchewan. $NaYF_4$@$NaGdF_4$ NCs in THF were deposited directly onto Au foil. After the solvent evaporated, the samples were quickly transferred to the experimental chamber. No charging effects were observed at low particle concentration. An example of a recorded XPS spectrum (hv = 450 eV), before background subtraction, is shown in Fig. 2.8a. Photoelectron spectra of elements Y^{3+} 3*d* (core) and Gd^{3+} 4*d* (shell) were recorded at increasing excitation photon energies, thus gradually increasing the sampling depth. Each spectrum was subject to a polynomial background subtraction, and the peak areas were corrected for the photoionization cross sections. The intensity ratio of Y^{3+} 3*d* to Gd^{3+} 4*d* core levels is shown in Fig. 2.8b. Changing the photon energy changes the kinetic energy range. An increase in the ratio is observed at increasing kinetic energies, demonstrating that Gd^{3+} lies predominantly on the surface of the particles. The ratio increase is caused by the screening of the Y^{3+} intensity by the surrounding Gd^{3+} atoms at low kinetic energies. Although the former characterizations, such as HAADF-STEM, EDS, etc., are all can be used to confirm the formation of the core@shell structure, XPS showed another conclusive proof for the characterization of the core@shell structure.

2.5.2.3 X-ray Diffraction

X-ray diffraction (XRD) is a tool used for identifying the atomic and molecular structure of a crystal, in which the crystalline atoms cause a beam of incident X-rays

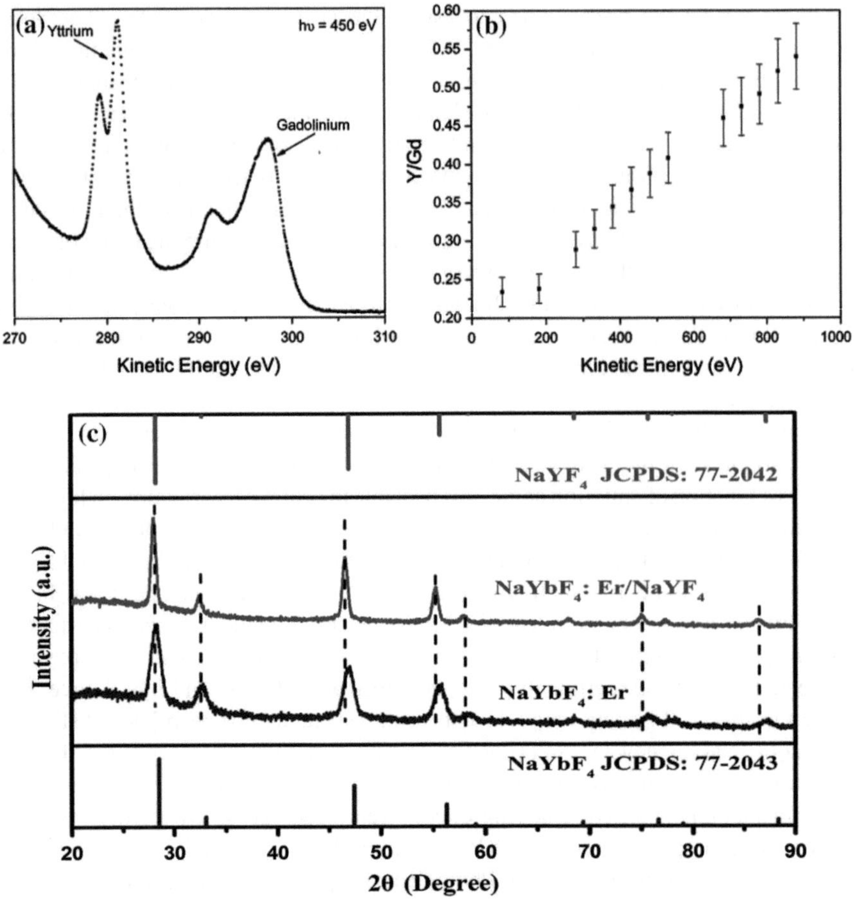

Fig. 2.8 **a** X-ray photoelectron spectroscopy of the Y^{3+} $3d$ and Gd^{3+} $4d$ core levels of $NaYF_4$@$NaGdF_4$ core@shell nanocrystals before subtraction of a polynomial background, at a photon energy of ca. 450 eV. **b** Intensity ratio of the Y^{3+} $3d$ to Gd^{3+} $4d$ core levels as a function of photoelectron kinetic energy. **c** Typical X-ray diffractions of $NaYbF_4$:Er and $NaYbF_4$:Er/$NaYF_4$ nanocrystals. **a**, **b** Reprinted with the permission from Ref. [89]. Copyright 2009 American Chemical Society. **c** Reprinted with the permission from Ref. [164]. Copyright 2013 American Chemical Society

to diffract into many specific directions. By measuring the angles and intensities of these diffracted beams, a crystallographer can produce a three-dimensional picture of the density of electrons within the crystal. From this electron density, the mean positions of the atoms in the crystal can be determined, as well as their chemical bonds, their disorder, and various other information. Furthermore, the XRD can also be used to evaluate the crystal size of the samples by using the Scherrer equation, which is limited to nanoscale particles and not applicable to grains larger than about 0.1–0.2 μm.

$$D = \frac{K\lambda}{B\cos\theta} \qquad (2.32)$$

where D is the mean size of the crystalline domains, which may be smaller or equal to the grain size; K is a dimensionless shape factor, with a value close to unity. The shape factor has a typical value of about 0.89, but varies with the actual shape of the crystallite. λ is the X-ray wavelength, which depends on the X-ray sources. B is the line broadening at half the maximum intensity (FWHM), after subtracting the instrumental line broadening, in radians. θ is the Bragg angle, in radians. It is important to realize that the Scherrer formula provides a lower bound on the particle size. The reason for this is that a variety of factors can contribute to the width of a diffraction peak besides instrumental effects and crystallite size; the most important of these are usually inhomogeneous strain and crystal lattice imperfections. The following sources of peak broadening are as follows: dislocations, stacking faults, twinning, microstresses, grain boundaries, sub-boundaries, coherency strain, chemical heterogeneities, and crystallite smallness. (Some of those and other imperfections may also result in peak shift, peak asymmetry, anisotropic peak broadening, or affect peak shape.)

Figure 2.8c shows the typical XRD patterns of the $NaYbF_4$:Er core nanocrystals and corresponding $NaYbF_4$:Er@$NaYF_4$ core@shell nanocrystals [164]. It is found that both the bare core and core@shell nanocrystals exhibit a cubic crystal structure. With the growth of $NaYF_4$ shells, the diffraction peaks shift to lower angles consistent with the bigger lattice constant for $NaYF_4$, compared with $NaYbF_4$ crystals. In addition, the diffraction peaks become narrow gradually with the growth of $NaYF_4$ shells, indicating that the crystalline domain size is increased for the core/shell nanocrystals, providing direct evidence of the epitaxial growth of the shell.

2.6 Summary and Perspectives

This chapter primarily summarizes the formation mechanism of the crystallization (including nucleation and growth) for monodisperse nanocrystals, various synthetic procedures for the upconversion nanoparticles, and the optical, chemical, and structural characterizations of upconversion nanoparticles.

As aforementioned, thermolysis, Ostwald-ripening, and hydro-thermal strategies are three general approaches to produce lanthanide-doped upconversion nanoparticles. However, the quality of the same resulting upconversion nanoparticles (the same size, same shape, and same composition ratios) produced by these approaches lacks comparison, because most of the equipment used to characterize the optical properties of upconversion nanoparticles are based on the home-made instruments. Standard measurement facilities and protocols are highly required to allow the quantitative comparison of optical properties.

Although the synthetic nanochemistries described in this chapter are able to produce monodisperse upconversion nanoparticles with a diameter of 20–100 nm, it remains a challenge to produce smaller but efficient upconversion nanoparticles. The clinical translation requires that imaging agents injected into the human body should be completely cleared in a reasonable amount of time. Renal clearance is the main clearance route for imaging agents without biodegradation into biologically benign components, which requires the hydrodynamic size of nanomaterials to be smaller than 10 nm with optimized surface. Hence, developing new or modified synthetic strategies to prepare sub-10-nm upconversion nanoparticles is of crucial importance in the future. Furthermore, such an ultrasmall size of upconversion nanoparticles is always associated with a significant decrease of the luminescent efficiency due to the extremely high “surface-to-volume” ratio. Thus, nanochemistry or nanostructure dealing with ultrasmall-size-related quenching mechanisms also needs our attention.

Last but not the least, through the development of upconversion nanoparticles, the structures of the nanoparticles become increasingly advanced and ingenious. Thus, the structure, especially the core@shell structure, becomes more and more complicated, indicating that the traditional synthesis method hardly fulfills the practical needs. Therefore, the end goal is the development of the new synthesis approach which can fulfill the requirements for the fabrication of complicated multi-shell-structured upconversion nanoparticles.

References

1. Li, X., Shen, H., Niu, J., Zhang, Y., Wang, H., Li, L.S.: Columnar self-assembly of Cu_2S hexagonal nanoplates induced by Tin(IV)-X complex as inorganic surface ligand. J. Am. Chem. Soc. **132**, 12778–12779 (2010)
2. Li, X., Niu, J.Z., Shen, H., Xu, W., Wang, H., Li, L.S.: Shape controlled synthesis of tadpole-like and heliotrope seed-like $AgInS_2$ nanocrystals. CrystEngComm **12**, 4410–4415 (2010)
3. Li, X., Shen, H., Li, S., Niu, J.Z., Wang, H., Li, L.S.: Investigation on type-II Cu_2S-CdS core/shell nanocrystals: synthesis and characterization. J. Mater. Chem. **20**, 923–928 (2010)
4. Li, X., Si, H., Niu, J.Z., Shen, H., Zhou, C., Yuan, H., Wang, H., Ma, L., Li, L.S.: Size-controlled syntheses and hydrophilic surface modification of Fe_3O_4, Ag, and Fe_3O_4/Ag heterodimer nanocrystals. Dalton Trans. **39**, 10984–10989 (2010)
5. Li, X., Wang, M., Shen, H., Zhang, Y., Wang, H., Li, L.S.: Inorganic Sn-X-complex-induced 1D, 2D, and 3D copper sulfide superstructures from anisotropic hexagonal nanoplate building blocks. Chem. Eur. J. **17**, 10357–10364 (2011)
6. Rogach, A.L., Talapin, D.V., Shevchenko, E.V., Kornowski, A., Haase, M., Weller, H.: Organization of matter on different size scales: monodisperse nanocrystals and their superstructures. Adv. Funct. Mater. **12**, 653–664 (2002)
7. Shen, H., Wang, H., Chen, X., Niu, J.Z., Xu, W., Li, X.M., Jiang, X.-D., Du, Z., Li, L.S.: Size- and shape-controlled synthesis of CdTe and PbTe nanocrystals using tellurium dioxide as the tellurium precursor. Chem. Mater. **22**, 4756–4761 (2010)
8. Sun, S., Ting, C.T., Wu, C.I.: The normal function of a speciation gene, Odysseus, and its hybrid sterility effect. Science **305**, 81–83 (2004)
9. Hyeon, T.: Chemical synthesis of magnetic nanoparticles. Chem. Commun. 927–934 (2003)

10. Park, J., Joo, J., Kwon, S.G., Jang, Y., Hyeon, T.: Synthesis of monodisperse spherical nanocrystals. Angew. Chem. Int. Ed. **46**, 4630–4660 (2007)
11. Bai, X., Song, H., Pan, G., Lei, Y., Wang, T., Ren, X., Lu, S., Dong, B., Dai, Q., Fan, L.: Size-dependent upconversion luminescence in Er^{3+}/Yb^{3+}-codoped nanocrystalline yttria: saturation and thermal effects. J. Phys. Chem. C **111**, 13611–13617 (2007)
12. Mai, H.-X., Zhang, Y.-W., Sun, L.-D., Yan, C.-H.: Highly efficient multicolor up-conversion emissions and their mechanisms of monodisperse $NaYF_4$:Yb, Er core and core/shell-structured nanocrystals. J. Phys. Chem. C **111**, 13721–13729 (2007)
13. Chen, G., Yang, C., Prasad, P.N.: Nanophotonics and nanochemistry: controlling the excitation dynamics for frequency up- and down-conversion in lanthanide-doped nanoparticles. Acc. Chem. Res. **46**, 1474–1486 (2013)
14. Wang, F., Liu, X.: Upconversion multicolor fine-tuning: visible to near-infrared emission from lanthanide-doped $NaYF_4$ nanoparticles. J. Am. Chem. Soc. **130**, 5642–5643 (2008)
15. Wang, F., Chatterjee, D.K., Li, Z., Zhang, Y., Fan, X., Wang, M.: Synthesis of polyethylenimine/$NaYF_4$ nanoparticles with upconversion fluorescence. Nanotechnology **17**, 5786–5791 (2006)
16. Li, Z., Zhang, Y.: Monodisperse silica-coated polyvinylpyrrolidone/$NaYF_4$ nanocrystals with multicolor upconversion fluorescence emission. Angew. Chem. Int. Ed. **45**, 7732–7735 (2006)
17. Zeng, J.H., Su, J., Li, Z.H., Yan, R.X., Li, Y.D.: Synthesis and upconversion luminescence of hexagonal-phase $NaYF_4$:Yb, Er^{3+}, phosphors of controlled size and morphology. Adv. Mater. **17**, 2119–2123 (2005)
18. Yi, G.S., Chow, G.M.: Colloidal LaF_3:Yb,Er,LaF_3:Yb,Ho and LaF_3:Yb,Tm nanocrystals with multicolor upconversion fluorescence. J. Mater. Chem. **15**, 4460–4464 (2005)
19. Yi, G.S., Lu, H.C., Zhao, S.Y., Yue, G., Yang, W.J., Chen, D.P., Guo, L.H.: Synthesis, characterization, and biological application of size-controlled nanocrystalline $NaYF_4$:Yb, Er infrared-to-visible up-conversion phosphors. Nano Lett. **4**, 2191–2196 (2004)
20. Heer, S., Kompe, K., Gudel, H.U., Haase, M.: Highly efficient multicolour upconversion emission in transparent colloids of lanthanide-doped $NaYF_4$ nanocrystals. Adv. Mater. **16**, 2102–2105 (2004)
21. Heer, S., Lehmann, O., Haase, M., Gudel, H.U.: Blue, green, and red upconversion emission from lanthanide-doped $LuPO_4$ and $YbPO_4$ nanocrystals in a transparent colloidal solution. Angew. Chem. Int. Ed. **42**, 3179–3182 (2003)
22. Stouwdam, J.W., van Veggel, F.: Near-infrared emission of redispersible Er^{3+}, Nd^{3+}, and Ho^{3+} doped LaF_3 nanoparticles. Nano Lett. **2**, 733–737 (2002)
23. Boyer, J.-C., Cuccia, L.A., Capobianco, J.A.: Synthesis of colloidal upconverting $NaYF_4$: Er^{3+}/Yb^{3+} and Tm^{3+}/Yb^{3+}monodisperse nanocrystals. Nano Lett. **7**, 847–852 (2007)
24. Mai, H.X., Zhang, Y.W., Si, R., Yan, Z.G., Sun, L.D., You, L.P., Yan, C.H.: High-quality sodium rare-earth fluoride nanocrystals: controlled synthesis and optical properties. J. Am. Chem. Soc. **128**, 6426–6436 (2006)
25. Boyer, J.C., Vetrone, F., Cuccia, L.A., Capobianco, J.A.: Synthesis of colloidal upconverting $NaYF_4$ nanocrystals doped with Er^{3+}, Yb^{3+} and Tm^{3+}, Yb^{3+} via thermal decomposition of lanthanide trifluoroacetate precursors. J. Am. Chem. Soc. **128**, 7444–7445 (2006)
26. Zhang, Y.W., Sun, X., Si, R., You, L.P., Yan, C.H.: Single-crystalline and monodisperse LaF_3 triangular nanoplates from a single-source precursor. J. Am. Chem. Soc. **127**, 3260–3261 (2005)
27. Si, R., Zhang, Y.W., You, L.P., Yan, C.H.: Rare-earth oxide nanopolyhedra, nanoplates, and nanodisks. Angew. Chem. Int. Ed. **44**, 3256–3260 (2005)
28. Wang, L., Li, P., Li, Y.: Down- and up-conversion luminescent nanorods. Adv. Mater. **19**, 3304–3307 (2007)
29. Zhang, F., Zhao, D.: Synthesis of uniform rare earth fluoride ($NaMF_4$) nanotubes by in situ ion exchange from their hydroxide [$M(OH)_3$] parents. ACS Nano **3**, 159–164 (2009)

30. Zhang, F., Li, J., Shan, J., Xu, L., Zhao, D.: Shape, size, and phase-controlled rare-earth fluoride nanocrystals with optical up-conversion properties. Chem. Eur. J. **15**, 11010–11019 (2009)
31. Wang, G., Peng, Q., Li, Y.: Upconversion luminescence of monodisperse $CaF_2:Yb^{3+}/Er^{3+}$ nanocrystals. J. Am. Chem. Soc. **131**, 14200–14201 (2009)
32. Schaefer, H., Ptacek, P., Eickmeier, H., Haase, M.: Synthesis of hexagonal Yb^{3+}, Er^{3+}-doped $NaYF_4$ nanocrystals at low temperature. Adv. Funct. Mater. **19**, 3091–3097 (2009)
33. Li, Z., Zhang, Y.: An efficient and user-friendly method for the synthesis of hexagonal-phase $NaYF_4$:Yb, Er/Tm nanocrystals with controllable shape and upconversion fluorescence. Nanotechnology **19**, 345606 (2008)
34. Zhang, F., Wan, Y., Yu, T., Zhang, F., Shi, Y., Xie, S., Li, Y., Xu, L., Tu, B., Zhao, D.: Uniform nanostructured arrays of sodium rare-earth fluorides for highly efficient multicolor upconversion luminescence. Angew. Chem. Int. Ed. **46**, 7976–7979 (2007)
35. Wang, L., Li, Y.: Controlled synthesis and luminescence of lanthanide doped $NaYF_4$ nanocrystals. Chem. Mater. **19**, 727–734 (2007)
36. Wang, L., Li, Y.: $Na(Y_{1.5}Na_{0.5})F_6$ single-crystal nanorods as multicolor luminescent materials. Nano Lett. **6**, 1645–1649 (2006)
37. Wang, X., Zhuang, J., Peng, Q., Li, Y.D.: A general strategy for nanocrystal synthesis. Nature **437**, 121–124 (2005)
38. Venkatramu, V., Falcomer, D., Speghini, A., Bettinelli, M., Jayasankar, C.K.: Synthesis and luminescence properties of Er^{3+}-doped $Lu_3Ga_5O_{12}$ nanocrystals. J. Lumin. **128**, 811–813 (2008)
39. Yang, K., Zheng, F., Wu, R., Li, H., Zhang, X.: Upconversion luminescent properties of YVO_4: Yb^{3+}, Er^{3+} nano-powder by sol-gel method. J. Rare Earths **24**, 162–166 (2006)
40. Patra, A., Friend, C.S., Kapoor, R., Prasad, P.N.: Fluorescence upconversion properties of Er^{3+}-doped TiO_2 and $BaTiO_3$ nanocrystallites. Chem. Mater. **15**, 3650–3655 (2003)
41. Patra, A., Friend, C.S., Kapoor, R., Prasad, P.N.: Upconversion in $Er^{3+}:ZrO_2$ nanocrystals. J. Phys. Chem. B **106**, 1909–1912 (2002)
42. Qin, X., Yokomori, T., Ju, Y.: Flame synthesis and characterization of rare-earth (Er^{3+}, Ho^{3+}, and Tm^{3+}) doped upconversion nanophosphors. Appl. Phys. Lett. **90**(7), 073104 (2007)
43. Li, X., Zhang, F., Zhao, D.: Lab on upconversion nanoparticles: optical properties and applications engineering via designed nanostructure. Chem. Soc. Rev. doi:10.1039/C4CS00163J (2014)
44. Li, X., Zhang, F., Zhao, D.: Highly efficient lanthanide upconverting nanomaterials: progresses and challenges. Nano Today **8**, 643–676 (2013)
45. Wang, G., Peng, Q., Li, Y.: Lanthanide-doped nanocrystals: synthesis, optical-magnetic properties, and applications. Acc. Chem. Res. **44**, 322–332 (2011)
46. Haase, M., Schäfer, H.: Upconverting nanoparticles. Angew. Chem. Int. Ed. **50**, 5808–5829 (2011)
47. Wang, F., Liu, X.: Recent advances in the chemistry of lanthanide-doped upconversion nanocrystals. Chem. Soc. Rev. **38**, 976–989 (2009)
48. Zhou, J., Liu, Z., Li, F.: Upconversion nanophosphors for small-animal imaging. Chem. Soc. Rev. **41**, 1323–1349 (2012)
49. Mader, H.S., Kele, P., Saleh, S.M., Wolfbeis, O.S.: Upconverting luminescent nanoparticles for use in bioconjugation and bioimaging. Curr. Opin. Chem. Biol. **14**, 582–596 (2010)
50. Cheng, L., Wang, C., Liu, Z.: Upconversion nanoparticles and their composite nanostructures for biomedical imaging and cancer therapy. Nanoscale **5**, 23–27 (2013)
51. Chatterjee, D.K., Gnanasammandhan, M.K., Zhang, Y.: Small upconverting fluorescent nanoparticles for biomedical applications. Small **6**, 2781–2795 (2010)
52. Auzel, F.: Upconversion and anti-stokes processes with f and d ions in solids. Chem. Rev. **104**, 139–173 (2004)
53. Fischer, L.H., Harms, G.S., Wolfbeis, O.S.: Upconverting nanoparticles for nanoscale thermometry. Angew. Chem. Int. Ed. **50**, 4546–4551 (2011)

54. Chen, G., Qju, H., Prasad, P.N., Chen, X.: Upconversion nanoparticles: design, nanochemistry, and applications in theranostics. Chem. Rev. **114**, 5161–5214 (2014)
55. Sun, X., Zhang, Y.-W., Du, Y.-P., Yan, Z.-G., Si, R., You, L.-P., Yan, C.-H.: From trifluoroacetate complex precursors to monodisperse rare-earth fluoride and oxyfluoride nanocrystals with diverse shapes through controlled fluorination in solution phase. Chem. Eur. J. **13**, 2320–2332 (2007)
56. Johnson, N.J.J., Korinek, A., Dong, C., van Veggel, F.C.J.M.: Self-focusing by Ostwald ripening: a strategy for layer-by-layer epitaxial growth on upconverting nanocrystals. J. Am. Chem. Soc. **134**, 11068–11071 (2012)
57. Li, Z., Zhang, Y., Jiang, S.: Multicolor core/shell-structured upconversion fluorescent nanoparticles. Adv. Mater. **20**, 4765–4769 (2008)
58. Qian, H.-S., Zhang, Y.: Synthesis of hexagonal-phase core-shell $NaYF_4$ nanocrystals with tunable upconversion fluorescence. Langmuir **24**, 12123–12125 (2008)
59. Zheng, W., Zhou, S., Chen, Z., Hu, P., Liu, Y., Tu, D., Zhu, H., Li, R., Huang, M., Chen, X.: Sub-10 nm lanthanide-doped CaF_2 nanoprobes for time-resolved luminescent biodetection. Angew. Chem. Int. Ed. **52**, 6671–6676 (2013)
60. Lamer, V.K., Dinegar, R.H.: Theory, production and mechanism of formation of monodispersed hydrosols. J. Am. Chem. Soc. **72**, 4847–4854 (1950)
61. Murray, C.B., Kagan, C., Bawendi, M.: Synthesis and characterization of monodisperse nanocrystals and close-packed nanocrystal assemblies. Annu. Rev. Mater. Sci. **30**, 545–610 (2000)
62. Peng, Z.A., Peng, X.G.: Nearly monodisperse and shape-controlled CdSe nanocrystals via alternative routes: nucleation and growth. J. Am. Chem. Soc. **124**, 3343–3353 (2002)
63. Peng, X.G., Wickham, J., Alivisatos, A.P.: Kinetics of II-VI and III-V colloidal semiconductor nanocrystal growth: "focusing" of size distributions. J. Am. Chem. Soc. **120**, 5343–5344 (1998)
64. Vossmeyer, T., Katsikas, L., Giersig, M., Popovic, I.G., Diesner, K., Chemseddine, A., Eychmuller, A., Weller, H.: CdS nanoclusters: synthesis, characterization, size dependent oscillator strength, temperature shift of the excitonic transition energy, and reversible absorbance shift. J. Phys. Chem. **98**, 7665–7673 (1994)
65. Spanhel, L., Anderson, M.A.: Semiconductor clusters in the sol-gel process: quantized aggregation, gelation, and crystal growth in concentrated zinc oxide colloids. J. Am. Chem. Soc. **113**, 2826–2833 (1991)
66. Alivisatos, A.P.: Perspectives on the physical chemistry of semiconductor nanocrystals. J. Phys. Chem. **100**, 13226–13239 (1996)
67. Zhang, H.Z., Banfield, J.F.: Understanding polymorphic phase transformation behavior during growth of nanocrystalline aggregates: insights from TiO_2. J. Phys. Chem. B **104**, 3481–3487 (2000)
68. Murray, C.B., Norris, D.J., Bawendi, M.G.: Synthesis and characterization of nearly monodisperse CdE (E = sulfur, selenium, tellurium) semiconductor nanocrystallites. J. Am. Chem. Soc. **115**, 8706–8715 (1993)
69. Jana, N.R., Gearheart, L., Murphy, C.J.: Evidence for seed-mediated nucleation in the chemical reduction of gold salts to gold nanoparticles. Chem. Mater. **13**, 2313–2322 (2001)
70. Yu, H., Gibbons, P.C., Kelton, K.F., Buhro, W.E.: Heterogeneous seeded growth: a potentially general synthesis of monodisperse metallic nanoparticles. J. Am. Chem. Soc. **123**, 9198–9199 (2001)
71. Wilcoxon, J.P., Provencio, P.P.: Heterogeneous growth of metal clusters from solutions of seed nanoparticles. J. Am. Chem. Soc. **126**, 6402–6408 (2004)
72. Park, J., Lee, E., Hwang, N.M., Kang, M.S., Kim, S.C., Hwang, Y., Park, J.G., Noh, H.J., Kini, J.Y., Park, J.H., Hyeon, T.: One-nanometer-scale size-controlled synthesis of monodisperse magnetic iron oxide nanoparticles. Angew. Chem. Int. Ed. **44**, 2872–2877 (2005)

73. Talapin, D.V., Rogach, A.L., Kornowski, A., Haase, M., Weller, H.: Highly luminescent monodisperse CdSe and CdSe/ZnS nanocrystals synthesized in a hexadecylamine-trioctylphosphine oxide-trioctylphospine mixture. Nano Lett. **1**, 207–211 (2001)
74. Hambrock, J., Becker, R., Birkner, A., Weiss, J., Fischer, R.A.: A non-aqueous organometallic route to highly monodispersed copper nanoparticles using [Cu(OCH(Me) $CH_2NMe_2)_2$]. Chem. Commun. 68–69 (2002)
75. Jana, N.R., Peng, X.G.: Single-phase and gram-scale routes toward nearly monodisperse Au and other noble metal nanocrystals. J. Am. Chem. Soc. **125**, 14280–14281 (2003)
76. Park, J., An, K.J., Hwang, Y.S., Park, J.G., Noh, H.J., Kim, J.Y., Park, J.H., Hwang, N.M., Hyeon, T.: Ultra-large-scale syntheses of monodisperse nanocrystals. Nat. Mater. **3**, 891–895 (2004)
77. Seo, W.S., Jo, H.H., Lee, K., Park, J.T.: Preparation and optical properties of highly crystalline, colloidal, and size-controlled indium oxide nanoparticles. Adv. Mater. **15**, 795–797 (2003)
78. Yu, W.W., Falkner, J.C., Yavuz, C.T., Colvin, V.L.: Synthesis of monodisperse iron oxide nanocrystals by thermal decomposition of iron carboxylate salts. Chem. Commun. 2306–2307 (2004)
79. Joo, J., Na, H.B., Yu, T., Yu, J.H., Kim, Y.W., Wu, F.X., Zhang, J.Z., Hyeon, T.: Generalized and facile synthesis of semiconducting metal sulfide nanocrystals. J. Am. Chem. Soc. **125**, 11100–11105 (2003)
80. Yi, G.-S., Chow, G.-M.: Water-soluble $NaYF_4$:Yb, Er(Tm)/$NaYF_4$/polymer core/shell/shell nanoparticles with significant enhancement of upconversion fluorescence. Chem. Mater. **19**, 341–434 (2007)
81. Reiss, H.: The growth of uniform colloidal dispersions. J. Chem. Phys. **19**, 482–487 (1951)
82. Talapin, D.V., Rogach, A.L., Haase, M., Weller, H.: Evolution of an ensemble of nanoparticles in a colloidal solution: theoretical study. J. Phys. Chem. B **105**, 12278–12285 (2001)
83. Talapin, D.V., Rogach, A.L., Shevchenko, E.V., Kornowski, A., Haase, M., Weller, H.: Dynamic distribution of growth rates within the ensembles of colloidal II-VI and III-V semiconductor nanocrystals as a factor governing their photoluminescence efficiency. J. Am. Chem. Soc. **124**, 5782–5790 (2002)
84. Chen, G., Shen, J., Ohulchanskyy, T.Y., Patel, N.J., Kutikov, A., Li, Z., Song, J., Pandey, R.K., Agren, H., Prasad, P.N., Han, G.: (alpha-$NaYbF_4$:Tm^{3+})/CaF_2 core/shell nanoparticles with efficient near-infrared to near-infrared upconversion for high-contrast deep tissue bioimaging. ACS Nano **6**, 8280–8287 (2012)
85. Zhan, Q., Qian, J., Liang, H., Somesfalean, G., Wang, D., He, S., Zhang, Z., Andersson-Engels, S.: Using 915 nm laser excited Tm^{3+}/Er^{3+}/Ho^{3+}-doped $NaYbF_4$ upconversion nanoparticles for in vitro and deeper in vivo bioimaging without overheating irradiation. ACS Nano **5**, 3744–3757 (2011)
86. Chen, G., Ohulchanskyy, T.Y., Liu, S., Law, W.-C., Wu, F., Swihart, M.T., Agren, H., Prasad, P.N.: Core/shell $NaGdF_4$:Nd^{3+}/$NaGdF_4$ nanocrystals with efficient near-infrared to near-infrared downconversion photoluminescence for bioimaging applications. ACS Nano **6**, 2969–2977 (2012)
87. Vetrone, F., Naccache, R., Mahalingam, V., Morgan, C.G., Capobianco, J.A.: The active-core/active-shell approach: a strategy to enhance the upconversion luminescence in lanthanide-doped nanoparticles. Adv. Funct. Mater. **19**, 2924–2929 (2009)
88. Naccache, R., Vetrone, F., Mahalingam, V., Cuccia, L.A., Capobianco, J.A.: Controlled synthesis and water dispersibility of hexagonal phase $NaGdF_4$:Ho^{3+}/Yb^{3+} nanoparticles. Chem. Mater. **21**, 717–723 (2009)
89. Abel, K.A., Boyer, J.-C., van Veggel, F.C.J.M.: Hard proof of the $NaYF_4$/$NaGdF_4$ nanocrystal core/shell structure. J. Am. Chem. Soc. **131**, 14644–14645 (2009)
90. Ye, X., Collins, J.E., Kang, Y., Chen, J., Chen, D.T.N., Yodh, A.G., Murray, C.B.: Morphologically controlled synthesis of colloidal upconversion nanophosphors and their shape-directed self-assembly. Proc. Natl. Acad. Sci. **107**, 22430–22435 (2010)

91. Komban, R., Klare, J.P., Voss, B., Nordmann, J., Steinhoff, H.-J., Haase, M.: An electron paramagnetic resonance spectroscopic investigation on the growth mechanism of $NaYF_4$:Gd nanocrystals. Angew. Chem. Int. Ed. **51**, 6506–6510 (2012)
92. Mahalingam, V., Vetrone, F., Naccache, R., Speghini, A., Capobianco, J.A.: Colloidal Tm^{3+}/ Yb^{3+}-doped $LiYF_4$ nanocrystals: multiple luminescence spanning the UV to NIR regions via low-energy excitation. Adv. Mater. **21**, 4025–4028 (2009)
93. Wang, J., Wang, F., Xu, J., Wang, Y., Liu, Y., Chen, X., Chen, H., Liu, X.: Lanthanide-doped $LiYF_4$ nanoparticles: synthesis and multicolor upconversion tuning. C. R. Chim. **13**, 731–736 (2010)
94. Chen, G., Ohulchanskyy, T.Y., Kachynski, A., Agren, H., Prasad, P.N.: Intense visible and near-infrared upconversion photoluminescence in colloidal $LiYF_4$:Er^{3+} nanocrystals under excitation at 1490 nm. ACS Nano **5**, 4981–4986 (2011)
95. Zhou, H.-P., Zhang, Y.-W., Mai, H.-X., Sun, X., Liu, Q., Song, W.-G., Yan, C.-H.: Spontaneous organization of uniform CeO_2 nanoflowers by 3D oriented attachment in hot surfactant solutions monitored with an in situ electrical conductance technique. Chem. Eur. J. **14**, 3380–3390 (2008)
96. Qiu, H., Chen, G., Fan, R., Cheng, C., Hao, S., Chen, D., Yang, C.: Tuning the size and shape of colloidal cerium oxide nanocrystals through lanthanide doping. Chem. Commun. **47**, 9648–9650 (2011)
97. Chen, G., Qiu, H., Fan, R., Hao, S., Tan, S., Yang, C., Han, G.: Lanthanide-doped ultrasmall yttrium fluoride nanoparticles with enhanced multicolor upconversion photoluminescence. J. Mater. Chem. **22**, 20190–20196 (2012)
98. Du, Y.P., Sun, X., Zhang, Y.W., Yan, Z.G., Sun, L.D., Yan, C.H.: Uniform alkaline earth fluoride nanocrystals with diverse shapes grown from thermolysis of metal trifluoroacetates in hot surfactant solutions. Cryst. Growth Des. **9**, 2013–2019 (2009)
99. Teng, X., Zhu, Y., Wei, W., Wang, S., Huang, J., Naccache, R., Hu, W., Tok, A.I.Y., Han, Y., Zhang, Q., Fan, Q., Huang, W., Capobianco, J.A., Huang, L.: Lanthanide-doped Na_xScF_{3+x} nanocrystals: crystal structure evolution and multicolor tuning. J. Am. Chem. Soc. **134**, 8340–8343 (2012)
100. Yi, G.S., Chow, G.M.: Synthesis of hexagonal-phase $NaYF_4$:Yb, Er and $NaYF_4$:Yb, Tm nanocrystals with efficient up-conversion fluorescence. Adv. Funct. Mater. **16**, 2324–2329 (2006)
101. Du, Y.-P., Zhang, Y.-W., Sun, L.-D., Yan, C.-H.: Luminescent monodisperse nanocrystals of lanthanide oxyfluorides synthesized from trifluoroacetate precursors in high-boiling solvents. J. Phys. Chem. C **112**, 405–415 (2008)
102. Du, Y.-P., Zhang, Y.-W., Sun, L.-D., Yan, C.-H.: Atomically efficient synthesis of self-assembled monodisperse and ultrathin lanthanide oxychloride nanoplates. J. Am. Chem. Soc. **131**, 3162–3163 (2009)
103. Niu, W., Wu, S., Zhang, S.: Utilizing the amidation reaction to address the "cooperative effect" of carboxylic acid/amine on the size, shape, and multicolor output of fluoride upconversion nanoparticles. J. Mater. Chem. **21**, 10894–10902 (2011)
104. Chen, G., Ohulchanskyy, T.Y., Kumar, R., Agren, H., Prasad, P.N.: Ultrasmall monodisperse $NaYF_4$:Yb^{3+}/Tm^{3+} nanocrystals with enhanced near-infrared to near-infrared upconversion photoluminescence. ACS Nano **4**, 3163–3168 (2010)
105. Si, R., Zhang, Y.-W., Zhou, H.-P., Sun, L.-D., Yan, C.-H.: Controlled-synthesis, self-assembly behavior, and surface-dependent optical properties of high-quality rare-earth oxide nanocrystals. Chem. Mater. **19**, 18–27 (2007)
106. Cao, Y.C.: Synthesis of square gadolinium-oxide nanoplates. J. Am. Chem. Soc. **126**, 7456–7457 (2004)
107. Zhao, F., Yuan, M., Zhang, W., Gao, S.: Monodisperse lanthanide oxysulfide nanocrystals. J. Am. Chem. Soc. **128**, 11758–11759 (2006)
108. Ding, Y., Gu, J., Ke, J., Zhang, Y.-W., Yan, C.-H.: Sodium doping controlled synthesis of monodisperse lanthanide oxysulfide ultrathin nanoplates guided by density functional calculations. Angew. Chem. Int. Ed. **50**, 12330–12334 (2011)

109. Ding, Y., Gu, J., Zhang, T., Yin, A.-X., Yang, L., Zhang, Y.-W., Yan, C.-H.: Chemoaffinity-mediated synthesis of $NaRES_2$-based nanocrystals as versatile nano-building blocks and durable nano-pigments. J. Am. Chem. Soc. **134**, 3255–3264 (2012)
110. Wang, F., Sun, L.-D., Gu, J., Wang, Y.-F., Feng, W., Yang, Y., Wang, J., Yan, C.-H.: Selective heteroepitaxial nanocrystal growth of rare earth fluorides on sodium chloride: synthesis and density functional calculations. Angew. Chem. Int. Ed. **51**, 8796–8799 (2012)
111. Zhou, H.-P., Zhang, C., Yan, C.-H.: Controllable assembly of diverse rare-earth nanocrystals via the langmuir-blodgett technique and the underlying size- and symmetry-dependent assembly kinetics. Langmuir **25**, 12914–12925 (2009)
112. Wang, H.Z., Uehara, M., Nakamura, H., Miyazaki, M., Maeda, H.: Synthesis of well-dispersed Y_2O_3:Eu nanocrystals and self-assembled nanodisks using a simple non-hydrolytic route. Adv. Mater. **17**, 2506–2509 (2005)
113. Shan, J., Qin, X., Yao, N., Ju, Y.: Synthesis of monodisperse hexagonal $NaYF_4$: Yb, Ln (Ln = Er, Ho and Tm) upconversion nanocrystals in TOPO. Nanotechnology **18** (2007)
114. Shan, J., Kong, W., Wei, R., Yao, N., Ju, Y.: An investigation of the thermal sensitivity and stability of the beta-$NaYF_4$:Yb, Er upconversion nanophosphors. J. Appl. Phys. **107** (2010)
115. Shan, J., Ju, Y.: Controlled synthesis of lanthanide-doped $NaYF_4$ upconversion nanocrystals via ligand induced crystal phase transition and silica coating. Appl. Phys. Lett. **91** (2007)
116. He, M., Huang, P., Zhang, C., Chen, F., Wang, C., Ma, J., He, R., Cui, D.: A general strategy for the synthesis of upconversion rare earth fluoride nanocrystals via a novel OA/ionic liquid two-phase system. Chem. Commun. **47**, 9510–9512 (2011)
117. Li, P., Peng, Q., Li, Y.: Dual-mode luminescent colloidal spheres from monodisperse rare-earth fluoride nanocrystals. Adv. Mater. **21**, 1945–1948 (2009)
118. Chen, D., Huang, P., Yu, Y., Huang, F., Yang, A., Wang, Y.: Dopant-induced phase transition: a new strategy of synthesizing hexagonal upconversion $NaYF_4$ at low temperature. Chem. Commun. **47**, 5801–5803 (2011)
119. Chen, Z., Chen, H., Hu, H., Yu, M., Li, F., Zhang, Q., Zhou, Z., Yi, T., Huang, C.: Versatile synthesis strategy for carboxylic acid-functionalized upconverting nanophosphors as biological labels. J. Am. Chem. Soc. **130**, 3023–3029 (2008)
120. Nguyen, T.-D., Dinh, C.-T., Do, T.-O.: Shape- and size-controlled synthesis of monoclinic ErOOH and cubic Er_2O_3 from micro- to nanostructures and their upconversion luminescence. ACS Nano **4**, 2263–2273 (2010)
121. Qiu, H., Chen, G., Sun, L., Hao, S., Han, G., Yang, C.: Ethylenediaminetetraacetic acid (EDTA)-controlled synthesis of multicolor lanthanide doped $BaYF_5$ upconversion nanocrystals. J. Mater. Chem. **21**, 17202–17208 (2011)
122. Yang, S., Gao, L.: Controlled synthesis and self-assembly of CeO_2 nanocubes. J. Am. Chem. Soc. **128**, 9330–9331 (2006)
123. Chen, D., Yu, Y., Huang, F., Huang, P., Yang, A., Wang, Y.: Modifying the size and shape of monodisperse bifunctional alkaline-earth fluoride nanocrystals through lanthanide doping. J. Am. Chem. Soc. **132**, 9976–9978 (2010)
124. Zhu, X., Zhang, Q., Li, Y., Wang, H.: Redispersible and water-soluble LaF_3:Ce, Tb nanocrystals via a microfluidic reactor with temperature steps. J. Mater. Chem. **18**, 5060–5062 (2008)
125. Li, S., Xie, T., Peng, Q., Li, Y.: Nucleation and growth of CeF_3 and $NaCeF_4$ nanocrystals. Chem. Eur. J. **15**, 2512–2517 (2009)
126. Chen, D., Yu, Y., Huanga, F., Wang, Y.: Phase transition from hexagonal LnF_3 (Ln = La, Ce, Pr) to cubic $Ln_{0.8}M_{0.2}F_{2.8}$ (M = Ca, Sr, Ba) nanocrystals with enhanced upconversion induced by alkaline-earth doping. Chem. Commun. **47**, 2601–2603 (2011)
127. Chen, D., Lei, L., Zhang, R., Yang, A., Xu, J., Wang, Y.: Intrinsic single-band upconversion emission in colloidal Yb/Er(Tm):$Na_3Zr(Hf)F_7$ nanocrystals. Chem. Commun. **48**, 10630–10632 (2012)
128. Wang, G., Peng, Q., Li, Y.: Synthesis and upconversion luminescence of BaY_2F_8:Yb^{3+}/Er^{3+} nanobelts. Chem. Commun. **46**, 7528–7529 (2010)

129. Schaefer, H., Ptacek, P., Zerzouf, O., Haase, M.: Synthesis and optical properties of KYF_4/Yb, Er nanocrystals, and their surface modification with undoped KYF_4. Adv. Funct. Mater. **18**, 2913–2918 (2008)
130. Shi, F., Wang, J., Zhai, X., Zhao, D., Qin, W.: Facile synthesis of beta-$NaLuF_4$:Yb/Tm hexagonal nanoplates with intense ultraviolet upconversion luminescence. CrystEngComm **13**, 3782–3787 (2011)
131. Peng, Y., Xing, M.-M., Luo, X.-X., Wang, L.-Q.: Synthesis and characteristic research of nanoparticles KY_3F_{10}:Yb, RE (RE = Er, Ho, Tm) by thermal decomposition. Acta Phys. Sin-Ch. Ed. **61** (2012)
132. Vetrone, F., Mahalingam, V., Capobianco, J.A.: Near-infrared-to-blue upconversion in colloidal $BaYF_5$:Tm^{3+}, Yb^{3+} nanocrystals. Chem. Mater. **21**, 1847–1851 (2009)
133. Wang, Y.-F., Sun, L.-D., Xiao, J.-W., Feng, W., Zhou, J.-C., Shen, J., Yan, C.-H.: Rare-earth nanoparticles with enhanced upconversion emission and suppressed rare-earth-ion leakage. Chem. Eur. J. **18**, 5558–5564 (2012)
134. Byrappa, K., Adschiri, T.: Hydrothermal technology for nanotechnology. Prog. Cryst. Growth Ch **53**, 117–166 (2007)
135. Haes, A.J., Van Duyne, R.P.: A nanoscale optical blosensor: sensitivity and selectivity of an approach based on the localized surface plasmon resonance spectroscopy of triangular silver nanoparticles. J. Am. Chem. Soc. **124**, 10596–10604 (2002)
136. Wang, L., Qin, W., Liu, Z., Zhao, D., Qin, G., Di, W., He, C.: Improved 800 nm emission of Tm^{3+} sensitized by Yb^{3+} and Ho^{3+} in beta-$NaYF_4$ nanocrystals under 980 nm excitation. Opt. Express **20**, 7602–7607 (2012)
137. Jiang, W., Kim, B.Y.S., Rutka, J.T., Chan, W.C.W.: Nanoparticle-mediated cellular response is size-dependent. Nat. Nanotechnol. **3**, 145–150 (2008)
138. Shaner, N.C., Steinbach, P.A., Tsien, R.Y.: A guide to choosing fluorescent proteins. Nat. Methods **2**, 905–909 (2005)
139. Yang, D., Kang, X., Shang, M., Li, G., Peng, C., Li, C., Lin, J.: Size and shape controllable synthesis and luminescent properties of $BaGdF_5$: Ce^{3+}/Ln^{3+} (Ln = Sm, Dy, Eu, Tb) nano/submicrocrystals by a facile hydrothermal process. Nanoscale **3**, 2589–2595 (2011)
140. Liu, Z.-J., Song, X.-X., Tang, Q.: Development of PEGylated $KMnF_3$ nanoparticles as a T-1-weighted contrast agent: chemical synthesis, in vivo brain MR imaging, and accounting for high relaxivity. Nanoscale **5**, 5073–5079 (2013)
141. Yan, R.X., Li, Y.D.: Down/up conversion in Ln^{3+}-doped YF_3 nanocrystals. Adv. Funct. Mater. **15**, 763–770 (2005)
142. Lemyre, J.L., Ritcey, A.M.: Synthesis of lanthanide fluoride nanoparticles of varying shape and size. Chem. Mater. **17**, 3040–3043 (2005)
143. Li, C., Yang, J., Yang, P., Lian, H., Lin, J.: Hydrothermal synthesis of lanthanide fluorides LnF_3 (Ln = La to Lu) nano-/microcrystals with multiform structures and morphologies. Chem. Mater. **20**, 4317–4326 (2008)
144. Fu, Z., Zheng, H., He, E., Gao, W., Li, G.: Enhancement of red emission by co-dopant Ln^{3+} ions in Eu^{3+}:LaOF nanoparticles. Sci. China Phys. Mech. Astron. **56**, 928–932 (2013)
145. Wang, F., Han, Y., Lim, C.S., Lu, Y., Wang, J., Xu, J., Chen, H., Zhang, C., Hong, M., Liu, X.: Simultaneous phase and size control of upconversion nanocrystals through lanthanide doping. Nature **463**, 1061–1065 (2010)
146. Liu, Q., Sun, Y., Yang, T., Feng, W., Li, C., Li, F.: Sub-10 nm hexagonal lanthanide-doped $NaLuF_4$ upconversion nanocrystals for sensitive bioimaging in vivo. J. Am. Chem. Soc. **133**, 17122–17125 (2011)
147. Li, C., Quan, Z., Yang, P., Yang, J., Lian, H., Lin, J.: Shape controllable synthesis and upconversion properties of $NaYbF_4$/$NaYbF_4$:Er^{3+} and YbF_3/YbF_3:Er^{3+} microstructures. J. Mater. Chem. **18**, 1353–1361 (2008)
148. Ostrowski, A.D., Chan, E.M., Gargas, D.J., Katz, E.M., Han, G., Schuck, P.J., Milliron, D.J., Cohen, B.E.: Controlled synthesis and single-particle imaging of bright, sub-10 nm lanthanide-doped upconverting nanocrystals. ACS Nano **6**, 2686–2692 (2012)

149. Johnson, N.J.J., Oakden, W., Stanisz, G.J., Prosser, R.S., van Veggel, F.C.J.M.: Size-tunable, ultrasmall $NaGdF_4$ nanoparticles: insights into their T1 MRI contrast enhancement. Chem. Mater. **23**, 3714–3722 (2011)
150. Zheng, W., Zhou, S., Chen, Z., Hu, P., Liu, Y., Tu, D., Zhu, H., Li, R., Huang, M., Chen, X.: Sub-10 nm lanthanide-doped CaF_2 nanoprobes for time-resolved luminescent biodetection. Angew. Chem. Int. Ed. **52**, 6671–6676 (2013)
151. Chen, D., Yu, Y., Huang, F., Lin, H., Huang, P., Yang, A., Wang, Z., Wang, Y.: Lanthanide dopant-induced formation of uniform sub-10 nm active-core/active-shell nanocrystals with near-infrared to near-infrared dual-modal luminescence. J. Mater. Chem. **22**, 2632–2640 (2012)
152. Xu, L., Yu, Y., Li, X., Somesfalean, G., Zhang, Y., Gao, H., Zhang, Z.: Synthesis and upconversion properties of monoclinic Gd_2O_3:Er^{3+} nanocrystals. Opt. Mater. **30**, 1284–1288 (2008)
153. X-x, Luo, W-h, Cao: Ethanol-assistant solution combustion method to prepare La_2O_2S:Yb, Pr nanometer phosphor. J. Alloy. Compd. **460**, 529–534 (2008)
154. Vetrone, F., Boyer, J.C., Capobianco, J.A., Speghini, A., Bettinelli, M.: Significance of Yb^{3+} concentration on the upconversion mechanisms in codoped Y_2O_3:Er^{3+}, Yb^{3+} nanocrystals. J. Appl. Phys. **96**, 661–667 (2004)
155. Boyer, J.-C., van Veggel, F.C.J.M.: Absolute quantum yield measurements of colloidal $NaYF_4$: Er^{3+}, Yb^{3+} upconverting nanoparticles. Nanoscale **2**, 1417–1419 (2010)
156. Sarkar, S., Meesaragandla, B., Hazra, C., Mahalingam, V.: Sub-5 nm Ln^{3+}-doped $BaLuF_5$ nanocrystals: a platform to realize upconversion via interparticle energy transfer (IPET). Adv. Mater. **25**, 856–860 (2013)
157. Wong, H.-T., Vetrone, F., Naccache, R., Chan, H.L.W., Hao, J., Capobianco, J.A.: Water dispersible ultra-small multifunctional $KGdF_4$:Tm^{3+}, Yb^{3+} nanoparticles with near-infrared to near-infrared upconversion. J. Mater. Chem. **21**, 16589–16596 (2011)
158. Wang, F., Deng, R., Wang, J., Wang, Q., Han, Y., Zhu, H., Chen, X., Liu, X.: Tuning upconversion through energy migration in core-shell nanoparticles. Nat. Mater. **10**, 968–973 (2011)
159. Zhang, F., Che, R., Li, X., Yao, C., Yang, J., Shen, D., Hu, P., Li, W., Zhao, D.: Direct imaging the upconversion nanocrystal core/shell structure at the subnanometer level: shell thickness dependence in upconverting optical properties. Nano Lett. **12**, 2852–2858 (2012)
160. Wang, F., Wang, J., Liu, X.: Direct evidence of a surface quenching effect on size-dependent luminescence of upconversion nanoparticles. Angew. Chem. Int. Ed. **49**, 7456–7460 (2010)
161. Abel, K.A., Boyer, J.-C., Andrei, C.M., van Veggel, F.C.J.M.: Analysis of the shell thickness distribution on $NaYF_4$/$NaGdF_4$ core/shell nanocrystals by EELS and EDS. J. Phys. Chem. L. **2**, 185–189 (2011)
162. Li, X., Wang, R., Zhang, F., Zhou, L., Shen, D., Yao, C., Zhao, D.: Nd^{3+} sensitized up/down converting dual-mode nanomaterials for efficient in-vitro and in-vivo bioimaging excited at 800 nm. Sci. Rep. **3**, 3536 (2013)
163. Huang, P., Zheng, W., Zhou, S., Tu, D., Chen, Z., Zhu, H., Li, R., Ma, E., Huang, M., Chen, X.: Lanthanide-doped $LiLuF_4$ upconversion nanoprobes for the detection of disease biomarkers. Angew. Chem. Int. Ed. **53**, 1252–1257 (2014)
164. Li, X., Shen, D., Yang, J., Yao, C., Che, R., Zhang, F., Zhao, D.: Successive layer-by-layer strategy for multi-shell epitaxial growth: shell thickness and doping position dependence in upconverting optical properties. Chem. Mater. **25**, 106–112 (2013)
165. Dong, C., Korinek, A., Blasiak, B., Tomanek, B., van Veggel, F.C.J.M.: Cation exchange: a facile method to make $NaYF_4$:Yb, Tm-$NaGdF_4$ core-shell nanoparticles with a thin, tunable, and uniform shell. Chem. Mater. **24**, 1297–1305 (2012)
166. Li, X., Wang, R., Zhang, F., Zhao, D.: Engineering homogeneous doping in single nanoparticle to enhance upconversion efficiency. Nano Lett. **14**, 3634–3639 (2014)

167. Li, J.J., Wang, Y.A., Guo, W.Z., Keay, J.C., Mishima, T.D., Johnson, M.B., Peng, X.G.: Large-scale synthesis of nearly monodisperse CdSe/CdS core/shell nanocrystals using air-stable reagents via successive ion layer adsorption and reaction. J. Am. Chem. Soc. **125**, 12567–12575 (2003)
168. Dong, C., van Veggel, F.C.J.M.: Cation exchange in lanthanide fluoride nanoparticles. ACS Nano **3**, 123–130 (2009)
169. Liu, Q., Sun, Y., Li, C., Zhou, J., Li, C., Yang, T., Zhang, X., Yi, T., Wu, D., Li, F.: F-18-labeled magnetic-upconversion nanophosphors via rare-earth cation-assisted ligand assembly. ACS Nano **5**, 3146–3157 (2011)
170. Xu, C.T., Svenmarker, P., Liu, H., Wu, X., Messing, M.E., Wallenberg, L.R., Andersson-Engels, S.: High-resolution fluorescence diffuse optical tomography developed with nonlinear upconverting nanoparticles. ACS Nano **6**, 4788–4795 (2012)
171. Wen, H., Zhu, H., Chen, X., Hung, T.F., Wang, B., Zhu, G., Yu, S.F., Wang, F.: Upconverting near-infrared light through energy management in core-shell-shell nanoparticles. Angew. Chem. Int. Ed. **125**, 13661–13665 (2013)

Chapter 3
Upconversion Luminescence of Lanthanide Ion-Doped Nanocrystals

Abstract The upconversion luminescence is one of the most significant parameters of upconversion nanoparticles. It is very important for various practical applications, such as bio-imaging, bio-sensing, multiplexed encoding, and so on. This chapter focuses on the upconversion luminescence of the lanthanide-doped nanocrystals, including the emission color tunability and the enhancement of the upconversion luminescence of upconversion nanocrystals.

Keywords Single-band upconversion luminescence · Multi-color emission · Fluorescence resonance energy transfer · Luminescence resonance energy transfer · Upconversion efficiency · Core@shell structure · Lanthanide activators · Upconversion nanocrystals · Quantum dots · Organic dyes

3.1 Introduction

Photon upconversion has been known and studied over a long period of time and is generally based on bulk materials. An extensive review of the early work on bulk materials was given by Auzel in 2004, and there are also other reviews on specific topics of upconversion nanoparticles [1–12]. It was not until recently that upconversion nanoparticles became available for biomedical applications due to the success achieved in producing highly efficient upconversion nanoparticles. Obtaining high efficiency is of particular importance, as it is the basic optical function of upconversion nanoparticles to perform high-performance theranostic application. On the other hand, manipulation of upconversion emission color is of particular importance for multiplexed bio-imaging and multiplexed assays. Multiplexed imaging enables real-time tracking of multiple targets such as proteins and genes, which are important for bio-analysis, clinical diagnosis, as well as for the therapeutic treatments [13–15]. Multiplexed assays entail parallel detection of multiple analytes, which enables

Xiaomin Li and Fan Zhang contributed together to this chapter.

F. Zhang, *Photon Upconversion Nanomaterials*,
Nanostructure Science and Technology, DOI 10.1007/978-3-662-45597-5_3

genotyping, DNA sequencing, and high-throughput screening of combinatorial libraries [16–19]. Therefore, this chapter will focus on the upconversion luminescence of the lanthanide-doped nanocrystals, including the emission color tunability and the enhancement of the upconversion luminescence of upconversion nanocrystals.

3.2 Upconversion Emission Color Tunability

The ability to fine-tune upconversion color output for Ln^{3+}-doped NPs is highly desirable for their applications in multiplexed biological labeling, imaging, and therapeutics. Recently, there has been an increasing focus on the preparation of Ln^{3+}-doped NPs with tunable upconversion emissions from UV to visible and NIR by means of different strategies. The emission colors of Ln^{3+}-doped upconversion nanoparticles can be manipulated by the control of either emission peaks or their relative emission intensities through tuning the host–dopant combinations, the dopant concentrations, the nanoparticles phase, size, shape, crystallinity, surface capping ligands, and so on.

3.2.1 Multi-color Emission Using Different Host–Dopant Combinations

Because each lanthanide ion has its own set of energy level structures, different lanthanide ions can produce distinct sharp emission peaks, thus covering collectively a broad spectrum, from NIR to the UV range (Table 3.1). It has been demonstrated that most of RE ions from La to Lu can exhibit upconversion, including Pr [20–24], Sm [25, 26], Eu [25, 26], Tb [25, 26], Dy [25, 26], Ho [27–29], Er [27, 30–32], and Tm [33, 34]. In addition, the luminescence from a given lanthanide ion is almost independent of different host materials due to the weak coupling of the *f–f* transition with the local crystal field provided by the host lattice. Thus, these features enable production of a range of selective color output by selecting different lanthanide dopants or using their combinations. It is worth pointing out that the phonon distribution from the host lattice and the surface ligands can produce a pronounced effect on the color output on the upconversion nanoparticles, because it can adjust the population between two close energy levels of a given lanthanide ion through multi-phonon-assisted non-radiative processes. Generally, a fixed lanthanide ion or its combination with others can produce a similar set of emission peaks in a given type of host material. The most common activators in upconversion nanoparticles are generally restricted to Er^{3+}, Tm^{3+}, and Ho^{3+} ions, characterized by their ladder-like energy levels. Interestingly, all of these lanthanide ions can be efficiently sensitized by the Yb^{3+} ion whose absorption is

Table 3.1 Typical dopant ions, the corresponding major emissions and energy transitions for making multi-colored upconversion nanocrystals reported in the literatures

Activators	Major emissions (nm)	Energy transitions	References
Pr^{3+}	489, 526, 548, 618, 652, 670, 732, 860	$^3P_0 \rightarrow {}^3H_4$, $^1I_6 \rightarrow {}^3H_5$, $^3P_0 \rightarrow {}^3H_5$, $^3P_0 \rightarrow {}^3H_6$, $^3P_0 \rightarrow {}^3F_2$, $^3P_1 \rightarrow {}^3F_3$, $^3P_0 \rightarrow {}^3F_4$, $^1I_6 \rightarrow {}^1G_4$	[20–24]
Sm^{3+}	555, 590	$^4G_{5/2} \rightarrow {}^6H_{5/2}$, $^4G_{5/2} \rightarrow {}^6H_{7/2}$	[25, 26]
Eu^{3+}	590, 615, 690	$^5D_0 \rightarrow {}^7F_1$, $^5D_0 \rightarrow {}^7F_2$, $^5D_0 \rightarrow {}^7F_4$	[25, 26]
Tb^{3+}	490, 540, 580, 615	$^5D_4 \rightarrow {}^7F_6$, $^5D_4 \rightarrow {}^7F_5$, $^5D_4 \rightarrow {}^7F_4$, $^5D_4 \rightarrow {}^7F_3$	[25, 26]
Dy^{3+}	570	$^4F_{9/2} \rightarrow {}^6H_{13/2}$	[25, 26]
Ho^{3+}	542, 645, 658	$^5S_2 \rightarrow {}^5I_8$, $^5F_5 \rightarrow {}^5I_8$	[27–29]
Er^{3+}	411, 523, 542, 656,	$^2H_{9/2} \rightarrow {}^4I_{15/2}$, $^2H_{11/2} \rightarrow {}^4I_{15/2}$, $^4S_{3/2} \rightarrow {}^4I_{15/2}$, $^4F_{9/2} \rightarrow {}^4I_{15/2}$	[27, 30–32]
Tm^{3+}	294, 345, 368, 450, 475, 650, 700, 800	$^1I_6 \rightarrow {}^3H_6$, $^1I_6 \rightarrow {}^3F_4$, $^1D_2 \rightarrow {}^3H_6$, $^1D_2 \rightarrow {}^3F_4$, $^1G_4 \rightarrow {}^3H_6$, $^1G_4 \rightarrow {}^3F_4$, $^3F_3 \rightarrow {}^3H_6$, $^3H_4 \rightarrow {}^3H_6$	[33, 34]

much stronger and located around 975 nm for the $^2F_{7/2} \rightarrow {}^2F_{5/2}$ transition. The ability of the sensitizer Yb^{3+} ion to efficiently transfer its absorption energy to Er^{3+}, Tm^{3+}, and Ho^{3+} ions enables a single-wavelength excitation (975 nm) of upconversion nanoparticles doped with these lanthanide ions or their various combinations for multi-color emissions (Fig. 3.1a). The Er^{3+} ions mainly have three upconversion emission bands, including two green emission bands at around 525/545 nm originated from the transitions $^2H_{11/2}/^4S_{3/2} \rightarrow {}^4I_{15/2}$ and a red emission band at around 660 nm originated from the transition $^4F_{9/2} \rightarrow {}^4I_{15/2}$ [5, 12]. The Ho^{3+} ions have two main upconversion bands of green and red emission at 541 and 647 nm, corresponding to the transitions $^5S_2/^5F_4 \rightarrow {}^5I_8$ and $^5F_5 \rightarrow {}^5I_8$, respectively [35, 36]. The main upconversion band of the Tm^{3+} ions is in the NIR range at 800 nm originated from the $^3H_4 \rightarrow {}^3H_6$ transition. This NIR upconversion band is located within the "optical transparency window" for biological tissues, in which both light absorption and scattering are significantly reduced [12]. This feature makes Tm^{3+}-doped upconversion nanoparticles particularly interesting for imaging deep-lying tissue in rodents. The Tm^{3+} ions have another three upconversion emission bands at around 479, 450, and 350 nm, generated by the transitions of $^1G_4 \rightarrow {}^3H_6$, $^1D_2 \rightarrow {}^3F_4$, and $^1D_2 \rightarrow {}^3H_6$, respectively [37]. The selection of different emitters or combinations of lanthanide dopants is a straightforward approach to tune the upconversion multi-color emission [12]. The first demonstration of efficient multi-color upconversion emission in colloidal lanthanide-doped nanoparticles was reported by Haase and coworkers. They utilized Yb^{3+}/Er^{3+} and Yb^{3+}/Tm^{3+} codoped $NaYF_4$ nanoparticles to yield strong yellow and blue emissions, respectively [30]. Similar observations have also been reported by Capobianco et al. in $NaYF_4$ nanoparticles [38]. Later, Nann et al. demonstrated four color emissions from $NaYbF_4{:}Tm^{3+}$ (blue), $NaYbF_4{:}Ho^{3+}$ (green), $NaYbF_4{:}Er^{3+}$

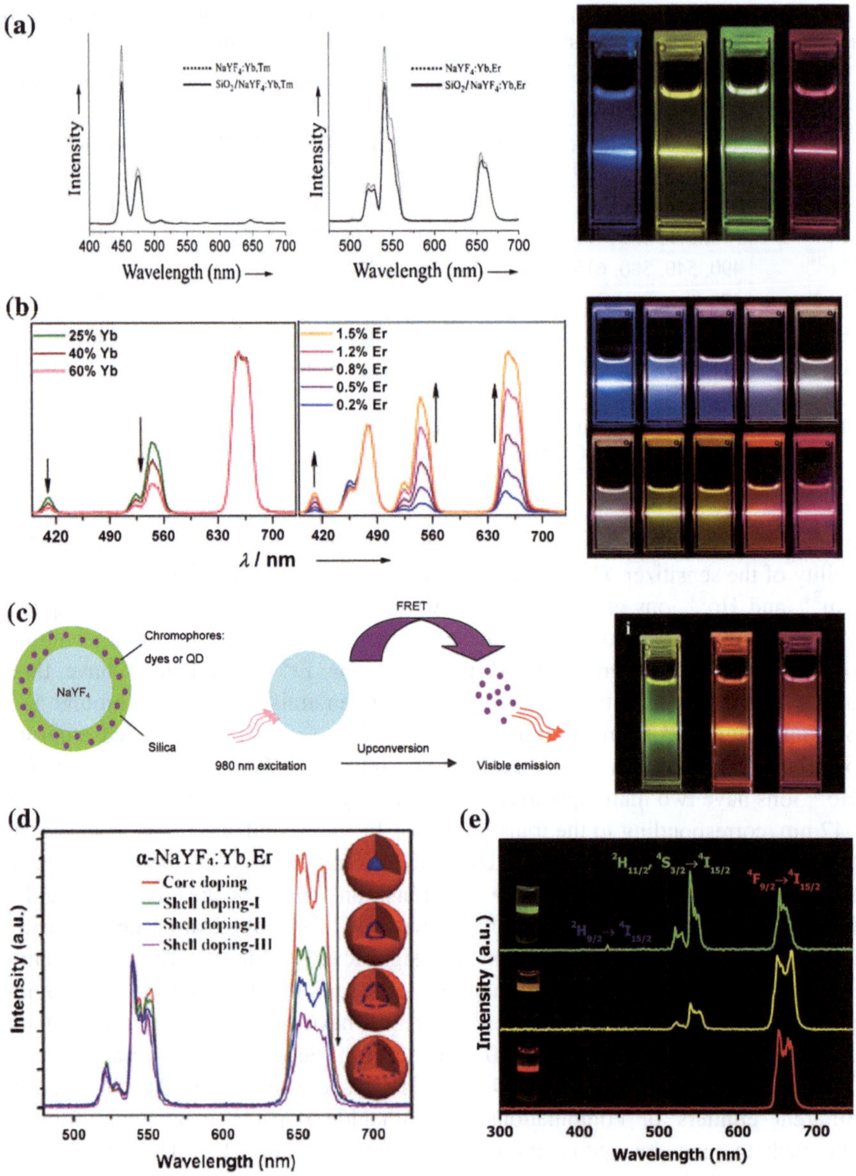
(a)
NaYF4:Yb,Tm
SiO2/NaYF4:Yb,Tm
Intensity
400 450 500 550 600 650 700
Wavelength (nm)
NaYF4:Yb,Er
SiO2/NaYF4:Yb,Er
Intensity
500 550 600 650 700
Wavelength (nm)
(b)
25% Yb
40% Yb
60% Yb
1.5% Er
1.2% Er
0.8% Er
0.5% Er
0.2% Er
420 490 560 630 700 420 490 560 630 700
λ / nm
(c)
Chromophores:
dyes or QD
NaYF4
Silica
FRET
980 nm excitation
Upconversion
Visible emission
i
(d)
α-NaYF4:Yb,Er
Core doping
Shell doping-I
Shell doping-II
Shell doping-III
Intensity (a.u.)
500 550 600 650 700
Wavelength (nm)
(e)
Intensity (a.u.)
300 400 500 600 700
Wavelength (nm)

◀ **Fig. 3.1** **a** Emission spectra of $NaYF_4$:Yb,Tm (*left*) and $NaYF_4$:Yb,Er (*middle*) nanospheres with and without silica coating. Photographs of $NaYF_4$@SiO_2 nanospheres in hexane under excitation of NIR laser (*right*): total upconversion luminescence of $NaYF_4$:Yb,Tm nanospheres (*blue*), total upconversion luminescence of $NaYF_4$:Yb,Er nanospheres (*yellow green*), and luminescence passing through *red* (*green*) or *green* (*red*) filters. **b** Room temperature upconversion emission spectra and emission colors of $NaYF_4$:Yb,Er (25–60/2 mol%) and $NaYF_4$:Yb,Tm,Er (20/0.2/0.2–1.5 mol%) nanoparticles in ethanol solutions. **c** Schematic drawing of FRET-based multi-color $NaYF_4$@SiO_2 NIR-to-vis upconversion nanospheres and photographs of total fluorescence of FITC-doped $NaYF_4$:Yb,Tm@SiO_2 nanospheres, TRITC-doped $NaYF_4$:Yb,Er@SiO_2 nanospheres, and QD605-doped $NaYF_4$:Yb,Tm@SiO_2 nanospheres. **d** Luminescent spectra of the Er-based α-phase core@shell nanoparticles with different doping positions. **e** Room temperature upconversion emission spectra of $NaYF_4$:Yb/Er (18/2 mol%) nanocrystals with 0, 5, and 30 mol% Mn^{2+} dopant ions dispersed in cyclohexane. **a** Reprinted with the permission from Ref. [43]. Copyright 2008 Wiley-VCH Verlag GmbH & Co. KGaA. **b** Reprinted with the permission from Ref. [39]. Copyright 2008 American Chemical Society. **c** Reprinted with the permission from Ref. [43]. Copyright 2008 Wiley-VCH Verlag GmbH & Co. KGaA. **d** Reprinted with the permission from Ref. [44]. Copyright 2013 American Chemical Society. **e** Reprinted with the permission from Ref. [45]. Copyright 2012 Wiley-VCH Verlag GmbH & Co. KGaA

(red), and $NaYF_4$:Yb^{3+} (magenta) nanoparticles, respectively [29]. Wang and Liu [39] reported a general and versatile approach to fine-tune the upconversion emission in a broad range of color output by a single-wavelength excitation at 980 nm. By selecting the Er^{3+} and/or Tm^{3+} as an activator and the Yb^{3+} as the sensitizer, the color output in lanthanide-doped cubic $NaYF_4$ can be finely tuned by variation of the concentration of lanthanide dopants. Recently, it has been established by Chen et al. [40] that the color output in YF_3 nanoparticles doped with different lanthanide dopants and combinations can be manipulated by adjusting the concentration of Yb^{3+}. Chan et al. [41] used combinatorial screening of multiply doped $NaYF_4$ nanoparticles to identify a series of doubly and triply doped upconverting nanoparticles that exhibit narrow and spectrally pure emission spectra at various visible wavelengths. It is known that blue, green, and red are three primary colors that can be combined to produce a palette of colors. The simultaneous use of green and red color through Er^{3+} or Ho^{3+} and the blue emission through Tm^{3+} can yield white upconversion emission in upconversion nanoparticles [42]. Given the broad range of emission wavelengths from lanthanide dopants, an appropriate selection of different lanthanide dopants or their combinations can generate a large library of emission spectra in the NIR-UV spectral region that are particularly useful for multiplexed labeling and encoding.

3.2.2 Doping Concentration-Induced Multi-color Emission

Apart from the emission wavelength tuning, the multi-color upconversion luminescence can also be realized by adjusting the relative intensity of multiple emission peaks. The doping concentration, which determines relative amount of dopants in the nanocrystals as well as the average distance between neighboring dopants, has a strong influence on the optical properties, especially on the relative intensity of emission peaks. Many groups have reported the realization of multi-color upconversion via the tuning of the average distance between neighboring dopants [39, 44, 46, 47]. For example, Liu et al. developed a general and versatile approach to tune the upconversion emission of α-$NaYF_4$:Yb,Er nanoparticles [39]. By precise control of the emission intensity balance through adjusting the different combinations of Yb,Tm and/or Yb,Er dopants, and doping concentration, the upconversion luminescence emissions could be deliberately tuned from VIS to NIR under 980-nm excitation (Fig. 3.1b). Our group showed that multi-color fine-tuning could be alternatively achieved via the three-component ion-doped upconversion nanoparticles system ($NaYF_4$:Yb/Ho/Tm) in a multi-photon emission process [15]. The codoped $NaYF_4$:20%Yb/0.2%Tm system exhibits a blue color emission, resulting from $^1D_2 \rightarrow {}^3F_4$ and $^1G_4 \rightarrow {}^3H_6$. By adding a second emitter (Ho^{3+}) with different concentrations to the system, the relative intensity ratio of the dual emissions can be precisely controlled. For example, variations in the Ho^{3+} concentration (0.2–2.0 %) lead to prominent changes in the green ($^2H_{11/2}/^4S_{3/2} \rightarrow {}^4I_{15/2}$) and red ($^4F_{9/2} \rightarrow {}^4I_{15/2}$) spectral region. Consequently, the adjustable balance of emission intensities enabled the tri-doped particle system to display tunable color output from blue to green. Similar phenomenon has also been observed in $NaYF_4$:Yb/Er [48], Y_2O_3:Yb/Er [49], ZrO_2:Yb/Er [50], Y_2O_3:Yb,Ho,Tm [51], etc. Most recently, Chan et al. [41] used combinatorial screening of multiply doped $NaYF_4$ nanoparticles to identify a series of doubly and triply doped upconversion nanoparticles that exhibit narrow and spectrally pure emission spectra at various visible wavelengths. They also elucidated the energy transfer mechanism giving rise to spectrally pure emission using the comprehensive kinetic model developed by their group. In conclusion, given the broad range of emission wavelengths from lanthanide dopants, an appropriate selection of different lanthanide dopants or their combinations can generate a large library of emission spectra in NIR-UV spectral region that are particularly useful for multiplexed labeling and encoding. As mentioned above, the tuning of doping concentration is still relied on the tunability of the reactant ratios (which can be regarded as macroscopic doping concentration). More recently, our group reported that the multi-color upconversion emission could be deliberately tuned through the doping position control, suggesting a facile route for development of novel upconversion multiplexed probes [44]. When the same amount of dopants (Fig. 3.1d) is doped in the different position of one consistent-sized nanoparticle, the local relative concentration of dopants can be increased as the doping position is gradually transferred from outer shell to inner core. Due to efficient cross-relaxation processes among Er^{3+} ions at high doping levels, the red/green emission ratio can be increased with increasing the local relative concentration of Er and Yb.

3.2.3 *Tuning Upconversion Emission Using FRET or LRET*

Fluorescence resonance energy transfer (FRET) is a mechanism that causes energy transfer from a donor fluorophore to an acceptor fluorophore through non-radiative dipole–dipole coupling [52–57]. When utilizing the upconversion nanoparticle as an energy donor and the dye or quantum dots as an energy acceptor, it offers a larger freedom for upconverted emission wavelengths than the one produced merely by the lanthanide ions. Indeed, Zhang et al. reported that an expanded range of multi-color emission can be obtained from the core/shell $NaYF_4$:Yb^{3+}/Er^{3+} (Tm^{3+})@silica nanostructures compared to that of organic dyes or quantum dots which are encapsulated into the silica shell (Fig. 3.1c) [43]. For each type of lanthanide-doped upconversion nanoparticles, a wide range of multiple colors can be produced through the FRET from the upconversion nanoparticles to the organic dyes or quantum dots that produce new emission colors. A new multiplexed NIR in vivo imaging has also recently been shown using NIR quantum dots and $NaYF_4$:Yb^{3+}/Tm^{3+} nanoparticles [58]. Instead of utilizing the emitting dye to produce additional upconversion color, the emission bands of upconversion can also be selectively quenched to produce varied color outputs via utilizing the strong absorption of unremitting dye or nanoparticles. For example, Wolfbeis et al. reported tuning of dual upconversion emission from either the $NaYF_4$:Yb^{3+}/Er^{3+} or the $NaYF_4$:Yb^{3+}/Tm^{3+} nanoparticles for ratio metric optical encoding by introducing various amounts of amino-reactive organic dye on the upconversion nanoparticle surface to selectively quench an intended upconversion band [59]. Alternatively, gold nanoparticles can also be utilized as a FRET quencher to quench the green emission peak in $NaYF_4$:Yb^{3+}/Er^{3+} [48, 60]. It is worth noting that the time-resolved downconversion FRET process has been established to detect trace amounts of interested analytes using lanthanide-doped nanoparticles that exhibit longtime photoluminescence as energy donors [61–64]. This suggests that the FRET upconversion mechanism can have important implications for sensing applications. Luminescence resonance energy transfer (LRET) is a radiative process where light emitted by the donor is absorbed by acceptor molecules. The difference between LRET and FRET resides in that the energy transfer process in LRET is radiative, whereas it is non-radiative for FRET. Liu et al. reported on multicolor in vivo imaging of upconversion nanocrystals with their emission tuned by LRET to dyes, indicating the effectiveness of LRET for expanding upconversion emissions [65]. In addition, upconversion LRET has been used for sensitive detection of CN– in water and for lection recognition [66, 67]. However, until now, both FRET and LRET mechanisms are utilized without clarification. Discerning these two mechanisms is essential for selecting the type of energy acceptor molecules, which can be accomplished by measuring the time-resolved spectroscopy of the donor emission.

3.2.4 Single-Band Upconversion Luminescence

In comparison with multi-color upconversion nanoparticles, nanoparticles featuring a single-band red upconversion luminescence with a high chromatic purity are more preferable to serving as ideal optical bio-labels for deep tissue imaging in vivo, since the red emission in the range of 600–700 nm falls within the "optical window" of the biological tissues that can afford deep tissue penetration [36]. Therefore, much effort has been devoted to realizing single-band upconversion red emission by rational control of the multi-peak emission profiles of Ln^{3+} [68, 69]. For example, Liu and Wang reported a high red-to-green emission ratio with increasing the doping concentration of Yb^{3+} in Yb/Er codoped $NaYF_4$ NPs [39]. In addition, as a result of the efficient energy transfer between Mn^{2+} and doped Ln^{3+} ions, single-band optical features can be easily achieved in Yb^{3+}/Er^{3+} and Yb^{3+}/Tm^{3+} codoped upconversion nanocrystals that contain Mn^{2+} ions (Fig. 3.1e) [69, 45, 70]. One of the most representative examples of this approach was reported by Zhao and coworkers, who developed a facile strategy to manipulate green and red upconversion emissions in hexagonal $NaYF_4$:Yb/Er nanocrystals by controlling the Mn^{2+} content [45].

The doping of Mn^{2+} ions into the lattice of $NaYF_4$ nanocrystals somewhat disturbed the transition possibilities between green and red emissions of Er^{3+} and thus facilitated the occurrence of red upconversion emission, resulting in an emission color output from green to red. In contrast, Liu and coworkers described another strategy to generate pure single-band upconversion emission in the red and NIR spectral regions, which was mainly based on the use of $KMnF_3$ as the host material with Ln^{3+} (Yb/Er,Yb/Ho and Yb/Tm) dopants homogeneously incorporated into the host lattice [69]. Furthermore, it was observed that the single-band feature found in the $KMnF_3$ nanocrystals was independent of the dopant concentration, pump power, and temperature, which differs markedly from what is observed in multi-color upconversion nanocrystals and makes them particularly attractive as an ideal optical bio-probe for deep tissue imaging without the constraints associated with multi-peak upconversion nanocrystals. Except the introducing of Mn^{2+}, Gao's group also reported the strong single-band red upconversion emission ($\sim$669 nm) with core/shell structured YOF:Yb,Er/YOF nanoparticles under 980-nm excitation [71]. More recently, Wang and coworkers also produced intense intrinsic single-band red (or NIR) upconversion emissions of Er^{3+} (or Tm^{3+}) ions in $Na_3Zr(Hf)F_7$ NPs doped with Yb^{3+}/Er^{3+} (or Tm^{3+}) ions, which originated from the Ln^{3+} clusters formed in $Na_3Zr(Hf)F_7$ host materials [68]. Furthermore, Zhao et al. also reported enhanced red emission from $GdF_3:Yb^{3+},Er^{3+}$ upconversion nanoparticles by Li^+ doping [72]. Unfortunately, up to now, only the single-band red (650–670 nm) and NIR ($\sim$800 nm) upconversion emissions have been realized. More single-band upconversion emissions with different colors and high chromatic purities are desirable for the perspective applications, such as the multiplexed intracellular cell detection and immunohistochemical staining without the constraints associated with conventional multi-peak upconversion nanoparticles.

3.2.5 Other Ways to Tune Upconversion Emission

Tuning upconversion emission by inter-particle energy transfer or antenna effect. Mahalingam and coworkers have reported the first observation of upconversion via interparticle energy transfer (IPET) [73]. A colloidal mixture containing ~5-nm $BaLuF_5$:Tm^{3+} and $BaLuF_5$:Yb^{3+} nanoparticles showed intense blue upconversion emission upon excitation at 980 nm. In contrast, a solution containing only $BaLuF_5$:Tm^{3+} or $BaLuF_5$:Yb^{3+} nanoparticles did not emit the luminescence, clearly displaying the energy transfer from the Yb^{3+}-doped nanoparticles to the Tm^{3+}-doped nanoparticles. This result has important applications, as a range of color output can be produced using similar mechanisms. Moreover, energy transfer between two nanoparticles is strongly dependent on the distance between them. This feature is quite similar to the FRET process, which can measure distance and detect molecular interactions in an excitation donor/acceptor pair that has found wide applications in biology and chemistry. However, further works in characterizing and understanding the basic energy transfer mechanism is required, particularly on the interactions between two single nanoparticles. Recently, energy transfers from dye molecules to $NaYF_4$:Yb^{3+}/Er^{3+} upconversion nanoparticles have been established to significantly improve the intensity of upconversion green emission [74]. This process utilizes an organic dye as an antenna to sensitize small-sized upconversion nanoparticles due to the fact that the absorption of the dye is orders of magnitude stronger than the absorption by the lanthanide ion, and a large amount of lanthanide ions are located at or around the nanoparticle surface due to its unique high "surface-to-volume" ratio produced by the nanoscale dimension. Despite the achieved high upconversion luminescence intensity from dye-sensitized upconversion nanoparticles, the net upconversion quantum yields in this system remain relatively low (~0.1 %). Further work on increasing the upconversion quantum yields in dye-sensitized upconversion nanoparticles will be of value. In addition, although no multi-color upconversion emissions are reported in this work, the antenna mechanism can be easily extended to other lanthanide ions or combinations to produce efficient multi-color emissions.

Tuning upconversion emission using cross-relaxation processes. The cross-relaxation process is a typical energy transfer process in which one ion will transfer part of its excited energy to the other ion. The cross-relaxation process is a fundamental consequence of ion–ion interaction; its efficiency is strongly dependent on the dopant concentration. The cross-relaxation process can be judiciously used to tune the color output in upconversion nanocrystals by enhancing emission from one excited level, while quenching it from another. Chen et al. [75] demonstrated for the first time that tuning upconversion nanocrystals from green to red can be realized in monodispersed cubic $NaYF_4$:Yb^{3+}/Ho^{3+} nanoparticles via tri-doping with Ce^{3+} ions under 970-nm diode laser excitation. Two cross-relaxation processes of $^5S_2/^5F_4(Ho) + {}^2F_{5/2}(Ce) \rightarrow {}^5F_5(Ho) + {}^2F_{7/2}(Ce)$ and $^5I_6(Ho) + {}^2F_{5/2}(Ce) \rightarrow {}^5I_7(Ho) + {}^2F_{7/2}(Ce)$ between the Ho^{3+} and Ce^{3+} ions were employed to select the dominant upconversion pathways to tune the radiation from green to red. Capobianco et al. utilized the cross-relaxation between the two Er^{3+} ions to reduce the green upconversion emission ($^2H_{11/2}/^4S_{3/2} \rightarrow {}^4I_{15/2}$) while increasing the red upconversion emission ($^4F_{9/2} \rightarrow {}^4I_{15/2}$) [76]. Moreover, they utilized the cross-relaxations between Tm^{3+} and Dy^{3+} to preferentially suppress high-energy upconverted emissions of Tm^{3+} in $Yb^{3+}/Tm^{3+}/Dy^{3+}$ codoped $LiYF_4$ colloidal nanoparticles [77]. The cross-relaxation process generally is the main reason for the well-known "concentration quenching mechanism." Caution should be exercised to avoid any decrease of the PL intensity while tuning the color output.

Tuning upconversion emission using ligand effects. The use of the ligand is essential for controlling the size and the shape of resulting upconversion nanoparticles using chemical reaction in a solution phase. However, it is found by Wu et al. that multi-color emissions from $NaYF_4:Yb^{3+}/Er^{3+}$ or $NaYF_4:Yb^{3+}/Tm^{3+}$ nanoparticles could be obtained by changing the ratio of two ligands, octadecylamine and the oleamide, coordinated to the nanoparticle surface [78]. They also found that the multi-color output of N-octadecyloleamide-coated $NaYF_4:Yb^{3+}/Er^{3+}$ nanoparticles can be subtly tuned by a "cooperative effect" resulting from the amidation reaction of the carboxylic acid group of oleic acid and the amine group of oleylamine [79]. Tuning the emission color by a "cooperative effect" might arise from two effects: (i) the change of the size and shape of resulting nanoparticles and (ii) the change of vibrational modes at the interface between the nanoparticle surface and the ligands. A direct measurement of the change of the phonon density distribution of upconversion nanoparticles can be utilized to illustrate the details of the mechanism for the variation of color output.

Tuning upconversion emission using size- and shape-induced surface effects. Several groups have examined the size- and shape-dependent optical properties of the Ln^{3+}-doped upconversion nanoparticles. Song and coworkers observed that the relative upconversion intensity of the red to the green increased gradually with decreasing particle size in the Y_2O_3:1%Er^{3+},4%Yb^{3+} nanoparticles [80]. Yan et al. reported the size-dependent upconversion emission of $NaYF_4:Yb^{3+}/Er^{3+}$ nanoparticles dispersed in solutions [81]. Multi-color upconversion emissions (green, yellow, and red) can be observed when decreasing the upconversion nanoparticles size under excitation with a 980-nm laser. A gradual decrease in the blue photoluminescence efficiency in $NaYF_4:Yb^{3+}/Tm^{3+}$ nanoparticles has been observed by decreasing the nanoparticle size [82]. It is known that the quantum confinement effect is not available for lanthanide ions that are doped into upconversion nanoparticles [83]. Thus, unlike quantum dots, the size-dependent emission cannot be interpreted by quantum mechanics. Alternatively, as the size of the upconversion nanoparticles decreases, the surface-induced (defects, ligands, solvents) effect becomes more prominent, which can produce more efficient multi-phonon-assisted non-radiative relaxations to modify the relative population among close-lying excited states or to deactivate upconversion luminescence. Liu et al. have investigated upconversion luminescence of $NaGdF_4:Yb^{3+}/Tm^{3+}$ nanoparticle of varying size (15, 20, and 30 nm). Despite a large disparity in the particle size, the emission spectra are essentially identical when coated with an inert shell of $NaGdF_4$ [84]. This result provided for the first time direct evidence to support surface-related effects accounting for the size-dependent behavior of upconversion luminescence. In addition, Zhang et al. and Murray et al. reported that $NaYF_4:Yb^{3+}/Er^{3+}/Tm^{3+}$ nanoplates, nanospheres, and nanoellipses or $NaYF_4:Yb^{3+},Er^{3+}/Ho^{3+}$ nanorods, nanoplates, and nanoprisms emitted varied upconversion luminescence [85, 86]. Although detailed explanations are not provided in both works, we believe that the varied color might arise from shape-related surface mechanisms that produce varied phonon influence on the lanthanide ions in upconversion nanoparticles.

3.3 Strategies for Enhancement of Upconversion Efficiency

Upconversion nanoparticles have attracted a great deal of attention in biological applications due to their unique upconversion luminescence, such as high chemical stability, low toxicity, and high signal-to-noise ratio [15, 47, 51, 87–94]. These features suggest that upconversion nanoparticles are becoming a new generation platform to achieve a reliable performance in the context of highly complex bio-imaging applications. Despite the gains, improvements are still needed to optimize optical properties for potential practical applications. The most important parameter that needs improvement is the luminescent efficiency of upconversion nanoparticles, which is still much lower compared with that of the bulk materials (3−5 %) [2, 10, 44, 95, 96]. In this section, we focus on the strategies for the enhancement of the upconversion efficiency.

3.3.1 Selection of Novel Host Matrix

The choice of the host lattice determines the distance between dopant ions, their relative spatial position, their coordination numbers, and type of anions surrounding the dopant [12]. The properties of the host lattice and its interaction with the dopant ions, therefore, have a strong influence on the upconversion process. Furthermore, the host materials generally require close lattice matches with dopant ions and have low phonon energies.

The phonon-induced non-radiative process is the main loss mechanism for up-conversion emissions; the luminescence efficiency is quite sensitive to the distribution of phonon density of states in upconversion nanoparticles. A non-radiative process involves multi-phonon-assisted relaxations whereby the energy difference between the higher and the lower energy level is converted into many lattice phonons [83]. A good indicator of how many phonons may be needed is the cutoff phonon energy of the host lattice. The host materials investigated in nanophosphors encompass oxides (e.g., Y_2O_3 and ZrO_2) [97, 98], fluorides (e.g., $NaYF_4$, $NaGdF_4$, and $NaLuF_4$) [99], vanadates [100], oxysulfide [101], and oxyfluorides or oxychlorides [102], etc.

Table 3.2 Phonon energy of commonly used matrices for rare earth ions

Material	Phonon energy (cm^{-1})	References
Phosphate glass	1,200	[4]
Silica glass	1,100	[4]
Fluoride glass	550	[4]
Chalcogenide glass	400	[4]
Y_2O_3	550	[97]
ZrO_2	500	[98]
$NaYF_4$	350	[99]
LaF_3	300	[4]
$LaCl_3$	240	[4]
YVO_4	890	[100]
Y_2O_2S	520	[101]
GdOCl	500	[102]
$LaPO_4$	1,050	[4]

The cutoff phonon energy of the different host materials is summarized in Table 3.2. Generally, the larger the number of phonons needed to convert the excitation energy into phonon energy, the lower the efficiency of the non-radiative process. Hence, to enhance the emission efficiency by reducing non-radiative rate, it is preferred that the lanthanide ions incorporated into a dielectric host have very low-frequency phonons. Among investigated hosts, fluoride materials have the lowest phonon cutoff energy and generally exhibit the highest upconversion efficiency due to the minimization of non-radiative losses in the intermediate states or the emitting states. The crystal field of a host material has a profound effect on the upconversion efficiency of lanthanide emitters, as the local crystal symmetry surrounding a lanthanide ion strongly influences its optical properties. A less symmetric crystal phase is generally favorable for the upconversion efficiency, as the intermixing of the lanthanide ion's *f* states with higher electronic configurations can be more manifested. For example, hexagonal $NaYF_4$:Yb^{3+}/Er^{3+} microcrystals exhibit visible upconversion PL, which is 4.4 times higher than its cubic counterparts [96], and monoclinic ZrO_2 nanoparticles emit higher upconversion luminescence than the tetragonal phase nanoparticles [98]. Selection of an appropriate host lattice with a low phonon cutoff energy and low crystal field symmetry is therefore important to realize an efficient upconversion process.

As all trivalent rare earth ions exhibit similar ionic size and chemical properties, their inorganic compounds are ideal host materials for upconverting lanthanide dopant ions. In addition, alkaline earth ions (Ca^{2+}, Sr^{2+}, and Ba^{2+}) and some transition metal ions (Zr^{4+} and Ti^{4+}) also exhibit ionic size close to lanthanide ions [5]. Therefore, inorganic compounds containing these ions are also frequently used as host materials for upconversion processes [103, 104]. However, lanthanide doping in these nanocrystals is always accompanied by the formation of crystal defects such as cation vacancies, the interstitial anions which are used to maintain charge neutrality, and so on. To maintain a single crystal phase of the host for efficient upconversion luminescence, the dopant concentration should be stringently controlled. Semiconductor nanocrystals such as ZnS have also been proposed as upconversion host materials [105–107]. However, it is still debatable whether lanthanide ions in these nanocrystals are mainly located on the outermost layer of the nanocrystals due to largely mismatched ionic size between the host and dopant ions or incorporated homogeneously in the host lattice. To date, a series of host materials such as $LiYF_4$ [34, 108– 110], $NaYF_4$ [32, 38, 111–113], $NaGdF_4$ [114, 115], $NaLuF_4$ [116], $BaYF_5$ [117, 118], KY_3F_{10} [119], $BaGdF_5$ [120], YOF [71], LuOF [121], LaOF [122], GdOF [121, 122], CaF_2 [123, 124], SrF_2 [123, 124], and BaF_2 [123] have been developed and reviewed in many papers [1–12].

Generally speaking, the main intra-4*f* electronic-dipole transitions of rare earth ions are forbidden by the quantum mechanical selection rules, which, however, can be broken by the local crystal field (host) of the rare earth ions due to the capability to intermix their *f* states with higher electronic configurations [10, 125, 126]. Consequently, tailoring the host crystal field of rare earth ions and thereby the radiation parameters might become an alternative general strategy to increase the upconversion luminescence for various host materials doped with diverse rare earth ions. Li^+ ion has the smallest cationic radius in the periodical table of elements, which is favorable for the movement and site occupation in the host lattice. These advantages make them attractive for the use in tailoring of the host lattice local crystal field. It has been demonstrated that Li^+ ion can be doped easily into the host lattice substitutionally or interstitially, and this can break the symmetry of the crystal field around the rare earth ions, leading to the enhancement of upconversion luminescence intensity [126–129]. For example, Zhang group reported a 25 times enhancement of the visible upconversion luminescence in Y_2O_3:Yb,Er nanoparticles by codoping with Li^+ ions in the host [126]. Cai et al. reported that green and red luminescence intensities of upconversion nanocrystals were enhanced by about 47 and 23 times in $Na_xLi_{1-x}GdF_4$:Yb,Er [127]. Wang and Nann [129] have gained more than a 30-fold increase in luminescence intensity for $Na_xLi_{1-x}YF_4$:Yb,Er upconversion nanocrystals (Fig. 3.2a). Most recently, the Chen group had developed a unique strategy for the synthesis of novel $LiLuF_4$:Ln^{3+}-based core@shell upconversion nanoparticles with typically high absolute upconversion quantum yield of 5.0 and 7.6 % for Er^{3+} and Tm^{3+}, respectively [130].

As is known, a common strategy for tuning upconversion nanocrystals is to control the doping concentration or species of lanthanide ions [15, 25, 39]. However, the phenomenon of concentration quenching of the excited state at high doping levels poses a significant constraint. Thus, lanthanide ions have to be stringently kept at a relatively low concentration to minimize luminescence quenching. Liu et al. described a new class of KYb_2F_7 host nanocrystals adopting an orthorhombic crystallographic structure in which the lanthanide ions are distributed in arrays of tetrad clusters (Fig. 3.2b) [131]. Importantly, this unique arrangement enables the preservation of excitation energy within the sublattice domain of host and effectively minimizes the migration of excitation energy to defects, even in stoichiometric compounds with a high Yb^{3+} content (calculated as 98 mol%). It means that the KYb_2F_7 can not only serve as a host matrix, but also serve as a sensitizer to absorb and preserve the energy. This allows the generation

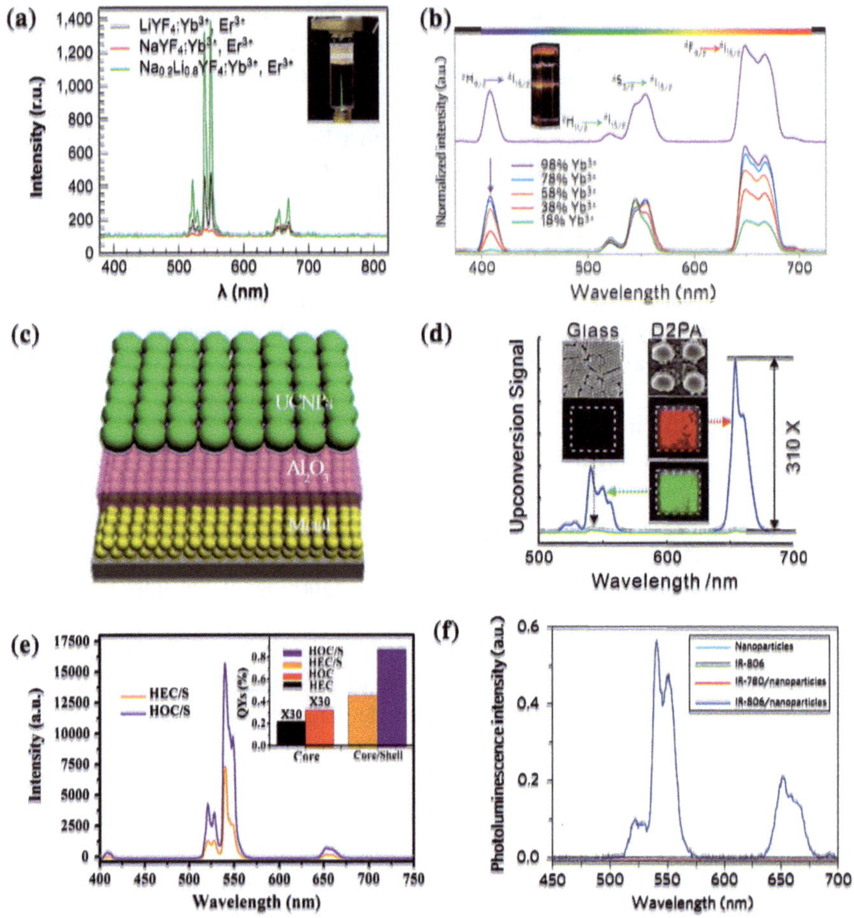

Fig. 3.2 **a** Upconversion luminescence spectra of $NaYF_4$:Yb,Er,$LiYF_4$:Yb,Er and $Na_xLi_{1-x}YF_4$:Yb,Er nanocrystals in chloroform excited by a 980-nm laser diode. **b** The upconversion luminescence of the KYb_2F_7-based upconversion nanoparticles. **c** Schematic structure of a spin cast layer of Au or Ag nanoparticles, a thin Al_2O_3 layer, and a monolayer of upconversion nanoparticles. **d** Enhanced upconversion luminescence on the glass and optimized D2PA substrate. **e** Upconversion emission spectra of $NaGdF_4$:Yb,Er/$NaYF_4$ HEC/S and HOC/S nanoparticles under identical experimental conditions. *Inset* shows the absolute quantum yields of the HEC, HOC, HEC/S, and HOC/S nanoparticles. **f** Steady-state emission spectra of β-$NaYF_4$:Yb,Er nanoparticles, IR-806, β-$NaYF_4$:Yb,Er nanoparticles/IR-780, and β-$NaYF_4$:Yb,Er nanoparticles/IR-806 in $CHCl_3$. **a** Reprinted with the permission from Ref. [129]. Copyright 2009 American Chemical Society. **b** Reprinted with the permission from Ref. [131]. Copyright 2014 Nature Publishing Group. **c** Reprinted with the permission from Ref. [132]. Copyright 2012 American Chemical Society. **d** Reprinted with the permission from Ref. [133]. Copyright 2012 Wiley-VCH Verlag GmbH & Co. KGaA. **e** Reprinted with the permission from Ref. [134]. Copyright 2014 American Chemical Society. **f** Reprinted with the permission from Ref. [74]. Copyright 2012 Nature Publishing Group

of an unusual four-photon-promoted violet upconversion emission from Er^{3+} with an intensity that is more than eight times higher than previously reported. The results highlighted the approach to enhance upconversion through energy clustering at the sublattice level of host may provide new opportunities for light-triggered biological reactions and PDT.

In general, the final crystal structure and grain size are quite related to the parameters of host nanoparticles. As summarized in Table 3.3, the size of current synthesized host nanoparticles doped with lanthanide ions is typically in the range of 20–50 nm. However, the optimal size for bio-imaging probes is reported to be less than 10 nm, so that the nanoparticles can be efficiently cleared from the body [112, 135, 136]. Thus, one of important goals still need to be achieved is to obtain small (<10 nm) lanthanide-doped nanoparticles with intense upconversion luminescence. In regard to this, much effort has been input to synthesize ultrasmall nanoparticles with intense upconversion luminescence and many great achievements have been obtained [64, 112, 116, 135]. For example, the Prasad group has reported the synthesis of monodispersed α-$NaYF_4$ host matrix-based upconversion nanoparticles with ultrasmall size (7–10 nm) and high upconversion efficiency [112]. The intensity

Table 3.3 Typical lanthanide-doped host matrixes with different sizes, crystal structures, and morphologies

Host: dopant	Size (nm)	Crystal structure	Morphology	References
$YLiF_4$:Yb,Tm	~80	Tetragonal	Diamond	[34]
$NaYF_4$:Yb,Er	13.6	Cubic	Polyhedron	[138]
$NaYF_4$:Yb,Er	187 × 71	Hexagonal	Hexagonal plates	[138]
$NaYF_4$:Yb,Er	21 ± 0.5	Hexagonal	Spheres	[43]
$NaYF_4$:Yb,Er	10.5 ± 0.7	Hexagonal	Spheres	[32]
$NaYF_4$:Yb,Er	4.5–15	Hexagonal	Spheres	[91]
$NaYF_4$:Yb,Tm	~7	Cubic	Spheres	[112]
$NaGdF_4$:Yb,Er	20, 31, 41	Cubic	Cube	[139]
$NaGdF_4$:Yb,Tm	10, 15, 25	Hexagonal	Spheres	[84]
$NaGdF_4$	2.5–8	Hexagonal	Hexagonal	[137]
$NaLuF_4$:Yb,Er	18.9	Cubic	Spheres	[116]
$NaLuF_4$:Gd, Yb,Tm	7.81–17	Hexagonal	Spheres	[116]
$BaLuF_5$	<5	Cubic	Dot	[73]
$KGdF_4$:Yb,Tm	~3.7	Cubic	Irregular	[135]
KYb_2F_7:Er	109 × 26 × 12	Orthorhombic	Ice lolly stick	[131]
$LiLuF_4$:Yb,Er	28 ± 1.5	Tetragonal	Rhomboid	[130]
$KMnF_3$:Yb,Er	~30	Cubic	Cube	[69]
YOF:Yb,Er	15 ± 0.4	Cubic	Spheres	[71]
CaF_2:Yb,Er	3.8 ± 0.5	Cubic	Cube	[104]
YF_3:Yb,Er	~3.7	Orthorhombic	Irregular	[40]

of NIR emission could increase up to 43 times along with an increase in the relative content of Yb^{3+} ions from 20 to 100 %. The resultant ultrasmall $NaYbF_4$:Tm^{3+} nanocrystals had NIR upconversion luminescence, which was 3.6 times more intense than that from $NaYF_4$:Yb,Tm nanocrystals with a size of 25–30 nm reported previously. Chen et al. [40] presented a simple and reproducible thermolysis approach to synthesize 3.7-nm-sized lanthanide-doped upconversion nanoparticles using YF_3 as a host matrix. The upconversion luminescence intensity of these ultrasmall nanoparticles was found to be about 10 times brighter than that of CaF_2 host with a size of ~4.5 nm, which were reported to be one of the smallest efficient upconversion nanoparticles. Cohen group also described the controlled synthesis of β-phase $NaYF_4$:Yb,Er nanocrystals with a protein size from 4.5 to 15 nm. The core@shell β-phase $NaYF_4$:Yb,Er@$NaYF_4$ nanoparticles less than 10 nm exhibited higher luminescence efficiency (0.18 ± 0.01 %) than the core nanoparticles with 37 nm in size (0.14 ± 0.01 %) [91]. Most recently, Liu et al. [116] synthesized the sub-10-nm hexagonal $NaLuF_4$ host nanoparticles, which displayed bright upconversion luminescence with a quantum yield of 0.47 ± 0.06 %. Mahalingam et al. showed the upconversion process via inter-particle energy transfer in lanthanide-doped $BaLuF_5$ matrix nanocrystals with a sub-5 nm size for the first time [73]. Van Veggel and coworkers reported the synthesis of monodisperse 2.5–8.0-nm $NaGdF_4$ host nanoparticles by optimizing the reaction parameters associated with both the nucleation and growth phases in a high-boiling binary solvent mixture [137]. Most recently, Chen group developed a strategy based on sodium codoping for the synthesis of monodisperse and ultrasmall CaF_2:Ln^{3+} (Ln = Ce,Tb; Yb,Er; Yb,Tm) upconversion nanoparticles through a coprecipitation route. The resulting sub-10-nm CaF_2:Ln^{3+} core-only and core@shell nanoparticles with tunable shell thickness yielded an enhanced upconversion luminescence [104].

3.3.2 Plasmonic-Enhanced Upconversion Luminescence

It is well known that metal nanostructures (i.e., plasmonic substrates) can efficiently collect light and enhance the intensity due to surface plasmon resonances (SPR). These effects have been widely used for enhancing various optical processes, such as Raman scattering and downconversion luminescence [140–144]. Therefore, the second strategy to enhance the upconversion luminescence is according to the SPR effect of metal [132, 145–156]. The optical properties of a photoluminescence material positioned in close proximity to colloidal metal nanoparticles are affected by the near-field electrodynamic environment [157]. The modified fields cause an enhancement or quenching of the photoluminescence relative to the native state, which depends sensitively on the distance between the luminescent material and metal surface. A judicious balance of the two factors needs to be worked out for plasmonic modulation of luminescence. Quenching would predominantly occur if the luminescent nanoparticle is in direct contact with metallic surface. Conversely, the enhancement of the luminescence occurs on moving the nanoparticle away from the metal surface and is maximized at some critical distance.

It is well known that SPR of metallic structures can enhance the fluorescence from adjacent fluorophores (dyes or quantum dots), when the distance between the metallic structure and the fluorophore is appropriate. Similarly, SPR can also be used to enhance the efficiency of luminescence of upconversion nanocrystals. The guiding principle is that the luminescence efficiency of the nanoparticles can be enhanced by tuning the SPR peak to the excitation/emission wavelength. Recently, a number of nanostructured systems of metal cores, spacer molecules, and photoluminescence materials have been examined. The distance dependency of photoluminescence enhancement and quenching has so far been observed with organic dyes [157], quantum dots [158], and rare earth complexes with organic molecules [159], polyelectrolytes [157], or silica [158] as spacers. In 2010, we reported a facile method to fabricate metal core-enhanced Er^{3+}-doped upconversion nanoparticles based on a core@spacer@shell approach. $Ag@SiO_2@Y_2O_3$:Er nanocomposites with different SiO_2 spacer thickness were synthesized [47]. It was found that the upconversion luminescence enhancements or quenching was dependent on the dielectric silica spacer thickness. The average photoluminescence intensity could reach a maximum value of 4 times that from the pure Y_2O_3:Er shell at an optimal spacer thickness of 30 nm. Almost at the same time, the Duan group reported a new approach for the modulation of upconversion luminescence through plasmonic interactions between upconversion nanoparticles and gold nanostructures [145]. The attachment of gold nanoparticles onto upconversion nanoparticles could more than double the upconversion luminescence intensity. The quantitative analysis of the confocal images showed that an intensity enhancement factor of approximately 2.6 could be achieved. This enhancement was at least partly attributed to surface-plasmon-coupled emission, which can increase the radiative decay rate and emission efficiency. Interestingly, they also found that the formation of a gold shell could significantly suppress the emission because of considerable scattering of excitation irradiation.

The SPR peaks of the above-introduced metal nanoparticles are confined to the visible region at around 500–550 nm, which overlap with the green emission, and therefore, a selective enhancement occurs for this emission peak only. Although gold nanoshells have been coated on the nanocrystals by the Duan group, it greatly suppressed the emission from the nanocrystals due to the strong scattering of the excitation irradiation. It is predicted that simultaneous enhancement for all the upconversion emission peaks can be achieved if the SPR peak is tuned to NIR region and made resonant with the absorption of upconversion nanoparticles at 980 nm thereby substantially increasing the excitation flux via local field enhancement effect. Recently, Priyam et al. have reported a poly-(amino acid) templated gold nanoshell encapsulation of silica-coated $NaYF_4$:Yb,Er upconversion nanoparticles and tuned the SPR peaks in a wide wavelength range from visible to NIR region with highly intense single plasmon peak. The luminescence was dramatically transformed from a quenching effect into an enhancement effect [155]. The photoluminescence intensities of the three upconversion emission peaks at 660 (red), 545 (green), and 410 nm (blue) could be enhanced for 2.1, 2.6, and 3.3 times, respectively.

It has been demonstrated that upconversion luminescence can be enhanced not only by the metal nanoparticles but also by the plasmonic metal microarrays. Most recently, Zhang et al. [133] have investigated a plasmon-enhanced upconversion luminescence of $NaYF_4$:Yb^{3+},Er^{3+} nanocrystals on a 3D plasmonic antenna architecture: the disk-coupled dots-on-pillar antenna array (D2PA) (Fig. 3.2d). They optimized the D2PA structure by tuning the pillar height (h), and consequently the resonance frequency of the substrate, and found that, when h = 75 nm, the resonance frequency matched the pumping laser's frequency and the enhancement factor of upconversion luminescence reached its maximum, 310-fold. Furthermore, Kagan and coworkers demonstrated amplification of luminescence for hexagonal $NaYF_4$ doped with Yb^{3+}, Er^{3+} or Yb^{3+}, Tm^{3+} by close proximity to metal nanoparticles (Fig. 3.2c) [132]. They presented a configuration in which close-packed monolayers of upconversion nanoparticles were separated from a dense multilayer of metal nanoparticles (Au or Ag) by a nanometer scale oxide. Metal-enhanced upconversion luminescence was found to be dependent on the thickness of the oxide spacer layer and the type of metal nanoparticles with enhancements of up to 5.2-fold proximal to Au and of up to 45-fold proximal to Ag nanoparticles.

Most previous works had clearly shown plasmonic-enhanced upconversion luminescence, but some works also reported on plasmonic-induced quenching of upconversion luminescence [46, 160, 161]. Time-resolved measurements did reveal that both the excitation and the emission processes were influenced by the coupling to plasmon resonance in the metallic nanospheres. However, these results could not determine which enhancement and/or quenching mechanism was occurring. The plasmonic-induced luminescence enhancement was known to result from the increase of the radiative rate and/or the excitation intensity by the SPR of metallic nanostructures, while the quenching of the PL stems from resonance energy transfer from the fluorophore to the metallic structure or the absorption of the emitted light by the metallic nanoparticle [162, 163]. For the Stokes-shifted photoluminescence of a fluorophore, the measurement of intrinsic quantum yield of the fluorophore can determine which mechanism takes places and how much it contributes; the increased radiative rate can increase the intrinsic quantum yield, while the increased excitation field is independent of the intrinsic quantum yield of the fluorophore. However, for the upconversion luminescence, the measurement of upconversion luminescence quantum yield becomes ineffective for such purpose, as the quantum yield of upconversion luminescence is overall a nonlinear process and is strongly dependent on the excitation density. To unravel the complex mechanisms of plasmonic-induced upconversion luminescence, a direct measurement of the quantum yield of the Stokes-shifted luminescence (where the upconversion luminescence shares the same peak position) of the lanthanide activator is helpful. Such a measurement can quantify the amount of gain that is due to the increased radiative rate produced by the plasmonic coupling. In addition, a measurement of the enhancement factors of different upconversion luminescence bands is constructive to understand the enhancement mechanism, as the plasmonic-increased local excitation field generally results in more pronounced effects on a multi-photon involved upconversion luminescence band, when utilizing a low excitation power density that can not produce a saturation effect. It is known that the plasmonic-induced quenching effect only occurs efficiently when the upconversion nanocrystals and metallic structures are placed in direct contact or at a quite close distance. Thus, it is imperative to introduce a spacer layer between the upconversion nanoparticles and the metallic structure to avoid the quenching effect. Moreover, the ability to precisely control the thickness of the spacer provides an opportunity to maximize the upconversion luminescence under NIR light excitation. Theoretical investigations on the interaction of plasmonic metallic structures of varying shapes and sizes with upconversion nanoparticles are needed in the future to provide the guidance on chemical preparations of an optimized plasmonic-enhanced upconversion nanocrystals system for theranostic applications.

3.3.3 Fabrication of Highly Efficient Lanthanide Upconversion Nanoparticles by Homogeneous Doping

As mentioned above, doping concentration is also considered to be one of the most important factors influencing the optical properties of the upconversion nanocrystals. For example, in order to get highest efficiency upconversion luminescence, the optimum doping concentration for Yb^{3+} and Er^{3+} codoped system is ~20 and ~2 %, respectively [43, 44, 91, 96, 114, 164, 165]. The higher or lower doping concentration does not seem to be beneficial to the promotion of the efficiency, which means that the higher doping concentration would induce the cross-relaxations between the dopants and the lower doping concentration would cause the reduction of the luminescent center [10, 96]. Up to now, the doping concentration of the products is usually denoted by the reactant ratios used in the experiment (which can be regarded as macroscopic doping concentration). It is still difficult to confirm whether the doping elements are homogeneously dispersed in the nanoparticles from inside to outside. In other words, the particles may possess quite different local doping concentrations at different positions of single nanoparticle even the macroscopic doping is maintained constant. The heterogeneous doping of the dopants will further induce a series of negative effects on the upconversion luminescence of the particles, such as the concentration quenching caused by cross-relaxations between the dopants, energy transfer barrier result from the spatial separation between the sensitizer and activator, and so on.

In 2014, the influence of spatial distributions and relative concentration of the dopants on the optical properties of upconversion nanoparticles were investigated for the first time in the single nanoparticle level by our group (Fig. 3.2e) [134]. We demonstrated that heterogeneous doping indeed existed in the upconversion nanoparticles prepared by the regular synthesis approaches reported previously [referred as heterogeneous doping core (HEC) nanoparticle]. The heterogeneous doping can induce the cross-relaxations and energy transfer barrier between dopant ions because of the local relative enrichment of the dopants. In order to avoid the deficiency, we developed a homogeneous doping strategy to fabricate the upconversion nanoparticles [i.e., homogeneous doping core (HOC) nanoparticle] based on the one-pot successive layer-by-layer (SLBL) method [44]. Most importantly, the overall upconversion emission intensity and lifetime of the HOC upconversion nanoparticles were remarkably enhanced compared with the HEC upconversion nanoparticles. The absolute quantum yield as high as 0.89 ± 0.05 % was realized for the $NaGdF_4$:Yb,Er/$NaYF_4$ upconversion nanoparticles by homogeneous doping, which is about two times higher than that of the HEC-based core/shell upconversion nanoparticles (0.47 ± 0.05 %). Furthermore, using homogeneous doping $NaGdF_4$:Yb,Tm nanoparticles as initial core, the multi-color emission intensity $NaGdF_4$:Yb,Tm/$NaGdF_4$:A (A = Tb^{3+},Eu^{3+}) could also exhibit 20–30 % enhancement compared with that of HEC-based nanostructure, suggesting that the homogeneous doping approach not only could be used for the efficiency enhancement of lanthanide ion-doped nanoparticles, but also provided a track for engineering the high-quality upconversion nanoparticles with advanced nanostructure and optical properties.

3.3.4 Tuning the Excitation Wavelength

NIR laser at a wavelength of about 980 nm is always used to excite upconversion nanoparticles for upconversion emission since the sensitizer ion Yb^{3+} has a high absorption cross section in its band, while 980 nm light has strong optical absorption of water and biological specimens [166]. This could degrade image resolution, limiting the imaging penetration depth of light and possibly resulting in substantial cell and tissue damage because of the overheating. Therefore, the excitation band tuning of upconversion nanoparticles into an appreciate range is also an important issue for improving the performance.

Recently, Zou et al. [74] reported the concept that an organic NIR dye (IR-806) could be used as an antenna to harvest NIR photons with a broad band (740–850 nm) for β-$NaYF_4$:Yb,Er nanoparticles (Fig. 3.2f). The overall upconversion by the dye-sensitized nanoparticles was dramatically enhanced (by a factor of $\sim$3,300) as a result of increased absorptivity and overall broadening of the absorption spectrum of the upconverter. Although the absolute upconversion efficiency was low, this composite shows higher efficiency compared with the bare upconversion nanoparticles under the 740–850-nm excitation. On the other hand, this paper provided a strategy to design some near-infrared dyes and upconversion nanoparticle composites to realize the high upconversion efficiency under the specific excitation with different wavelength. For example, the dye could absorb photons at 980 nm and then transfer the energy to upconversion nanoparticles; combined the 980 nm photons absorbed by the nanoparticles, the efficiency would be enhanced.

Prasad's group reported the intense upconversion photoluminescence in colloidal $LiYF_4:Er^{3+}$ nanocrystals under excitation with telecom wavelength at 1,490 nm [109]. The intensities of two- and three-photon anti-Stokes photoluminescence bands were higher than or comparable to that of the Stokes emission under excitation with a low power density in the range of 5–120 W/cm^2. The quantum yield of the photoluminescence (~1.2 ± 0.1 %) was almost four times higher than that (0.3 ± 0.1 %) of the lanthanide-doped nanocrystals reported previously under an excitation at ~980 nm [95]. Zhan et al. [167] also developed a new promising excitation approach to improve the performance of in vitro or in vivo imaging with a cost-effective laser at 915 nm instead of 980 nm light. This novel laser excitation method provided drastically less heating of the biological specimen and deep imaging penetration depth in animals or tissues due to the low water absorption of 915-nm laser. To sum it up, the excitation wavelength tuning may provide an alternative and significant way to improve the performance of upconversion nanoparticles.

3.4 Core@Shell Structure-Based Luminescence Engineering of the Upconversion Nanocrystals

In addition to lanthanide-doped core-only upconversion nanoparticles, much progress has also been made in the synthesis of upconversion nanoparticles with core@shell structures in the past decade. It should be noted that the small lattice mismatch between the core and shell materials is the important precondition for the fabrication of high-quality core@shell structures. In the upconversion nanoparticles "nanolab", the core@shell structure plays a very important role in the luminescence engineering of upconversion nanoparticles, because core@shell nanoparticles not only provide an excellent platform to combine different property or functional units into one single nanoparticle, but also offer an effective solution to problems associated with the core-only nanoparticles.

3.4.1 Enhancing Upconversion Luminescence with Core@Shell Structure

Nanosized lanthanide-doped upconversion nanoparticles normally have a high proportion of dopant ions distributed on the surface of particles. Luminescence is readily quenched by high-energy oscillators arising from surface defects and weakly bound impurities, ligands, and solvents due to lack of effective protection by the host lattice. Excitation energy of the interior ions can also be transferred to the surface through adjacent dopant ions and eventually dissipated non-radiatively. To make up the deficiencies, growing a shell with small lattice mismatch around the core to reduce non-radiative decay losses of the surface luminescence can provide an effective way to improve the optical properties of upconversion nanoparticles. In such core@shell structures, the dopant ions are confined in the interior core of the nanocrystals. The shell can effectively suppress energy loss on the crystal surface, leading to the enhanced luminescence efficiency. The first demonstration of the core@shell structured upconversion nanoparticles was reported by Lezhnina and coworkers [168], which was focused on EuF_3@GdF_3, GdF_3@EuF_3, LaF_3:Yb, Ho@LaF_3, and LaF_3:Nd@LaF_3 core@shell nanoparticles. In 2007, Chow et al. achieved ~7 and 29 times luminescence enhancement for $NaYF_4$:Yb,Er and $NaYF_4$:Yb,Tm nanocrystals after coating with $NaYF_4$ shells, respectively [169]. Since then, many core@shell nanocomposites have been prepared to improve the efficiency, such as $NaYF_4$:Yb,Er(Tm)@$NaGdF_4$ [165, 170, 171], $NaYF_4$:Yb,Er (Tm)@$NaYF_4$ [81, 169, 172], $NaGdF_4$:Yb,Tm(Er,Nd)@$NaGdF_4$ [84, 139, 173], $NaGdF_4$:Yb,Er(Tm,Ce,Tb)@$NaYF_4$ [44, 174], CeF_3:Tb@LaF_3 [175–177], KYF_4: Yb,Er@KYF_4 [178], YOF:Yb,Er@YOF [71], and so on.

Interestingly, Liu's group recently showed that the epitaxial growth of an optically inert $NaYF_4$ layer around a lanthanide-doped $NaGdF_4$@$NaGdF_4$ core@shell nanoparticle could effectively prevent surface quenching of excitation energy (Fig. 3.3a) [26]. To verify the effect of the activator emission enhancement by $NaYF_4$ layer protection, they further investigated a series of $NaGdF_4$:Yb, Tm@$NaGdF_4$:A@$NaGdF_4$ core@shell@shell nanoparticles (A = Eu, Tb, Dy, Sm). The results showed that Gd excitation energy could migrate through the Gd shell and dissipated through the trapping by the surface defects or solvent molecules. The photoluminescence studies showed that the Tb^{3+} emission in the core@shell@shell nanoparticles was essentially not improved. When Gd^{3+} ions in the outermost shell were gradually replaced by optically inert Y^{3+}, a steady enhancement of the activator emission was observed. Taken together, the results conclusively suggested that the suppression of surface quenching to Gd^{3+} ions was primarily responsible for the enhanced activator emissions.

The shell composites of the core@shell upconversion nanoparticles mentioned above are all host shells without dopants. Because of the simple and single property of the host shell, the enhancement of the luminescence efficiency is very limited. The design and fabrication of the upconversion nanoparticles with more efficient core@shell structure is desirable [114, 179, 180]. For example, Capobianco and coworkers showed a strong enhancement of the green and red emission bands of $NaGdF_4$:Yb,Er particles by growing the Yb^{3+} doped $NaGdF_4$ shells around cores (Fig. 3.3b) [114]. Except for the ETU process in the $NaGdF_4$:Yb,Er cores, additional energy transfer from excited Yb^{3+} ions in the shell to Er^{3+} ions in the core increased the overall efficiency of nanoparticles. The upconversion emission of the active-core/active-shell particles was more intense by a factor of about 3 (green) and 10 (red) compared to the standard core@shell particles with a doped core and an inert shell.

The fabrication of core@shell structures is a very effective strategy to improve the optical property of upconversion nanoparticles. However, only few papers reported a precise link between the shell thickness of core@shell nanoparticles and

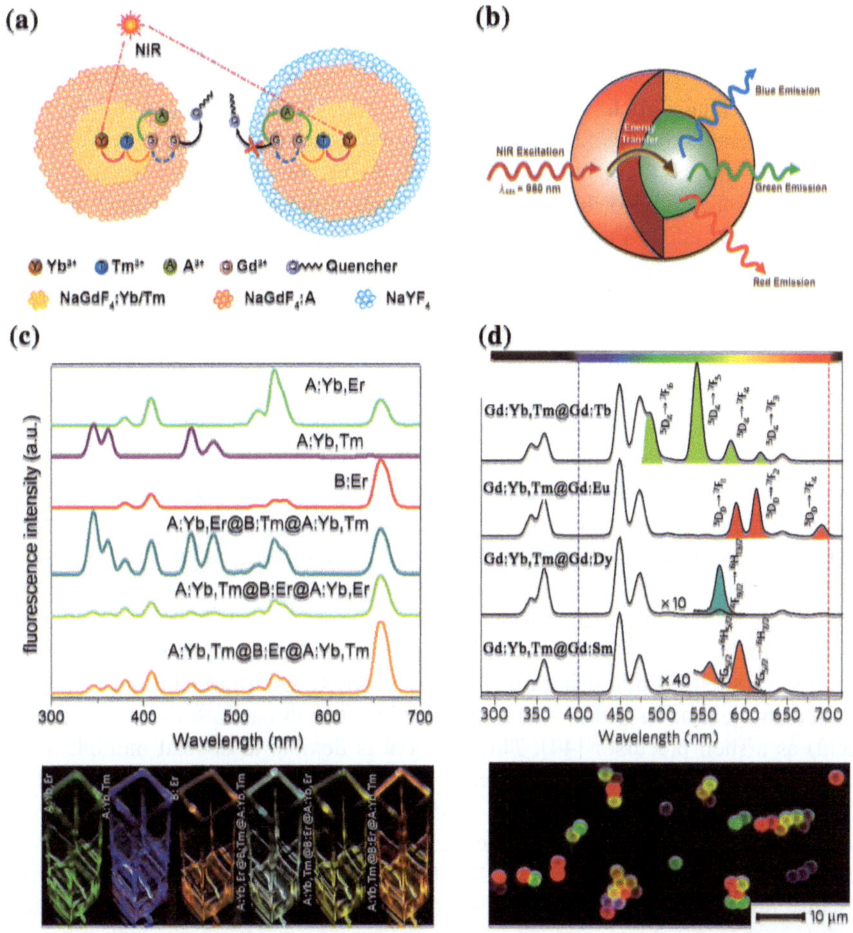

Fig. 3.3 **a** Schematic illustration of the energy transfer mechanism in the core@shell structured and $NaYF_4$-coated core@shell@shell structured nanoparticles. **b** General depiction of the $NaGdF_4$: Er^{3+}, Yb^{3+}@$NaGdF_4$:Yb^{3+} active-core@active-shell nanoparticles under 980-nm excitation. **c** Upconversion spectra and optical images of core@shell structure-based multi-color upconversion nanoparticles. Layer component of upconversion nanoparticles are as follows, respectively: A:Yb, Er; A:Yb,Tm; B:Er; A:Yb,Er@B:Tm@A:Yb,Tm; A:Yb,Tm@B:Er@A:Yb,Er; A:Yb,Tm@B: Er@A:Yb,Tm; $NaYF_4$ is defined as A, and $NaYbF_4$ is defined as B. **d** Emission spectra of the as-prepared $NaGdF_4$@$NaGdF_4$ core–shell nanoparticles doped with different activators (activator emissions are highlighted with color). Luminescence micrograph of polystyrene beads tagged with core–shell nanoparticles comprising $NaGdF_4$:Yb,Tm@$NaGdF_4$ (*blue*), NaGdF4:Yb,Tm@NaGdF4:Tb (*green*), $NaGdF_4$:Yb,Tm@$NaGdF_4$:Eu (*red*), and a binary mixture of $NaGdF_4$:Yb, Tm@$NaGdF_4$:Tb and $NaGdF_4$:Yb,Tm@$NaGdF_4$:Eu (*yellow*), respectively. **a** Reprinted with the permission from Ref. [26]. Copyright 2012 American Chemical Society. **b** Reprinted with the permission from Ref. [114]. Copyright 2009 Wiley-VCH Verlag GmbH & Co. KGaA. **c** Reprinted with the permission from Ref. [179]. Copyright 2013 Elsevier B.V. **d** Reprinted with the permission from Ref. [25]. Copyright 2011 Nature Publishing Group

their optical properties, mainly because the core@shell structures have only been inferred from indirect measurements. Recently, with the high-angle annular dark-field (HAADF) technology, our group and van Veggel et al. had shown that the core@shell nanoparticles synthesized by the most widely used seed-mediated heating-up method did not have a uniform shell, meaning that the core was partially covered by the shell [165, 171, 181]. Therefore, the core nanoparticles could not be passivated efficiently to decrease the core surface defects and insulated from the environmental effect. Furthermore, we also realized direct imaging of the $NaYF_4$:Yb,Er@$NaGdF_4$ core@shell structure at the subnanometer level for the first time with cryo-transmission electron microscopy (Cryo-TEM) and rigorous electron energy loss spectroscopy (EELS) under the low operation temperature (96 K) [165]. A strong linear link between the $NaGdF_4$ shell thickness and optical response of the hexagonal $NaYF_4$:Yb,Er@$NaGdF_4$ core@shell nanocrystals had been established. In order to avoid the growth of the heterogeneous shell, a reproducible multi-cycle batch method to grow host shells on $NaYF_4$:Yb,Er nanocrystals with monolayer (ML) thickness was demonstrated for the first time [165]. During the epitaxial growth of the host shell layer-by-layer, surface defects of the upconversion nanoparticle cores could be gradually passivated by the homogeneous shell deposition process, which results in the obvious enhancement in overall upconversion emission intensity and lifetime and more resistant to quenching by water molecules. Furthermore, we also developed a facile one-pot SLBL method to realize the successive shell coating using high boiling point Re-OA (rare earth chlorides dissolved in oleic acid) and Na-TFA-OA (sodium trifluoroacetate dissolved in oleic acid) as a shell precursor [44]. This protocol is flexible to deposit multiple shells uniformly by successive introduction of the shell precursor solutions instead of the lengthy multi-cycle batch operation [165]. As a result of the high-quality uniform shell and advanced core@shell structures, the upconversion luminescence efficiency can be improved up to 0.51 ± 0.08 %. Therefore, we believe that high-quality upconversion nanoparticles with more reasonable and advanced structures would be fabricated by simply changing amounts or species of the shell precursors with this SLBL method.

Actually, the most effective way to enhance upconversion intensity is increasing the number of emitters [10]. However, high doping levels of emitters lead to deleterious resonant energy transfer, resulting in quenching of excitation energy [5]. It has been extensively demonstrated that the optimal dopant concentration is ∼2 mol% for emitter Er^{3+}, as far as upconversion nanomaterial $NaYF_4$:Yb,Er is concerned [96], and no effort has yet been made on the breakthrough of this limitation. Recently, Kong and Zhang et al. demonstrated that the well-accepted upper limitation of the concentration quenching threshold could increase from 2 to 5 mol% for Er^{3+} by a designed spatial separation strategy of the emitter doping area [182]. The separated layer effectively minimized the energy transfer between the emitters in the inside and outside layers and largely lowered the possibility of trapping of excitation energy by defects, resulting in enhancement of the upconversion luminescence. Actually, the quenching effect with high doping levels of emitters had been demonstrated in 2008 by Qian and Zhang [180]. They reported the synthesis of hexagonal $NaYF_4$:Yb,Tm@$NaYF_4$:Yb,Er nanocrystals with Tm and Er doped in the core and shell, respectively. In contrast to the Tm/Er codoped $NaYF_4$ nanocrystals in which Tm luminescence was quenched, both Tm and Er emissions were observed from the core@shell nanocrystals. $NaYF_4$:Yb, Tm@$NaYF_4$:Yb,Er@$NaYF_4$:Yb,Tm nanocrystals with a sandwich structure were also synthesized, and the luminescence from Er and Tm was greatly enhanced. Furthermore, Liu group showed that the spatial confinement of dissimilar dopant ions within specific regions of the core@shell nanoparticles was critical for generating efficient energy migration-mediated upconversion (EMU) [25]. The core@shell structure was designed to eliminate cross-relaxation-induced energy loss due to direct interaction between the Yb/Tm pair and activator in their work. To verify the requirement of a core@shell structure separating Yb/Tm pair and activator (Eu, Tb, Dy, Sm), they prepared a series of $NaGdF_4$@$NaGdF_4$ core@shell nanoparticles with the core homogeneously doped with Yb^{3+}, Tm^{3+}, and Tb^{3+} at different doping levels. By comparing the emission spectra of these core@shell nanoparticles with the Yb/Tm pair and activators at separated layers, it was found that the homogenously doped nanoparticles with Tb^{3+}, Eu^{3+}, Dy^{3+}, and Sm^{3+} all exhibited weak emission intensities in stark contrast to strong emissions from the heterogeneously doped counterpart. The decreased emission intensity could be attributed to cross-relaxation between Tm^{3+} and Tb^{3+} (or Eu^{3+}, Dy^{3+}, Sm^{3+}). The suppressed emission peaks of Tm^{3+} centered at 360 and 450 nm also proved the interaction between Tm^{3+} and Tb^{3+} (or Eu^{3+}, Dy^{3+}, Sm^{3+}). It is noteworthy that the cross-relaxation between Tm^{3+} and Tb^{3+} could be slightly eliminated at a low Tb^{3+} dopant content. However, a low concentration of Tb^{3+} was also unfavorable for trapping the excitation energy of Gd^{3+}. As a result, a suppressed EMU process occurred with significant luminescence quenching.

Besides rare earth-based shells, the heterogeneous core@shell structures have also been investigated for the enhancement of the upconversion efficiency of the core nanoparticles. Wang et al. [183] reported ~300-fold enhancement of upconversion luminescence in $NaYF_4$:Ln^{3+}@CaF_2 heterogeneous core@shell nanoparticles with a size of 10–13 nm in comparison with the parent $NaYF_4$:Ln^{3+} nanoparticles. The hetero-shell of CaF_2 was chosen as well to coat $NaYbF_4$:Tm, leading to a 35-fold enhancement of NIR-to-NIR upconversion intensity and reaching a quantum yield as high as 0.6 % under an excitation power density of 0.3 W/cm^2 [184]. Han et al. reported on two orders of magnitude enhancement of NIR-to-UV upconversion luminescence in $NaYF_4$:Yb,Tm@CaF_2 core@shell nanoparticles compared to the core nanoparticles [185]. As compared to the typically used $NaYF_4$ shell, CaF_2 coating was found to be more effective in resisting quenching in aqueous medium to preserve the upconverting emissions and more efficient in suppression of lanthanide leakage.

3.4.2 Tuning Upconversion Emission Using Core@Shell Structures

The use of core@shell architectures with two or more lanthanide activators incorporated into the core and different shells is a typical strategy to tune the luminescence of upconversion nanoparticles, while maintaining the high luminescence efficiency [186]. This strategy also explores different combinations of lanthanide activators to produce multi-color emissions, but exercises a spatial isolation of these lanthanide activators using an appropriate core@shell structure. It eliminates deleterious cross-relaxations between the lanthanide activators and suppresses surface-related quenching mechanism in the core nanoparticles, thus yielding a range of tunable emission with high efficiency. Recently, Liu et al. have reported the tuning of upconversion properties for a rather wide range of activators using the combination of gadolinium sublattice-mediated energy migration and core@shell structural engineering, which would expand the range of applications for lanthanide-doped nanoparticles [25]. The results show that many lanthanide ions, including Tb^{3+}, Eu^{3+}, Dy^{3+}, and Sm^{3+}, without long-lived intermediary energy states can also serve as the activators to generate tunable upconversion emissions spanning the visible spectral region (Fig. 3.3d). Similarly, Chen group designed a novel strategy to fabricate uniform $NaGdF_4$ NPs that were composed of a $NaGdF_4$:Yb,Tm core and $NaGdF_4$:Eu shell in order to achieve dual-mode up/downconversion luminescence from identical Eu^{3+} ions. The use of such core@shell structures allowed the elimination of the deleterious cross-relaxations between doped Eu^{3+}, Yb^{3+}, and Tm^{3+} ions, which thereby yielded a 10 times enhancement in the upconversion luminescence of Eu^{3+} as compared with the triply doped core-only counterparts [187]. Zhang et al. reported on hexagonal phase core@shell structured $NaYF_4$:Yb,Tm@$NaYF_4$:Yb,Er [180] and sandwich structured core@shell@shell $NaYF_4$:Yb,Tm@$NaYF_4$:Yb,Er@$NaYF_4$:Yb, Tm,$NaYF_4$:Yb,Er@$NaYbF_4$:Tm@$NaYF_4$:Yb,Tm,$NaYF_4$:Yb,Tm@$NaYbF_4$:Er@Na YF_4:Yb,Er, and $NaYF_4$:Yb,Tm@$NaYbF_4$:Er@$NaYF_4$:Yb,Tm nanoparticles to produce multiple emission peaks and for multi-color cellular imaging (Fig. 3.3c) [179].

3.4.3 Tuning of the Excitation Wavelength

One major limitation of conventional upconversion nanoparticles is the low conversion efficiency especially under a low-intensity laser excitation, mainly due to the extremely weak absorption of sensitizer Yb^{3+} ion in NIR wavelength range [74, 188, 189]. Moreover, owing to the very narrow NIR absorption spectral window of upconversion nanoparticles, 980-nm laser is the most commonly used NIR excitation source to generate upconversion luminescence. However, the use of 980-nm NIR laser as an excitation source for biological applications has an intrinsic disadvantage: Water, as the most significant component of the animal and human body, exhibits a high photon absorbance around 980 nm [166]. The intensity of the laser beam would be overwhelmingly attenuated while diffusing in biological tissues, resulting in the limited penetration ability. Furthermore, the absorbed light energy would transform into thermal energy, often causing serious sample overheating in conjunction with substantial cell and tissue damage under continuous irradiation. Therefore, to make upconversion nanoparticles more suitable for biological applications, it is necessary to enhance the absorption of NIR light and optimize the excitation light wavelength into an appropriate range. In contrast to 980-nm NIR laser, the use of 800-nm NIR laser as the excitation source may both overcome the overheating issues and significantly improve the penetration depth for deep tissue imaging since the NIR region around 700–900 nm reaches a minima absorbance for all biomolecules [166]. Unfortunately, the upconversion luminescence of conventional upconversion nanoparticles could be hardly detected under excitation of 800 nm, due to the extremely low-absorption cross section of commonly used activators (Er^{3+}, Tm^{3+}, and Ho^{3+}) and sensitizer (Yb^{3+}) around 800 nm. To overcome this problem, it is essential to combine upconversion nanoparticles with other components which have large absorption around 800 nm. The Nd^{3+} ion could become a quite potential candidate as the sensitizer to realize the upconversion luminescence of upconversion nanoparticles, due to the intense absorption cross section of Nd^{3+} around 800 nm and efficient energy transfer from excited Nd^{3+} to Yb^{3+} ions. Therefore, assisted by the delicate design of the core@shell nanostructure, the excitation wavelength of upconversion nanoparticles can be blue-shifted to ~800 nm.

The intrinsic excitation wavelength of colloidal dispersible upconversion nanoparticles was first engineered by Han group according to the fabrication of cascade sensitization of Nd/Yb/Er (Tm) tri-doped core@shell β-$NaYF_4$ colloidal upconversion nanoparticles [190]. Nevertheless, the doping concentration of Nd in the tri-doped system must be constrained to a very low level (less than 1 %) to minimize the quenching of the excitation energy through efficient energy back transfer from activators to ^4I_J manifolds of Nd^{3+}. Thus, directly doping Nd^{3+} into upconversion nanoparticles always results in much lower upconversion luminescence under 800-nm irradiation than that under 980-nm irradiation. Afterward, Xie et al. developed a new type of $NaYF_4$:Yb,Nd,A@$NaYF_4$:Nd (A = Tm,Er,Ho) core@shell upconversion nanoparticles which could be effectively excited at 795 nm. The core@shell upconversion nanoparticles were synthesized through spatially confined doping of neodymium (Nd^{3+}) ions (Fig. 3.4a, b) [191]. The use of Nd^{3+} ions as sensitizers facilitated the energy transfer and photon upconversion of a series of lanthanide activators (Er^{3+}, Tm^{3+}, and Ho^{3+}) at a biocompatible excitation wavelength (795 nm) and also significantly minimized the overheating problem associated with 980-nm excitation. Importantly, the core@shell design enabled high-concentration doping of Nd^{3+} (~20 %) in the shell layer and thus markedly enhanced the upconversion luminescence from the activators, providing highly attractive luminescent biomarkers for bio-imaging without auto-fluorescence and concern of overheating. Similarly, to address the spatially separating the activator and Nd^{3+}, the Yan group also synthesized β-$NaGdF_4$:Yb,Er@$NaGdF_4$:Nd core@-shell nanoparticles. The new excitation band located at 808 nm could achieve high

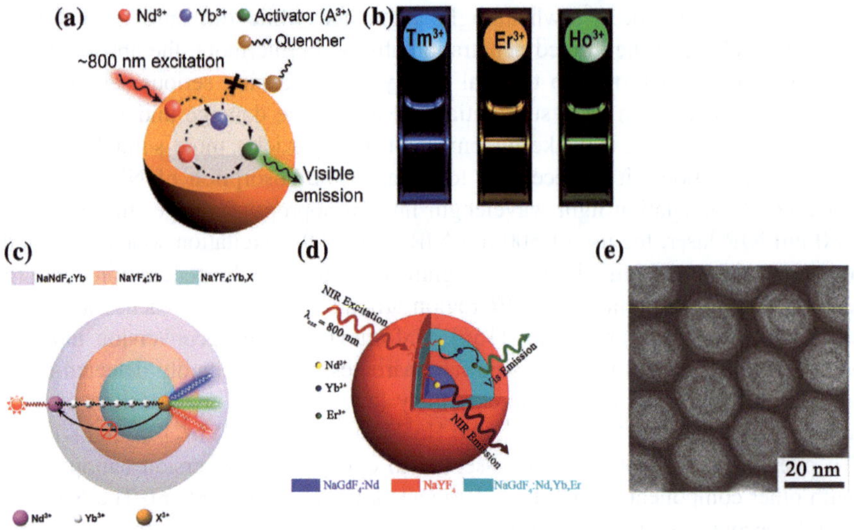

Fig. 3.4 **a**, **b** Schematic design and luminescence photographs of the $NaYF_4$:Yb,Nd,X@$NaYF_4$:Nd (X = Tm,Er,Ho) core@shell nanoparticles for photon upconversion under 800-nm excitation. **c** Schematic illustration of the proposed energy transfer mechanisms in the quenching-shield $NaYF_4$:Yb,X@$NaYF_4$:Yb@$NaNdF_4$:Yd (X = Tm,Er,Ho) sandwich structured upconversion nanoparticles upon 800-nm excitation. **d**, **e** General strategy to achieve the upconversion and downconversion dual-mode luminescence with multilayer $NGdF_4$:Nd@$NaYF_4$@$NaGdF_4$:Nd,Yb, Er@$NaYF_4$ Core@Shell1@Shell2@Shell3 nanoparticle and the corresponding HAADF-STEM image of the obtained multilayer nanoparticles. **a**, **b** Reprinted with the permission from Ref. [191]. Copyright 2013 American Chemical Society. **c** Reprinted with the permission from Ref. [194]. Copyright 2013 Wiley-VCH Verlag GmbH & Co. KGaA. **d**, **e** Reprinted with the permission from Ref. [195]. Copyright 2013 Nature Publishing Group

upconversion excitation efficiency similar to that at 980 nm, but with greatly minimized tissue overheating [192]. Wen et al. [193] described a novel design, based on nanostructural engineering to separate unwanted electronic transitions for constructing a new class of materials displaying tunable upconversion emissions spanning from UV to the visible spectral region by single-wavelength excitation at 808 nm. These results demonstrated that core@shell nanostructures could be harnessed to integrate and coordinate several distinct optical processes for realizing unusual upconversion of NIR light in the medical spectral window (about 800 nm) into tunable emissions spanning from UV to the visible spectral region. Most recently, Zhong et al. [194] fabricated a quenching-shield sandwich structure ($NaYF_4$:Yb,Er@$NaYF_4$:Yb@$NaNdF_4$:Yb) to eliminate the photon quenching for 800 nm excited upconversion luminescence of Nd^{3+}-sensitized nanoparticles (Fig. 3.4c). This unique nanostructure was essential to eliminate the deleterious cross-relaxation pathways between the activator and sensitizer by means of a precisely controlled transition layer ($NaYF_4$:Yb). The optimized emission intensity of the quenching-shield sandwich structure upon 800-nm excitation was even brighter than that at 980 nm with a low excitation power density, as a result of the higher absorption cross section of the sensitizer Nd^{3+} ion. Core@shell1@shell2@shell3 structured $NaGdF_4$:Nd@$NaYF_4$@$NaGdF_4$:Nd,Yb,Er@$NaYF_4$ nanocrystals were also well designed and synthesized in our group (Fig. 3.4d, e) [195]. Each of the parts assumed respective role and worked together to achieve dual-mode up/downconversion luminescence upon the low heat effect 800-nm excitation. Nd^{3+}, Yb^{3+}, and Er^{3+} tri-doped $NaGdF_4$:Nd,Yb,Er upconversion layers [NIR (800 nm)-to-Visible (540 nm)] with a constitutional efficient 800 nm excitable property were achieved for the in vitro bio-imaging with low auto-fluorescence and photodamage effects. Moreover, typical NIR (800 nm)-to-NIR (860–895 nm) DC luminescence of Nd^{3+} has also been realized with this designed nanostructure. Due to the low heat effect, high penetration depth of the excitation source, high efficiency of the DC luminescence, and high contrast in the in vivo DC imaging of a whole body nude mouse were all achieved.

3.5 Upconversion Nanocrystals Versus Quantum Dots and Dyes

As aforementioned, the lanthanide-doped nanocrystals can exhibit unique efficient anti-Stokes upconversion luminescence. Compared with the traditionally used downconversion fluorescent organic dyes and quantum dots, the advantages of the optical probes based on upconversion nanocrystals are sharp emission bandwidth, tunable emission, high photostability, and little background auto-fluorescence. In this section, we systematically compared the advantages and shortcomings between the upconversion nanocrystals and quantum dots and organic dyes from various aspects, such as luminescence mechanism, multi-color emission, luminescent stability, efficiency, ultrasmall nanocrystals fabrication, and so on.

As discussed in Chap. 1, the most efficient mechanisms behind lanthanide ion-based upconversion process are two-step excitation (excitation to the ground-state followed by excited-state absorption, GSA/ESA) [10] and subsequent energy transfer (ETU, also known as addition de photon par transfers d'energies, APTE effect) [10, 196]. GSA/ESA is a single-ion-based process, whereas ETU always involves at least two ions. GSA/ESA is observed for low doping concentration (<1 %). In the case of ETU, upconversion from a low to a high excited state is realized by energy transfer between two excited ions, possibly different types of ions. By comparison, the photoluminescence mechanism of semiconductor quantum dots is quite different from that of upconversion nanoparticles [197, 198]. As one of the most important parameters for semiconductors, the width of band gap is fixed, which is determined by its identity. However, the situation changes in the case of nanoscale semiconductor particles with small sizes (less than $\sim$10 nm) because of the regime of quantum confinement. It means that electronic excitations "feel" the presence of the particle boundaries and respond to changes in the particle size by adjusting their energy spectra. This phenomenon is known as the quantum size effect, and the nanoscale particles that exhibit the effect are often referred as quantum dots. For organic dyes, the photoluminescence mechanism can be explained by the Jablonski energy diagram [199, 200].

The ability to manipulate multi-color output of luminescence nanocrystals is particularly important for their applications in multiplexed biological labeling. For the upconversion nanocrystals, the multi-color emission can be easily realized by tuning the emission wavelength (selection of host–dopant combinations, using FRET or LRET effect, etc.) or relative intensity of the multiple emission peaks (tuning the doping concentration, size and shape effect, single-band upconversion luminescence, etc.), which has been discussed in the earlier portions of this chapter.

It has been demonstrated that the semiconductor materials could show the quantum size effect when the radius of the crystals is smaller than the Bohr radius. As the size of quantum dots decreases, the spacing between the valence band and conduction bands (i.e., the band gap) increases, leading to a blue shift of the emission wavelength. The quantum dots can be excited efficiently at any wavelength shorter than the emission peak, and the excitation wavelength has no influence on the characteristic narrow, symmetric spectrum properties of the emission [201]. By comparison, the multi-color emission of organic dyes can be accomplished by controlling organic fluorophores. Conventional dye molecules impose stringent requirements on optical systems used to make luminescence measurements. Their narrow excitation spectra make simultaneous excitation difficult in most cases. Furthermore, the broad emission spectra with a long tail at red wavelengths introduce spectral cross talk between different detection channels, making quantitation of the relative amounts of different probes difficult.

In summary, although both of quantum dots and organic dyes can exhibit the multi-color emission with different mechanisms, some obvious limitations must be considered in comparison with upconversion nanoparticles. Organic fluorescent dyes have different excitation wavelengths from UV to VIS, requiring multiple excitation lasers for the process, which increases the cost of the detection instrument. Furthermore, when the emission spectra overlap, radiative and/or non-radiative energy transfer may complicate the multi-color emissions. Quantum dots offer significant advantages over conventional organic fluorescent dyes because they are more photostable and have narrow emission line widths. However, most of the quantum dots are made of toxic materials and they "blink." Furthermore, difficulties in distinguishing between various emission signals when the emission spectra overlap are similar to those for organic dyes. For upconversion nanoparticles, the rare earth ions can be excited in the background free and tissue-transparent IR region instead of at UV and visible frequencies to give multi-color emissions. Due to the unique 4*f* orbital structure, spectra arising from lanthanide ions *f*–*f* transitions are narrow and insensitive to their environment, unlike those of quantum dots or organic dyes. Typically, trivalent rare earth ions in inorganic hosts have emission line widths of ~10–20 nm in visible region, which is about half of that for quantum dots (25–40 nm) and much narrower than that for organic dyes (30–50 nm) or transition metal ions (~100 nm) [202, 203]. This feature allows more resolvable bands to be packed into the same spectral bandwidth, which enables a large number of distinct combinations.

Until now, the low upconversion efficiencies of upconversion nanoparticles is still the most serious limitation for commercialization, which results in most applications being operated at high excitation densities. The efficiency of an upconverter is determined by many variables which influence the radiative, non-radiative, and energy transfer rates, e.g., phonon energies, spectral overlap between sensitizer emission and activator absorption, temperature, and defects that quench the emission. Very importantly, as upconversion is a nonlinear process, the efficiency strongly depends on excitation power. It has been summarized in Chap. 2 that most of the lanthanide-doped upconversion nanocrystals show very low quantum yields (<1 %). In comparison with relative low quantum yield of up-conversion nanocrystals, high efficiency is one of the most important characteristics of quantum dots, which gradually increased to nearly 100 % (up to 97 %) during past decades of rapid development [204]. However, the high efficiency is based on the downconversion photoluminescence. Most of excitation light sources for quantum dots are located in UV or VIS regions, which restrict the bio-applications due to the background auto-luminescence and the low penetration depth. Liu et al. compared in vivo imaging sensitivity of quantum dots with upconversion nano-particles side by side (Fig. 3.5) [205]. They found that the in vivo detection limit of upconversion nanoparticles to be at least one order of magnitude lower than that of the former, due to the auto-luminescence-free nature of the latter. Actually, quantum dots could also emit upconversion luminescence at high energies via two-photon or multi-photon absorption process [206–211], which was similar but different from the upconversion process of the upconversion nanoparticles. The multi-photon process could be characterized by the simultaneous absorption of two or more pump photons followed by the emission of the output radiation at a shorter wavelength than the pump one. Up to now, many papers have reported the multi-photon emission of the quantum dots. For example, Loh's group found that CdSe/AsS core/shell nanocrystals exhibited efficient two-photon fluorescence, which could be used in NIR-excited cellular labeling [208]. Liu's group observed bright two- and three-photon excitation luminescence from the Mn-doped ZnS quantum rods (QRs) using tunable NIR femtosecond laser [209]. More recently, Yu et al. also realized high resolution in vitro and in vivo imaging by combining three-photon excitation of ZnS nanocrystals and visible emission from Mn^{2+} dopants [206]. Although quantum dots could serve as the NIR-excited bio-labels, expensive high-intensity (10^6–10^9 W/cm^2) pulse lasers had to be introduced as an excitation source for the generation of a simultaneous two-photon process. In stark contrast, upconversion processes of lanthanide-doped upconversion nanocrystals could be induced by a low-power (1–10^3 W/cm^2) inexpensive continuous wave laser. Moreover, the efficiency of lanthanide-doped upconversion nanocrystals was remarkably higher than that of the multi-photon absorption processes of quantum dots.

Luminescence stability of the bio-probe is also considered to be one of the most important parameters in applications. In 2009, Wu et al. [212] reported that individual

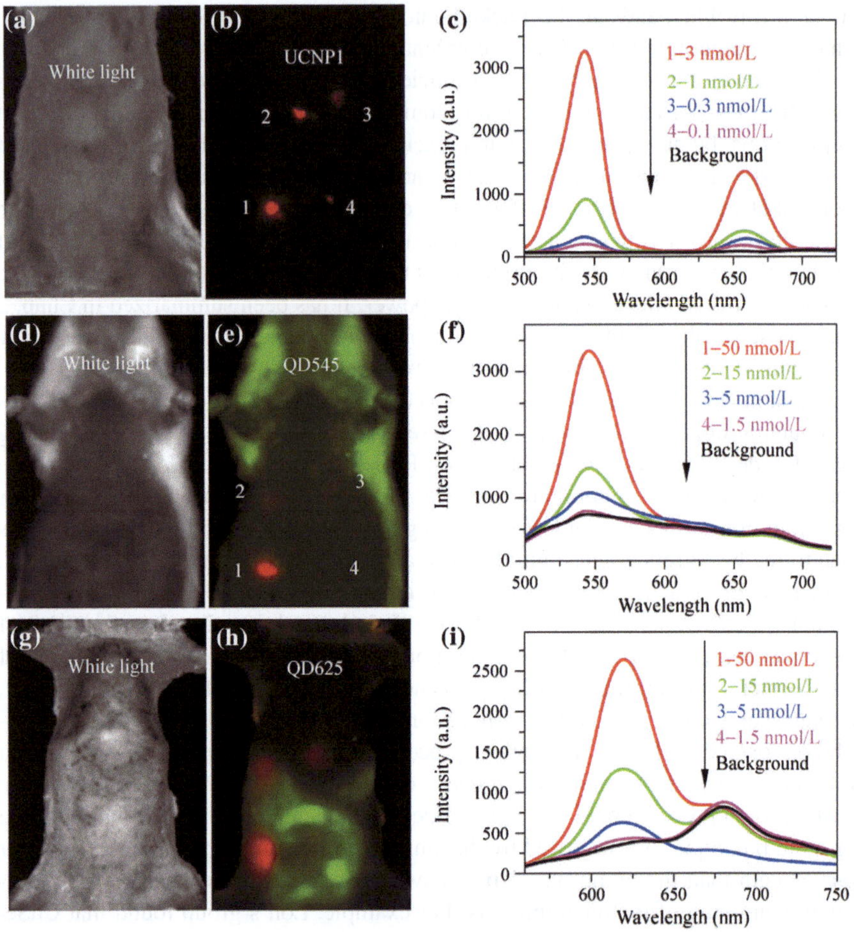

Fig. 3.5 Comparison of imaging sensitivities between upconversion nanoparticles and quantum dots. **a** A white light image of a mouse subcutaneously injected with various concentrations of upconversion nanoparticle. **b** An in vivo upconversion luminescence image of the injected mouse. **c** Upconversion emission spectra recorded at the injection sites. **d, g** White light images of mice subcutaneously injected with quantum dots. **e** Spectrally resolved luminescence images of QD545-injected mouse. **h** QD 625-injected mouse (*red* and *green* colors represent QD luminescence and auto-fluorescence, respectively). **f, i** Luminescence spectra recorded at the QD injection sites (the luminescence spectra of 1.5 nmol/L quantum dots were nearly identical to the background spectra). Reprinted with the permission from Ref. [205]. Copyright 2010 Springer

Yb^{3+}- and Er^{3+}-doped hexagonal phase $NaYF_4$ upconversion nanoparticles emitted bright anti-Stokes visible upconversion luminescence with exceptional photostability when being excited by a 980-nm continuous wave laser. The individual upconversion nanoparticle exhibited no on/off emission behavior or “blinking” down to the millisecond timescale and no loss of intensity following an hour of continuous excitation.

They also found that aqueous upconversion nanoparticles exhibited similar non-blinking behavior and resistance to bleaching as hydrophobic upconversion nanoparticles. Almost at the same time, Park et al. [139] also reported the single particle level measurements of upconversion luminescence with $NaGdF_4$:Er^{3+},Yb^{3+}/$NaGdF_4$ core/shell nanocrystals. The luminescence of the particle was extremely resistant to photobleaching and photoblinking. Our group also showed the photostability of the single $NaYF_4$:Yb,Ho,Tm@$NaYF_4$ core/shell upconversion nanoparticle. The $NaYF_4$:Yb,Ho,Tm@$NaYF_4$ did not exhibit photobleaching or photodamage after 1 h of continuous laser irradiation (Fig. 3.6a–e) [15]. Furthermore, no photoblinking could be determined when the bin time for each data point in emission intensity was reduced to 1 ms. The intensity at single particle level was preserved even after 6 h of continuous irradiation by a continuous wave laser at 980 nm, indicating an excellent stability of $NaYF_4$:Yb,Ho,Tm@$NaYF_4$ core/shell particles against photobleaching. Furthermore, Li et al. also demonstrated that upconversion nanoparticles showed some distinctive advantages, such as the low level of photobleaching, the elimination of background interference from either endogenous fluorophores, or colabeled fluorescent probes [213]. As shown in Fig. 3.6f, under the simultaneous and high-power illumination provided by CW lasers at 405, 543, and 980 nm for 400 s, the luminescence intensity still remained at 96 % of its initial value at the end of the test, whereas the intensity of organic dyes DiI and DAPI decreased rapidly to 2.5 and 1 %, respectively, demonstrating the higher photostability of the upconversion nanoparticles compared with the dyes. For biological labeling applications, lanthanide-doped nanocrystals should not only exhibit high luminescence efficiency, but also have surface wetting characteristics that are compatible with biomolecules or biomolecular assemblies. It is worth noting that the luminescence would undergo a quenching process when upconversion nanoparticles were transferred from oil phase to aqueous solution. This phenomenon should be due to the high-energy stretching vibration (ca. 3,500 cm^{-1}) effect. Water molecules are regarded as surface oscillators that significantly quench the luminescence of rare earth dopants [44, 84, 165, 214].

In comparison with upconversion nanoparticles, quantum dots also show better performance in resisting photobleaching. Wu et al. [215] designed an experiment to monitor the luminescence intensity changes of the quantum dots and organic dyes, Alexa Fluor 488, simultaneously in the same cell with the same excitation wavelength for 3 min. The results showed that regardless of which target was stained with QD–streptavidin, the fluorescent signal was very stable against photobleaching. In contrast, Alexa 488 was bleached quickly. They also compared the photostability of QD 608–streptavidin and Alexa 488–streptavidin with specimens mounted with glycerol or the anti-fade mounting medium Vectashield. Without an anti-fade medium, the luminescence intensity of Alexa 488 decreased to 50 % of the initial intensity at 10 s and to 10 % at 60 s. With the protection of anti-fade reagents, Alexa 488 retained ~80 % at 60 s and 55 % at the end of 3 min of illumination. For QD 608–streptavidin, the intensity was 104 % of the initial intensity for glycerol-mounted specimens and 97 % for anti-fade medium-mounted specimens at the end of the illumination. The results indicate that even with

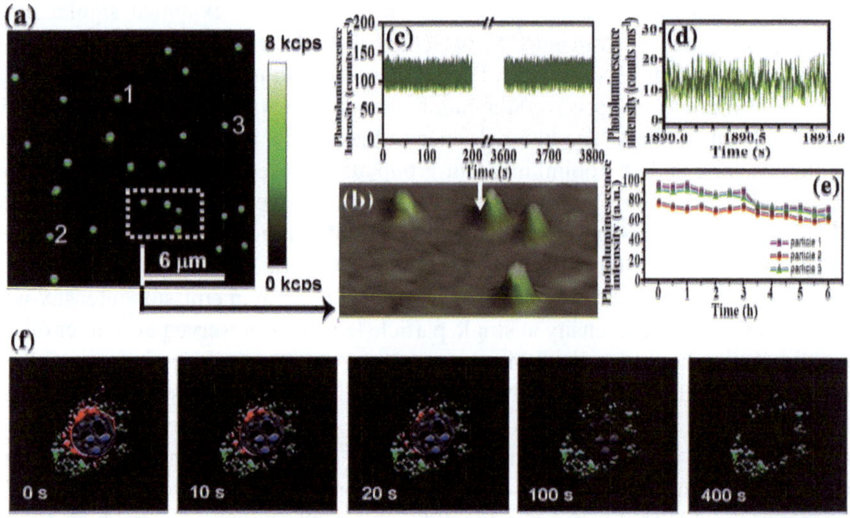

Fig. 3.6 **a**, **b** Confocal upconverted luminescent image of the individual $NaYF_4$:Yb,Ho, Tm@$NaYF_4$ upconversion nanoparticles, laser power density is about 8×10^6 W/cm^2. **c** The time trace of emission intensity from a single upconversion nanoparticles under continuous laser illumination for more than 1 h, suggesting the durable photostability [15]. **d** The zoom-in time trace and histogram of emission intensity showing non-blinking behavior of single upconversion nanoparticles. **e** The longtime trace of emission intensity from three individual upconversion nanoparticles under continuous laser illumination for more than 6 h, suggesting a photostability behavior. **f** Simultaneous excitations provided by CW lasers at 405, 543, and 980 nm with powers of ~1.6, 0.13, and 19 mW in the focal plane, respectively. **a–e** Reprinted with the permission from Ref. [15]. Copyright 2011 Wiley-VCH Verlag GmbH & Co. KGaA. **f** Reprinted with the permission from Ref. [213]. Copyright 2009 American Chemical Society

anti-bleaching protection, Alexa 488–streptavidin is still much less photostable than the QD–streptavidin.

Although quantum dots show higher stability in comparison with organic dyes, photoblinking is still one of the most serious limitations [204, 216, 217]. The photoluminescence from quantum dots is defined by large intensity fluctuations, and it turns "on" and "off" intermittently even under continuous photoexcitation [216]. Despite over a decade of research, completely non-blinking is still a big challenge for quantum dots, and an understanding of the blinking phenomenon remains elusive. In 2008, Mahler et al. [217] synthesized novel CdSe/CdS core/shell quantum dots with thick crystalline shells, 68 % of which did not blink when being observed individually at 33 Hz for 5 min. About 1 year later, Wang et al. [218] first reported ternary core/shell CdZnSe/ZnSe semiconductor nanocrystals that individually exhibited continuous, non-blinking photoluminescence. One fly in the ointment was that the luminescence of the quantum dots showed a multi-peaked spectral, resulting in some limitations to a certain extent in the practical use. Although some papers reported the synthesis of the high-quality quantum dots with suppressed blinking [204], the high particle uniformity, high photoluminescence efficiency, narrow and symmetric emission spectral line shapes and non-photoblinking were still difficult to obtain.

In conclusion, both upconversion nanoparticles and quantum dots have much higher luminescence stabilities than organic dyes. Quantum dots have ultrahigh downconversion efficiency, but always accompanied by photoblinking. In contrast, upconversion nanoparticles show relative higher upconversion efficiency than the two-photon quantum dots with non-blinking property. According to an ancient Chinese saying, "sometimes a foot may prove short while an inch may prove long," as evident, upconversion nanoparticles and quantum dots should show their respective advantages in different applications.

3.6 Summary and Perspectives

In this chapter, the optical properties of upconversion lanthanide ion-doped nanocrystals, including emission color tunability, enhancement of upconversion efficiency, core@shell structure-based luminescence engineering, and the optical properties comparison between upconversion nanocrystals and quantum dots and dyes, were summarized. Despite so many successes, the optical properties of the upconversion nanocrystals are still in need of more optimization to fulfill the requirements of the practical applications.

For the multi-color emission, although great efforts have been dedicated to achieve tunable color output, it is still a challenge to develop a general protocol to precisely tune the color output of upconversion nanoparticles over a broad spectral range, while still maintaining high efficiency. The core@shell structure to tune upconversion seems promising, but surface-related deactivations have to be addressed. In addition, the range of upconverted emission is still limited, unable to cover all of the panchromatic range. Future opportunities for tuning upconversion emission can be divided in the following directions: (1) tuning upconversion using core@shell structures with minimized surface-related deactivations. Spatial isolation of lanthanide emitters is necessary to avoid cross-relaxations to maintain high dopant concentration and high luminescence efficiency. Meanwhile, the shell is able to effectively suppress surface-related deactivations. (2) Combining two or more existing color-tuning approaches (e.g., control the energy migration by designing the advanced core@shell or multi-shell structure; coupling upconversion mechanisms to other types of emitters through FRET or LRET, etc.) may open totally new design criteria for achieving wide tunability at higher efficiency. (3) It can be seen that most of the upconversion luminescence obtained from the designed nanostructures is multi-peak emission, which means that the color is not pure enough for some special applications. Single-band emission nanomaterials can be realized by controlling the energy migration of the lanthanide dopants.

Despite recent advances in improving the efficiency of upconversion nanoparticles, the insufficient brightness of upconversion nanoparticles is still the major reason that limits their popularity in biology. The insufficient brightness arises from two parts: (1) the low extinction coefficient of upconversion nanoparticles, which is an intrinsic feature of *4f–4f* optical transitions in lanthanide ions. To alleviate the extinction limit of trivalent lanthanide ions, one important way is to utilize an antenna effect from dyes, plasmons, quantum dots, or other doping elements with a strong absorption to sensitize the upconversion nanoparticles. (2) Another source of the lack of brightness is from the inadequate quantum yield of upconversion nanoparticles especially at the low excitation density. Manipulation of the local environment around the lanthanide dopants by delicate design and optimization of the structure of the upconversion nanoparticles may be a promising direction to enhance the quantum yield.

For core@shell structure upconversion nanocrystals, the structures of the nanoparticles become more and more advanced and ingenious. Thus, the structure, especially the core@shell structure, becomes more and more complicated, indicating that the traditional synthesis method hardly fulfills the practical needs. The future for core@shell nanocrystals lies in the development of the new synthesis approach which can fulfill the requirements for the fabrication of complicated multi-shell structured upconversion nanoparticles. Lastly, standard measurement facilities and protocols are highly required to allow the quantitative comparison of optical properties for the samples with different structures or prepared with different methods.

References

1. Li, X., Zhang, F., Zhao, D.: Lab on upconversion nanoparticles: optical properties and applications engineering via designed nanostructure. Chem. Soc. Rev. doi:10.1039/C4CS00163J (2014)
2. Li, X., Zhang, F., Zhao, D.: Highly efficient lanthanide upconverting nanomaterials: progresses and challenges. Nano Today **8**, 643–676 (2013)
3. Wang, G., Peng, Q., Li, Y.: Lanthanide-doped nanocrystals: synthesis, optical-magnetic properties, and applications. Acc. Chem. Res. **44**, 322–332 (2011)
4. Haase, M., Schaefer, H.: Upconverting nanoparticles. Angew. Chem. Int. Ed. **50**, 5808–5829 (2011)
5. Wang, F., Liu, X.: Recent advances in the chemistry of lanthanide-doped upconversion nanocrystals. Chem. Soc. Rev. **38**, 976–989 (2009)
6. Zhou, J., Liu, Z., Li, F.: Upconversion nanophosphors for small-animal imaging. Chem. Soc. Rev. **41**, 1323–1349 (2012)
7. Mader, H.S., Kele, P., Saleh, S.M., Wolfbeis, O.S.: Upconverting luminescent nanoparticles for use in bioconjugation and bioimaging. Curr. Opin. Chem. Biol. **14**, 582–596 (2010)
8. Cheng, L., Wang, C., Liu, Z.: Upconversion nanoparticles and their composite nanostructures for biomedical imaging and cancer therapy. Nanoscale **5**, 23–27 (2013)
9. Chatterjee, D.K., Gnanasammandhan, M.K., Zhang, Y.: Small upconverting fluorescent nanoparticles for biomedical applications. Small **6**, 2781–2795 (2010)
10. Auzel, F.: Upconversion and anti-stokes processes with f and d ions in solids. Chem. Rev. **104**, 139–173 (2004)
11. Fischer, L.H., Harms, G.S., Wolfbeis, O.S.: Upconverting nanoparticles for nanoscale thermometry. Angew. Chem. Int. Ed. **50**, 4546–4551 (2011)

12. Chen, G., Qju, H., Prasad, P.N., Chen, X.: Upconversion nanoparticles: design, nanochemistry, and applications in theranostics. Chem. Rev. **114**, 5161–5214 (2014)
13. Wilson, R., Cossins, A.R., Spiller, D.G.: Encoded microcarriers for high-throughput multiplexed detection. Angew. Chem. Int. Ed. **45**, 6104–6117 (2006)
14. Finkel, N.H., Lou, X., Wang, C., He, L.: Barcoding the microworld. Anal. Chem. **76**, 352A–359A (2004)
15. Zhang, F., Shi, Q., Zhang, Y., Shi, Y., Ding, K., Zhao, D., Stucky, G.D.: Fluorescence upconversion microbarcodes for multiplexed biological detection: nucleic acid encoding. Adv. Mater. **23**, 3775–3779 (2011)
16. Gorris, H.H., Wolfbeis, O.S.: Photon-upconverting nanoparticles for optical encoding and multiplexing of cells, biomolecules, and microspheres. Angew. Chem. Int. Ed. **52**, 3584–3600 (2013)
17. Meldal, M., Christensen, S.F.: Microparticle matrix encoding of beads. Angew. Chem. Int. Ed. **49**, 3473–3476 (2010)
18. Hardenbol, P., Yu, F.L., Belmont, J., MacKenzie, J., Bruckner, C., Brundage, T., Boudreau, A., Chow, S., Eberle, J., Erbilgin, A., Falkowski, M., Fitzgerald, R., Ghose, S., Iartchouk, O., Jain, M., Karlin-Neumann, G., Lu, X.H., Miao, X., Moore, B., Moorhead, M., Namsaraev, E., Pasternak, S., Prakash, E., Tran, K., Wang, Z.Y., Jones, H.B., Davis, R.W., Willis, T.D., Gibbs, R.A.: Highly multiplexed molecular inversion probe genotyping: over 10,000 targeted SNPs genotyped in a single tube assay. Genome Res. **15**, 269–275 (2005)
19. Klostranec, J.M., Xiang, Q., Farcas, G.A., Lee, J.A., Rhee, A., Lafferty, E.I., Perrault, S.D., Kain, K.C., Chan, W.C.W.: Convergence of quantum dot barcodes with microfluidics and signal processing for multiplexed high-throughput infectious disease diagnostics. Nano Lett. **7**, 2812–2818 (2007)
20. Ming, C., Song, F., Yan, L.: Spectroscopic study and green upconversion of Pr^{3+}/Yb^{3+}-codoped $NaY(WO_4)_2$ crystal. Opt. Commun. **286**, 217–220 (2013)
21. Pelle, F., Dhaouadi, M., Michely, L., Aschehoug, P., Toncelli, A., Veronesi, S., Tonelli, M.: Spectroscopic properties and upconversion in $Pr^{3+}:YF_3$ nanoparticles. Phys. Chem. Chem. Phys. **13**, 17453–17460 (2011)
22. Maciel, G.S., Guimaraes, R.B., Barreto, P.G., Carvalho, I.C.S., Rakov, N.: The influence of Yb^{3+} doping on the upconversion luminescence of Pr^{3+} in aluminum oxide based powders prepared by combustion synthesis. Opt. Mater. **31**, 1735–1740 (2009)
23. Naccache, R., Vetrone, F., Speghini, A., Bettinelli, M., Capobianco, J.A.: Cross-relaxation and upconversion processes in Pr^{3+} singly doped and Pr^{3+}/Yb^{3+} codoped nanocrystalline $Gd_3Ga_5O_{12}$: the sensitizer/activator relationship. J. Phys. Chem. C **112**, 7750–7756 (2008)
24. Joshi, C., Rai, S.B.: Structural, thermal, and optical properties of Pr^{3+}/Yb^{3+} co-doped oxyhalide tellurite glasses and its nano-crystalline parts. Solid State Sci. **14**, 997–1003 (2012)
25. Wang, F., Deng, R., Wang, J., Wang, Q., Han, Y., Zhu, H., Chen, X., Liu, X.: Tuning upconversion through energy migration in core–shell nanoparticles. Nat. Mater. **10**, 968–973 (2011)
26. Su, Q., Han, S., Xie, X., Zhu, H., Chen, H., Chen, C.-K., Liu, R.-S., Chen, X., Wang, F., Liu, X.: The effect of surface coating on energy migration-mediated upconversion. J. Am. Chem. Soc. **134**, 20849–20857 (2012)
27. Liu, C., Chen, D.: Controlled synthesis of hexagon shaped lanthanide-doped LaF_3 nanoplates with multicolor upconversion fluorescence. J. Mater. Chem. **17**, 3875–3880 (2007)
28. Qin, X., Yokomori, T., Ju, Y.: Flame synthesis and characterization of rare-earth (Er^{3+}, Ho^{3+}, and Tm^{3+}) doped upconversion nanophosphors. Appl. Phys. Lett. **90**, 073104 (2007)
29. Ehlert, O., Thomann, R., Darbandi, M., Nann, T.: A four-color colloidal multiplexing nanoparticle system. ACS Nano **2**, 120–124 (2008)
30. Heer, S., Kompe, K., Gudel, H.U., Haase, M.: Highly efficient multicolour upconversion emission in transparent colloids of lanthanide-doped $NaYF_4$ nanocrystals. Adv. Mater. **16**, 2102–2105 (2004)

31. Heer, S., Lehmann, O., Haase, M., Gudel, H.U.: Blue, green, and red upconversion emission from lanthanide-doped $LuPO_4$ and $YbPO_4$ nanocrystals in a transparent colloidal solution. Angew. Chem. Int. Ed. **42**, 3179–3182 (2003)
32. Yi, G.S., Chow, G.M.: Synthesis of hexagonal-phase $NaYF_4$:Yb,Er and $NaYF_4$:Yb,Tm nanocrystals with efficient up-conversion fluorescence. Adv. Funct. Mater. **16**, 2324–2329 (2006)
33. Zhang, X., Yang, P., Li, C., Wang, D., Xu, J., Gai, S., Lin, J.: Facile and mass production synthesis of β-$NaYF_4$:Yb^{3+},Er^{3+}/Tm^{3+} 1D microstructures with multicolor up-conversion luminescence. Chem. Commun. **47**, 12143–12145 (2011)
34. Mahalingam, V., Vetrone, F., Naccache, R., Speghini, A., Capobianco, J.A.: Colloidal Tm^{3+}/Yb^{3+}-doped $LiYF_4$ nanocrystals: multiple luminescence spanning the UV to NIR regions via low-energy excitation. Adv. Mater. **21**, 4025–4028 (2009)
35. Zhang, F., Wan, Y., Yu, T., Zhang, F., Shi, Y., Xie, S., Li, Y., Xu, L., Tu, B., Zhao, D.: Uniform nanostructured arrays of sodium rare-earth fluorides for highly efficient multicolor upconversion luminescence. Angew. Chem. Int. Ed. **46**, 7976–7979 (2007)
36. Liu, Y., Tu, D., Zhu, H., Chen, X.: Lanthanide-doped luminescent nanoprobes: controlled synthesis, optical spectroscopy, and bioapplications. Chem. Soc. Rev. **42**, 6924–6958 (2013)
37. Suyver, J., Grimm, J., Van Veen, M., Biner, D., Krämer, K., Güdel, H.: Upconversion spectroscopy and properties of $NaYF_4$ doped with Er^{3+}, Tm^{3+} and/or Yb^{3+}. J. Lumin. **117**, 1–12 (2006)
38. Boyer, J.C., Vetrone, F., Cuccia, L.A., Capobianco, J.A.: Synthesis of colloidal upconverting $NaYF_4$ nanocrystals doped with Er^{3+}, Yb^{3+} and Tm^{3+}, Yb^{3+} via thermal decomposition of lanthanide trifluoroacetate precursors. J. Am. Chem. Soc. **128**, 7444–7445 (2006)
39. Wang, F., Liu, X.: Upconversion multicolor fine-tuning: visible to near-infrared emission from lanthanide-doped $NaYF_4$ nanoparticles. J. Am. Chem. Soc. **130**, 5642–5643 (2008)
40. Chen, G., Qiu, H., Fan, R., Hao, S., Tan, S., Yang, C., Han, G.: Lanthanide-doped ultrasmall yttrium fluoride nanoparticles with enhanced multicolor upconversion photoluminescence. J. Mater. Chem. **22**, 20190–20196 (2012)
41. Chan, E.M., Han, G., Goldberg, J.D., Gargas, D.J., Ostrowski, A.D., Schuck, P.J., Cohen, B. E., Milliron, D.J.: Combinatorial discovery of lanthanide-doped nanocrystals with spectrally pure upconverted emission. Nano Lett. **12**, 3839–3845 (2012)
42. Chen, G.Y., Liu, Y., Zhang, Y.G., Somesfalean, G., Zhang, Z.G., Sun, Q., Wang, F.P.: Bright white upconversion luminescence in rare-earth-ion-doped Y_2O_3 nanocrystals. Appl. Phys. Lett. **91**, 133103 (2007)
43. Li, Z., Zhang, Y., Jiang, S.: Multicolor core/shell-structured upconversion fluorescent nanoparticles. Adv. Mater. **20**, 4765–4769 (2008)
44. Li, X., Shen, D., Yang, J., Yao, C., Che, R., Zhang, F., Zhao, D.: Successive layer-by-layer strategy for multi-shell epitaxial growth: shell thickness and doping position dependence in upconverting optical properties. Chem. Mater. **25**, 106–112 (2013)
45. Tian, G., Gu, Z., Zhou, L., Yin, W., Liu, X., Yan, L., Jin, S., Ren, W., Xing, G., Li, S., Zhao, Y.: Mn^{2+} dopant-controlled synthesis of $NaYF_4$:Yb/Er upconversion nanoparticles for in vivo imaging and drug delivery. Adv. Mater. **24**, 1226–1231 (2012)
46. Li, Z., Wang, L., Wang, Z., Liu, X., Xiong, Y.: Modification of $NaYF_4$:Yb,Er@SiO_2 nanoparticles with gold nanocrystals for tunable green-to-red upconversion emissions. J. Phys. Chem. C **115**, 3291–3296 (2011)
47. Zhang, F., Braun, G.B., Shi, Y., Zhang, Y., Sun, X., Reich, N.O., Zhao, D., Stucky, G.: Fabrication of Ag@SiO_2@Y_2O_3:Er nanostructures for bioimaging: tuning of the upconversion fluorescence with silver nanoparticles. J. Am. Chem. Soc. **132**, 2850–2851 (2010)
48. Wang, L.Y., Yan, R.X., Hao, Z.Y., Wang, L., Zeng, J.H., Bao, J., Wang, X., Peng, Q., Li, Y.D.: Fluorescence resonant energy transfer biosensor based on upconversion-luminescent nanoparticles. Angew. Chem. Int. Ed. **44**, 6054–6057 (2005)

49. Vetrone, F., Boyer, J.C., Capobianco, J.A., Speghini, A., Bettinelli, M.: Significance of Yb3+ concentration on the upconversion mechanisms in codoped Y_2O_3:Er^{3+},Yb^{3+} nanocrystals. J. Appl. Phys. **96**, 661–667 (2004)
50. Chen, G.Y., Zhang, Y.G., Somesfalean, G., Zhang, Z.G., Sun, Q., Wang, F.P.: Two-color upconversion in rare-earth-ion-doped ZrO_2 nanocrystals. Appl. Phys. Lett. **89**, 163105 (2006)
51. Zhang, F., Haushalter, R.C., Haushalter, R.W., Shi, Y., Zhang, Y., Ding, K., Zhao, D., Stucky, G.D.: Rare-earth upconverting nanobarcodes for multiplexed biological detection. Small **7**, 1972–1976 (2011)
52. Medintz, I.L., Uyeda, H.T., Goldman, E.R., Mattoussi, H.: Quantum dot bioconjugates for imaging, labelling and sensing. Nat. Mater. **4**, 435–446 (2005)
53. Nagai, T., Ibata, K., Park, E.S., Kubota, M., Mikoshiba, K., Miyawaki, A.: A variant of yellow fluorescent protein with fast and efficient maturation for cell-biological applications. Nat. Biotechnol. **20**, 87–90 (2002)
54. Shaner, N.C., Campbell, R.E., Steinbach, P.A., Giepmans, B.N.G., Palmer, A.E., Tsien, R.Y.: Improved monomeric red, orange and yellow fluorescent proteins derived from *Discosoma* sp red fluorescent protein. Nat. Biotechnol. **22**, 1567–1572 (2004)
55. Maier, S.A., Kik, P.G., Atwater, H.A., Meltzer, S., Harel, E., Koel, B.E., Requicha, A.A.G.: Local detection of electromagnetic energy transport below the diffraction limit in metal nanoparticle plasmon waveguides. Nat. Mater. **2**, 229–232 (2003)
56. Giepmans, B.N.G., Adams, S.R., Ellisman, M.H., Tsien, R.Y.: The fluorescent toolbox for assessing protein location and function. Science **312**, 217–224 (2006)
57. Weiss, S.: Fluorescence spectroscopy of single biomolecules. Science **283**, 1676–1683 (1999)
58. Jeong, S., Won, N., Lee, J., Bang, J., Yoo, J., Kim, S.G., Chang, J.A., Kim, J., Kim, S.: Multiplexed near-infrared in vivo imaging complementarily using quantum dots and upconverting $NaYF_4$:Yb^{3+}, Tm^{3+} nanoparticles. Chem. Commun. **47**, 8022–8024 (2011)
59. Gorris, H.H., Ali, R., Saleh, S.M., Wolfbeis, O.S.: Tuning the dual emission of photon-upconverting nanoparticles for ratiometric multiplexed encoding. Adv. Mater. **23**, 1652–1655 (2011)
60. Wang, L.Y., Li, Y.D.: Green upconversion nanocrystals for DNA detection. Chem. Commun. 2557–2559 (2006)
61. Liu, Y., Zhou, S., Tu, D., Chen, Z., Huang, M., Zhu, H., Ma, E., Chen, X.: Amine-functionalized lanthanide-doped zirconia nanoparticles: optical spectroscopy, time-resolved fluorescence resonance energy transfer biodetection, and targeted imaging. J. Am. Chem. Soc. **134**, 15083–15090 (2012)
62. Ju, Q., Tu, D., Liu, Y., Li, R., Zhu, H., Chen, J., Chen, Z., Huang, M., Chen, X.: Amine-functionalized lanthanide-doped $KGdF_4$ nanocrystals as potential optical/magnetic multimodal bioprobes. J. Am. Chem. Soc. **134**, 1323–1330 (2012)
63. Tu, D., Liu, L., Ju, Q., Liu, Y., Zhu, H., Li, R., Chen, X.: Time-resolved FRET biosensor based on amine-functionalized lanthanide-doped $NaYF_4$ nanocrystals. Angew. Chem. Int. Ed. **50**, 6306–6310 (2011)
64. Zheng, W., Zhou, S., Chen, Z., Hu, P., Liu, Y., Tu, D., Zhu, H., Li, R., Huang, M., Chen, X.: Sub-10 nm lanthanide-doped CaF_2 nanoprobes for time-resolved luminescent biodetection. Angew. Chem. Int. Ed. **52**, 6671–6676 (2013)
65. Cheng, L., Yang, K., Shao, M., Lee, S.-T., Liu, Z.: Multicolor in vivo imaging of upconversion nanoparticles with emissions tuned by luminescence resonance energy transfer. J. Phys. Chem. C **115**, 2686–2692 (2011)
66. Yao, L., Zhou, J., Liu, J., Feng, W., Li, F.: Iridium-complex-modified upconversion nanophosphors for effective LRET detection of cyanide anions in pure water. Adv. Funct. Mater. **22**, 2667–2672 (2012)
67. Bogdan, N., Vetrone, F., Roy, R., Capobianco, J.A.: Carbohydrate-coated lanthanide-doped upconverting nanoparticles for lectin recognition. J. Mater. Chem. **20**, 7543–7550 (2010)

68. Chen, D., Lei, L., Zhang, R., Yang, A., Xu, J., Wang, Y.: Intrinsic single-band upconversion emission in colloidal Yb/Er(Tm):$Na_3Zr(Hf)F_7$ nanocrystals. Chem. Commun. **48**, 10630–10632 (2012)
69. Wang, J., Wang, F., Wang, C., Liu, Z., Liu, X.: Single-band upconversion emission in lanthanide-doped $KMnF_3$ nanocrystals. Angew. Chem. Int. Ed. **50**, 10369–10372 (2011)
70. Zeng, J.H., Xie, T., Li, Z.H., Li, Y.: Monodispersed nanocrystalline fluoroperovskite upconversion phosphors. Cryst. Growth Des. **7**, 2774–2777 (2007)
71. Yi, G., Peng, Y., Gao, Z.: Strong red-emitting near-infrared-to-visible upconversion fluorescent nanoparticles. Chem. Mater. **23**, 2729–2734 (2011)
72. Yin, W., Zhao, L., Zhou, L., Gu, Z., Liu, X., Tian, G., Jin, S., Yan, L., Ren, W., Xing, G., Zhao, Y.: Enhanced red emission from GdF_3:Yb^{3+},Er^{3+} upconversion nanocrystals by Li^+ doping and their application for bioimaging. Chem. Eur. J. **18**, 9239–9245 (2012)
73. Sarkar, S., Meesaragandla, B., Hazra, C., Mahalingam, V.: Sub-5 nm Ln^{3+}-doped $BaLuF_5$ nanocrystals: a platform to realize upconversion via interparticle energy transfer (IPET). Adv. Mater. **25**, 856–860 (2013)
74. Zou, W., Visser, C., Maduro, J.A., Pshenichnikov, M.S., Hummelen, J.C.: Broadband dye-sensitized upconversion of near-infrared light. Nat. Photonics **6**, 560–564 (2012)
75. Chen, G., Liu, H., Somesfalean, G., Liang, H., Zhang, Z.: Upconversion emission tuning from green to red in Yb^{3+}/Ho^{3+}-codoped $NaYF_4$ nanocrystals by tridoping with Ce^{3+} ions. Nanotechnology **20**, 385704 (2009)
76. Vetrone, F., Boyer, J.C., Capobianco, J.A., Speghini, A., Bettinelli, M.: Concentration-dependent near-infrared to visible upconversion in nanocrystalline and bulk Y_2O_3:Er^{3+}. Chem. Mater. **15**, 2737–2743 (2003)
77. Mahalingam, V., Naccache, R., Vetrone, F., Capobianco, J.A.: Preferential suppression of high-energy upconverted emissions of Tm^{3+} by Dy^{3+} ions in Tm^{3+}/Dy^{3+}/Yb^{3+}-doped $LiYF_4$ colloidal nanocrystals. Chem. Commun. **47**, 3481–3483 (2011)
78. Niu, W., Wu, S., Zhang, S.: A facile and general approach for the multicolor tuning of lanthanide-ion doped $NaYF_4$ upconversion nanoparticles within a fixed composition. J. Mater. Chem. **20**, 9113–9117 (2010)
79. Niu, W., Wu, S., Zhang, S.: Utilizing the amidation reaction to address the “cooperative effect” of carboxylic acid/amine on the size, shape, and multicolor output of fluoride upconversion nanoparticles. J. Mater. Chem. **21**, 10894–10902 (2011)
80. Bai, X., Song, H., Pan, G., Lei, Y., Wang, T., Ren, X., Lu, S., Dong, B., Dai, Q., Fan, L.: Size-dependent upconversion luminescence in Er^{3+}/Yb^{3+}-codoped nanocrystalline yttria: saturation and thermal effects. J. Phys. Chem. C **111**, 13611–13617 (2007)
81. Mai, H.-X., Zhang, Y.-W., Sun, L.-D., Yan, C.-H.: Highly efficient multicolor up-conversion emissions and their mechanisms of monodisperse $NaYF_4$:Yb,Er core and core/shell-structured nanocrystals. J. Phys. Chem. C **111**, 13721–13729 (2007)
82. Yin, A., Zhang, Y., Sun, L., Yan, C.: Colloidal synthesis and blue based multicolor upconversion emissions of size and composition controlled monodisperse hexagonal $NaYF_4$:Yb,Tm nanocrystals. Nanoscale **2**, 953–959 (2010)
83. Chen, G., Yang, C., Prasad, P.N.: Nanophotonics and nanochemistry: controlling the excitation dynamics for frequency up- and down-conversion in lanthanide-doped nanoparticles. Acc. Chem. Res. **46**, 1474–1486 (2013)
84. Wang, F., Wang, J., Liu, X.: Direct evidence of a surface quenching effect on size-dependent luminescence of upconversion nanoparticles. Angew. Chem. Int. Ed. **49**, 7456–7460 (2010)
85. Li, Z., Zhang, Y.: An efficient and user-friendly method for the synthesis of hexagonal-phase $NaYF_4$:Yb,Er/Tm nanocrystals with controllable shape and upconversion fluorescence. Nanotechnology **19**, 345606 (2008)
86. Ye, X., Collins, J.E., Kang, Y., Chen, J., Chen, D.T.N., Yodh, A.G., Murray, C.B.: Morphologically controlled synthesis of colloidal upconversion nanophosphors and their shape-directed self-assembly. Proc. Natl. Acad. Sci. **107**, 22430–22435 (2010)

87. Jin, J., Gu, Y.-J., Man, C.W.-Y., Cheng, J., Xu, Z., Zhang, Y., Wang, H., Lee, V.H.-Y., Cheng, S.H., Wong, W.-T.: Polymer-coated $NaYF_4$:Yb^{3+},Er^{3+} upconversion nanoparticles for charge-dependent cellular imaging. ACS Nano **5**, 7838–7847 (2011)
88. Zhang, F., Braun, G.B., Pallaoro, A., Zhang, Y., Shi, Y., Cui, D., Moskovits, M., Zhao, D., Stucky, G.D.: Mesoporous multifunctional upconversion luminescent and magnetic "nanorattle" materials for targeted chemotherapy. Nano Lett. **12**, 61–67 (2012)
89. Dai, Y., Ma, Pa, Cheng, Z., Kang, X., Zhang, X., Hou, Z., Li, C., Yang, D., Zhai, X., Lin, J.: Up-conversion cell imaging and pH-induced thermally controlled drug release from $NaYF_4$:Yb^{3+}/Er^{3+}@hydrogel core-shell hybrid microspheres. ACS Nano **6**, 3327–3338 (2012)
90. Chen, Z., Chen, H., Hu, H., Yu, M., Li, F., Zhang, Q., Zhou, Z., Yi, T., Huang, C.: Versatile synthesis strategy for carboxylic acid-functionalized upconverting nanophosphors as biological labels. J. Am. Chem. Soc. **130**, 3023–3029 (2008)
91. Ostrowski, A.D., Chan, E.M., Gargas, D.J., Katz, E.M., Han, G., Schuck, P.J., Milliron, D.J., Cohen, B.E.: Controlled synthesis and single-particle imaging of bright, sub-10 nm lanthanide-doped upconverting nanocrystals. ACS Nano **6**, 2686–2692 (2012)
92. Bogdan, N., Vetrone, F., Ozin, G.A., Capobianco, J.A.: Synthesis of ligand-free colloidally stable water dispersible brightly luminescent lanthanide-doped upconverting nanoparticles. Nano Lett. **11**, 835–840 (2011)
93. Qian, H.S., Guo, H.C., Ho, P.C.-L., Mahendran, R., Zhang, Y.: Mesoporous-silica-coated up-conversion fluorescent nanoparticles for photodynamic therapy. Small **5**, 2285–2290 (2009)
94. Xu, C.T., Svenmarker, P., Liu, H., Wu, X., Messing, M.E., Wallenberg, L.R., Andersson-Engels, S.: High-resolution fluorescence diffuse optical tomography developed with nonlinear upconverting nanoparticles. ACS Nano **6**, 4788–4795 (2012)
95. Boyer, J.-C., van Veggel, F.C.J.M.: Absolute quantum yield measurements of colloidal $NaYF_4$:Er^{3+}, Yb^{3+} upconverting nanoparticles. Nanoscale **2**, 1417–1419 (2010)
96. Kramer, K.W., Biner, D., Frei, G., Gudel, H.U., Hehlen, M.P., Luthi, S.R.: Hexagonal sodium yttrium fluoride based green and blue emitting upconversion phosphors. Chem. Mater. **16**, 1244–1251 (2004)
97. Martinez, A., Morales, J., Diaz-Torres, L.A., Salas, P., De la Rosa, E., Oliva, J., Desirena, H.: Green and red upconverted emission of hydrothermal synthesized Y_2O_3:Er^{3+}-Yb^{3+} nanophosphors using different solvent ratio conditions. Mater. Sci. Eng. B **174**, 164–168 (2010)
98. Patra, A., Friend, C.S., Kapoor, R., Prasad, P.N.: Effect of crystal nature on upconversion luminescence in Er^{3+}:ZrO_2 nanocrystals. Appl. Phys. Lett. **83**, 284–286 (2003)
99. Shalav, A., Richards, B.S., Trupke, T., Kramer, K.W., Gudel, H.U.: Application of $NaYF_4$:Er^{3+} up-converting phosphors for enhanced near-infrared silicon solar cell response. Appl. Phys. Lett. **86**, 013505 (2005)
100. Mialon, G., Tuerkcan, S., Dantelle, G., Collins, D.P., Hadjipanayi, M., Taylor, R.A., Gacoin, T., Alexandrou, A., Boilott, J.-P.: High up-conversion efficiency of YVO_4:Yb,Er nanoparticles in water down to the single-particle level. J. Phys. Chem. C **114**, 22449–22454 (2010)
101. Kumar, G.A., Pokhrel, M., Martinez, A., Dennis, R.C., Villegas, I.L., Sardar, D.K.: Synthesis and spectroscopy of color tunable Y_2O_2S:Yb^{3+},Er^{3+} phosphors with intense emission. J. Alloy. Compd. **513**, 559–565 (2012)
102. Li, Y., Wei, X., Yin, M.: Synthesis and upconversion luminescent properties of Er^{3+} doped and Er^{3+}-Yb^{3+} codoped GdOCl powders. J. Alloy. Compd. **509**, 9865–9868 (2011)
103. Wang, G., Peng, Q., Li, Y.: Upconversion luminescence of monodisperse CaF_2:Yb^{3+}/Er^{3+} nanocrystals. J. Am. Chem. Soc. **131**, 14200–14201 (2009)
104. Zheng, W., Zhou, S., Chen, Z., Hu, P., Liu, Y., Tu, D., Zhu, H., Li, R., Huang, M., Chen, X.: Sub-10 nm lanthanide-doped CaF_2 nanoprobes for time-resolved luminescent biodetection. Angew. Chem. Int. Ed. **125**, 6803–6813 (2013)
105. Bol, A.A., van Beek, R., Meijerink, A.: On the incorporation of trivalent rare earth ions in II-VI semiconductor nanocrystals. Chem. Mater. **14**, 1121–1126 (2002)

106. Patra, A., Friend, C.S., Kapoor, R., Prasad, P.N.: Upconversion in Er^{3+}:ZrO_2 nanocrystals. J. Phys. Chem. B **106**, 1909–1912 (2002)
107. Patra, A., Friend, C.S., Kapoor, R., Prasad, P.N.: Fluorescence upconversion properties of Er^{3+}-doped TiO_2 and $BaTiO_3$ nanocrystallites. Chem. Mater. **15**, 3650–3655 (2003)
108. Mahalingam, V., Naccache, R., Vetrone, F., Capobianco, J.A.: Sensitized Ce^{3+} and Gd^{3+} ultraviolet emissions by Tm^{3+} in colloidal $LiYF_4$ nanocrystals. Chem. Eur. J. **15**, 9660–9663 (2009)
109. Chen, G., Ohulchanskyy, T.Y., Kachynski, A., Agren, H., Prasad, P.N.: Intense visible and near-infrared upconversion photoluminescence in colloidal $LiYF_4$:Er^{3+} nanocrystals under excitation at 1490 nm. ACS Nano **5**, 4981–4986 (2011)
110. Zhang, B.F., Frigoli, M., Angiuli, F., Vetrone, F., Capobianco, J.A.: Photoswitching of bis-spiropyran using near-infrared excited upconverting nanoparticles. Chem. Commun. **48**, 7244–7246 (2012)
111. Boyer, J.-C., Cuccia, L.A., Capobianco, J.A.: Synthesis of colloidal upconverting $NaYF_4$:Er^{3+}/Yb^{3+} and Tm^{3+}/Yb^{3+} monodisperse nanocrystals. Nano Lett. **7**, 847–852 (2007)
112. Chen, G., Ohulchanskyy, T.Y., Kumar, R., Agren, H., Prasad, P.N.: Ultrasmall monodisperse $NaYF_4$:Yb^{3+}/Tm^{3+} nanocrystals with enhanced near-infrared to near-infrared upconversion photoluminescence. ACS Nano **4**, 3163–3168 (2010)
113. Kumar, R., Nyk, M., Ohulchanskyy, T.Y., Flask, C.A., Prasad, P.N.: Combined optical and MR bioimaging using rare earth ion doped $NaYF_4$ nanocrystals. Adv. Funct. Mater. **19**, 853–859 (2009)
114. Vetrone, F., Naccache, R., Mahalingam, V., Morgan, C.G., Capobianco, J.A.: The active-core/active-shell approach: a strategy to enhance the upconversion luminescence in lanthanide-doped nanoparticles. Adv. Funct. Mater. **19**, 2924–2929 (2009)
115. Naccache, R., Vetrone, F., Mahalingam, V., Cuccia, L.A., Capobianco, J.A.: Controlled synthesis and water dispersibility of hexagonal phase $NaGdF_4$:Ho^{3+}/Yb^{3+} nanoparticles. Chem. Mater. **21**, 717–723 (2009)
116. Liu, Q., Sun, Y., Yang, T., Feng, W., Li, C., Li, F.: Sub-10 nm hexagonal lanthanide-doped $NaLuF_4$ upconversion nanocrystals for sensitive bioimaging in vivo. J. Am. Chem. Soc. **133**, 17122–17125 (2011)
117. Vetrone, F., Mahalingam, V., Capobianco, J.A.: Near-infrared-to-blue upconversion in colloidal $BaYF_5$:Tm^{3+}, Yb^{3+} nanocrystals. Chem. Mater. **21**, 1847–1851 (2009)
118. Zhang, C., Ma, P.A., Li, C., Li, G., Huang, S., Yang, D., Shang, M., Kang, X., Lin, J.: Controllable and white upconversion luminescence in $BaYF_5$:Ln^{3+} (Ln = Yb,Er,Tm) nanocrystals. J. Mater. Chem. **21**, 717–723 (2011)
119. Mahalingam, V., Vetrone, F., Naccache, R., Speghini, A., Capobianco, J.A.: Structural and optical investigation of colloidal Ln^{3+}/Yb^{3+} co-doped KY_3F_{10} nanocrystals. J. Mater. Chem. **19**, 3149–3152 (2009)
120. Yang, D., Li, C., Li, G., Shang, M., Kang, X., Lin, J.: Colloidal synthesis and remarkable enhancement of the upconversion luminescence of $BaGdF_5$:Yb^{3+}/Er^{3+} nanoparticles by active-shell modification. J. Mater. Chem. **21**, 5923–5927 (2011)
121. Sun, X., Zhang, Y.-W., Du, Y.-P., Yan, Z.-G., Si, R., You, L.-P., Yan, C.-H.: From trifluoroacetate complex precursors to monodisperse rare-earth fluoride and oxyfluoride nanocrystals with diverse shapes through controlled fluorination in solution phase. Chem. Eur. J. **13**, 2320–2332 (2007)
122. Du, Y.-P., Zhang, Y.-W., Sun, L.-D., Yan, C.-H.: Luminescent monodisperse nanocrystals of lanthanide oxyfluorides synthesized from trifluoroacetate precursors in high-boiling solvents. J. Phys. Chem. C **112**, 405–415 (2008)
123. Quan, Z., Yang, D., Yang, P., Zhang, X., Lian, H., Liu, X., Lin, J.: Uniform colloidal alkaline earth metal fluoride nanocrystals: nonhydrolytic synthesis and luminescence properties. Inorg. Chem. **47**, 9509–9517 (2008)
124. Du, Y.P., Sun, X., Zhang, Y.W., Yan, Z.G., Sun, L.D., Yan, C.H.: Uniform alkaline earth fluoride nanocrystals with diverse shapes grown from thermolysis of metal trifluoroacetates in hot surfactant solutions. Cryst. Growth Des. **9**, 2013–2019 (2009)

125. Weber, M.: Radiative and multiphonon relaxation of rare-earth ions in Y_2O_3. Phys. Rev. **171**, 283 (1968)
126. Chen, G., Liu, H., Liang, H., Somesfalean, G., Zhang, Z.: Upconversion emission enhancement in Yb^{3+}/Er^{3+}-codoped Y_2O_3 nanocrystals by tridoping with Li^+ ions. J. Phys. Chem. C **112**, 12030–12036 (2008)
127. Cheng, Q., Sui, J., Cai, W.: Enhanced upconversion emission in Yb^{3+} and Er^{3+} codoped $NaGdF_4$ nanocrystals by introducing Li^+ ions. Nanoscale **4**, 779–784 (2012)
128. Zhao, C., Kong, X., Liu, X., Tu, L., Wu, F., Zhang, Y., Liu, K., Zeng, Q., Zhang, H.: Li^+ ion doping: an approach for improving the crystallinity and upconversion emissions of $NaYF_4$: Yb^{3+}, Tm^{3+} nanoparticles. Nanoscale **5**, 8084–8089 (2013)
129. Wang, H.-Q., Nann, T.: Monodisperse upconverting nanocrystals by microwave-assisted synthesis. ACS Nano **3**, 3804–3808 (2009)
130. Huang, P., Zheng, W., Zhou, S., Tu, D., Chen, Z., Zhu, H., Li, R., Ma, E., Huang, M., Chen, X.: Lanthanide-doped $LiLuF_4$ upconversion nanoprobes for the detection of disease biomarkers. Angew. Chem. Int. Ed. **53**, 1252–1257 (2014)
131. Wang, J., Deng, R., MacDonald, M.A., Chen, B., Yuan, J., Wang, F., Chi, D., Hor, T.S.A., Zhang, P., Liu, G.: Enhancing multiphoton upconversion through energy clustering at sublattice level. Nat. Mater. **13**, 157–162 (2014)
132. Saboktakin, M., Ye, X., Oh, S.J., Hong, S.-H., Fafarman, A.T., Chettiar, U.K., Engheta, N., Murray, C.B., Kagan, C.R.: Metal-enhanced upconversion luminescence tunable through metal nanoparticle-nanophosphor separation. ACS Nano **6**, 8758–8766 (2012)
133. Zhang, W., Ding, F., Chou, S.Y.: Large enhancement of upconversion luminescence of $NaYF_4$: Yb^{3+}/Er^{3+} nanocrystal by 3D plasmonic nano-antennas. Adv. Mater. **24**, OP236–OP241 (2012)
134. Li, X., Wang, R., Zhang, F., Zhao, D.: Engineering homogeneous doping in single nanoparticle to enhance upconversion efficiency. Nano Lett. **14**, 3634–3639 (2014)
135. Wong, H.-T., Vetrone, F., Naccache, R., Chan, H.L.W., Hao, J., Capobianco, J.A.: Water dispersible ultra-small multifunctional $KGdF_4:Tm^{3+}$, Yb^{3+} nanoparticles with near-infrared to near-infrared upconversion. J. Mater. Chem. **21**, 16589–16596 (2011)
136. Longmire, M., Choyke, P.L., Kobayashi, H.: Clearance properties of nano-sized particles and molecules as imaging agents: considerations and caveats. Nanomedicine **3**, 703–717 (2008)
137. Johnson, N.J.J., Oakden, W., Stanisz, G.J., Prosser, R.S., van Veggel, F.C.J.M.: Size-tunable, ultrasmall $NaGdF_4$ nanoparticles: insights into their T1 MRI contrast enhancement. Chem. Mater. **23**, 3714–3722 (2011)
138. Mai, H.X., Zhang, Y.W., Si, R., Yan, Z.G., Sun, L.D., You, L.P., Yan, C.H.: High-quality sodium rare-earth fluoride nanocrystals: controlled synthesis and optical properties. J. Am. Chem. Soc. **128**, 6426–6436 (2006)
139. Park, Y.I., Kim, J.H., Lee, K.T., Jeon, K.-S., Na, H.B., Yu, J.H., Kim, H.M., Lee, N., Choi, S.H., Baik, S.-I., Kim, H., Park, S.P., Park, B.-J., Kim, Y.W., Lee, S.H., Yoon, S.-Y., Song, I.C., Moon, W.K., Suh, Y.D., Hyeon, T.: Nonblinking and nonbleaching upconverting nanoparticles as an optical imaging nanoprobe and T1 magnetic resonance imaging contrast agent. Adv. Mater. **21**, 4467–4471 (2009)
140. Xiong, Y.J., McLellan, J.M., Chen, J.Y., Yin, Y.D., Li, Z.Y., Xia, Y.N.: Kinetically controlled synthesis of triangular and hexagonal nanoplates of palladium and their SPR/SERS properties. J. Am. Chem. Soc. **127**, 17118–17127 (2005)
141. Lu, Y., Liu, G.L., Lee, L.P.: High-density silver nanoparticle film with temperature-controllable interparticle spacing for a tunable surface enhanced Raman scattering substrate. Nano Lett. **5**, 5–9 (2005)
142. Talley, C.E., Jackson, J.B., Oubre, C., Grady, N.K., Hollars, C.W., Lane, S.M., Huser, T.R., Nordlander, P., Halas, N.J.: Surface-enhanced Raman scattering from individual Au nanoparticles and nanoparticle dimer substrates. Nano Lett. **5**, 1569–1574 (2005)
143. Chan, Y.-H., Chen, J., Wark, S.E., Skiles, S.L., Son, D.H., Batteas, J.D.: Using patterned arrays of metal nanoparticles to probe plasmon enhanced luminescence of CdSe quantum dots. ACS Nano **3**, 1735–1744 (2009)

144. Munechika, K., Chen, Y., Tillack, A.F., Kulkarni, A.P., Plante, I.J.-L., Munro, A.M., Ginger, D.S.: Spectral control of plasmonic emission enhancement from quantum dots near single silver nanoprisms. Nano Lett. **10**, 2598–2603 (2010)
145. Zhang, H., Li, Y., Ivanov, I.A., Qu, Y., Huang, Y., Duan, X.: Plasmonic modulation of the upconversion fluorescence in $NaYF_4$:Yb/Tm hexaplate nanocrystals using gold nanoparticles or nanoshells. Angew. Chem. Int. Ed. **49**, 2865–2868 (2010)
146. Li, Z.Q., Chen, S., Li, J.J., Liu, Q.Q., Sun, Z., Wang, Z.B., Huang, S.M.: Plasmon-enhanced upconversion fluorescence in $NaYF_4$:Yb/Er/Gd nanorods coated with Au nanoparticles or nanoshells. J. Appl. Phys. **111**(1), 014310 (2012)
147. Schietinger, S., Aichele, T., Wang, H.-Q., Nann, T., Benson, O.: Plasmon-enhanced upconversion in single $NaYF_4$:Yb^{3+}/Er^{3+} codoped nanocrystals. Nano Lett. **10**, 134–138 (2010)
148. Yuan, P., Lee, Y.H., Gnanasammandhan, M.K., Guan, Z., Zhang, Y., Xu, Q.-H.: Plasmon enhanced upconversion luminescence of $NaYF_4$:Yb,Er@SiO_2@Ag core-shell nanocomposites for cell imaging. Nanoscale **4**, 5132–5137 (2012)
149. Deng, W., Sudheendra, L., Zhao, J., Fu, J., Jin, D., Kennedy, I.M., Goldys, E.M.: Upconversion in $NaYF_4$:Yb,Er nanoparticles amplified by metal nanostructures. Nanotechnology **22**(32), 325604 (2011)
150. Li, C., Zhu, Y., Zhang, X., Yang, X., Li, C.: Metal-enhanced fluorescence of carbon dots adsorbed Ag@SiO_2 core-shell nanoparticles. RSC Adv. **2**, 1765–1768 (2012)
151. Zhao, P., Zhu, Y., Yang, X., Fan, K., Shen, J., Li, C.: Facile synthesis of upconversion luminescent mesoporous Y_2O_3:Er microspheres and metal enhancement using gold nanoparticles. RSC Adv. **2**, 10592–10597 (2012)
152. Verhagen, E., Kuipers, L., Polman, A.: Enhanced nonlinear optical effects with a tapered plasmonic waveguide. Nano Lett. **7**, 334–337 (2007)
153. Sudheendra, L., Ortalan, V., Dey, S., Browning, N.D., Kennedy, I.M.: Plasmonic enhanced emissions from cubic $NaYF_4$:Yb:Er/Tm nanophosphors. Chem. Mater. **23**, 2987–2993 (2011)
154. Fujii, M., Nakano, T., Imakita, K., Hayashi, S.: Upconversion luminescence of Er and Yb codoped $NaYF_4$ nanoparticles with metal shells. J. Phys. Chem. C **117**, 1113–1120 (2013)
155. Priyam, A., Idris, N.M., Zhang, Y.: Gold nanoshell coated $NaYF_4$ nanoparticles for simultaneously enhanced upconversion fluorescence and darkfield imaging. J. Mater. Chem. **22**, 960–965 (2012)
156. Liu, N., Qin, W., Qin, G., Jiang, T., Zhao, D.: Highly plasmon-enhanced upconversion emissions from Au@beta-$NaYF_4$:Yb,Tm hybrid nanostructures. Chem. Commun. **47**, 7671–7673 (2011)
157. Schneider, G., Decher, G., Nerambourg, N., Praho, R., Werts, M.H.V., Blanchard-Desce, M.: Distance-dependent fluorescence quenching on gold nanoparticles ensheathed with layer-by-layer assembled polyelectrolytes. Nano Lett. **6**, 530–536 (2006)
158. Liu, N., Prall, B.S., Klimov, V.I.: Hybrid gold/silica/nanocrystal-quantum-dot superstructures: synthesis and analysis of semiconductor-metal interactions. J. Am. Chem. Soc. **128**, 15362–15363 (2006)
159. Nabika, H., Deki, S.: Enhancing and quenching functions of silver nanoparticles on the luminescent properties of europium complex in the solution phase. J. Phys. Chem. B **107**, 9161–9164 (2003)
160. Qian, L.P., Zhou, L.H., Too, H.-P., Chow, G.-M.: Gold decorated $NaYF_4$:Yb,Er/$NaYF_4$/silica (core/shell/shell) upconversion nanoparticles for photothermal destruction of BE(2)-C neuroblastoma cells. J. Nanopart. Res. **13**, 499–510 (2011)
161. Liu, S., Chen, G., Ohulchanskyy, T.Y., Swihart, M.T., Prasad, P.N.: Facile synthesis and potential bioimaging applications of hybrid upconverting and plasmonic $NaGdF_4$:Yb^{3+},Er^{3+}/silica/gold nanoparticles. Theranostics **3**, 275–281 (2013)
162. Lakowicz, J.R.: Radiative decay engineering: biophysical and biomedical applications. Anal. Biochem. **298**, 1–24 (2001)

163. Dulkeith, E., Morteani, A.C., Niedereichholz, T., Klar, T.A., Feldmann, J., Levi, S.A., van Veggel, F., Reinhoudt, D.N., Moller, M., Gittins, D.I.: Fluorescence quenching of dye molecules near gold nanoparticles: radiative and nonradiative effects. Phys. Rev. Lett. **89**, 203002 (2002)
164. Johnson, N.J.J., Korinek, A., Dong, C., van Veggel, F.C.J.M.: Self-focusing by ostwald ripening: a strategy for layer-by-layer epitaxial growth on upconverting nanocrystals. J. Am. Chem. Soc. **134**, 11068–11071 (2012)
165. Zhang, F., Che, R., Li, X., Yao, C., Yang, J., Shen, D., Hu, P., Li, W., Zhao, D.: Direct imaging the upconversion nanocrystal core/shell structure at the subnanometer level: shell thickness dependence in upconverting optical properties. Nano Lett. **12**, 2852–2858 (2012)
166. Weissleder, R.: A clearer vision for in vivo imaging. Nat. Biotechnol. **19**, 316–317 (2001)
167. Zhan, Q., Qian, J., Liang, H., Somesfalean, G., Wang, D., He, S., Zhang, Z., Andersson-Engels, S.: Using 915 nm laser excited $Tm^{3+}/Er^{3+}/Ho^{3+}$-doped $NaYbF_4$ upconversion nanoparticles for in vitro and deeper in vivo bioimaging without overheating irradiation. ACS Nano **5**, 3744–3757 (2011)
168. Lezhnina, M.M., Jüstel, T., Kätker, H., Wiechert, D.U., Kynast, U.H.: Efficient luminescence from rare-earth fluoride nanoparticles with optically functional shells. Adv. Funct. Mater. **16**, 935–942 (2006)
169. Yi, G.-S., Chow, G.-M.: Water-soluble $NaYF_4$:Yb,Er(Tm)/$NaYF_4$/polymer core/shell/shell nanoparticles with significant enhancement of upconversion fluorescence. Chem. Mater. **19**, 341–343 (2007)
170. Guo, H., Li, Z., Qian, H., Hu, Y., Muhammad, I.N.: Seed-mediated synthesis of $NaYF_4$:Yb, Er/$NaGdF_4$ nanocrystals with improved upconversion fluorescence and MR relaxivity. Nanotechnology **21**(12), 125602 (2010)
171. Dong, C., Korinek, A., Blasiak, B., Tomanek, B., van Veggel, F.C.J.M.: Cation exchange: a facile method to make $NaYF_4$:Yb,Tm-$NaGdF_4$ core-shell nanoparticles with a thin, tunable, and uniform shell. Chem. Mater. **24**, 1297–1305 (2012)
172. Wang, Y., Tu, L., Zhao, J., Sun, Y., Kong, X., Zhang, H.: Upconversion luminescence of beta-$NaYF_4$:Yb^{3+}, Er^{3+}@beta-$NaYF_4$ core/shell nanoparticles: excitation power, density and surface dependence. J. Phys. Chem. C **113**, 7164–7169 (2009)
173. Chen, G., Ohulchanskyy, T.Y., Liu, S., Law, W.-C., Wu, F., Swihart, M.T., Agren, H., Prasad, P.N.: Core/shell $NaGdF_4$:Nd^{3+}/$NaGdF_4$ nanocrystals with efficient near-infrared to near-infrared downconversion photoluminescence for bioimaging applications. ACS Nano **6**, 2969–2977 (2012)
174. Boyer, J.-C., Gagnon, J., Cuccia, L.A., Capobianco, J.A.: Synthesis, characterization, and spectroscopy of $NaGdF_4$:Ce^{3+}, Tb^{3+}/$NaYF_4$ core/shell nanoparticles. Chem. Mater. **19**, 3358–3360 (2007)
175. Li, C., Liu, X., Yang, P., Zhang, C., Lian, H., Lin, J.: LaF_3, CeF_3, CeF_3:Tb^{3+}, and CeF_3:Tb^{3+}@LaF_3 (core-shell) nanoplates: hydrothermal synthesis and luminescence properties. J. Phys. Chem. C **112**, 2904–2910 (2008)
176. Wang, Z.L., Quan, Z.W., Jia, P.Y., Lin, C.K., Luo, Y., Chen, Y., Fang, J., Zhou, W., O'Connor, C.J., Lin, J.: Facile synthesis and photoluminescent properties of redispersible CeF_3, CeF_3:Tb^{3+}, and CeF_3:Tb^{3+}/LaF_3 (core/shell) nanoparticles. Chem. Mater. **18**, 2030–2037 (2006)
177. Gai, S., Yang, P., Li, X., Li, C., Wang, D., Dai, Y., Lin, J.: Monodisperse CeF_3, CeF_3:Tb^{3+}, and CeF_3:Tb^{3+}@LaF_3 core/shell nanocrystals: synthesis and luminescent properties. J. Mater. Chem. **21**, 14610–14615 (2011)
178. Schaefer, H., Ptacek, P., Zerzouf, O., Haase, M.: Synthesis and optical properties of KYF_4/Yb,Er nanocrystals, and their surface modification with undoped KYF_4. Adv. Funct. Mater. **18**, 2913–2918 (2008)
179. Dou, Q., Idris, N.M., Zhang, Y.: Sandwich-structured upconversion nanoparticles with tunable color for multiplexed cell labeling. Biomaterials **34**, 1722–1731 (2013)
180. Qian, H.-S., Zhang, Y.: Synthesis of hexagonal-phase core-shell $NaYF_4$ nanocrystals with tunable upconversion fluorescence. Langmuir **24**, 12123–12125 (2008)

181. Abel, K.A., Boyer, J.-C., Andrei, C.M., van Veggel, F.C.J.M.: Analysis of the shell thickness distribution on $NaYF_4/NaGdF_4$ core/shell nanocrystals by EELS and EDS. J. Phys. Chem. L. **2**, 185–189 (2011)
182. Liu, X., Kong, X., Zhang, Y., Tu, L., Wang, Y., Zeng, Q., Li, C., Shi, Z., Zhang, H.: Breakthrough in concentration quenching threshold of upconversion luminescence via spatial separation of the emitter doping area for bio-applications. Chem. Commun. **47**, 11957–11959 (2011)
183. Wang, Y.-F., Sun, L.-D., Xiao, J.-W., Feng, W., Zhou, J.-C., Shen, J., Yan, C.-H.: Rare-earth nanoparticles with enhanced upconversion emission and suppressed rare-earth-ion leakage. Chem. Eur. J. **18**, 5558–5564 (2012)
184. Chen, G.Y., Shen, J., Ohulchanskyy, T.Y., Patel, N.J., Kutikov, A., Li, Z.P., Song, J., Pandey, R.K., Agren, H., Prasad, P.N., Han, G.: (alpha-$NaYbF_4$:Tm^{3+})/CaF_2 core/shell nanoparticles with efficient near-infrared to near-infrared upconversion for high-contrast deep tissue bioimaging. ACS Nano **6**, 8280–8287 (2012)
185. Shen, J., Chen, G., Ohulchanskyy, T.Y., Kesseli, S.J., Buchholz, S., Li, Z., Prasad, P.N., Han, G.: Upconversion: tunable near infrared to ultraviolet upconversion luminescence enhancement in (α-$NaYF_4$:Yb,Tm)/CaF_2 core/shell nanoparticles for in situ real-time recorded biocompatible photoactivation (small 19/2013). Small **9**, 3213–3217 (2013)
186. Kar, A., Patra, A.: Impacts of core–shell structures on properties of lanthanide-based nanocrystals: crystal phase, lattice strain, downconversion, upconversion and energy transfer. Nanoscale **4**, 3608 (2012)
187. Liu, Y., Tu, D., Zhu, H., Li, R., Luo, W., Chen, X.: A strategy to achieve efficient dual-mode luminescence of Eu^{3+} in lanthanides doped multifunctional $NaGdF_4$ nanocrystals. Adv. Mater. **22**, 3266–3271 (2010)
188. Xie, X., Liu, X.: Photonics: upconversion goes broadband. Nat. Mater. **11**, 842–843 (2012)
189. Gu, Z., Yan, L., Tian, G., Li, S., Chai, Z., Zhao, Y.: Recent advances in design and fabrication of upconversion nanoparticles and their safe theranostic applications. Adv. Mater. **25**, 3758–3779 (2013)
190. Shen, J., Chen, G., Vu, A.M., Fan, W., Bilsel, O.S., Chang, C.C., Han, G.: Engineering the upconversion nanoparticle excitation wavelength: cascade sensitization of tri-doped upconversion colloidal nanoparticles at 800 nm. Adv. Opt. Mater. **1**, 644–650 (2013)
191. Xie, X., Gao, N., Deng, R., Sun, Q., Xu, Q.-H., Liu, X.: Mechanistic investigation of photon upconversion in Nd^{3+}-sensitized core-shell nanoparticles. J. Am. Chem. Soc. **135**, 12608–12611 (2013)
192. Wang, Y.-F., Liu, G.-Y., Sun, L.-D., Xiao, J.-W., Zhou, J.-C., Yan, C.-H.: Nd^{3+}-sensitized upconversion nanophosphors: efficient in vivo bioimaging probes with minimized heating effect. ACS Nano **7**, 7200–7206 (2013)
193. Wen, H., Zhu, H., Chen, X., Hung, T.F., Wang, B., Zhu, G., Yu, S.F., Wang, F.: Upconverting near-infrared light through energy management in core-shell-shell nanoparticles. Angew. Chem. Int. Ed. **125**, 13661–13665 (2013)
194. Zhong, Y., Tian, G., Gu, Z., Yang, Y., Gu, L., Zhao, Y., Ma, Y., Yao, J.: Elimination of photon quenching by a transition layer to fabricate a quenching-shield sandwich structure for 800 nm excited upconversion luminescence of Nd^{3+}-sensitized nanoparticles. Adv. Mater. **26**, 2831–2837 (2013)
195. Li, X., Wang, R., Zhang, F., Zhou, L., Shen, D., Yao, C., Zhao, D.: Nd^{3+} sensitized up/down converting dual-mode nanomaterials for efficient in-vitro and in-vivo bioimaging excited at 800 nm. Sci. Rep. **3**, 3536 (2013)
196. Auzel, F.E.: Materials and devices using double-pumped-phosphors with energy transfer. Proc. IEEE **61**, 758–786 (1973)
197. Bawendi, M.G., Steigerwald, M.L., Brus, L.E.: The quantum mechanics of larger semiconductor clusters ("quantum dots"). Annu. Rev. Phys. Chem. **41**, 477–496 (1990)
198. Wang, Y., Herron, N.: Nanometer-sized semiconductor clusters: materials synthesis, quantum size effects, and photophysical properties. J. Phys. Chem. **95**, 525–532 (1991)
199. Jablonski, A.: Efficiency of anti-stokes fluorescence in dyes. Nature **131**, 839–840 (1933)

200. Jablonski, A.: About the mechanism of photo-luminescence of dye phosphors. Z. Angew. Phys. **94**, 38–46 (1935)
201. Michalet, X., Pinaud, F.F., Bentolila, L.A., Tsay, J.M., Doose, S., Li, J.J., Sundaresan, G., Wu, A.M., Gambhir, S.S., Weiss, S.: Quantum dots for live cells, in vivo imaging, and diagnostics. Science **307**, 538–544 (2005)
202. Bruchez, M., Moronne, M., Gin, P., Weiss, S., Alivisatos, A.P.: Semiconductor nanocrystals as fluorescent biological labels. Science **281**, 2013–2016 (1998)
203. Chan, W.C.W., Nie, S.M.: Quantum dot bioconjugates for ultrasensitive nonisotopic detection. Science **281**, 2016–2018 (1998)
204. Chen, O., Zhao, J., Chauhan, V.P., Cui, J., Wong, C., Harris, D.K., Wei, H., Han, H.-S., Fukumura, D., Jain, R.K., Bawendi, M.G.: Compact high-quality CdSe-CdS core-shell nanocrystals with narrow emission linewidths and suppressed blinking. Nat. Mater. **12**, 445–451 (2013)
205. Cheng, L., Yang, K., Zhang, S., Shao, M., Lee, S., Liu, Z.: Highly-sensitive multiplexed in vivo imaging using PEGylated upconversion nanoparticles. Nano Res. **3**, 722–732 (2010)
206. Yu, J.H., Kwon, S.-H., Petrasek, Z., Park, O.K., Jun, S.W., Shin, K., Choi, M., IlPark, Y., Park, K., Na, H.B., Lee, N., Lee, D.W., Kim, J.H., Schwille, P., Hyeon, T.: High-resolution three-photon biomedical imaging using doped ZnS nanocrystals. Nat. Mater. **12**, 359–366 (2013)
207. Maestro, L.M., Ramirez-Hernandez, J.E., Bogdan, N., Capobianco, J.A., Vetrone, F., Garcia Sole, J., Jaque, D.: Deep tissue bio-imaging using two-photon excited CdTe fluorescent quantum dots working within the biological window. Nanoscale **4**, 298–302 (2012)
208. Wang, J., Lin, M., Yan, Y., Wang, Z., Ho, P.C., Loh, K.P.: CdSe/AsS core-shell quantum dots: preparation and two-photon fluorescence. J. Am. Chem. Soc. **131**, 11300–11301 (2009)
209. Deng, Z., Tong, L., Flores, M., Lin, S., Cheng, J.-X., Yan, H., Liu, Y.: High-quality manganese-doped zinc sulfide quantum rods with tunable dual-color and multiphoton emissions. J. Am. Chem. Soc. **133**, 5389–5396 (2011)
210. Fu, Y., Han, T.T., Luo, Y., Agren, H.: Multiphoton excitation of quantum dots by ultrashort and ultraintense laser pulses. Appl. Phys. Lett. **88**, 221114 (2006)
211. Fu, Y., Han, T.T., Agren, H., Lin, L., Chen, P., Liu, Y., Tang, G.Q., Wu, J., Yue, Y., Dai, N.: Design of semiconductor CdSe core ZnS/CdS multishell quantum dots for multiphoton applications. Appl. Phys. Lett. **90**, 173102 (2007)
212. Wu, S., Han, G., Milliron, D.J., Aloni, S., Altoe, V., Talapin, D.V., Cohen, B.E., Schuck, P.J.: Non-blinking and photostable upconverted luminescence from single lanthanide-doped nanocrystals. Proc. Natl. Acad. Sci. **106**, 10917–10921 (2009)
213. Yu, M., Li, F., Chen, Z., Hu, H., Zhan, C., Yang, H., Huang, C.: Laser scanning up-conversion luminescence microscopy for imaging cells labeled with rare-earth nanophosphors. Anal. Chem. **81**, 930–935 (2009)
214. Johnson, N.J.J., Sangeetha, N.M., Boyer, J.-C., van Veggel, F.C.J.M.: Facile ligand-exchange with polyvinylpyrrolidone and subsequent silica coating of hydrophobic upconverting beta-$NaYF_4$:Yb^{3+}/Er^{3+} nanoparticles. Nanoscale **2**, 771–777 (2010)
215. Wu, X.Y., Liu, H.J., Liu, J.Q., Haley, K.N., Treadway, J.A., Larson, J.P., Ge, N.F., Peale, F., Bruchez, M.P.: Immunofluorescent labeling of cancer marker Her2 and other cellular targets with semiconductor quantum dots. Nat. Biotechnol. **21**, 41–46 (2003)
216. Nirmal, M., Dabbousi, B.O., Bawendi, M.G., Macklin, J.J., Trautman, J.K., Harris, T.D., Brus, L.E.: Fluorescence intermittency in single cadmium selenide nanocrystals. Nature **383**, 802–804 (1996)
217. Mahler, B., Spinicelli, P., Buil, S., Quelin, X., Hermier, J.-P., Dubertret, B.: Towards non-blinking colloidal quantum dots. Nat. Mater. **7**, 659–664 (2008)
218. Wang, X., Ren, X., Kahen, K., Hahn, M.A., Rajeswaran, M., Maccagnano-Zacher, S., Silcox, J., Cragg, G.E., Efros, A.L., Krauss, T.D.: Non-blinking semiconductor nanocrystals. Nature **459**, 686–689 (2009)

Chapter 4
Upconversion Nanoparticle-Based Nanocomposites

Abstract In recent years, the development of multi-functional nanomaterials with fantastic physical, chemical, and biological properties has become an attractive research topic, demonstrating high potential in biomedicine, catalysis, energy conversion, water treatment adsorbents, and so on. As one of the most important luminescence nanomaterials, upconversion nanoparticles (UCNPs) were also used to fabricate multi-functional nanocomposites (UCNPs-X). In this chapter, we summarize recent advanced upconversion nanoparticles-based nanocomposites, including UCNPs-mSiO_2 (mSiO_2 = mesoporous SiO_2), UCNPs-MNPs (MNPs = magnetic nanoparticles), UCNPs-metal, and UCNPs-semiconductor nanocomposites.

Keywords Upconversion nanoparticles · Multi-functional nanomaterials · UCNPs-based nanocomposites · UCNP-mSiO_2 · Magnetic nanoparticles · UCNPs-metal nanocomposites · UCNPs-semiconductor nanocomposites · Sol-gel method

4.1 Introduction

In recent years, the development of multi-functional nanomaterials with fantastic physical, chemical, and biological properties has become an attractive research topic, demonstrating high potential in biomedicine, catalysis, energy conversion, water treatment adsorbents, and so on [1–7]. An important research direction in nanoparticle synthesis is the expansion from single-component nanoparticles to multi-component hybrid nanostructures with discrete domains of different materials arranged in a controlled fashion. The advantages of multi-component structures lie in three aspects. The first is to realize multi-functionality: Different functionalities can be integrated, with the dimension and material parameters of the individual components independently optimized. Such multi-component structures are

Xiaomin Li and Fan Zhang contributed together to this chapter.

F. Zhang, *Photon Upconversion Nanomaterials*,
Nanostructure Science and Technology, DOI 10.1007/978-3-662-45597-5_4

commonly formed through seeded-growth of one material on seed particles of another, yielding core@shell, core@satellite peanut-, dumbbell-, flower-like composite nanoparticles, etc. [8–15]. Other approaches include embedding of different inorganic materials together within silica spheres, or within polymer particles or capsules [16]. In this chapter, we summarize recent advanced UCNPs-based nanocomposites, including UCNPs-mSiO_2 (mSiO_2 = mesoporous SiO_2), UCNPs-MNPs (MNPs = magnetic nanoparticles), UCNPs-metal, and UCNPs-semiconductor nanocomposites. In this chapter, we mainly focus on the synthesis of the nanocomposites; the details of the corresponding properties and applications of the nanocomposites are summarized in the following chapters.

4.2 UCNP-mSiO_2

Silanization is a popular technique for surface modification of nanoparticles since silica is highly biocompatible. The ease of surface modification (e.g., –COOH, –NH_2, –SH, etc.) allows for bio-labeling, drug targeting, and delivery. Furthermore, the silica-coated functional nanoparticles can be protected by the SiO_2 shell against the influence of physiological conditions and outside environments. Two routes can be adopted to coat silica, depending on the polarity of the capping ligands on the nanoparticle surface [17, 18]. One is the Stöber method, which can be utilized to coat silica on hydrophilic upconversion nanoparticles [19–23]. Another route is the reverse microemulsion (water-in-oil) method to realize silica coating onto upconversion nanoparticles with hydrophobic capping ligands [24–28]. Both of two methods can be considered as the division of sol-gel method. The sol-gel method is a typical technique for the fabrication of UCNPs-SiO_2 nanocomposites that does not require complicated procedures or instruments. The representative sol-gel process involves inorganic precursors [tetraethyl orthosilicate (TEOS), metal salts, metal-alkoxides, etc.] that upon reaction with water undergo hydrolysis and condensation, leading to the formation of 3D oxide networks. The Stöber method, invented by Stöber in the 1960s, is one of the most well-known representative examples of the sol-gel process for the synthesis of uniform colloidal silica spheres [19]. This discovery profoundly influenced the synthesis of novel core@shell-structured materials with a variety of components, sizes, and properties in which silica can serve as either a core or a shell [29, 30]. Alternatively, Osseo-Asare and Arriagada discovered a novel water-in-oil reverse-microemulsion method for the synthesis of silica nanoparticles with more uniform size [31]. Via the microemulsion method (using the surfactant Trix-100 or Igepal CO-520/cyclohexane/aqueous solution), a thin and dense silica layer was coated onto the surface of upconversion nanoparticles to form highly uniform and monodisperse core@shell-structured UCNPs@SiO_2 nanospheres [32, 33]. The surface of SiO_2 can be readily functionalized with diverse groups such as amines, carboxyl, or thiols, enabling other nanoparticles or photosensitive molecules to connect. For example, simply replacing a part of TEOS with (3-aminopropyl)triethoxysilane (APTES) allows the surface of the obtained

UCNPs@SiO_2 to be modified with –NH_2 [34]. With the help of the surface functional group, the "alien species" can easily be combined with upconversion nanoparticles. Folic acid (FA), which has been widely used as a targeting agent, can be easily conjugated with the amine-functionalized UCNPs@SiO_2 nanocomposites to realize targeted imaging or delivery. With the help of SiO_2 shells, the scope of potential applications of upconversion nanomaterials is extremely extended. For example, Zhang and coworkers first demonstrated the potential of upconversion nanoparticles-based gene delivery [35]. Anti-Her2 antibody and folic acid-modified $NaYF_4$@SiO_2 nanoparticles were prepared and used for luminescence imaging and targeted delivery of siRNA. The results showed that GL3 siRNA was delivered to SK-BR-3 cells with Her2 receptor by using the nanoparticles, and their delivery could be tracked. A FRET system based on ultrafine silica-coated $NaYF_4$ upconversion nanoparticles was designed by the same group to study the intracellular fate of siRNA in live cells [36]. A FRET phenomenon between the upconversion nanoparticles donor and siRNA intercalating dye BOBO-3 acceptor was simultaneously used to study their intracellular uptake, release, and bio-stability. Chien et al. [37] formulated UCNPs@SiO_2 nanocomposites as the NIR-triggered targeting and drug delivery vehicles that successfully deliver in vitro and in vivo for NIR photocontrolled targeting, bio-imaging, and chemotherapy. To achieve phototargeting, the tumor-homing agent (i.e., folic acid) has been constructed as a photoresponsive molecule. For the chemotherapeutic effect, the anti-tumor drug doxorubicin is thiolated on the surface of SiO_2 to form a disulfide bond that can be cleaved by lysosomal enzymes within the cells. The caged upconversion nanoparticles can serve as a platform for the improvement of selective targeting and possible reduction of adverse side effects from chemotherapy. Zhai et al. [38] reported a novel design based on a combination of UCNPs@SiO_2 nanoparticles and polyoxometalates, for electrically controlled fluorescence switches and sensitive detection of antioxidants in aqueous solution. In addition, the surface conjugation of organic groups or guest molecules, small nanoparticles, can also be anchored on the surface of the UCNPs@SiO_2 nanocomposites according to the surface function groups. For example, Shi and coworkers have demonstrated that the obtained UCNPs@SiO_2 can be integrated with other functional building blocks such as Au nanoparticles (for CT imaging) and ultrasmall CuS (for photothermal therapy) to form multi-functional UCNPs@SiO_2@Au or UCNPs@SiO_2@CuS nanoparticles, which display their respective properties of each component so as to achieve multimodal bio-imaging and synergetic therapy (radiotherapy and photothermal ablation) [39, 40].

As mentioned above, the "alien species" are mainly anchored on the outer surface of UCNPs@SiO_2 nanocomposites, indicating that only a small amount of "alien species" is introduced due to the limited area of outer surface of UCNPs@SiO_2 nanocomposites. A modified Stöber method is used to coat the functional nanoparticles with mesoporous silica (mSiO_2). The formation of mesoporous silica is nearly the same as the synthesis of dense SiO_2 (as mentioned above) except for the presence of the structure-directing agent (i.e., cetyltrimethylammonium bromide, CTAB) [22]. In general, there are mainly two routes to get the nanocomposites. The first route is based on coating mesoporous silica (mSiO_2) layers on the surface of UCNPs@SiO_2

to form the UCNPs@SiO_2@$mSiO_2$ core@shell@shell structure (Fig. 4.1a) [22, 41–43]. The second route is based on overcoating with $mSiO_2$ layer on the surface of upconversion nanoparticles directly in one step (Fig. 4.1f) [44–46]. Mesoporous silica possesses intriguing properties including good compatibility, porous structures with high surface area (providing reservoirs for loading various guest molecules), tunable pore size (offering the selectivity for adsorption and

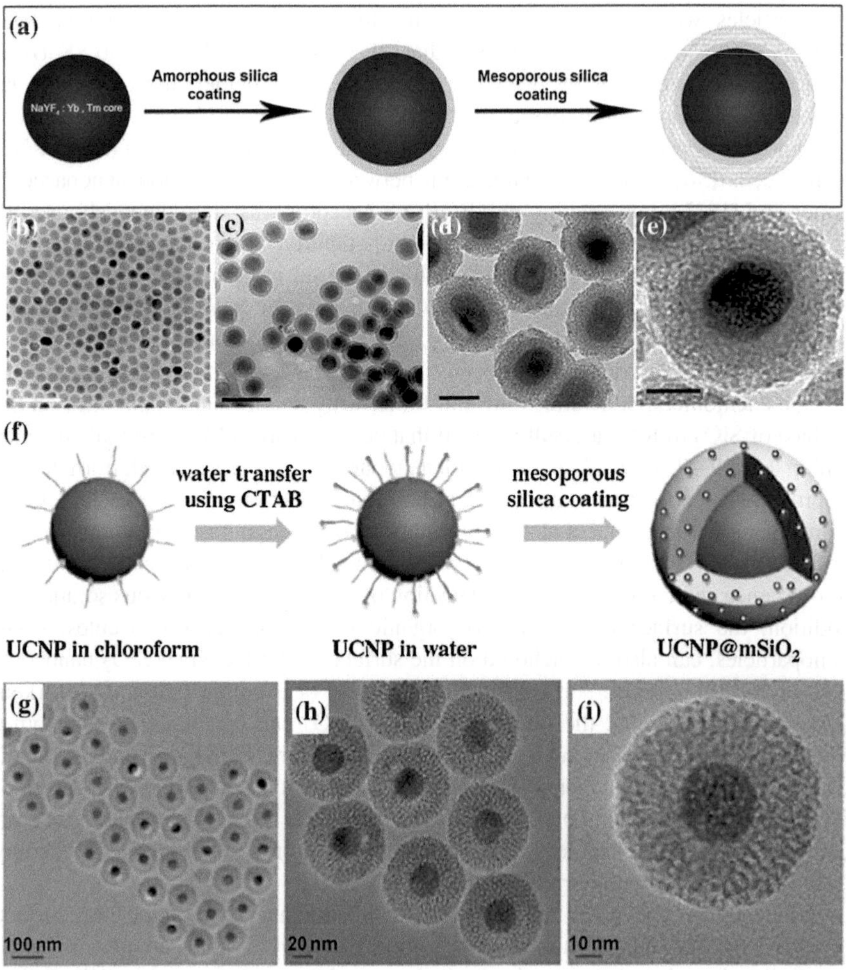

Fig. 4.1 The synthetic routes for mesoporous silica overcoating. **a** Two-step approach for the fabrication of the UCNPs@SiO_2@$mSiO_2$ nanocomposite. **b–e** TEM images of upconversion nanoparticles (**b**), UCNPs@SiO_2 (**c**), UCNPs@SiO_2@$mSiO_2$ nanocomposites (**d**, **e**). **f** One-step method for the $mSiO_2$ overcoating on the hydrophobic surface of upconversion nanoparticles. **g–i** TEM images of the UCNPs@$mSiO_2$ nanocomposite. **a–e** Reprinted with the permission from Ref. [43]. Copyright 2012 National Academy of Sciences. **f–i** Reprinted with the permission from Ref. [44]. Copyright 2012 Wiley-VCH Verlag GmbH & Co. KGaA

controllability of the release of the restricted nanoparticles), and ease of surface functionalization (providing active sites for linking other biological molecules). As such, if upconversion nanoparticles combine with silica or mesoporous silica, a novel core@shell-structured nanomaterial for drug delivery can be obtained. By reasonable core@shell-structured design and different synthesis strategies, upconversion nanoparticles can be assembled on, encapsulated within, or combined with other specific nanoparticles inside and on the surface of silica or mesoporous silica for different application purposes. In this section, dramatic efforts have been devoted to the fabrication of a series of UCNPs-$mSiO_2$ nanocomposites through sol-gel chemistry and the corresponding applications [18].

4.2.1 Two-Step Sol-gel Method for the Fabrication of UCNPs-mSiO₂ Nanocomposites

In order to produce a mesoporous silica layer, there are two commonly used structure-directing agents: (1) surfactants such as CTAB and (2) organosilanes such as octadecyltrimethoxysilane (C18TMS). The former can form ordered mesopores while the latter can form disordered and worm-like ones. In an earlier study via a two-step sol-gel reaction, Lin et al. fabricated multi-functional nanocomposites using Fe_3O_4 nanospheres as the core that were subsequently coated with dense silica and ordered mesoporous silica, which was further functionalized by the deposition of $NaYF_4$:Yb/Er (Tm) upconversion nanoparticles in the mesoporous channels [47]. The resultant composite nanomaterials Fe_3O_4@SiO_2@$mSiO_2$@UCNPs exhibit high magnetization and bright upconversion emission under 980-nm laser excitation. The intermediate solid SiO_2 as a protective matrix plays an important role in protecting the upconversion luminescence from quenching by the inner black Fe_3O_4 core, while a $mSiO_2$ shell can be used to load drug model ibuprofen. More importantly, the drug release amount can be monitored by the change in the upconversion emission intensity. This class of multi-functional system seems to have the potential to track the drug delivery process based on its magnetic and luminescent properties. Following this concept, instead of coating upconversion nanoparticles at the surface of $mSiO_2$, upconversion nanoparticles as an inner core were encapsulated by $mSiO_2$ to form Gd_2O_3:Er@SiO_2@$mSiO_2$ [48]. In addition, Sun et al. reported the one-pot self-assembly of multi-functional mesoporous nanoprobes with magnetic nanoparticles (MNPs) and hydrophobic upconversion nanoparticles, in which CTAB-stabilized upconversion nanoparticles (positively charged) were self-assembled onto Fe_3O_4@SiO_2 (negatively charged) such as core-satellites via electrostatic and van der Waals interactions. Then, the subsequent cocondensation of TEOS and CTAB resulted in the formation of the outer mesoporous silica layer [49]. Apart from CTAB, C18TMS is a common directing agent for the synthesis of disordered mesopore structures. Zhang et al. and Lin's group reported the synthesis of UCNPs@SiO_2@$mSiO_2$ nanoparticles using surfactant C18TMS as a pore generator via two-step sol-gel reactions [33, 41].

4.2.2 One-Step Sol-gel Method for the Fabrication of UCNPs-mSiO_2 Nanocomposites

Despite its success, there are some problems to be addressed with the two-step sol-gel method. For example, one of the biggest shortcomings in the two-step sol-gel reaction is that the fabrication of these materials typically requires intermediate coatings of solid silica followed by further deposition of a mesoporous silica layer, which involves complicated multi-step procedures. To overcome these hurdles, the direct coating of mesoporous silica on the surface of a single upconversion nanoparticle via a facile and general strategy is greatly demanded. In 2006, Heyon's group reported a typical method for the direct encapsulation of hydrophobic inorganic nanoparticles with a mesoporous silica shell [50]. In this method, CTAB not only acts as a capping and phase-transfer agent, but also as the template for the formation of mesoporous structures in the silica sol-gel reaction. This outstanding work provides a new opportunity to directly coat diverse hydrophobic nanoparticles possessing different compositions, shapes, and sizes with mesoporous silica shells [51, 52]. Enlightened by these studies, Lin's group and Shi's group have independently reported the synthesis of uniform and monodisperse UCNPs@mSiO_2 nanocomposites using a one-step sol-gel process [44, 45, 53, 54]. The typical synthesis procedure for UCNPs@mSiO_2 nanocomposites is as following [53]. Hydrophobic upconversion nanoparticles were first transferred from the organic phase to the aqueous phase by using CTAB as a secondary surfactant. CTAB-terminated upconversion nanoparticles can then act as the seed for the coating of the mesoporous silica shell via the sol-gel process. Finally, the surfaces of the UCNPs@mSiO_2 nanospheres were modified with polyethylene glycol (PEG) in order to enhance the nanocarrier dispersion and long-term stability under physiological conditions, prolong the circulation time of the nanocarrier in blood, and facilitate preferential accumulation at the tumor sites by the enhanced permeation and retention (EPR) effect [55]. Thus, in this kind of composite nanomaterial, the core imparts it with luminescence and/or magnetic properties for simultaneous upconversion nanoparticles and magnetic resonance imaging (MRI), whereas the mesoporous shell allows it to be loaded with an anticancer drug. The average diameter of nanospheres is determined to be below 100 nm, which is within the acceptable size range for biomedical applications in vivo.

In addition, CTAB-stabilized upconversion nanoparticles can be employed to construct pH-responsive drug delivery nanocarriers by using poly(acrylicacid) (PAA), a biodegradable superabsorbent, as a nanoreactor and template. Recently, Wang and coworkers fabricated a novel and unique multi-functional concentric-UCNPs@mSiO_2@PAA core@double shell nanostructure [56]. The formation of eccentric PAA shells should be related to the change in the interfacial energy between PAA, UCNPs@mSiO_2 nanoparticles, and the solvent, likely resulting in a minimum interfacial energy. Such material has two special characteristics: ultrahigh drug storage capacity and sensitive pH-responsive drug release properties. The drug loading capacity in eccentric-(concentric-UCNPs@mSiO_2)@PAA was 2 mg of DOX per 1 mg nanomaterials. This is because PAA, with an abundance of carboxyl groups, not only effectively loads the positively charged drugs by electrostatic interactions but also exhibits a pH-responsive performance. In subsequent work, the same researchers further improved the experimental protocol by taking into account that PAA in the outer layer is unstable in water and prone to swelling. Special eccentric UCNPs@PAA@SiO_2 core@shell nanoclusters consisting of a single $NaYF_4$:Yb/Er/Gd upconversion nanoparticle as the core, PAA as the intermediate layer, and eccentric SiO_2 as the outer layer were successfully manufactured [57]. In brief, PAA molecules were self-assembled around CTAB-UCNPs to obtain eccentric UCNP-s@PAA core@shell nanospheres. During this process, the resulting eccentric PAA shell is like a "reservoir" that absorbs and retains water molecules inside its net structure because PAA is a polymer with a high water-absorbent capacity. The hydrolysis reaction of TEOS is subsequently confined to the PAA network by the sol-gel process, leading to the formation of eccentric UCNPs@PAA-SiO_2 core@shell nanoclusters. Likewise, eccentric UCNPs@PAA-SiO_2 nanomaterial has good capability for drug loading; however, its stability in water has been greatly improved compared to the former.

4.2.3 Yolk–Shell UCNPs-mSiO_2 Nanocomposites

Apart from the usage of pore making agents discussed above, upconversion nanoparticles-based nanoparticles obtained by sol-gel reactions can be further etched to produce yolk–shell (or rattle-type) drug delivery nanocarriers [58–61]. Yolk-shell nanostructures have a unique interstitial hollow space between the core and mesoporous shell and are attractive as the new generation drug delivery nanoplatform due to the greatly enhanced drug-loading capacity [62–64]. One of the most straightforward synthesis methodologies for yolk-shell nanostructures is the surface-protected etching method. This approach was first reported by Yin and coworkers to prepare hollow mesoporous SiO_2, where polyvinylpyrrolidone (PVP) was used as the protecting layer and was coated with solid SiO_2 spheres. Subsequent selective etching of the interior SiO_2 was followed, using NaOH as the etchant by virtue of the structural difference between the core and shell of SiO_2 [65]. In general, the selection of appropriate etchants and surface-protecting agents is critical to obtain high-quality yolk-shell-structured nanomaterials [66]. The most frequently used etchants include NaOH, HF, $NaCO_3$, and even hot water, while the surface-protecting agents are PVP, poly(dimethyldiallylammonium chloride), polyethyleneimine (PEI), etc.

Upconversion nanoparticles-based yolk-shell structures are composed of core@shell and hollow structures with a special core@void@shell configuration. For instance, Wang et al. employed PEI-bearing positively charged networks as surface-protecting agents to fabricate upconversion nanoparticles-based yolk-shell structures, in which hot water acted as an etchant [59]. Li and coworkers reported a rattle-structured Fe_3O_4@void@$NaLuF_4$:Yb/Er nanostructure that can provide the ternary modality for magnetic resonance (MR), CT, and upconversion luminescence imaging [60]. Bu and coworkers fabricated a multi-functional rattle-structured nanotheranostic with a movable upconversion nanoparticle core consisting of an outer mesoporous SiO_2 shell and a hollow cavity between them [61]. Hydrophobic $NaYF_4$:Yb/Er@$NaGdF_4$ (denoted as Gd-UCNP) nanoparticles were coated with double dense SiO_2 layers using a two-step sol-gel method to form Gd-UCNPs@d1-SiO_2@d2-SiO_2. In order to ensure that the reaction would take place exclusively inside the inner SiO_2 layer, the outer SiO_2 layer was coated with PVP to protect it against etching. Then, a milder etchant of hot water was used to create hollow cavities because it can dissolve the colloidal SiO_2 shell by breaking the internal Si–O–Si bonds at a controllable rate. Using a "surface-protected hot water etching" strategy, the intermediate SiO_2 layer was selectively etched away to leave behind a hollow cavity inside the thin porous SiO_2 shell, eventually producing yolk-shell-structured UCNPs-SiO_2 nanocomposites. It is noteworthy that traditional alkaline or acidic etchants did not work owing to the difficulty in controlling the etching rates. This design can achieve two major goals: (i) A Gd-UCNP core can act as a luminescence/ MRI dual-mode imaging probe for locating tumors in vivo; and (ii) the hollow cavity and porous shell can load drugs such as cisplatin for localized therapy via synergetic chemo-/radiotherapy. Most recently, they also reported the synthesis of multi-functional $NaYF_4$:Yb/Tm@$NaGdF_4$ core/hollow mesoporous silica shell nanoparticles (UCNP@hmSiO_2) and their application in simultaneous NIR-triggered drug delivery and in situ quantitative drug release monitoring by both upconverted luminescence and longitudinal relaxation time ($T1$) MR imaging in real time [67]. The synthesis procedure is nearly the same as their former work, except that UCNP@dSiO_2@m-SiO_2 nanoparticles were used as the initial particles to conduct the "surface-protected etching" process (Fig. 4.2g–n). Then, hot water was chosen as a mild etchant to selectively remove the dense silica layer at a controllable rate, with the PVP protected porous silica shell being unaffected. Zhao et al. [68] also reported the yolk-shell upconversion nanocages of UCNPs@SiO_2 (Fig. 4.2a–f), but the method was not based on the "surface-protected etching." As following, the OA ligand on the up-conversion nanoparticles surface was removed and dispersed in the mixed solution of surfactant laurylsulfonatebetaine (LSB)/sodium dodecyl benzenesulfonate (SDBS) to form a stock solution. The stock solution was kept at 40 °C, then, triethanolamine, APS, and TEOS were successively added to the resulting solution and stirred for another 1 h. Finally, the yolk-shell upconversion nanocages were obtained. It can be seen that the procedure is easier than the multi-step "surface-protected etching" route.

Electrospinning is a cost-effective and versatile method for preparing one-dimensional (1D) materials including polymers, inorganic materials, and hybrid compounds [69, 70]. When inorganic rare earth luminescent nanoparticles are dispersed into polymer hybrid precursors, various 1D rare earth luminescent materials with multiform morphologies such as fiber, wire, belt, and tube can be achieved readily via the electrospinning route, which has potential applications in fluorescent lamps and color displays [71–73]. Lin's group has demonstrated that the electrospinning route is also an effective approach to prepare UCNP-mSiO_2 multi-functional materials, which can act as promising drug carriers for biomedical applications [74–79]. In their early work, hydrophilic $NaYF_4$:Yb/Er upconversion nanoparticles were directly dispersed in electrospinning solution containing TEOS. Then, porous

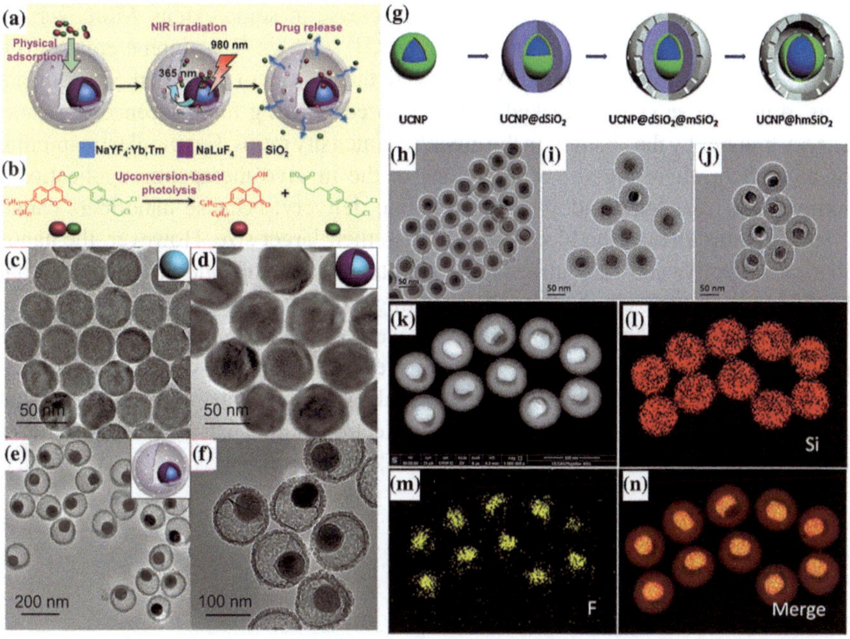

Fig. 4.2 **a**, **b** Schematic illustration of the NIR-regulated PDD and the photolysis of the pro-drug under upconversion luminescence from the yolk-shell UCNPs@$mSiO_2$ nanocomposite. TEM images of the $NaYF_4$:Yb,Tm (**c**), $NaYF_4$:Yb,Tm@$NaLuF_4$ (**d**), and yolk-shell UCNPs@$mSiO_2$ nanocomposites (**e**, **f**). **g** The formation process of the hollow structure UCNPs@$mSiO_2$ nanocomposites. TEM images of **b** UCNP@SiO_2, **c** UCNP@SiO_2@$mSiO_2$, and **d** hollow structure UCNPs@$mSiO_2$ nanocomposite. **k–n** STEM images and corresponding elemental mappings of the hollow structure UCNPs@$mSiO_2$ nanocomposites. **a–f** Reprinted with the permission from Ref. [68]. Copyright 2014 Wiley-VCH Verlag GmbH & Co. KGaA. **g–n** Reprinted with the permission from Ref. [67]. Copyright 2011 Wiley-VCH Verlag GmbH & Co. KGaA

SiO_2 fiber-decorated upconversion nanoparticles (UCNPs@SiO_2 fibers) were obtained after high temperature annealing, which served as a carrier to the loading and releasing of drugs. Moreover, by fastidiously controlling experimental conditions, UCNPs@SiO_2 tubes were successfully fabricated [77]. Most recently, they adopted a novel and ingenious architecture design to obtain a multi-functional dual drug delivery system via the electrospinning technique by combining the advantages of inorganic materials (UCNPs@$mSiO_2$) and organic materials (polymer) [76]. The working principle of their strategy is shown as follows. Firstly, DOX delivery carrier UCNPs@$mSiO_2$ nanospheres were fabricated according to a phase transfer-assisted surfactant-templating coating process. Subsequently, the as-obtained DOX-UCNPs@$mSiO_2$ nanoparticles were mixed with electrospinning solution including poly(ε-caprolactone)-gelatin and another anti-inflammatory drug, indomethacin, so as to form dual drug-loaded composite fibers, denoted as UCNPs@$mSiO_2$@fibers. The two drugs exhibited release behaviors in vitro that presented distinct release properties. Optimum drug release is a sustained and long-term process that can effectively solve the problem of drug burst release to some extent. Moreover, the upconversion luminescent intensity ratios of $^2H_{11/2}/^4S_{3/2}$–$^4I_{15/2}$ (green emission) to $^4F_{9/2}$–$^4I_{15/2}$ (red emission) from Er^{3+} vary with the amount of DOX in the system, and thus, drug release can be tracked and monitored utilizing luminescence resonance energy transfer by the changes in the green/red intensity ratios. For the electrospinning of nanofibers, there are some limitations in the in vivo therapeutic application if intravenous injection is adopted for systemic delivery, because nanofibers cannot effectively circulate in the bloodstream due to their larger size. However, the nanofibers are effective and very attractive for topical treatment of solid tumors and wound healing due to their characteristics, including extremely high-specific surface and porosity, and excellent air permeability as well as surface wettability. Nanofibrous scaffolds can maintain an appropriately moist environment for wounds by facilitating oxygen permeation and allowing fluid accumulation, effectively protecting the wound from bacterial penetration [80–83]. As such, electrospinning nanofibers are promising materials for facilitating wound healing and skin regeneration.

4.3 UCNPs-MNPs

In the past two decades, the synthetic MNPs and related nanoparticles have played an important role in many bio-application fields [84–90]. Some forms of these nanoparticles are now clinically used for improving MR contrast and have also been used in magnetic targeting drugs delivery system. Furthermore, the rapid development in the synthetic methods for preparing MNPs have allowed researchers to synthetically control the important features of these probes, such as size, magnetic dopants, magneto-crystalline phases, and surface states, resulting in the discovery of "designed" MNPs. These innovative nanoparticle probes exhibit superior magnetism and MR contrast effects which have been shown to be better than that of conventional MR contrast agents [91–94]. According to the composition of the MNPs, they can be classified into three major types: single metal oxide (iron oxide, manganese oxide, cobalt oxide, etc.), dopant MNPs [MFe_2O_4, M = Mn, Ni, Co], and metal alloy (FeCo, FePt, etc.). MNPs high-quality synthetic MNPs are typically prepared through the thermal decomposition of metal-complex precursors in hot non-hydrolytic organic solution containing surfactants. Thermal decomposition of the precursors generate monomers and their aggregation above a supersaturation level induces nucleation and subsequent nanoparticles growth [95–103]. During these stages, by tuning growth parameters, such as monomer concentration, crystalline phase of the nuclei, choice of solvent and surfactants, and surface energy growth temperature and time, it is possible to control the size, composition, and magneto-crystalline phase of MNPs.

Rapid development in nanotechnology has allowed the engineering of various functional nanomaterials with multiple discrete function-related components integrated in one nanoparticle for applications in multimodality bio-applications to circumvent the limitations of each single mode application. Conventional chemotherapy is commonly used in the clinical management of many cancers, often with severe side effects due to the cytotoxicity of the agents and the genetic heterogeneity of the tumors, that is, their phenotypic drug sensitivity or resistivity [104]. Nanoengineered drug delivery systems could improve the clinical performance of conventional chemotherapeutic agents and reduce overall toxicity by enhancing the specificity of the drug's delivery through tumor targeting and making elevated local dosages possible [105, 106]. A major goal in nanomedicine is the coherent implementation of multi-functional platforms within a single-targeted nanodelivery system that would simultaneously perform diagnosis, targeted delivery, and efficient therapy (theranostics) [16, 107–111]. Along these lines, luminescent and MNPs have been used as bio-labeling, contrast agents, and for MRI and targeted drug delivery, leading recently to major advances in bio-applications. Until recently, the multi-functional luminescent-magnetic materials have been based on quantum dots, organic molecules, and polymers [107–109]. The biological applications of these downconversion luminescent materials have at times been restricted due to their limited tissue-penetration depth, low signal-to-background ratio given the significant auto-fluorescence from biological tissues, in some cases by chemical instability, and by uncertain toxicity [112]. An alternative that avoids some of these limitations is near-infrared to UV-vis upconversion nanoparticles. Here, the luminescent materials emit visible light upon NIR-light excitation making them superior bio-imaging agents to systems employing downconversion luminescent materials. While not wholly free from tissue-penetration limits, upconversion materials absorbing at NIR wavelengths greatly improve penetration depth relative to downconversion materials, thereby also decreasing photodamage to tissue especially at high light fluxes. Upconversion materials also typically exhibit high photo and chemical stability allowing long integration times to be used and increasing signal-to-noise due to an essentially dark upconversion background [113–116]. On the other hand, to benefit from their magnetic properties, the magnetic materials can serve as MRI and targeted drug delivery agents. So, with the upconversion-magnetic hybrid nanostructured materials, we will certainly accomplish multimodality in bio-applications to circumvent the limitations of each single mode application.

In this section, we describe some useful designs of multi-functional upconversion luminescent-magnetic nanoparticles (UCNPs-MNPs), which have been attracting increased research effort because of their easily accessible multimodality. The integration of upconversion nanoparticles and MNPs forms multi-functional structures with both luminescence and magnetic properties, which lead to the demonstration of intracellular manipulation of nanoparticles and a promising candidate for dual-functional molecular imaging (i.e., combining MRI and optical imaging). Furthermore, the multi-functional nanomaterials emit visible luminescence upon NIR excitation and can be manipulated using an external magnetic field, for example, as a contrast agent for a specific target, making them attractive for a variety of biological applications. For example, we can use them as in vitro drug release assays and in vivo anti-tumor agents.

Generally, there are three kinds of strategies to fabricate multi-functional UCNPs-MNPs [117]. The first strategy is SiO_2-assisted synthesis of the multi-functional UCNPs-MNPs nanocomposites [58, 60, 118–120]. The second approach is based on combining upconversion nanoparticles and magnetic nanocrystals assisted by cross-linker [1, 121–123]. The third way is based on the seed-mediated growth method [124–126].

4.3.1 SiO_2-Assisted Synthesis of UCNPs-MNPs Multi-functional Nanocomposites

Silanization is a popular technique for surface modification of nanoparticles since silica is highly biocompatible with a surface that is easy to modify, which would allow for bio-labeling, drug targeting, and delivery. Furthermore, the SiO_2-coated functional nanoparticles can be protected by the SiO_2 shell against the influence of physiological conditions and outside environments, which will show more stable properties. So, with assistance of the advantages of SiO_2, SiO_2-assisted method is considered to be one of the most effective routes to synthesize the UCNPs-MNPs multi-functional nanocomposites. For example, Zhang et al. [58] fabricated nanorattles consisting of hydrophilic rare-earth-doped $NaYF_4$ shells each containing a loose magnetic nanoparticle through an ion-exchange process (Fig. 4.3a). The inner

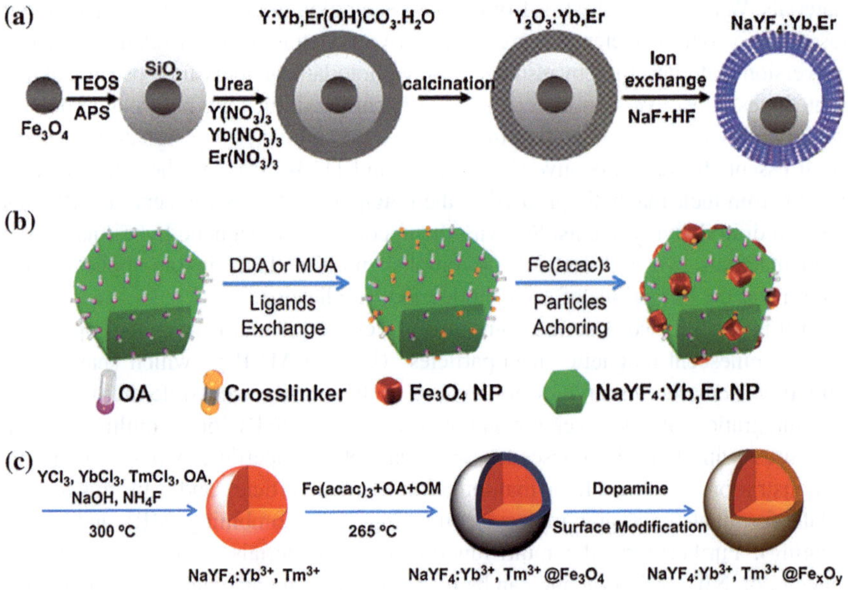

Fig. 4.3 **a** Synthetic procedure for the drug-loaded $Fe_3O_4@SiO_2@\alpha\text{-}NaYF_4$/Yb,Er nanorattles (DOX-MUC-F-NR). **b** Cross-linker anchoring strategy for bi-functional $Fe_3O_4/NaYF_4$:Yb,Er heteronanoparticles. **c** Schematic representation of the synthetic routine of the water-soluble $NaYF_4:Yb^{3+},Tm^{3+}@Fe_xO_y$ nanocrystals. **a** Reprinted with the permission from Ref. [58]. Copyright 2012 American Chemical Society. **b** Reprinted with the permission from Ref. [121]. Copyright 2010 The Royal Society of Chemistry. **c** Reprinted with the permission from Ref. [124]. Copyright 2011 Elsevier B.V

magnetic Fe_3O_4 nanoparticles are coated with a SiO_2 layer to avoid iron leaching in acidic biological environments. This multi-functional mesoporous nanostructure with both upconversion luminescent and magnetic properties has excellent water dispersibility and a high drug-loading capacity. The Fe_3O_4@SiO_2 nanoparticles were synthesized from monodispersed ~20-nm diameter Fe_3O_4 nanocrystals as a core, using a reverse-microemulsion method. The Fe_3O_4@SiO_2 nanospheres are then coated with a layer of Y/Yb, Er(OH)$CO_3 \cdot H_2O$ via a homogeneous precipitation method. Thermal treatment at high temperature transformed the amorphous Y/Yb, Er(OH)$CO_3 \cdot H_2O$ into the cubic phase of Y_2O_3. The Fe_3O_4@SiO_2@α-$NaYF_4$/Yb,Er magnetic upconversion fluoride nanorattles (MUC-F-NR) are formed via ion-exchange of MUC-O-NS in the presence of HF and NaF solution. Finally, mesoporous multi-functional upconversion luminescent and magnetic "nanorattle" was obtained. The nanorattles emit visible luminescence upon NIR excitation and have very good superparamagnetic property, making it an attractive system for a variety of biological applications. Furthermore, the BET surface area and the pore size are calculated to be ~73 m^2/g and 4.8 nm, which offer possibilities for further magnetic-targeted drug delivery. This method could also be extended to the synthesis of core/shell Fe_3O_4@$NaLuF_4$:Yb,Er/Tm nanostructure [60]. Hu et al. [120] presents a facile method for the fabrication of core-shell-structured $NaYF_4$:Yb,Er/Tm@SiO_2@Fe_3O_4 nanoparticles. In this approach, when the EDTA-capped $NaYF_4$:Yb,Er/Tm nanoparticles were mixed with the oleic acid-modified magnetic Fe_3O_4 nanoparticles in isopropyl alcohol, followed by addition of TEOS and ammonium hydroxide, core shell $NaYF_4$:Yb,Er/Tm@SiO_2@Fe_3O_4 nanoparticles were directly yielded; neither surface modification nor high temperature was needed. The obtained nanoparticles have very good superparamagnetic and luminescent properties. Liu et al. [118] have also reported a new type of monodisperse core/shell nanoparticles with both upconversion luminescence nanocrystals and superparamagnetic (SPM) nanocrystals encapsulated in silica shells (SiO_2/UC-SPM). The obtained composite nanoparticles are capable of emitting pure eye-visible luminescence upon NIR excitation and being driven by an external magnetic field to a specific target, making them attractive for a variety of biomedical applications, such as bio-imaging, drug targeting, and bio-separation. The synthesis method is based on the simple and more convenient one-pot reverse-microemulsion method. However, owing to the lack of mesoporosity, it is limited as a practical drug delivery system.

4.3.2 Cross-Linker-Assisted Synthesis of UCNPs-MNPs Multi-functional Nanocomposites

In addition to the SiO_2-assisted synthesis method, cross-linker-assisted method has also been developed [1, 121–123]. Cheng et al. [1] described a novel class of multi-functional nanocomposites based on upconversion nanoparticles with combined optical and magnetic properties. The multi-functional nanocomposites were prepared by layer-by-layer (LBL) self-assembly. Ultrasmall superparamagnetic dopamine

(DA)-modified iron oxide (Fe_3O_4) nanoparticles (IONPs) are adsorbed on the surface of poly(acrylic acid) (PAA)-modified $NaYF_4$-based UCNP by electrostatic attraction to form a UCNP-IONP complex, on top of which a thin gold shell is formed by seed-induced reduction growth. The layer of IONPs between upconversion nanoparticles and the Au shell not only affords multi-functional nanocomposites magnetic properties but also significantly reduces the luminescence quenching effects of the gold nanostructure to upconversion nanoparticles. The primary force between the PAA-modified upconversion nanoparticles and dopamine-modified magnetic particles is from the condensation reaction between the carboxyl and amino group. Shen and coworkers reported the $Fe_3O_4/NaYF_4$:Yb,Er heteronanoparticle synthesized by a novel ligand anchoring strategy (Fig. 4.3b) [121]. Fe_3O_4 nanoparticles are localized at the surface of upconversion nanoparticles assisted by cross-linkers. Compared with other chemical linkage methods for hybrid nanostructures, this cross-linker anchoring strategy comprises the ligand exchange process, crystallization of Fe_3O_4 nanoparticles, and integration into heteronanoparticles in a sequential one-pot reaction. The process is facile and effective for construction of heteronanoparticles in a solvo-thermal system. This approach is not limited by lattice mismatch for different functional components and is applicable to quantum dots and noble metal nanoparticles, etc. The heteronanoparticles well preserve the upconversion luminescence and room-temperature superparamagnetism, which enables both effective optical detection and magnetic trapping. They can be fabricated into luminescent micropatterns with magnetic manipulation. Xu et al. [123] demonstrate a facile synthetic method to prepare a novel nanocomposite for multi-model imaging and imaging combined drug delivery. They can encapsulate hydrophobic upconversion nanoparticles together with iron oxide (IO) nanoparticles by using an amphiphilic block copolymer, poly(styrene-block-allylalcohol) (PS_{16}-b-PAA_{10}), via a microemulsion method, obtaining an UC-IO@Polymer multi-functional nanocomposite system. By further introducing a fluorescent Squaraine (SQ) dye during sample preparation, the obtained UC-IO@Polymer-SQ could be used for triple-modal upconversion luminescence/downconversion optical/MR imaging in vitro and in vivo. Most recently, Chen et al. fabricated monodispersed biocompatible Yb/Er or Yb/Tm-doped β-$NaGdF_4$ upconversion phosphors using polyelectrolytes to prevent irreversible particle aggregation during conversion of the precursor, $Gd_2O(CO_3)_2 \cdot H_2O$:Yb/Er or Yb/Tm, to β-$NaGdF_4$:Yb/Er or Yb/Tm. The polyelectrolyte on the outer surface of nanophosphors also provided an amine tag for PEGylation. This method is also employed to fabricate PEGylated magnetic upconversion phosphors with Fe_3O_4 as the core and β-$NaGdF_4$ as a shell [127].

4.3.3 Seed-Mediated Growth Method for the Synthesis of UCNPs-MNPs Multi-functional Nanocomposites

When considering seed-mediated growth, the energy of the reaction plays an important role. The energy barrier for the homogeneous nucleation of a material is typically quite large; it is rare in nanoparticles systems that particles will spontaneously nucleate and grow in solution without a driving force such as a supersaturated solution, a strong reducing agent, or the use of high temperatures to overcome the energy barrier toward nucleation. Attempts to synthesize hybrid nanoparticles make use of this energy barrier or perhaps more accurately the lower energy barrier for the heterogeneous nucleation of a second phase onto a seed particle. In comparison with the methods mentioned above, the seed-mediated growth method should be simpler and show a more stable hybrid nanostructure than the cross-linker assisted and SiO_2-assisted method. On the other hand, the size of the obtained core/shell hybrid nanostructure can be easily controlled below 100 nm, which increases the biocompatibility of the materials. So, we can extend this seed-mediated growth method to the synthesis of the UCNPs-MNPs multi-functional nanocomposites [124–126].

Xia et al. [124] synthesized new core/shell-structured NaYF4:Yb^{3+},Tm^{3+}@Fe_xO_y nanoparticles, with 20-nm Yb^{3+}, Tm^{3+} codoped $NaYF_4$ nanoparticles as a core, and 5-nm Fe_xO_y nanoparticles as a shell. These core/shell nanoparticles exhibit excellent near-infrared upconversion luminescence emission at 800 nm under excitation by a continuous-wave 980 nm laser and superparamagnetic properties with a saturation magnetization of ~12 emu/g (Fig. 4.3c). Similarly, Zhong et al. [125] introduced a core@shell-structured nanoparticles composed of magnetite core and upconversion luminescent shell by using the seed-mediated growth method. Each of the components in the nanocomposites displays a unique function. Fe_3O_4 as the core endows nanoparticles with magnetic property, and the outer $NaGdF_4$:Yb/Er shell improves the luminescent efficiency. Although the composite obtained by using seed-mediated growth method gains both upconversion luminescence and magnetic properties, to date, core/shell structures based on upconversion and MNPs have not been certified from direct measurements. So, more direct evidence should be provided on the characterizations of the obtained UCNPs-MNPs multi-functional materials synthesized by this method in the future. Besides the core@shell structure, Liu et al. reported the Fe_3O_4/$NaYF_4$ heterosubmicrorods with a good bi-functional of superparamagnetic and upconversion luminescence properties were successfully fabricated via a very facile hydrothermal method. In contrast with most previous works, this method is facile to perform, with less cost and higher yield, promising to be more suitable for industrial production. Particularly, by TEM, we found that morphologies of the heterosubmicrorods were pictured as many Fe_3O_4 nanoparticles tightly attaching to the surface of $NaYF_4$ submicrorods [128].

4.3.4 Applications of UCNPs-MNPs Nanocomposites

Zijlmans et al. [129] were the first to exploit lanthanide-doped phosphors' upconverting properties to study biological recognition events. Since then, there has been substantial contribution to the development of biological labeling strategies that rely on the use of upconversion nanoparticles. Synthetic MNPs are emerging as versatile probes in biomedical applications, especially in the area of magnetic resonance imaging [130–135] and drug delivery [85, 136–140]. So,when considering the applications of the UCNPs-MNPs multi-functional nanocomposites, the strategies have been shown to be particularly attractive in multimodal imaging and magnetic-targeted drug delivery. Because the multimodal imaging is comprehensively summarized in Chap. 6, in this section, we focus on magnetic-targeted drug delivery using the UCNPs-MNPs nanocomposites.

One major problem in current chemotherapy is the drug dosage given to patients. A low dosage is ineffective in the treatment of a tumor, whereas a high dosage of chemo-therapeutics is intolerable for patients due to toxicity and side effects. Therefore, a new design for anticancer drugs is desirable to increase the local effective therapeutic concentration. Along this line, combining MNPs with the pharmaceutical carriers, conspired with upconversion luminescent nanoparticles as bio-labeling agents, the UCNPs-MNPs multi-functional nanocomposites are considered to be one of the most effective ways to increase the therapeutic effectiveness.

Most recently, our group reported a concentric nanostructure with both upconversion luminescence and magnetic properties [58]. Importantly, the nanoparticle additionally contains a hollow volume that can accommodate a drug payload. The material was used as a drug delivery system to load the anticancer drug. Cytotoxicity and cell imaging experiments demonstrate the nanorattle materials to be biocompatible and easily endocytosed by cells. Intracellular drug release was also observed. The multi-functional nanorattles show advantages for in vivo enhancement of therapy efficacy by reducing the systemic toxicity of anti-tumor drugs both through protection of the drug during circulation and using magnetic fields to target the material to the lesion. Furthermore, with such a coreshell architecture surface coatings tailored for tumor targeting are conceivable, making it possible to engineer additional functionalities into this nanosystem. In addition to the triple-modal imaging of the polymer-encapsulated UCNPs-Fe_3O_4 nanocomposites, which was developed by the Liu group, a chemotherapy drug, doxorubicin, is also loaded into the nanocomposite, forming an UC-IO@Polymer-DOX complex, which enables novel imaging-guided and magnetic-targeted drug delivery [123]. Besides the magnetic-guided chemotherapy, the UCNPs-MNPs nanocomposites can also be used in the targeted photothermal therapy. For example, Cheng et al. described a novel class of multi-functional nanocomposites based on upconversion nanoparticles with combined optical and magnetic properties useful in multimodality imaging. The multi-functional nanocomposites were prepared by layer-by-layer (LBL) self-assembly. Ultrasmall superparamagnetic iron oxide (Fe_3O_4) nanoparticles (IONPs) are adsorbed on the surface of a $NaYF_4$-based UCNP by electrostatic attraction to form an UCNP–IONP complex, on top of which a thin gold shell is formed by seed-induced reduction growth. The layer of IONPs between upconversion nanoparticles and the Au shell not only affords multi-functional nanocomposites magnetic properties but also significantly reduces the luminescence quenching effects of the gold nanostructure on the upconversion nanoparticles. The UCNP–IONP–Au multi-functional nanocomposites are then coated by PEG to impart stability in physiological solutions used for in vitro targeted upconversion luminescence, MR, and dark-field light scattering imaging under the magnetic field. The surface plasmon resonance absorption contributed by the gold shell in multi-functional nanocomposites is utilized for dual-targeted (folic acid targeted and magnetic targeted) photothermal ablation of cancer cells.

4.4 UCNPs-Metals

It has been mentioned in Chap. 3 that metal nanostructures (i.e., plasmonics substrates) can efficiently collect light and enhance the intensity due to surface plasmon resonances (SPR). These effects have been widely used for enhancing various optical processes, such as Raman scattering and downconversion luminescence [141–145]. So, in this section, we focus on the fabrication of UCNPs-metal nanocomposites, which is an important way to enhance the upconversion luminescence according to the SPR effect of metals [146–152]. Because the metal-enhanced upconversion luminescence has been summarized in the former chapter, we just focus on the fabrication or synthesis of the UCNPs-metal nanocomposites. The interaction of upconversion nanocrystals with the metallic structures has been fabricated in three ways [153]: (i) Upconversion nanocrystals of $NaYF_4$ doped with lanthanide ion pairs of Yb^{3+}/Er^{3+} or Yb^{3+}/Tm^{3+} were deposited on a metallic film of metal islands [154], dense metal nanoparticles (Au or Ag) [146, 155], or 3D plasmonic antenna [156, 157]. (ii) Metallic nanoparticles are attached or self-assembled onto the surface of lanthanide-doped fluoride, oxide, or vanadate upconversion nanoparticles [147, 148, 152, 158–161]. (iii) core/shell structures of metallic/silica/UCNPs, UCNPs/silica/metallic, metallic/UCNPs, and UCNPs/metallic [150, 151, 162, 163].

(i) **Upconversion nanocrystals of $NaYF_4$ doped with lanthanide ion pairs of Yb^{3+}/Er^{3+} or Yb^{3+}/Tm^{3+} are deposited on a metallic film of metal islands, dense metal nanoparticles (Au or Ag), or 3D plasmonic antenna**. For example, Sun et al. [164] fabricated ~30-nm β-$NaYF_4$:Yb,Er upconversion nanoparticles on a gold pyramid array (Fig. 4.4a). They used simple chemical etching techniques to pattern the silicon substrate. When the surface of a [100]-oriented silicon wafer is exposed to a solution of KOH, anisotropic etching can lead to gold pyramidal divets. We formed such patterns by coating a wafer with gold, selectively exposing the wafer with photolithography (with 2-μm periodicity circular pattern), and immersing the substrate in KOH. The divets were then coated with ~250 nm of gold film by thermal evaporation, which was removed with epoxy to produce gold pyramid arrays. The gold pyramids were smooth, highly reproducible, and exhibited sharp tips. The upconversion nanoparticles were dispersed on this gold pyramid array substrate. Similarly, Zhang et al. [156] utilized a novel 3D plasmonic architecture, the disk-coupled dots-on-pillar antenna array (D2PA), to enhance the upconversion luminescence of $NaYF_4:Yb^{3+}/Er^{3+}$ codoped nanocrystals. The D2PA structure is a 3D nanocavity array composed of Au nanodisks on top of periodic dielectric pillars, an Au backplane

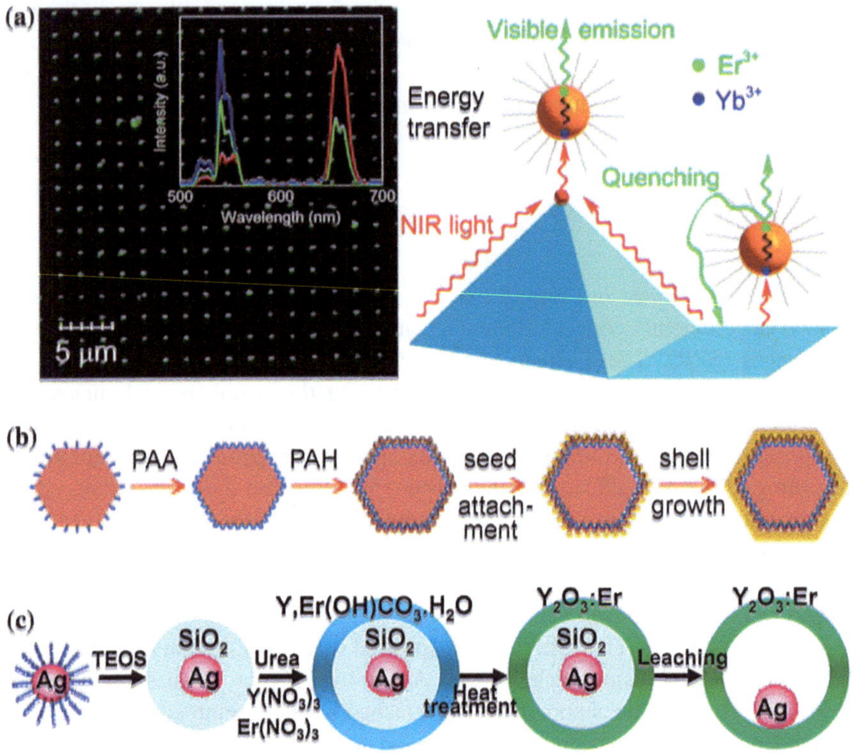

Fig. 4.4 **a** Confocal images for the *green* emission on 29-nm $NaYF_4:Yb^{3+}/Er^{3+}$ particles (*left*) (*inset* upconversion emission spectra of 29-nm β-$NaYF_4:Yb^{3+}/Er^{3+}$ nanoparticles on the gold pyramid, flat gold, and glass substrate, respectively). Schematic of the energy transfer, upconversion, and quenching processes on the *top* and *bottom* of the gold pyramid substrate (*right*). **b** Illustration of surface functionalization, gold nanocrystals attachment, and gold nanoshell growth on the upconversion nanoparticles. **c** Synthetic procedure of the $Ag@SiO_2@Y_2O_3$:Er nanocomposites. **a** Reprinted with the permission from Ref. [164]. Copyright 2014 American Chemical Society. **b** Reprinted with the permission from Ref. [147]. Copyright 2010 Wiley-VCH Verlag GmbH & Co. KGaA. **c** Reprinted with the permission from Ref. [162]. Copyright 2010 American Chemical Society

at the feet of the pillars, and dense Au nanodots on the sidewalls. The D2PA substrates were fabricated by the nanoimprint method, which allows us to mass produce nanostructures with a sub-10 nm precision at the wafer scale. In brief, SiO_2 nanopillars were first patterned with nanoimprint and reactive ion etching (RIE), and then Au nanodisks, backplane, and nanodots on the nanopillar sidewall were all formed within one step of Au evaporation. Many structural parameters that affect the resonance properties of the D2PA can be controlled and tuned precisely in the fabrication (e.g., pillar radius, height, etc.). The nanocrystal solution was diluted by 10 times in hexane and then was spin-coated on the D2PA substrate (4,000 rpm for 1 min), resulting in a submonolayer coverage. Most of the nanocrystals were attached on the sidewall of the glass pillars and Au nanodisks. Besides the above-mentioned metal array, the multilayer metal nanoparticles super lattices can also be used to fabricate the UCNPs-metal nanocomposites. For example, Saboktakin et al. [146] presented a configuration in which close-packed monolayers of upconversion nanoparticles are separated from a dense multilayer of metal nanoparticles super lattices (Au or Ag) by a nanometer-scale oxide grown by atomic layer deposition. To enhance the emission of these upconversion nanoparticles, they prepared spherical Au and Ag nanoparticles, 5 nm in diameter with standard deviation of $\sigma = 5\ \%$. To fabricate the UCNPs-spacer-metal nanoparticles structures, dense Au or Ag nanoparticles films were formed on glass substrates by spin-coating. By using solutions with identical concentrations of metal nanoparticles deposited under the same spin-coating conditions, the particle density of the films stays the same from sample to sample. A thin layer of Al_2O_3 was subsequently deposited using atomic layer deposition (ALD). This deposition method provides an accurate thickness of the dielectric spacer on the nanometer scale, which allows precise control over the distance between upconversion nanoparticles and metal nanoparticles in our structure. Next, a large-area monolayer of upconversion nanoparticles was deposited on top of the oxide layer. They use this geometry to macroscopically study the role of the spacer layer thickness on the enhancement process through steady-state photoluminescence measurements and investigate the nature of the enhancement by time-resolved photoluminescence rise and decay time experiments.

(ii) **Metallic nanoparticles are attached or self-assembled onto the surface of lanthanide-doped upconversion nanoparticles**. The surface modification of the as-synthesized upconversion nanoparticles with new ligands is implemented to make them dispersible in aqueous phase and carry positive charges [147, 148, 152, 160], which is important to attract the negatively charged metallic nanoparticles to the surface of upconversion nanoparticles in water or for in situ growth of gold nanorods [161]. For example, Duan et al. reported a new approach for the fabrication of the UCNPs-metal nanocomposites (Fig. 4.4b) [147]. They used hydrophobic upconversion nanoparticles as the initial seed. The attachment of gold nanoparticles and the growth of gold nanoshells are typically carried out in aqueous solution, therefore requiring the dispersion of the upconversion nanoparticles in water. To this end, two surface-modification steps have been performed for gold seed attachment and gold shell growth. Firstly, a ligand-exchange process was carried out by using PAA as a multidentate ligand that displaces the original hydrophobic ligands on the nanoparticle surface. The resulting PAA-coated nanoparticles are typically negatively charged. To facilitate the subsequent attachment of the negatively charged gold nanoparticles, an additional layer of poly(allylamine hydrochloride) (PAH) was coated onto the nanoparticle surface to render them positively charged. To attach the gold nanoparticles onto the upconversion nanoparticles surface, negatively charged gold nanoparticles were prepared separately and then mixed with the aqueous solution of upconversion nanoparticles in the desired ratio and allowed to age for a controlled time. Gold shell growth was then carried out by introducing additional gold precursors and reductant into the solution. Similarly, Liu et al. [152] also synthesized the Au@β-NaYF4:Yb,Tm hybrid nanostructures in two steps by using trisodium citrate as a multi-role surfactant. In the first step, we employed a hydrothermal method to synthesize $NaYF_4$:Yb, Tm nanoparticles by using trisodium citrate as a capping ligand to control the morphology. Then in the second step, trisodium citrate acted both as an Au^{3+} reducing regent and a coupling agent. During the preparation process, the added chloroauric acid was reduced by the trisodium citrate, and then, the in situ formed Au nanoparticles were attached on the surfaces of β-$NaYF_4$:Yb,Tm upconversion nanoparticles. In addition, Kannan et al. introduced gold nanorods decoration on $NaYF_4$:Yb/Tm upconversion nanoparticles by functionalizing the upconversion nanoparticles with polyamidoamine generation 1 (PAMAM G1) dendrimer, followed by a single-step seed-mediated growth of long-range gold nanorods to enhance "biological window" upconversion emission [161]. Dong et al. [165] present the synthesis of core-shell-structured hexagonal-phase $NaYF_4$:Yb^{3+},Er^{3+}@Ag nanoparticles and their unique bio-functional properties. The approach for the preparation of multi-functional $NaYF_4$:Yb^{3+},Er^{3+}@Ag nanoparticles is as following. Small, well-crystallized hexagonal $NaYF_4$:Yb^{3+},Er^{3+} nanoparticles were generated by thermal decomposition in an oleylamine system. Then, thioglycolic acid (TGA) absorbed on to the surface of upconversion nanoparticles as the ligands and water-soluble hexagonal-phase $NaYF_4$:Yb^{3+},Er^{3+} nanoparticles were then obtained. After the addition of sodium citrate solution, the surface of nanoparticles became negative and could adsorb the metal ions. A silver shell was further formed on the surface of $NaYF_4$:Yb^{3+},Er^{3+} upconversion nanoparticles through the ascorbic acid (AA) reduction. Besides the water solvable upconversion nanoparticles, the upconversion nanoparticles dispersed in organic phase can also be directly assembled with silver nanowires or gold nanoparticles synthesized in organic phase to investigate the plasmonic effect [166]. Feng et al. [166] used a solvent evaporation method to assemble $NaYF_4$:Yb,Er nanocrystals and Ag nanowires layer-by-layer. A drop of $NaYF_4$:Yb,Er hexane colloid solution was deposited on a silicon substrate to cover the whole plate by a capillary tube to form a dense layer of $NaYF_4$:Yb,Er nanocrystals. After drying it in under ambient condition to fully evaporate hexane, a drop of Ag nanowire solution was deposited to cover part of the $NaYF_4$:Yb,Er nanocrystal layer. Similarly, Ag or Au nanocrystals can be positioned at various points of the same silicon plate to form a composite assembly, which makes it easier for comparing among different areas when considering the influence of various metal nanocrystals.

(iii) **By core/shell structures of metallic/silica/UCNPs, UCNPs/silica/metallic**. The core/shell design is of particular importance for theranostic applications, as it provides a novel platform to integrate plasmonics and upconversion luminescence at the nanoscale. Moreover, the distance between the metallic nanoparticles and the upconversion nanoparticles can be precisely controlled by the thickness of the spacer silica shell. For example, we reported a facile method to fabricate metal core-enhanced Er^{3+} upconversion luminescence based on a core/

spacer/shell approach [162]. The procedure for creating $Ag@SiO_2@Y_2O_3$:Er nanostructure is illustrated as Fig. 4.4c. Firstly, water-soluble Ag nanoparticles were synthesized in the presence of polyvinylpyrrolidone by heating ethylene glycol and $AgNO_3$. Through a modified Stöber method, the Ag nanoparticles were coated with a silica layer to obtain $Ag@SiO_2$ composite. The uniform $Ag@SiO_2$ spheres were then coated with a layer of Y, Er(OH)$CO_3 \cdot H_2O$ through a homogeneous precipitation method from yttrium, erbium nitrate, and urea at 90 °C for 2 h. After heat treatment at 700 °C for 3 h, the amorphous Y:Er(OH) $CO_3 \cdot H_2O$ layer transforms into cubic phase Y_2O_3:Er. In this procedure, the size of the Ag nanoparticles, SiO_2 thickness, and the thickness of the upconversion layer can be controlled by changing the reaction parameters. Priyam et al. [151] also developed a strategy in which the $NaYF_4$:Yb,Er upconversion nanoparticles are first coated with silica and then a continuous gold nanoshell is grown via a poly(amino acid)-templated approach. The poly(amino acid), poly-L-histidine (PLH), is positively charged at neutral pH and electrostatically adsorbed onto the negatively charged silica-coated $NaYF_4$:Yb,Er upconversion nanoparticles. The amount of PLH required for complete surface coverage is optimized. The key feature of the PLH is that its histidine moieties can strongly complex with Au^{3+} ions. Thus, Au^{3+} ions are densely packed all along the poly(amino acid) chains, and subsequently, ultrathin gold nanoshells are produced.

4.5 UCNPs-Semiconductors

Over the past few years, many efforts have been made for the fabrication and control of size and shape of semiconductor nanostructures, because of their importance in catalysis, electronics, photonics, optoelectronics, sensing, and potential applications in nanodevices. One of the most important characters of semiconductors is the ability to absorb the light with a specific range of wavelengths, which is corresponding to their band gap. The absorbed light can be converted to electricity, luminescence, or heat. Combined with the unique optical properties of upconversion nanoparticles, some UCNPs-semiconductor nanocomposites have been developed to extend the potential applications of upconversion nanoparticles.

Xiao et al. [39] constructed a multi-functional core/satellite nanocomposite by decorating ultrasmall CuS semiconductor nanoparticles onto the surface of a silica-coated upconversion nanoparticles (Fig. 4.5a–e). These multi-functional nanocomposites were prepared by a simple water-in-oil reverse-microemulsion strategy and subsequent electrostatic adsorption technique between positively charged UCNP@SiO_2-NH_2 and negatively charged citrate-stabilized CuS nanoparticles. Such core/satellite nanocomposite of 45 nm in diameter on average could produce significant amounts of cytotoxic heat upon 980 nm laser excitation and simultaneously could serve as radiosensitizers to generate dose-enhancement effects of radiotherapy (RT) by the high Z elements (Yb, Gd, and Er) contained in upconversion nanoparticles, which would result in remarkable in vitro cell damage and

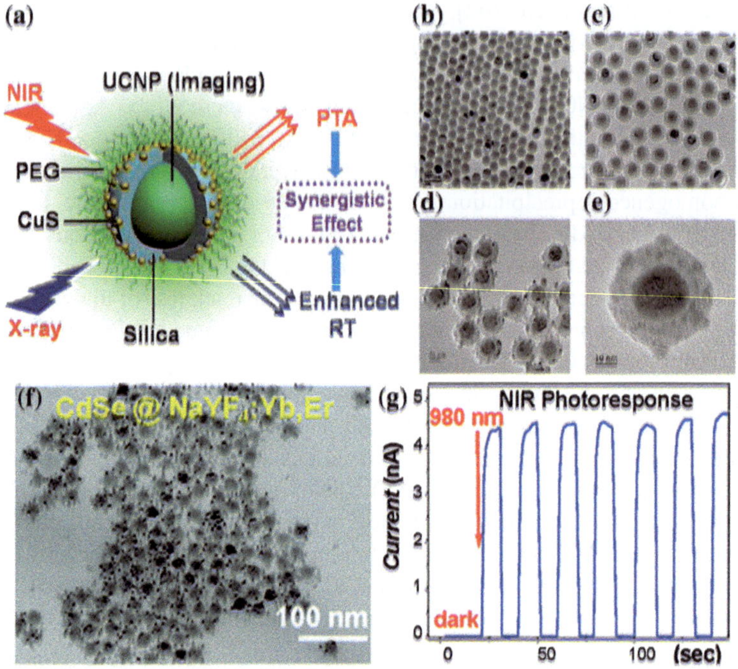

Fig. 4.5 **a** Schematic of UCNPs-CuS nanocomposites for enhanced RT/PTA synergistic therapy. **b–e** TEM images of **b** UCNPs, **c** UCNP@SiO_2–NH_2, and **d**, **e** UCNPs-CuS nanocomposites at different magnifications. **f**, **g** TEM image of the UCNPs-CdSe nanocomposites and NIR photoresponse ability of the nanocomposite. **a–e** Reprinted with the permission from Ref. [39]. Copyright 2013 American Chemical Society. **f**, **g** Reprinted with the permission from Ref. [169]. Copyright 2010 American Chemical Society

in vivo tumor regression. Complete eradication of tumor tissue without late recurrence could be achieved by the radiotherapy/photothermal ablation (RT/PTA) synergistic therapeutic effect. A simulation experiment demonstrates that this enhanced RT could assist PTA in eliminating deep-seated tumors. The in vivo systemic toxicity of nanocomposites examined by hematological and histological analysis in major organs did not show any signs of adverse effect to mice within 1 month. Furthermore, upconversion nanoparticles itself as a natural imaging nanoprobe endows these nanocomposites with several imaging modalities, such as upconversion luminescence, magnetic resonance, and computer tomography (CT), which indicates the great potential for an early cancer diagnosis and multimodal image-guided tumorous therapeutic alliance in the future.

Liu group reported a novel design, based on a combination of lanthanide-doped upconversion nanoparticles and MnO_2 semiconductor nanosheets, for rapid, selective detection of glutathione in aqueous solutions and living cells [167]. In their synthesis, the oleic acid ligand capped on the surface of core-shell upconversion nanoparticles was oxidized to afford azelaic acid by using a technique developed by Li and coworkers [168]. This modification provides hydrophilic nanoparticles without notable effects on the resulting particle shape, size, and optical property. The as-prepared hydrophilic core-shell $NaYF_4$:Yb/Tm@$NaYF_4$ nanoparticles were then used to direct the growth of MnO_2 nanosheets by addition of an aqueous $KMnO_4$ solution in the presence of 2-(*N*-morpholino)ethanesulfonic acid (MES) buffer at pH 6. $KMnO_4$ was reduced by MES to form amorphous MnO_2 nanosheets. The formation of UCNPs/MnO_2 nanosheet assemblies is dominated by electrostatic interactions.

Yan et al. [169] used a seeded-growth method to synthesize heterostructures consisting of CdSe quantum dots attached to $NaYF_4$:Yb,Er upconversion nanoparticles (Fig. 4.5f, g). Briefly, oleic acid-capped $NaYF_4$:Yb,Er upconversion nanoparticles prepared by Capobianco's method [170] were used to direct the growth of CdSe from cadmium stearate and Bu3PSe precursors in the presence of oleylamine. Use of two dissimilar ligands (oleylamine for CdSe and oleic acid for $NaYF_4$) is essential for linking the two materials. This might be due to electrostatic attraction between oppositely charged (acid and amine-capped) nanoparticles. The resulting material consisted of a $NaYF_4$:Yb,Er core decorated with CdSe quantum dots. The average sizes of the $NaYF_4$:Yb,Er and CdSe nanocrystals were estimated to be 21.6 ± 1.2 nm and 7.1 ± 0.9 nm, respectively. There is a strong attachment between the two parts of the heterostructures, these can be broken apart only by prolonged ultrasonication. Based on the new nanoheterostructure, the upconverted NIR photons are used to create excitons to generate charge carriers in CdSe-based films, leading to a reversible and stable NIR photoconductivity switch. This is the first example of the use of upconverting nanocrystals in electronic devices. The concept is potentially applicable in photovoltaics to harvest photons with subband gap energies.

Photocatalysis is an environmentally friendly technique to eliminate toxic organic substances in air and water. The process is initiated by an irradiation of semiconductors with a suitable light source. The subsequently formed free radicals with strong oxidizing ability could mineralize organic pollutants to CO_2 and H_2O. TiO_2 is considered to be an efficient photocatalyst for its appropriate electronic band structure, biological and chemical inertness, strong oxidizing power and long-term stability. However, TiO_2 possesses a wide band gap of 3.2 eV, allowing it only to be activated by UV light. Therefore, the modification of TiO_2 to reduce the band gap and make it sensitive to the longer-wavelength light is one of the most important objectives in related realms, where various methods have been proposed, such as substitutional doping elements and combining TiO_2 with organic dyes or narrow gap semiconductors quantum dots. Unfortunately, a large fraction of the NIR in sunlight remains untapped for photocatalysis. Combining the advantages of upconversion nanoparticles and the excellent photocatalysis properties of TiO_2, upconversion nanoparticles might serve as an intermediate which, after absorbing NIR light, can transfer energy to TiO_2 to generate strongly oxidative holes and reductive electrons. So, the fabrication of UCNPs-semiconductors nanocomposites is also very important for the photocatalysis. For example, Zhang et al. synthesized the lanthanide-doped $NaYF_4$@TiO_2 core-shell nanorods with highly crystalline and tunable TiO_2 shells under mild conditions without calcinations [171]. The microcomposite with a moderate TiO_2 shell shows excellent photocatalytic activity under near-infrared irradiation. Typically, $NaYF_4$:Yb,Tm microrods were first hydrothermally synthesized based on previous reports [172]. Then, the rods were washed with acid to remove the attached hydrophobic ligands. In the second step, the fluoride rods were added to a polyvinylpyrolidone solution in water and ethanol. Then, the Ti precursor was dropped into the solution under vigorous stirring. Finally, the amorphous TiO_2 was crystallized under hydrothermal treatment. By varying the ratio of fluoride: Ti, microcomposites with different shell thicknesses could be obtained. Two key participants were involved in the hydrothermal combination above: (a) TiF_4 was chosen as a substitute for alkoxides as it mildly hydrolyzes in hydrothermal conditions, (b) PVP was coated on the surface of the fluoride rods and stabilizes them in the solution, ensuring uniform nucleation of TiO_2 nanocrystals on the rods. Ji et al. also prepared the hexagonal $NaYF_4$:Yb^{3+},Er^{3+} nanoparticles supported on TiO_2 nanobelts by using two-step procedures of ion-exchangeable process and hydrothermal treatment: Layered titanate nanobelts were first ion-exchanged with Y^{3+}, Yb^{3+}, and Er^{3+} cations to produce titanate nanobelts with these cations, and then, the product nanobelts in NaF solution were treated under hydrothermal condition to transform into anatase TiO_2 nanobelts supported with β-$NaYF_4$:Yb^{3+},Er^{3+} nanoparticles [173]. Besides the TiO_2, Li et al. [174] reported that the nanocomposites of $NaYF_4$:Yb,Tm-CdS can also be used a novel near-infrared-driven photocatalyst. The synthesis process for the NIR photocatalyst is very simple. CdS, a well-known low-band gap semiconductor (about 2.5 eV), are pretreated with thioglycolic acid molecules ($SHCH_2COOH$). The SH groups are for self-stabilization, and the COOH groups functionalize the surface. The surfaces of (β-phase) $NaYF_4$:Yb,Tm microcrystals are activated by the treatment with mercaptoethanolmolecules ($SHCH_2CH_2OH$). The strong interaction between the –SH groups and lanthanide ions would leave many OH groups on the surface of the microcrystals. The subsequent reaction between mercaptoethanol and thioglycolic acid forms a stable $NaYF_4$:Yb,Tm/CdS composite.

For conventional solar cells, one of the major energy loss mechanisms is the transmission of subband gap photons. The use of upconversion materials provides a solution to the transmission loss by converting two subband gap photons into one above-band gap photon. Indeed, upconversion nanoparticles have been coupled to a wide range of solar cells including crystalline Si [175], amorphous Si thin-film [176], organic [177], and dye-sensitized solar cells (DSSCs) [178]. So, the combinations between the upconversion nanoparticles and semiconductor are also very important for the solar energy applications, which will be further illustrated in Chap. 11. In this section, we just focus on the fabrication methods of the UCNPs-semiconductor composites in the solar cells. Su et al. [179] reported a rational design and fabrication of interconnected porous network of three-dimensional heteronanostructured photoanodes for enhanced near-infrared light harvesting. In a typical experiment, monodispersed polystyrene microspheres of $\sim$510 nm in diameter were self-assembled onto fluorine-doped tin oxide substrates to form sacrificial opal templates by an evaporation-induced convective deposition method. These polystyrene opal photonic crystals form a face-centered cubic lattice structure with a hierarchical long-range order. Subsequently, TiO_2 infiltrated through the opal templates by an atomic layer deposition method. The TiO_2-infiltrated opal templates were then subjected to pyrolysis at 450 °C to thermally remove the polystyrene spheres, leaving behind the TiO_2 inverse opal photonic crystals with an anatase phase. The photonic crystals show wide pore interconnectivity and hierarchically ordered periodicity. These are important attributes for the subsequent infiltration of upconversion nanoparticles, sensitization of quantum dots, and penetration of electrolyte throughout the entire photoanode. The infiltration of the nanoparticles was made possible through the capillary effect of the porous network under high vacuum, which allowed homogeneous distribution of the nanoparticles on the surface of the inverse opal. Prior to the final coating of quantum dots sensitizers, a thin layer of TiO_2 was deposited onto the nanoparticles by an atomic layer deposition technique. After the deposition, an annealing step was required to crystallize TiO_2 but the temperature must be controlled to prevent the transformation of upconversion nanoparticles from hexagonal to cubic phase. Besides the UCNPs-TiO_2 photonic crystals, β-$NaYF_4$:Er^{3+},Yb^{3+}@SiO_2@TiO_2 hexagonal submicroprisms and β-$NaYF_4$:Yb^{3+},Er^{3+}/SiO_2/TiO_2 submicroplates were also synthesized by Liang et al. [180, 181]. Firstly, β-$NaYF_4$:Yb^{3+},Er^{3+} plates or prisms were synthesized by hydrothermal treatment. Then, the SiO_2 and TiO_2 shells were coated on the surface of the plates or prisms by using TEOS and tetrabutyl titanate as precursors under base condition and then the mixture was hydrothermally treated. Polyvinyl-pyrrolidone-K30 was used as the active agent in both SiO_2 and TiO_2 coating processes. Finally, the obtained double shell β-$NaYF_4$:Yb^{3+},Er^{3+}/SiO_2/TiO_2 plates were mixed with ethanol, water, and acetylacetone to prepare a paste and used for the fabrication of the photoanode.

Graphene has recently attracted dramatic interest from researchers working in physics, chemistry, biology, materials, and modeling, due to its superior mechanical, optical, and electronic properties, as well as its excellent controllability and manipulability compared to its analogs [182, 183]. So, besides the above-mentioned semiconductor materials, graphene can also be used to combine with the upconversion nanoparticles. And the obtained UCNPs-graphene nanocomposites can exhibit some new excellent properties in many application fields. For example, a new nanocomposite consisting of upconversion material (YF_3:Yb^{3+},Tm^{3+}), TiO_2 (P25), and graphene has been prepared by Ren et al. [184] and shown to be an advanced sunlight-activated photocatalyst. During the facile hydrothermal method, the reduction of graphene oxide and loading of YF_3:Yb^{3+},Tm^{3+} and P25 were achieved simultaneously, and the functionalities of each part were integrated together. Li et al. also reported the $NaYF_4$:Er^{3+}/Yb^{3+}-graphene composites prepared by growing $NaYF_4$:E^{3+}/Yb^{3+} nanoparticles in interlayers of expanded graphite, accompanied with the simultaneous exfoliation of expanded graphite under polyvinylpyrrolidone-assisted hydrothermal conditions. The obtained $NaYF_4$:Er^{3+}/Yb^{3+}-graphene composites can emit bright green upconversion emissions under 980-nm excitation. In addition, the $NaYF_4$:Er^{3+}/Yb^{3+}-graphene composites were chosen to design TiO_2-$NaYF_4$:Er^{3+}/Yb^{3+}-graphene composite photoanodes, which have the ability to improve the efficiencies of the solar cells. Furthermore, the UCNPs-graphene nanocomposites can also be fabricated aimed to some bio-applications. For example, Liu et al. constructed a highly sensitive bio-sensing platform by using UCNPs-graphene nanocomposites. In their synthesis, the surfaces of the upconversion nanoparticles are modified with carboxylic acid groups by using PAA [185]. Amine-containing single-stranded DNA (ssDNA) is covalently bound to the carboxylic acid groups on the upconversion nanoparticles surface with EDC/NHS-assisted standard procedures. Then, the ssDNA-UCNPs can be brought in close proximity to the graphene oxide surface through the strong non-covalent π–π stacking interaction between ssDNA and graphene oxide. Since there is a good overlap between luminescence emission of upconversion nanoparticles and absorption spectrum of graphene oxide (GO), the highly efficient luminescence quenching of ssDNA-UCNPs can be considered as a direct consequence of efficient FRET between upconversion nanoparticles and GO. Most recently, Wang et al. [186] reported a multi-functional nanoplatform, which is developed by covalently grafting core@shell-structured upconversion nanoparticles with nanographene oxide (NGO) via PEG, and then loading phthalocyanine (ZnPc) on the surface of NGO. The obtained UCNPs-NGO/ZnPc nanocomposites are not only used as upconversion luminescence imaging probes of cells and whole-body animals with high contrast for diagnosis, but also can generate cytotoxic singlet oxygen under light excitation for photodynamic therapy (PDT), as well as rapidly and efficiently convert the 808 nm laser energy into thermal energy for photothermal therapy. In their synthesis, the upconversion nanoparticles were converted into hydrophilic UCNPs-NH_2 via a ligand exchanging process using poly(allylamine) (PAAm) as surface coating agent to replace the original oleic acid ligands. In order to endow the obtained upconversion nanoparticles with multimodal therapeutic functions, NGO as both PTT agent and carriers for PS was covalently grafted with upconversion nanoparticles via a carbodiimide cross-linking reaction between the amino group of UCNPs-NH_2 and the carboxyl group of NGO-PEG. NGO-PEG was obtained by covalently grafting amino carboxylic PEG onto NGO sheet, and the left carboxylic groups at the PEG terminals were available for fluorescent probes. Yan et al. [187] also fabricated the biocompatible and flexible GO/UCNPs hybrid film for optical pH sensing. The nanocomposites films is prepared by mixing GO with polyethylenimine (PEI)-functionalized upconversion nanoparticles and then vacuum filtrating the resulting hybrid dispersion. Due to the plenty of oxygen-containing groups, the GO nanosheets process negative charge when dispersed in water over a wide range of pH values. Thus, positively charged PEI-$NaYF_4$:Yb,Er nanoparticles can be readily loaded onto GO nanosheets via electrostatic interaction.

4.6 Summary and Perspectives

In this chapter, we have described recent developments in the synthesis and applications of upconversion nanoparticles-based multi-functional nanoparticles. This research continues to be a vibrant and growing interdisciplinary field. In the near future, there are a number of areas of research that are particularly promising, but concerted efforts are still required for success: (1) The reported strategies to synthesize the upconversion nanoparticles-based nanocomposites are often accompanied by a reduction of luminescent efficiency due to unwanted quenching between different domains of the nanocomposites. So, in future research, attention must be paid to the synthesis of high-efficiency upconversion luminescence multi-functional nanocomposites. (2) Despite the fact that some successful results have been obtained, the development of multi-functional nanoprobes for multimodal imaging is still in its early stages. It is still a major challenge to design a multi-functional nanoprobe for simultaneous bio-imaging covering optical imaging, MRI, CT, and other kinds of monitoring. In addition, active-targeted multimodal imaging and drug delivery system still needs to be developed to realize the disease detection and therapy simultaneously. (3) The brightness of upconversion nanoparticles is still unsatisfactory (especially in the application of solar cells), in particular, under a low excitation density of 10^{-2} to 10^{-1} W/cm^{-2}. On the other hand, most of the absorption bands of the upconversion nanoparticles are located at around 980 ± 50 nm, which is too narrow for the solar cell applications.

References

1. Cheng, L., Yang, K., Li, Y., Chen, J., Wang, C., Shao, M., Lee, S.-T., Liu, Z.: Facile preparation of multifunctional upconversion nanoprobes for multimodal imaging and dual-targeted photothermal therapy. Angew. Chem. Int. Ed. **50**, 7385–7390 (2011)
2. Zhang, R., Wu, C., Tong, L., Tang, B., Xu, Q.-H.: Multifunctional core-shell nanoparticles as highly efficient imaging and photosensitizing agents. Langmuir **25**, 10153–10158 (2009)
3. Deng, Y., Cai, Y., Sun, Z., Liu, J., Liu, C., Wei, J., Li, W., Liu, C., Wang, Y., Zhao, D.: Multifunctional mesoporous composite microspheres with well-designed nanostructure: a highly integrated catalyst system. J. Am. Chem. Soc. **132**, 8466–8473 (2010)
4. Rolison, D.R., Long, J.W., Lytle, J.C., Fischer, A.E., Rhodes, C.P., McEvoy, T.M., Bourga, M.E., Lubers, A.M.: Multifunctional 3D nanoarchitectures for energy storage and conversion. Chem. Soc. Rev. **38**, 226–252 (2009)
5. Wang, H., Sun, L., Li, Y., Fei, X., Sun, M., Zhang, C., Li, Y., Yang, Q.: Layer-by-layer assembled Fe_3O_4@C@CdTe core/shell microspheres as separable luminescent probe for sensitive sensing of Cu^{2+} Ions. Langmuir **27**, 11609–11615 (2011)
6. Liu, L., Li, B., Qin, R., Zhao, H., Ren, X., Su, Z.: Synthesis and characterization of new bifunctional nanocomposites possessing upconversion and oxygen-sensing properties. Nanotechnology **21**, (2010)
7. Li, M.-J., Chen, Z., Yam, V.W.-W., Zu, Y.: Multi functional ruthenium(II) polypyridine complex-based core-shell magnetic silica nanocomposites: magnetism, luminescence, and electrochemiluminescence. ACS Nano **2**, 905–912 (2008)
8. Shi, W.L., Zeng, H., Sahoo, Y., Ohulchanskyy, T.Y., Ding, Y., Wang, Z.L., Swihart, M., Prasad, P.N.: A general approach to binary and ternary hybrid nanocrystals. Nano Lett. **6**, 875–881 (2006)
9. Gu, H.W., Zheng, R.K., Zhang, X.X., Xu, B.: Facile one-pot synthesis of bifunctional heterodimers of nanoparticles: a conjugate of quantum dot and magnetic nanoparticles. J. Am. Chem. Soc. **126**, 5664–5665 (2004)

10. Yu, H., Chen, M., Rice, P.M., Wang, S.X., White, R.L., Sun, S.H.: Dumbbell-like bifunctional Au-Fe_3O_4 nanoparticles. Nano Lett. **5**, 379–382 (2005)
11. Khon, E., Mereshchenko, A., Tarnovsky, A.N., Acharya, K., Klinkova, A., Hewa-Kasakarage, N.N., Nemitz, I., Zamkov, M.: Suppression of the plasmon resonance in Au/CdS colloidal nanocomposites. Nano Lett. **11**, 1792–1799 (2011)
12. Lopez-Lorente, A.I., Simonet, B.M., Valcarcel, M.: Analytical potential of hybrid nanoparticles. Anal. Bioanal. Chem. **399**, 43–54 (2011)
13. Sitt, A., Salant, A., Menagen, G., Banin, U.: Highly emissive nano rod-in-rod heterostructures with strong linear polarization. Nano Lett. **11**, 2054–2060 (2011)
14. Shevchenko, E.V., Bodnarchuk, M.I., Kovalenko, M.V., Talapin, D.V., Smith, R.K., Aloni, S., Heiss, W., Alivisatos, A.P.: Gold/iron oxide core/hollow-shell nanoparticles. Adv. Mater. **20**, 4323–4329 (2008)
15. Cozzoli, P.D., Pellegrino, T., Manna, L.: Synthesis, properties and perspectives of hybrid nanocrystal structures. Chem. Soc. Rev. **35**, 1195–1208 (2006)
16. Insin, N., Tracy, J.B., Lee, H., Zimmer, J.P., Westervelt, R.M., Bawendi, M.G.: Incorporation of iron oxide nanoparticles and quantum dots into silica microspheres. ACS Nano **2**, 197–202 (2008)
17. Li, X., Zhang, F., Zhao, D.: Lab on upconversion nanoparticles: optical properties and applications engineering via designed nanostructure. Chem. Soc. Rev. doi: 10.1039/C4CS00163J (2014)
18. Yang, D., Ma, P.a., Hou, Z., Cheng, Z., Li, C., Lin, J.: Current advances in lanthanide ion (Ln^{3+})-based upconversion nanomaterials for drug delivery. Chem. Soc. Rev. doi: 10.1039/C4CS00155A (2014)
19. Stober, W., Fink, A., Bohn, E.: Controlled growth of monodisperse silica spheres in the micron size range. J. Colloid Interface Sci. **26**, 62–69 (1968)
20. Sivakumar, S., Diamente, P.R., van Veggel, F.C.J.M.: Silica-coated Ln^{3+}-doped LaF_3 nanoparticles as robust down- and upconverting biolabels. Chem. Eur. J. **12**, 5878–5884 (2006)
21. Li, Z., Zhang, Y.: Monodisperse silica-coated polyvinylpyrrolidone/$NaYF_4$ nanocrystals with multicolor upconversion fluorescence emission. Angew. Chem. Int. Ed. **45**, 7732–7735 (2006)
22. Yang, J., Deng, Y., Wu, Q., Zhou, J., Bao, H., Li, Q., Zhang, F., Li, F., Tu, B., Zhao, D.: Mesoporous silica encapsulating upconversion luminescence rare-earth fluoride nanorods for secondary excitation. Langmuir **26**, 8850–8856 (2010)
23. Johnson, N.J.J., Sangeetha, N.M., Boyer, J.-C., van Veggel, F.C.J.M.: Facile ligand-exchange with polyvinylpyrrolidone and subsequent silica coating of hydrophobic upconverting beta-$NaYF_4$:Yb^{3+}/Er^{3+} nanoparticles. Nanoscale **2**, 771–777 (2010)
24. Li, Z., Zhang, Y., Shuter, B., Idris, N.M.: Hybrid lanthanide nanoparticles with paramagnetic shell coated on upconversion fluorescent nanocrystals. Langmuir **25**, 12015–12018 (2009)
25. Das, G.K., Heng, B.C., Ng, S.-C., White, T., Loo, J.S.C., D'Silva, L., Padmanabhan, P., Bhakoo, K.K., Selvan, S.T., Tan, T.T.Y.: Gadolinium oxide ultranarrow nanorods as multimodal contrast agents for optical and magnetic resonance imaging. Langmuir **26**, 8959–8965 (2010)
26. Hu, H., Xiong, L., Zhou, J., Li, F., Cao, T., Huang, C.: Multimodal-luminescence core-shell nanocomposites for targeted imaging of tumor cells. Chem. Eur. J. **15**, 3577–3584 (2009)
27. Rantanen, T., Jarvenpaa, M.-L., Vuojola, J., Arppe, R., Kuningas, K., Soukka, T.: Upconverting phosphors in a dual-parameter LRET-based hybridization assay. Analyst **134**, 1713–1716 (2009)
28. Li, Z., Guo, H., Qian, H., Hu, Y.: Facile microemulsion route to coat carbonized glucose on upconversion nanocrystals as high luminescence and biocompatible cell-imaging probes. Nanotechnology **21**, (2010)
29. Erathodiyil, N., Ying, J.Y.: Functionalization of inorganic nanoparticles for bioimaging applications. Accounts. Chem. Res. **44**, 925–935 (2011)

30. Guerrero-Martinez, A., Perez-Juste, J., Liz-Marzan, L.M.: Recent progress on silica coating of nanoparticles and related nanomaterials. Adv. Mater. **22**, 1182–1195 (2010)
31. Osseo-Asare, K., Arriagada, F.J.: Growth kinetics of nanosize silica in a nonionic water-in-oil microemulsion: a reverse micellar pseudophase reaction model. J. Colloid Interface Sci. **218**, 68–76 (1999)
32. Li, F., Li, C., Liu, X., Chen, Y., Bai, T., Wang, L., Shi, Z., Feng, S.: Hydrophilic, upconverting, multicolor, lanthanide-doped $NaGdF_4$ nanocrystals as potential multifunctional bioprobes. Chem. Eur. J. **18**, 11641–11646 (2012)
33. Liu, B., Li, C., Yang, D., Hou, Z., Pa, Ma., Cheng, Z., Lian, H., Huang, S., Lin, J.: Upconversion-luminescent core/mesoporous-silica-shell-structured beta-$NaYF_4:Yb^{3+}$, Er^{3+} @SiO_2@mSiO_2 composite nanospheres: fabrication and drug-storage/release properties. Eur. J. Inorg. Chem. **2014**, 1906–1913 (2014)
34. Bharali, D.J., Klejbor, I., Stachowiak, E.K., Dutta, P., Roy, I., Kaur, N., Bergey, E.J., Prasad, P.N., Stachowiak, M.K.: Organically modified silica nanoparticles: a nonviral vector for in vivo gene delivery and expression in the brain. Proc. Natl. Acad. Sci. **102**, 11539–11544 (2005)
35. Jiang, S., Zhang, Y., Lim, K.M., Sim, E.K.W., Ye, L.: NIR-to-visible upconversion nanoparticles for fluorescent labeling and targeted delivery of siRNA. Nanotechnology **20**, (2009)
36. Jiang, S., Zhang, Y.: Upconversion nanoparticle-based FRET system for study of siRNA in live cells. Langmuir **26**, 6689–6694 (2010)
37. Chien, Y.-H., Chou, Y.-L., Wang, S.-W., Hung, S.-T., Liau, M.-C., Chao, Y.-J., Su, C.-H., Yeh, C.-S.: Near-infrared light photocontrolled targeting, bioimaging, and chemotherapy with caged upconversion nanoparticles in vitro and in vivo. ACS Nano **7**, 8516–8528 (2013)
38. Zhai, Y., Zhu, C., Ren, J., Wang, E., Dong, S.: Multifunctional polyoxometalates-modified upconversion nanoparticles: integration of electrochromic devices and antioxidants detection. Chem. Commun. **49**, 2400–2402 (2013)
39. Xiao, Q., Zheng, X., Bu, W., Ge, W., Zhang, S., Chen, F., Xing, H., Ren, Q., Fan, W., Zhao, K., Hua, Y., Shi, J.: A core/satellite multifunctional nanotheranostic for in vivo imaging and tumor eradication by radiation/photothermal synergistic therapy. J. Am. Chem. Soc. **135**, 13041–13048 (2013)
40. Xing, H., Bu, W., Zhang, S., Zheng, X., Li, M., Chen, F., He, Q., Zhou, L., Peng, W., Hua, Y., Shi, J.: Multifunctional nanoprobes for upconversion fluorescence MR and CT trimodal imaging. Biomaterials **33**, 1079–1089 (2012)
41. Qian, H.S., Guo, H.C., Ho, P.C.-L., Mahendran, R., Zhang, Y.: Mesoporous-silica-coated up-conversion fluorescent nanoparticles for photodynamic therapy. Small **5**, 2285–2290 (2009)
42. Garcia, J.V., Yang, J., Shen, D., Yao, C., Li, X., Wang, R., Stucky, G.D., Zhao, D., Ford, P. C., Zhang, F.: NIR-triggered release of caged nitric oxide using upconverting nanostructured materials. Small **8**, 3800–3805 (2012)
43. Jayakumar, M.K.G., Idris, N.M., Zhang, Y.: Remote activation of biomolecules in deep tissues using near-infrared-to-UV upconversion nanotransducers. Proc. Natl. Acad. Sci. **109**, 8483–8488 (2012)
44. Liu, J., Bu, W., Zhang, S., Chen, F., Xing, H., Pan, L., Zhou, L., Peng, W., Shi, J.: Controlled synthesis of uniform and monodisperse upconversion core/mesoporous silica shell nanocomposites for bimodal imaging. Chem. Eur. J. **18**, 2335–2341 (2012)
45. Li, C., Yang, D., Pa, Ma., Chen, Y., Wu, Y., Hou, Z., Dai, Y., Zhao, J., Sui, C., Lin, J.: Multifunctional upconversion mesoporous silica nanostructures for dual modal imaging and in vivo drug delivery. Small **9**, 4150–4159 (2013)
46. Liu, J., Bu, W., Pan, L., Shi, J.: NIR-triggered anticancer drug delivery by upconverting nanoparticles with integrated azobenzene-modified mesoporous silica. Angew. Chem. Int. Ed. **52**, 4375–4379 (2013)

47. Gai, S., Yang, P., Li, C., Wang, W., Dai, Y., Niu, N., Lin, J.: Synthesis of magnetic, up-conversion luminescent, and mesoporous core-shell-structured nanocomposites as drug carriers. Adv. Funct. Mater. **20**, 1166–1172 (2010)
48. Xu, Z., Li, C., Pa, Ma., Hou, Z., Yang, D., Kang, X., Lin, J.: Facile synthesis of an up-conversion luminescent and mesoporous Gd_2O_3:Er^{3+}@nSiO_2@mSiO_2 nanocomposite as a drug carrier. Nanoscale **3**, 661–667 (2011)
49. Liu, Z., Sun, L., Li, F., Liu, Q., Shi, L., Zhang, D., Yuan, S., Liu, T., Qiu, Y.: One-pot self-assembly of multifunctional mesoporous nanoprobes with magnetic nanoparticles and hydrophobic upconversion nanocrystals. J. Mater. Chem. **21**, 17615–17618 (2011)
50. Kim, J., Lee, J.E., Lee, J., Yu, J.H., Kim, B.C., An, K., Hwang, Y., Shin, C.H., Park, J.G., Hyeon, T.: Magnetic fluorescent delivery vehicle using uniform mesoporous silica spheres embedded with monodisperse magnetic and semiconductor nanocrystals. J. Am. Chem. Soc. **128**, 688–689 (2006)
51. Lee, J.E., Lee, N., Kim, T., Kim, J., Hyeon, T.: Multifunctional mesoporous silica nanocomposite nanoparticles for theranostic applications. Accounts. Chem. Res. **44**, 893–902 (2011)
52. Li, C., Liu, J., Alonso, S., Li, F., Zhang, Y.: Upconversion nanoparticles for sensitive and in-depth detection of Cu^{2+} ions. Nanoscale **4**, 6065–6071 (2012)
53. Li, C., Hou, Z., Dai, Y., Yang, D., Cheng, Z., Ma, Pa, Lin, J.: A facile fabrication of upconversion luminescent and mesoporous core- shell structured beta-$NaYF_4$:Yb^{3+},Er^{3+} @mSiO_2 nanocomposite spheres for anti-cancer drug delivery and cell imaging. Biomaterials. Science **1**, 213–223 (2013)
54. Sun, L., Liu, T., Qiu, Y., Liu, J., Shi, L., Wolfbeis, O.S.: Direct formation of mesoporous upconverting core-shell nanoparticles for bioimaging of living cells. Microchim. Acta **181**, 775–781 (2014)
55. Li, L., Tang, F., Liu, H., Liu, T., Hao, N., Chen, D., Teng, X., He, J.: In vivo delivery of silica nanorattle encapsulated docetaxel for liver cancer therapy with low toxicity and high efficacy. ACS Nano **4**, 6874–6882 (2010)
56. Li, L., Liu, C., Zhang, L., Wang, T., Yu, H., Wang, C., Su, Z.: Multifunctional magnetic-fluorescent eccentric-(concentric-Fe_3O_4@SiO_2)@polyacrylic acid core-shell nanocomposites for cell imaging and pH-responsive drug delivery. Nanoscale **5**, 2249–2253 (2013)
57. Chen, L., Li, L., Zhang, L., Xing, S., Wang, T., Wang, Y.A., Wang, C., Su, Z.: Designed fabrication of unique eccentric mesoporous silica nanocluster-based core-shell nanostructures for pH-responsive drug delivery. ACS Appl. Mater. Interfaces **5**, 7282–7290 (2013)
58. Zhang, F., Braun, G.B., Pallaoro, A., Zhang, Y., Shi, Y., Cui, D., Moskovits, M., Zhao, D., Stucky, G.D.: Mesoporous multifunctional upconversion luminescent and magnetic "nanorattle" materials for targeted chemotherapy. Nano Lett. **12**, 61–67 (2012)
59. Zhang, L., Wang, T., Yang, L., Liu, C., Wang, C., Liu, H., Wang, Y.A., Su, Z.: General route to multifunctional uniform yolk/mesoporous silica shell nanocapsules: a platform for simultaneous cancer-targeted imaging and magnetically guided drug delivery. Chem. A Eur. J. **18**, 12512–12521 (2012)
60. Zhu, X., Zhou, J., Chen, M., Shi, M., Feng, W., Li, F.: Core–shell Fe_3O_4@$NaLuF_4$:Yb,Er/Tm nanostructure for MRI, CT and upconversion luminescence tri-modality imaging. Biomaterials **33**, 4618–4627 (2012)
61. Fan, W., Shen, B., Bu, W., Chen, F., Zhao, K., Zhang, S., Zhou, L., Peng, W., Xiao, Q., Xing, H., Liu, J., Ni, D., He, Q., Shi, J.: Rattle-structured multifunctional nanotheranostics for synergetic chemo-/radiotherapy and simultaneous magnetic/luminescent dual-mode imaging. J. Am. Chem. Soc. **135**, 6494–6503 (2013)
62. Zhu, Y., Ikoma, T., Hanagata, N., Kaskel, S.: Rattle-type Fe_3O_4@SiO_2 hollow mesoporous spheres as carriers for drug delivery. Small **6**, 471–478 (2010)
63. Kang, X., Cheng, Z., Yang, D., Ma, Pa, Shang, M., Peng, C., Dai, Y., Lin, J.: Design and synthesis of multifunctional drug carriers based on luminescent rattle-type mesoporous silica microspheres with a thermosensitive hydrogel as a controlled switch. Adv. Funct. Mater. **22**, 1470–1481 (2012)

64. Tang, F., Li, L., Chen, D.: Mesoporous silica nanoparticles: synthesis biocompatibility and drug delivery. Adv. Mater. **24**, 1504–1534 (2012)
65. Zhang, Q., Zhang, T., Ge, J., Yin, Y.: Permeable silica shell through surface-protected etching. Nano Lett. **8**, 2867–2871 (2008)
66. Fang, X., Zhao, X., Fang, W., Chen, C., Zheng, N.: Self-templating synthesis of hollow mesoporous silica and their applications in catalysis and drug delivery. Nanoscale **5**, 2205–2218 (2013)
67. Fischer, L.H., Harms, G.S., Wolfbeis, O.S.: Upconverting nanoparticles for nanoscale thermometry. Angew. Chem. Int. Ed. **50**, 4546–4551 (2011)
68. Zhao, L., Peng, J., Huang, Q., Li, C., Chen, M., Sun, Y., Lin, Q., Zhu, L., Li, F.: Near-infrared photoregulated drug release in living tumor tissue via yolk-shell upconversion nanocages. Adv. Funct. Mater. **24**, 363–371 (2014)
69. Wang, P., Wang, Y., Tong, L.: Functionalized polymer nanofibers: a versatile platform for manipulating light at the nanoscale. Light Sci. Appl. **2**, e102 (2013)
70. Ma, Q., Wang, J., Dong, X., Yu, W., Liu, G.: Electrospinning fabrication and characterization of magnetic-upconversion fluorescent bifunctional core–shell nanofibers. J. Nanopart. Res. **16**, 1–10 (2014)
71. Li, G., Hou, Z., Peng, C., Wang, W., Cheng, Z., Li, C., Lian, H., Lin, J.: Electrospinning derived one-dimensional LaOCl:Ln^{3+} (Ln = Eu/Sm,Tb,Tm) nanofibers, nanotubes and microbelts with multicolor-tunable emission properties. Adv. Funct. Mater. **20**, 3446–3456 (2010)
72. Dong, B., Song, H., Yu, H., Zhang, H., Qin, R., Bai, X., Pan, G., Lu, S., Wang, F., Fan, L., Dai, Q.: Upconversion properties, of Ln^{3+} doped $NaYF_4$/polymer composite fibers prepared by electrospinning. J. Phys. Chem. C **112**, 1435–1440 (2008)
73. Hou, Z., Yang, P., Li, C., Wang, L., Lian, H., Quan, Z., Lin, J.: Preparation and luminescence properties of YVO_4:Ln and Y(V, P)O_4:Ln (Ln = Eu^{3+}, Sm^{3+}, Dy^{3+}) nanofibers and microbelts by sol-gel/electrospinning process. Chem. Mater. **20**, 6686–6696 (2008)
74. Hou, Z., Li, C., Ma, P., Li, G., Cheng, Z., Peng, C., Yang, D., Yang, P., Lin, J.: Electrospinning preparation and drug-delivery properties of an up-conversion luminescent porous $NaYF_4$:Yb^{3+}, Er^{3+}@silica fiber nanocomposite. Adv. Funct. Mater. **21**, 2356–2365 (2011)
75. Liu, M., Liu, H., Sun, S., Li, X., Zhou, Y., Hou, Z., Lin, J.: Multifunctional hydroxyapatite/Na(Y/Gd)F_4:Yb^{3+}, Er^{3+} composite fibers for drug delivery and dual modal imaging. Langmuir **30**, 1176–1182 (2014)
76. Hou, Z., Li, X., Li, C., Dai, Y., Ma, Pa, Zhang, X., Kang, X., Cheng, Z., Lin, J.: Electrospun upconversion composite fibers as dual drugs delivery system with individual release properties. Langmuir **29**, 9473–9482 (2013)
77. Li, X., Hou, Z., Ma, Pa, Zhang, X., Li, C., Cheng, Z., Dai, Y., Lian, J., Lin, J.: Multifunctional $NaYF_4$:Yb/Er/Gd nanocrystal decorated SiO_2 nanotubes for anti-cancer drug delivery and dual modal imaging. RSC Adv. **3**, 8517–8526 (2013)
78. Hou, Z., Li, C., Ma, Pa, Cheng, Z., Li, X., Zhang, X., Dai, Y., Yang, D., Lian, H., Lin, J.: Up-conversion luminescent and porous $NaYF_4$:Yb^{3+},Er^{3+}@SiO_2 nanocomposite fibers for anti-cancer drug delivery and cell imaging. Adv. Funct. Mater. **22**, 2713–2722 (2012)
79. Hou, Z., Zhang, C., Li, C., Xu, Z., Cheng, Z., Li, G., Wang, W., Peng, C., Lin, J.: Luminescent porous silica fibers as drug carriers. Chem. A Eur. J. **16**, 14513–14519 (2010)
80. Nguyen, T., Ghosh, C., Hwang, S.-G., Tran, L., Park, J.: Characteristics of curcumin-loaded poly (lactic acid) nanofibers for wound healing. J. Mater. Sci. **48**, 7125–7133 (2013)
81. Choi, J.S., Leong, K.W., Yoo, H.S.: In vivo wound healing of diabetic ulcers using electrospun nanofibers immobilized with human epidermal growth factor (EGF). Biomaterials **29**, 587–596 (2008)
82. Bölgen, N., Vargel, İ., Korkusuz, P., Menceloğlu, Y.Z., Pişkin, E.: In vivo performance of antibiotic embedded electrospun PCL membranes for prevention of abdominal adhesions. J. Biomed. Mater. Res. B Appl. Biomater. **81B**, 530–543 (2007)

83. Chen, Z.G., Wang, P.W., Wei, B., Mo, X.M., Cui, F.Z.: Electrospun collagen–chitosan nanofiber: a biomimetic extracellular matrix for endothelial cell and smooth muscle cell. Acta Biomater. **6**, 372–382 (2010)
84. Hu, F., Wei, L., Zhou, Z., Ran, Y., Li, Z., Gao, M.: Preparation of biocompatible magnetite nanocrystals for in vivo magnetic resonance detection of cancer. Adv. Mater. **18**, 2553–2556 (2006)
85. Chen, F.-H., Zhang, L.-M., Chen, Q.-T., Zhang, Y., Zhang, Z.-J.: Synthesis of a novel magnetic drug delivery system composed of doxorubicin-conjugated Fe_3O_4 nanoparticle cores and a PEG-functionalized porous silica shell. Chem. Commun. **46**, 8633–8635 (2010)
86. Yu, M.K., Jeong, Y.Y., Park, J., Park, S., Kim, J.W., Min, J.J., Kim, K., Jon, S.: Drug-loaded superparamagnetic iron oxide nanoparticles for combined cancer imaging and therapy in vivo. Angew. Chem. Int. Ed. **47**, 5362–5365 (2008)
87. Jun, Y.W., Huh, Y.M., Choi, J.S., Lee, J.H., Song, H.T., Kim, S., Yoon, S., Kim, K.S., Shin, J.S., Suh, J.S., Cheon, J.: Nanoscale size effect of magnetic nanocrystals and their utilization for cancer diagnosis via magnetic resonance imaging. J. Am. Chem. Soc. **127**, 5732–5733 (2005)
88. Huh, Y.M., Jun, Y.W., Song, H.T., Kim, S., Choi, J.S., Lee, J.H., Yoon, S., Kim, K.S., Shin, J.S., Suh, J.S., Cheon, J.: In vivo magnetic resonance detection of cancer by using multifunctional magnetic nanocrystals. J. Am. Chem. Soc. **127**, 12387–12391 (2005)
89. Katagiri, K., Imai, Y., Koumoto, K., Kaiden, T., Kono, K., Aoshima, S.: Magnetoresponsive on-demand release of hybrid liposomes formed from Fe_3O_4 nanoparticles and thermosensitive block copolymers. Small **7**, 1683–1689 (2011)
90. Chen, H., Deng, C., Zhang, X.: Synthesis of Fe_3O_4@SiO_2@PMMA core-shell-shell magnetic microspheres for highly efficient enrichment of peptides and proteins for MALDI-ToF MS analysis. Angew. Chem. Int. Ed. **49**, 607–611 (2010)
91. Seo, W.S., Lee, J.H., Sun, X., Suzuki, Y., Mann, D., Liu, Z., Terashima, M., Yang, P.C., McConnell, M.V., Nishimura, D.G., Dai, H.: FeCo/graphitic-shell nanocrystals as advanced magnetic-resonance-imaging and near-infrared agents. Nat. Mater. **5**, 971–976 (2006)
92. Nasongkla, N., Bey, E., Ren, J., Ai, H., Khemtong, C., Guthi, J.S., Chin, S.-F., Sherry, A.D., Boothman, D.A., Gao, J.: Multifunctional polymeric micelles as cancer-targeted MRI-ultrasensitive drug delivery systems. Nano Lett. **6**, 2427–2430 (2006)
93. Song, H.T., Choi, J.S., Huh, Y.M., Kim, S., Jun, Y.W., Suh, J.S., Cheon, J.: Surface modulation of magnetic nanocrystals in the development of highly efficient magnetic resonance probes for intracellular labeling. J. Am. Chem. Soc. **127**, 9992–9993 (2005)
94. Jun, Y.-W., Lee, J.-H., Cheon, J.: Chemical design of nanoparticle probes for high-performance magnetic resonance imaging. Angew. Chem. Int. Ed. **47**, 5122–5135 (2008)
95. Hou, Y., Xu, Z., Sun, S.: Controlled synthesis and chemical conversions of FeO nanoparticles. Angew. Chem. Int. Ed. **46**, 6329–6332 (2007)
96. Xu, Z., Shen, C., Hou, Y., Gao, H., Sun, S.: Oleylamine as both reducing agent and stabilizer in a facile synthesis of magnetite nanoparticles. Chem. Mater. **21**, 1778–1780 (2009)
97. Tan, Y., Zhuang, Z., Peng, Q., Li, Y.: Room-temperature soft magnetic iron oxide nanocrystals: synthesis, characterization, and size-dependent magnetic properties. Chem. Mater. **20**, 5029–5034 (2008)
98. Li, X., Si, H., Niu, J.Z., Shen, H., Zhou, C., Yuan, H., Wang, H., Ma, L., Li, L.S.: Size-controlled syntheses and hydrophilic surface modification of Fe_3O_4, Ag, and Fe_3O_4/Ag heterodimer nanocrystals. Dalton Trans. **39**, 10984–10989 (2010)
99. Sun, S.H., Zeng, H., Robinson, D.B., Raoux, S., Rice, P.M., Wang, S.X., Li, G.X.: Monodisperse MFe_2O_4 (M = Fe, Co, Mn) nanoparticles. J. Am. Chem. Soc. **126**, 273–279 (2004)
100. Kim, D., Lee, N., Park, M., Kim, B.H., An, K., Hyeon, T.: Synthesis of uniform ferrimagnetic magnetite nanocubes. J. Am. Chem. Soc. **131**, 454–455 (2009)
101. Kovalenko, M.V., Bodnarchuk, M.I., Lechner, R.T., Hesser, G., Schaeffler, F., Heiss, W.: Fatty acid salts as stabilizers in size- and shape-controlled nanocrystal synthesis: the case of inverse spinel iron oxide. J. Am. Chem. Soc. **129**, 6352–6353 (2007)

102. Sun, S.H., Zeng, H.: Size-controlled synthesis of magnetite nanoparticles. J. Am. Chem. Soc. **124**, 8204–8205 (2002)
103. Park, J., An, K.J., Hwang, Y.S., Park, J.G., Noh, H.J., Kim, J.Y., Park, J.H., Hwang, N.M., Hyeon, T.: Ultra-large-scale syntheses of monodisperse nanocrystals. Nat. Mater. **3**, 891–895 (2004)
104. Nagasubramanian, R., Innocenti, F., Ratain, M.J.: Pharmacogenetics in cancer treatment. Annu. Rev. Med. Sel. Top. Clin. Sci. **54**, 437–452 (2003)
105. Allen, T.M., Cullis, P.R.: Drug delivery systems: entering the mainstream. Science **303**, 1818–1822 (2004)
106. Janib, S.M., Moses, A.S., MacKay, J.A.: Imaging and drug delivery using theranostic nanoparticles. Adv. Drug Deliv. Rev. **62**, 1052–1063 (2010)
107. Kim, J., Park, S., Lee, J.E., Jin, S.M., Lee, J.H., Lee, I.S., Yang, I., Kim, J.-S., Kim, S.K., Cho, M.-H., Hyeon, T.: Designed fabrication of multifunctional magnetic gold nanoshells and their application to magnetic resonance imaging and photothermal therapy. Angew. Chem. Int. Ed. **45**, 7754–7758 (2006)
108. Selvan, S.T., Patra, P.K., Ang, C.Y., Ying, J.Y.: Synthesis of silica-coated semiconductor and magnetic quantum dots and their use in the imaging of live cells. Angew. Chem. Int. Ed. **46**, 2448–2452 (2007)
109. Lim, S.J., An, B.K., Jung, S.D., Chung, M.A., Park, S.Y.: Photoswitchable organic nanoparticles and a polymer film employing multifunctional molecules with enhanced fluorescence emission and bistable photochromism. Angew. Chem. Int. Ed. **43**, 6346–6350 (2004)
110. Kim, J., Kim, H.S., Lee, N., Kim, T., Kim, H., Yu, T., Song, I.C., Moon, W.K., Hyeon, T.: Multifunctional uniform nanoparticles composed of a magnetite nanocrystal core and a mesoporous silica shell for magnetic resonance and fluorescence imaging and for drug delivery. Angew. Chem. Int. Ed. **47**, 8438–8441 (2008)
111. Salgueirino-Maceira, V., Correa-Duarte, M.A., Spasova, M., Liz-Marzan, L.M., Farle, M.: Composite silica spheres with magnetic and luminescent functionalities. Adv. Funct. Mater. **16**, 509–514 (2006)
112. Li, Z., Zhang, Y., Jiang, S.: Multicolor core/shell-structured upconversion fluorescent nanoparticles. Adv. Mater. **20**, 4765–4769 (2008)
113. van de Rijke, F., Zijlmans, H., Li, S., Vail, T., Raap, A.K., Niedbala, R.S., Tanke, H.J.: Up-converting phosphor reporters for nucleic acid microarrays. Nat. Biotechnol. **19**, 273–276 (2001)
114. Wang, F., Liu, X.: Recent advances in the chemistry of lanthanide-doped upconversion nanocrystals. Chem. Soc. Rev. **38**, 976–989 (2009)
115. Auzel, F.: Upconversion and anti-stokes processes with f and d ions in solids. Chem. Rev. **104**, 139–173 (2004)
116. Mai, H.X., Zhang, Y.W., Si, R., Yan, Z.G., Sun, L.D., You, L.P., Yan, C.H.: High-quality sodium rare-earth fluoride nanocrystals: controlled synthesis and optical properties. J. Am. Chem. Soc. **128**, 6426–6436 (2006)
117. Li, X., Zhao, D., Zhang, F.: Multifunctional upconversion-magnetic hybrid nanostructured materials: synthesis and bioapplications. Theranostics **3**, 292–305 (2013)
118. Liu, Z., Yi, G., Zhang, H., Ding, J., Zhang, Y., Xue, J.: Monodisperse silica nanoparticles encapsulating upconversion fluorescent and superparamagnetic nanocrystals. Chem. Commun. 694–696 (2008)
119. Mi, C., Zhang, J., Gao, H., Wu, X., Wang, M., Wu, Y., Di, Y., Xu, Z., Mao, C., Xu, S.: Multifunctional nanocomposites of superparamagnetic (Fe_3O_4) and NIR-responsive rare earth-doped up-conversion fluorescent ($NaYF_4$:Yb,Er) nanoparticles and their applications in biolabeling and fluorescent imaging of cancer cells. Nanoscale **2**, 1141–1148 (2010)
120. Hu, D., Chen, M., Gao, Y., Li, F., Wu, L.: A facile method to synthesize superparamagnetic and up-conversion luminescent $NaYF_4$:Yb, Er/Tm@SiO_2@Fe_3O_4 nanocomposite particles and their bioapplication. J. Mater. Chem. **21**, 11276–11282 (2011)

121. Shen, J., Sun, L.-D., Zhang, Y.-W., Yan, C.-H.: Superparamagnetic and upconversion emitting Fe_3O_4/$NaYF_4$:Yb,Er hetero-nanoparticles via a crosslinker anchoring strategy. Chem. Commun. **46**, 5731–5733 (2010)
122. Cheng, L., Yang, K., Li, Y., Zeng, X., Shao, M., Lee, S.-T., Liu, Z.: Multifunctional nanoparticles for upconversion luminescence/MR multimodal imaging and magnetically targeted photothermal therapy. Biomaterials **33**, 2215–2222 (2012)
123. Xu, H., Cheng, L., Wang, C., Ma, X., Li, Y., Liu, Z.: Polymer encapsulated upconversion nanoparticle/iron oxide nanocomposites for multimodal imaging and magnetic targeted drug delivery. Biomaterials **32**, 9364–9373 (2011)
124. Xia, A., Gao, Y., Zhou, J., Li, C., Yang, T., Wu, D., Wu, L., Li, F.: Core-shell $NaYF_4:Yb^{3+}$, Tm^{3+}@Fe_xO_y nanocrystals for dual-modality T-2-enhanced magnetic resonance and NIR-to-NIR upconversion luminescent imaging of small-animal lymphatic node. Biomaterials **32**, 7200–7208 (2011)
125. Zhong, C., Yang, P., Li, X., Li, C., Wang, D., Gai, S., Lin, J.: Monodisperse bifunctional Fe_3O_4@$NaGdF_4$:Yb/Er@$NaGdF_4$:Yb/Er core-shell nanoparticles. RSC Adv. **2**, 3194–3197 (2012)
126. Zhang, M., Shi, S., Meng, J., Wang, X., Fan, H., Zhu, Y., Wang, X., Qian, Y.: Preparation and characterization of near-infrared luminescent bifunctional core/shell nanocomposites. J. Phys. Chem. C **112**, 2825–2830 (2008)
127. Chen, H., Qi, B., Moore, T., Colvin, D.C., Crawford, T., Gore, J.C., Alexis, F., Mefford, O. T., Anker, J.N.: Synthesis of brightly PEGylated luminescent magnetic upconversion nanophosphors for deep tissue and dual MRI imaging. Small **10**, 160–168 (2014)
128. Liu, D., Zhao, D., Shi, F., Zheng, K., Qin, W.: Superparamagnetic and upconversion luminescent properties of Fe_3O_4/$NaYF_4$:Yb,Er hetero-submicro-rods. Mater. Lett. **85**, 1–3 (2012)
129. Zijlmans, H., Bonnet, J., Burton, J., Kardos, K., Vail, T., Niedbala, R.S., Tanke, H.J.: Detection of cell and tissue surface antigens using up-converting phosphors: a new reporter technology. Anal. Biochem. **267**, 30–36 (1999)
130. Fan, H.-M., Olivo, M., Shuter, B., Yi, J.-B., Bhuvaneswari, R., Tan, H.-R., Xing, G.-C., Ng, C.-T., Liu, L., Lucky, S.S., Bay, B.-H., Ding, J.: Quantum dot capped magnetite nanorings as high performance nanoprobe for multiphoton fluorescence and magnetic resonance imaging. J. Am. Chem. Soc. **132**, 14803–14811 (2010)
131. Cheong, S., Ferguson, P., Feindel, K.W., Hermans, I.F., Callaghan, P.T., Meyer, C., Slocombe, A., Su, C.-H., Cheng, F.-Y., Yeh, C.-S., Ingham, B., Toney, M.F., Tilley, R.D.: Simple synthesis and functionalization of iron nanoparticles for magnetic resonance imaging. Angew. Chem. Int. Ed. **50**, 4206–4209 (2011)
132. Malvindi, M.A., Greco, A., Conversano, F., Figuerola, A., Corti, M., Bonora, M., Lascialfari, A., Doumari, H.A., Moscardini, M., Cingolani, R., Gigli, G., Casciaro, S., Pellegrino, T., Ragusa, A.: Magnetic/silica nanocomposites as dual-mode contrast agents for combined magnetic resonance imaging and ultrasonography. Adv. Funct. Mater. **21**, 2548–2555 (2011)
133. Li, Z., Yi, P.W., Sun, Q., Lei, H., Zhao, H.L., Zhu, Z.H., Smith, S.C., Lan, M.B., Lu, G.Q.: Ultrasmall water-soluble and biocompatible magnetic iron oxide nanoparticles as positive and negative dual contrast agents. Adv. Funct. Mater. **22**, 2387–2393 (2012)
134. Tian, Y., Yu, B., Li, X., Li, K.: Facile solvothermal synthesis of monodisperse Fe_3O_4 nanocrystals with precise size control of one nanometre as potential MRI contrast agents. J. Mater. Chem. **21**, 2476–2481 (2011)
135. Hu, F., MacRenaris, K.W., Waters, E.A., Liang, T., Schultz-Sikma, E.A., Eckermann, A.L., Meade, T.J.: Ultrasmall, water-soluble magnetite nanoparticles with high relaxivity for magnetic resonance imaging. J. Phys. Chem. C **113**, 20855–20860 (2009)
136. Lee, J.E., Lee, N., Kim, H., Kim, J., Choi, S.H., Kim, J.H., Kim, T., Song, I.C., Park, S.P., Moon, W.K., Hyeon, T.: Uniform mesoporous dye-doped silica nanoparticles decorated with multiple magnetite nanocrystals for simultaneous enhanced magnetic resonance imaging, fluorescence imaging, and drug delivery. J. Am. Chem. Soc. **132**, 552–557 (2010)

137. Wu, H., Zhang, S., Zhang, J., Liu, G., Shi, J., Zhang, L., Cui, X., Ruan, M., He, Q., Bu, W.: A hollow-core, magnetic, and mesoporous double-shell nanostructure. in situ decomposition/reduction synthesis, bioimaging, and drug-delivery properties. Adv. Funct. Mater. **21**, 1850–1862 (2011)
138. Gong, X., Peng, S., Wen, W., Sheng, P., Li, W.: Design and fabrication of magnetically functionalized core/shell microspheres for smart drug delivery. Adv. Funct. Mater. **19**, 292–297 (2009)
139. Zhang, X.F., Mansouri, S., Clime, L., Ly, H.Q., Yahia, L.H., Veres, T.: Fe_3O_4-silica core-shell nanoporous particles for high-capacity pH-triggered drug delivery. J. Mater. Chem. **22**, 14450–14457 (2012)
140. Santra, S., Kaittanis, C., Grimm, J., Perez, J.M.: Drug/dye-loaded, multifunctional iron oxide nanoparticles for combined targeted cancer therapy and dual optical/magnetic resonance imaging. Small **5**, 1862–1868 (2009)
141. Xiong, Y.J., McLellan, J.M., Chen, J.Y., Yin, Y.D., Li, Z.Y., Xia, Y.N.: Kinetically controlled synthesis of triangular and hexagonal nanoplates of palladium and their SPR/SERS properties. J. Am. Chem. Soc. **127**, 17118–17127 (2005)
142. Lu, Y., Liu, G.L., Lee, L.P.: High-density silver nanoparticle film with temperature-controllable interparticle spacing for a tunable surface enhanced Raman scattering substrate. Nano Lett. **5**, 5–9 (2005)
143. Talley, C.E., Jackson, J.B., Oubre, C., Grady, N.K., Hollars, C.W., Lane, S.M., Huser, T.R., Nordlander, P., Halas, N.J.: Surface-enhanced Raman scattering from individual Au nanoparticles and nanoparticle dimer substrates. Nano Lett. **5**, 1569–1574 (2005)
144. Chan, Y.-H., Chen, J., Wark, S.E., Skiles, S.L., Son, D.H., Batteas, J.D.: Using patterned arrays of metal nanoparticles to probe plasmon enhanced luminescence of CdSe quantum dots. ACS Nano **3**, 1735–1744 (2009)
145. Munechika, K., Chen, Y., Tillack, A.F., Kulkarni, A.P., Plante, I.J.-L., Munro, A.M., Ginger, D.S.: Spectral control of plasmonic emission enhancement from quantum dots near single silver nanoprisms. Nano Lett. **10**, 2598–2603 (2010)
146. Saboktakin, M., Ye, X., Oh, S.J., Hong, S.-H., Fafarman, A.T., Chettiar, U.K., Engheta, N., Murray, C.B., Kagan, C.R.: Metal-enhanced upconversion luminescence tunable through metal nanoparticle-nanophosphor separation. ACS Nano **6**, 8758–8766 (2012)
147. Zhang, H., Li, Y., Ivanov, I.A., Qu, Y., Huang, Y., Duan, X.: Plasmonic modulation of the upconversion fluorescence in $NaYF_4$:Yb/Tm hexaplate nanocrystals using gold nanoparticles or nanoshells. Angew. Chem. Int. Ed. **49**, 2865–2868 (2010)
148. Schietinger, S., Aichele, T., Wang, H.-Q., Nann, T., Benson, O.: Plasmon-enhanced upconversion in single $NaYF_4$:Yb^{3+}/Er^{3+} codoped nanocrystals. Nano Lett. **10**, 134–138 (2010)
149. Verhagen, E., Kuipers, L., Polman, A.: Enhanced nonlinear optical effects with a tapered plasmonic waveguide. Nano Lett. **7**, 334–337 (2007)
150. Sudheendra, L., Ortalan, V., Dey, S., Browning, N.D., Kennedy, I.M.: Plasmonic enhanced emissions from cubic $NaYF_4$:Yb:Er/Tm nanophosphors. Chem. Mater. **23**, 2987–2993 (2011)
151. Priyam, A., Idris, N.M., Zhang, Y.: Gold nanoshell coated $NaYF_4$ nanoparticles for simultaneously enhanced upconversion fluorescence and darkfield imaging. J. Mater. Chem. **22**, 960–965 (2012)
152. Liu, N., Qin, W., Qin, G., Jiang, T., Zhao, D.: Highly plasmon-enhanced upconversion emissions from Au@beta-$NaYF_4$:Yb,Tm hybrid nanostructures. Chem. Commun. **47**, 7671–7673 (2011)
153. Chen, G., Qju, H., Prasad, P.N., Chen, X.: Upconversion nanoparticles: design, nanochemistry, and applications in theranostics. Chem. Rev. **114**, 5161–5214 (2014)
154. Zhang, H., Xu, D., Huang, Y., Duan, X.: Highly spectral dependent enhancement of upconversion emission with sputtered gold island films. Chem. Commun. **47**, 979–981 (2011)

155. Xu, W., Xu, S., Zhu, Y., Liu, T., Bai, X., Dong, B., Xu, L., Song, H.: Ultra-broad plasma resonance enhanced multicolor emissions in an assembled Ag/$NaYF_4$:Yb,Er nano-film. Nanoscale **4**, 6971–6973 (2012)
156. Zhang, W., Ding, F., Chou, S.Y.: Large enhancement of upconversion luminescence of $NaYF_4:Yb^{3+}/Er^{3+}$ nanocrystal by 3D plasmonic nano-antennas. Adv. Mater. **24**, OP236–OP241 (2012)
157. Paudel, H.P., Zhong, L., Bayat, K., Baroughi, M.F., Smith, S., Lin, C., Jiang, C., Berry, M. T., May, P.S.: Enhancement of near-infrared-to-visible upconversion luminescence using engineered plasmonic gold surfaces. J. Phys. Chem. C **115**, 19028–19036 (2011)
158. Tu, N., Wang, L.: Surface plasmon resonance enhanced upconversion luminescence in aqueous media for TNT selective detection. Chem. Commun. **49**, 6319–6321 (2013)
159. Li, Z.Q., Li, X.D., Liu, Q.Q., Chen, X.H., Sun, Z., Liu, C., Ye, X.J., Huang, S.M.: Core/shell structured $NaYF_4:Yb^{3+}/Er^{3+}/Gd^{3+}$ nanorods with Au nanoparticles or shells for flexible amorphous silicon solar cells. Nanotechnology **23** (2012)
160. Zhao, P., Zhu, Y., Yang, X., Fan, K., Shen, J., Li, C.: Facile synthesis of upconversion luminescent mesoporous Y_2O_3:Er microspheres and metal enhancement using gold nanoparticles. RSC Adv. **2**, 10592–10597 (2012)
161. Kannan, P., Rahim, F.A., Chen, R., Teng, X., Huang, L., Sun, H., Kim, D.-H.: Au nanorod decoration on $NaYF_4$:Yb/Tm nanoparticles for enhanced emission and wavelength-dependent biomolecular sensing. ACS Appl. Mater. Interfaces **5**, 3508–3513 (2013)
162. Zhang, F., Braun, G.B., Shi, Y., Zhang, Y., Sun, X., Reich, N.O., Zhao, D., Stucky, G.: Fabrication of Ag@SiO_2@Y_2O_3:Er nanostructures for bioimaging: tuning of the upconversion fluorescence with silver nanoparticles. J. Am. Chem. Soc. **132**, 2850–2851 (2010)
163. Xu, W., Chen, B., Yu, W., Zhu, Y., Liu, T., Xu, S., Min, X., Bai, X., Song, H.: The up-conversion luminescent properties and silver-modified luminescent enhancement of YVO_4: Yb^{3+},Er^{3+} NPs. Dalton Trans. **41**, 13525–13532 (2012)
164. Sun, Q.-C., Mundoor, H., Ribot, J.C., Singh, V., Smalyukh, I.I., Nagpal, P.: Plasmon-enhanced energy transfer for improved upconversion of infrared radiation in doped-lanthanide nanocrystals. Nano Lett. **14**, 101–106 (2014)
165. Dong, B., Xu, S., Sun, J., Bi, S., Li, D., Bai, X., Wang, Y., Wang, L., Song, H.: Multifunctional $NaYF_4:Yb^{3+}$, Er^{3+}@Agcore/shell nanocomposites: integration of upconversion imaging and photothermal therapy. J. Mater. Chem. **21**, 6193–6200 (2011)
166. Feng, W., Sun, L.-D., Yan, C.-H.: Ag nanowires enhanced upconversion emission of $NaYF_4$: Yb,Er nanocrystals via a direct assembly method. Chem. Commun. 4393–4395 (2009)
167. Deng, R., Xie, X., Vendrell, M., Chang, Y.-T., Liu, X.: Intracellular glutathione detection using MnO_2-nanosheet-modified upconversion nanoparticles. J. Am. Chem. Soc. **133**, 20168–20171 (2011)
168. Chen, Z., Chen, H., Hu, H., Yu, M., Li, F., Zhang, Q., Zhou, Z., Yi, T., Huang, C.: Versatile synthesis strategy for carboxylic acid-functionalized upconverting nanophosphors as biological labels. J. Am. Chem. Soc. **130**, 3023–3029 (2008)
169. Yan, C., Dadvand, A., Rosei, F., Perepichka, D.F.: Near-IR photoresponse in new up-converting CdSe/$NaYF_4$:Yb,Er nanoheterostructures. J. Am. Chem. Soc. **132**, 8868–8869 (2010)
170. Boyer, J.-C., Cuccia, L.A., Capobianco, J.A.: Synthesis of colloidal upconverting $NaYF_4:Er^{3+}/Yb^{3+}$ and Tm^{3+}/Yb^{3+} monodisperse nanocrystals. Nano Lett. **7**, 847–852 (2007)
171. Zhang, Y., Hong, Z.: Synthesis of lanthanide-doped $NaYF_4$@TiO_2 core-shell composites with highly crystalline and tunable TiO_2 shells under mild conditions and their upconversion-based photocatalysis. Nanoscale **5**, 8930–8933 (2013)
172. Wang, F., Han, Y., Lim, C.S., Lu, Y., Wang, J., Xu, J., Chen, H., Zhang, C., Hong, M., Liu, X.: Simultaneous phase and size control of upconversion nanocrystals through lanthanide doping. Nature **463**, 1061–1065 (2010)

173. Ji, T., Yang, F., Du, H., Guo, H., Yang, J.: Preparation and characterization of upconversion nanocomposite for beta-$NaYF_4$:Yb^{3+}, Er^{3+}-supported TiO_2 nanobelts. J. Rare Earth **28**, 529–533 (2010)
174. Li, C., Wang, F., Zhu, J., Yu, J.C.: $NaYF_4$ Yb,Tm/CdS composite as a novel near-infrared-driven photocatalyst. Appl. Catal. B Environ. **100**, 433–439 (2010)
175. Shalav, A., Richards, B.S., Trupke, T., Kramer, K.W., Gudel, H.U.: Application of $NaYF_4$: Er^{3+} up-converting phosphors for enhanced near-infrared silicon solar cell response. Appl. Phys. Lett. **86**, 013505 (2005)
176. de Wild, J., Meijerink, A., Rath, J.K., van Sark, W.G.J.H.M., Schropp, R.E.I.: Towards upconversion for amorphous silicon solar cells. Sol. Energ. Mater. Sol. C **94**, 1919–1922 (2010)
177. Liang, H., Chen, G., Li, L., Liu, Y., Qin, F., Zhang, Z.: Upconversion luminescence in Yb^{3+}/Tb^{3+}-codoped monodisperse $NaYF_4$ nanocrystals. Opt. Commun. **282**, 3028–3031 (2009)
178. Shan, G.-B., Demopoulos, G.P.: Near-infrared sunlight harvesting in dye-sensitized solar cells via the insertion of an upconverter-TiO_2 nanocomposite layer. Adv. Mater. **22**, 4373–4377 (2010)
179. Su, L.T., Karuturi, S.K., Luo, J., Liu, L., Liu, X., Guo, J., Sum, T.C., Deng, R., Fan, H.J., Liu, X., Tok, A.I.Y.: Photon upconversion in hetero-nanostructured photoanodes for enhanced near-infrared light harvesting. Adv. Mater. **25**, 1603–1607 (2013)
180. Liang, L., Liu, Y., Bu, C., Guo, K., Sun, W., Huang, N., Peng, T., Sebo, B., Pan, M., Liu, W.: Highly uniform, bifunctional core/double-shell-structured β-$NaYF_4$: Er^{3+},Yb^{3+}@SiO_2@TiO_2 hexagonal sub-microprisms for high-performance dye sensitized solar cells. Adv. Mater. **25**, 2174–2180 (2013)
181. Liang, L., Liu, Y., Zhao, X.-Z.: Double-shell β-$NaYF_4$:Yb^{3+}, Er^{3+}/SiO_2/TiO_2 submicroplates as a scattering and upconverting layer for efficient dye-sensitized solar cells. Chem. Commun. **49**, 3958–3960 (2013)
182. Taychatanapat, T., Watanabe, K., Taniguchi, T., Jarillo-Herrero, P.: Quantum hall effect and landau-level crossing of dirac fermions in trilayer graphene. Nat. Phys. **7**, 621–625 (2011)
183. Huang, X., Qi, X., Boey, F., Zhang, H.: Graphene-based composites. Chem. Soc. Rev. **41**, 666–686 (2012)
184. Ren, L., Qi, X., Liu, Y., Huang, Z., Wei, X., Li, J., Yang, L., Zhong, J.: Upconversion-P25-graphene composite as an advanced sunlight driven photocatalytic hybrid material. J. Mater. Chem. **22**, 11765–11771 (2012)
185. Liu, C., Wang, Z., Jia, H., Li, Z.: Efficient fluorescence resonance energy transfer between upconversion nanophosphors and graphene oxide: a highly sensitive biosensing platform. Chem. Commun. **47**, 4661–4663 (2011)
186. Wang, Y., Wang, H., Liu, D., Song, S., Wang, X., Zhang, H.: Graphene oxide covalently grafted upconversion nanoparticles for combined NIR mediated imaging and photothermal/photodynamic cancer therapy. Biomaterials **34**, 7715–7724 (2013)
187. Yan, L., Chang, Y.-N., Yin, W., Liu, X., Xiao, D., Xing, G., Zhao, L., Gu, Z., Zhao, Y.: Biocompatible and flexible graphene oxide/upconversion nanoparticle hybrid film for optical pH sensing. Phys. Chem. Chem. Phys. **16**, 1576–1582 (2014)

Chapter 5
Surface Modification and Bioconjugation of Upconversion Nanoparticles

Abstract One of the most important applications of upconversion nanoparticles (UCNPs) is bioapplication. A crucial issue facing the use of hydrophobic UCNPs for bioapplications is to render them water soluble and provide reactive groups for subsequent bioconjugation to various biomolecules. Surface modification can act as the bridge between the synthesis and biomedical applications of UCNPs. Surface modification of the UCNPs should not only improve the photostability of the nanoparticles with desirable interfacial properties, but also provide a potential platform for attaching biological molecules and other conjugated materials for various biomedical applications.

Keywords Surface modification · Hydrophilic surface · Upconversion nanoparticles · Bioconjugation · Biomedical application · Ligand exchange · Surface polymerization

5.1 Introduction

Surface modification is the act of augmenting the surface of a material with physical, chemical, or biological characteristics different from the ones originally found on that surface. This modification is usually made to solid materials, but it is possible to find examples of modification to the surface of specific liquids. The modification could be achieved by different methods with a view to altering a wide range of characteristics of the surface, such as roughness [1], hydrophilicity [2], surface charge [3], surface energy, biocompatibility [2, 4], and reactivity [5]. Surface modification could be classified through three main groups: physical modification (physical adsorption, Langmuir blodgett film), chemical modification (oxidation by strong acids, ozone treatment, chemisorption, and flame treatment), and radiation (glow discharge, corona discharge, photo activation (UV), laser, ion

Chi Yao and Fan Zhang contributed together to this chapter.

F. Zhang, *Photon Upconversion Nanomaterials*,
Nanostructure Science and Technology, DOI 10.1007/978-3-662-45597-5_5

beam, plasma immersion ion implantation, electron beam lithography, and γ-irradiation) [6].

Nanoparticle surface modification is an important part of the material surface modification. Nanoparticle research has extended into the biomedical field with recent in vitro and in vivo animal studies which demonstrate that quite a number of nanoparticles (NPs) need bio-stable and biocompatible surface with low toxicity, to guarantee the use of NPs as biolabels and drug delivery vehicles [7–10]. NPs possess cores with different chemical compositions, very high surface areas, a wide variety of surface groups, and facile surface functionalization. Based on these properties, it has been suggested that NPs can be tailored to a specific surface property to promote designated biomedical applications. While NPs have their own intrinsic properties, it is generally necessary in biological applications to impart additional properties or functions through physical or chemical coupling between a NP and one or more molecules [11, 12]. The converse is often true in the case of many synthetic materials, where the NP enables detection but requires conjugation to a biological moiety for targeting or bioactivity. The association of one or more biologically relevant molecules at the interface of a NP defines a NP-bioconjugate, and combines the unique optoelectronic or physicochemical properties of NP materials with biological activity such as selective binding [13–15]. Biomolecules of interest may include one or more of the following: peptides [16], proteins [17], and antibodies [18, 19]; enzymes and ribozymes [16, 20], oligonucleotides and aptamers [21, 22], carbohydrates [23], lipids [22], drugs or other biologically active small molecules [15], reporter molecules, or contrast agents (e.g., radiolabels, fluorescent dyes) [24]. NP-bioconjugates can be prepared via the formation of new chemical bonds between functional groups associated with a NP and the biomolecule or small molecule of interest. NPs may also offer the potential for association through dative or coordinate bonding, electrostatic interactions, and van der Waals interactions. Moreover, it can often be necessary to account for a second conjugate preparation, where the biomolecule of interest is also conjugated with a reporter molecule. This is frequently the case in biosensing applications that use NPs. In general, the structure of a NP-bioconjugate affects its function, and the controlled display of biomolecules on NPs is paramount in obtaining conjugates with well-defined and reproducible properties.

Usually, the synthesis process of NPs requires the participation of ligand molecules. The ligand molecules bound to the nanoparticle surface not only control the growth of the particles during synthesis, but also prevent the aggregation of the NPs. The repulsive force between particles can, in principle, be due to electrostatic repulsion, steric exclusion, or a hydration layer on the surface. Depending on the particle system, i.e., the core material and the solvent in which the particles are dispersed, the choice of the right ligand may yield stable particles. Like most of the synthesis methods for high-quality nanoparticles, synthetic approaches for UCNPs with the uniform size distributions usually involve the use of organic capping ligands and result in UCNPs that are hydrophobic as synthesized. Therefore, their

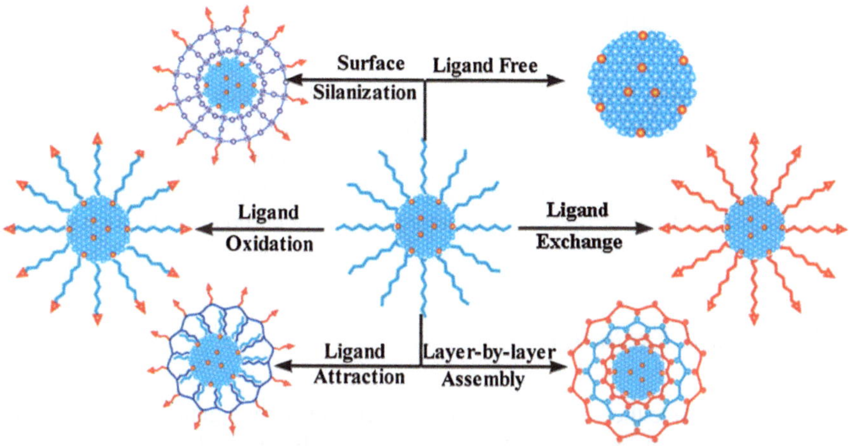

Fig. 5.1 Typical surface modification strategies used for making hydrophilic UCNPs with pendant functional groups. Reprinted with the permission from Ref. [87]. Copyright 2013 Royal Society of Chemistry

hydrophobic ligands compromise their water solubility and their compatibility with the biological milieu. Exploring surface engineering that renders these UCNPs hydrophilic and biocompatible has been essential for demonstrating their potential uses in various biomedical applications. This goal, in turn, is almost completely dependent on the properties of the surface modification approach designed to provide aqueous stability to the UCNPs and facilitate their integration within biological environment. To this end, a variety of surface modification strategies such as ligand exchange, ligand oxidation, ligand-free synthesis, ligand attraction, electrostatic layer-by-layer assembly, and surface silanization have been designed to transfer UCNPs with hydrophobic surfaces into water over the past decade (Fig. 5.1), as will be briefly overviewed in the following sections.

For biological applications, a surface modified-UCNP should exhibit high upconversion luminescence (UCL) efficiency, as well as have hydrophilic surface characteristics that are compatible with biomolecules or biomolecular assemblies such as live cells and bodies. If used in in vivo imaging or detection, low in vivo toxicity is an additional important issue [25]. Strategies to make water soluble and biocompatible UCNPs are guided by several criteria: (i) stability over a wide pH range and at high salt concentrations; (ii) easy functionalization; (iii) minimally perturb the inherent UCNPs properties or those of an attached biomolecule; and (iv) minimal undesired interaction with other molecules native to biological environments, that is, low nonspecific adsorption.

5.2 Ligand Engineering

Surface engineering is the sub-discipline of materials science which deals with the surface of solid matter. For the modification of hydrophobic UCNPs, a common approach of ligand engineering involves a ligand exchange reaction with inorganic acid or hydrophilic bifunctional molecules and involves a direct oxidation of the terminal group of native ligands to generate a pendant carboxylic functional group.

5.2.1 Ligand Exchange

In order to improve the stability of given nanoparticles, the ligand molecules on the surface can be exchanged by others that can possibly provide new properties or functionality to the particles [26]. In most cases, the incoming ligand molecule binds more strongly to the inorganic nanoparticle surface [27–30]. For ligand exchange, the new ligand molecules should have an affinity as strong as possible to the inorganic core in order to quickly and effectively replace the original surfactant molecules. In addition, the molecular geometry of the ligands in relation to the particle diameter is a factor that influences how densely the molecules are packed around the particles, which in turn influences ultimately the colloidal stability of the particles.

The coordination mode of ligand molecules (e.g., oleic acid (OA)) on the surface of UCNPs is shown in Fig. 5.2a. The carboxyl ends conjugate to Ln^{3+} with two C–O bonds, and this kind of force is quite strong. Naturally, the ligand molecules that are strongly bound to the nanoparticle surface or more tightly to each other will be less likely to detach from the particles' surface. That makes the ligand exchange process relatively difficult and time-consuming. Here are some examples of successful obtaining hydrophilic UCNPs by ligand exchange processes.

5.2.1.1 Organic Ligand-Free

The "organic ligand-free" approach can obtain water-dispersible organic ligand-free lanthanide-doped UCNPs [31]. This method is actually the ligand exchange between oleic acid molecules and inorganic ions (H_3O^+). This approach is based on removing the oleate ligand from the surface of oleate-capped Ln-UCNPs synthesized via thermal decomposition using a simple acid treatment process. This procedure would provide a great opportunity to engineer the surface properties of UCNPs by easily introducing various capping ligands and thus tuning their properties for required application. Importantly, the water-soluble organic ligand-free UCNPs would facilitate further conjunctions with electronegative groups such as −SH, −COOH, −NH2, −OH of hydrophilic and offer interesting opportunities for designing novel UCNPs-based materials for electronic or biomedical applications.

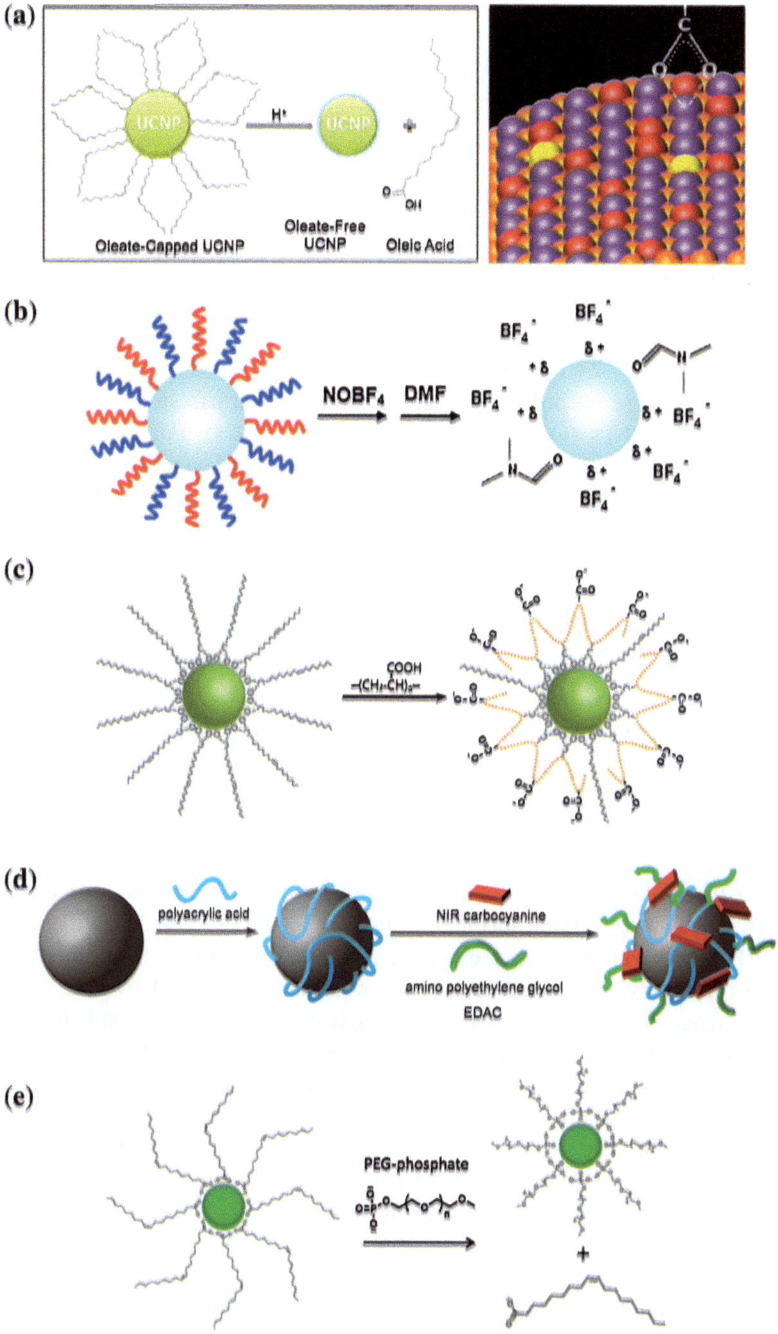
(a)
UCNP
H+
Oleate-Capped UCNP
Oleate-Free UCNP
Oleic Acid
C
O
(b)
NOBF4
DMF
BF4-
δ+
(c)
COOH
(d)
polyacrylic acid
NIR carbocyanine
amino polyethylene glycol
EDAC
(e)
PEG-phosphate
+

◀ **Fig. 5.2** **a** Schematic illustration of the ligand exchange process. The removal of the oleate (OA) capping ligand from the surface of the $NaYF_4:Er^{3+},Yb^{3+}$ nanoparticles. **b** Schematic illustration of the ligand-exchange process with $NOBF_4$. The general principle of the ligand exchange approach by using oleic acid coated iron oxide nanocrystals as a model system. **c** Ligand exchange with PAA in a polyol solvent such as DEG at high temperature makes these nanocrystals water soluble. **d** Surface modification of Y_2O_3 upconverting nanoparticles with polyacrylic acid, polyethylene glycol and NIR emitting fluorophores. **e** The replacement of oleate with a PEG-phosphate ligand as an efficient method for the generation of water-dispersible $NaYF_4$ UCNPs. **a** Reprinted with the permission from Ref. [31]. Copyright 2011 American Chemical Society. **b** Reprinted with the permission from Ref. [32]. Copyright 2010 American Chemical Society. **c** Reprinted with the permission from Ref. [48]. Copyright 2007 American Chemical Society. **d** Reprinted with the permission from Ref. [51]. Copyright 2009 Royal Society of Chemistry. **e** Reprinted with the permission from Ref. [52]. Copyright 2009 American Chemical Society

A facile acid–base reaction had been successfully applied to modify the oleate-capped $NaYF_4:Er^{3+},Yb^{3+}$ NPs (OA-capped Ln-UCNPs, Fig. 5.2a). For the oleate-capped Ln-UCNPs, the oleic acid molecules are chemically adsorbed on the Ln-UCNP through the coordination between the COO– group and the Ln(III) ions on the surface. The NPs were treated with HCl at pH 4, which protonates the oleate ligand resulting in the release of oleic acid from the surface. The oleate-free NPs show hydrophilic properties, which would indicate that the NPs are dispersed by electrostatic (electric double layer repulsion) stabilization forces. The scheme, which they propose for the acid–base reaction occurring at the surface of the Ln-UCNP is as the following:

$$\mathrm{LnOA} \xrightarrow[\mathrm{pH}\,4]{[}]\mathrm{HClLnOH} + \mathrm{OAH} \rightarrow [\mathrm{LnOH}_2^+]\cdots\mathrm{C1}^-$$

$$\mathrm{LnOA} \xrightarrow[\mathrm{pH}\,4]{[}]\mathrm{HClLnOH} + \mathrm{OAH} \xrightarrow[\mathrm{pH}\,7.4]{[}][\mathrm{LnO}^-]\cdots\mathrm{H_3O}^+$$

To investigate the effect of removing the oleate ligand from the surface of the UCNPs on the upconversion spectra, oleate-free $NaYF_4:Er^{3+},Yb^{3+}$ NPs dispersed in water (H_2O) as well as deuterated water (D_2O). Comparison of the three upconversion spectra reveals that the peak positions remain the same. However, there is a drastic change on two fronts, the overall upconversion intensity and the relative integrated red:green (R/G) emission ratio. The UCL is higher for the oleate-free nanoparticles dispersed in water compared to the oleate-capped nanoparticles in toluene while the overall UCL intensity is highest for the oleate-free nanoparticles dispersed in D_2O compared to those in H_2O. These results provide the evidence for the importance of the isotope effect, which reduce the nonradiative pathways since the OD oscillators are much less efficient in the multiphonon nonradiative deactivation process compared to OH.

A similar "organic ligand-free" strategy could be achieved by a ligand exchange using nitrosonium tetrafluoroborate ($NOBF_4$, Fig. 5.2b) [32]. $NOBF_4$ or diazonium tetra-fluoroborate compounds were employed to replace the original ligands, thus stabilizing the NPs in various polar, hydrophilic media such as *N,N*-dimethylformamide (DMF), dimethylsulfoxide (DMSO), or acetonitrile. When the dichloromethane solution of $NOBF_4$ was added to the NP dispersion in hexane, NPs were found to precipitate immediately after gentle shaking, indicating a dramatic change in NP solubility as a result of surface modification. After centrifugation to remove the supernatant, the precipitated NPs could be readily redispersed in various polar, hydrophilic solvents such as DMF, DMSO, or acetonitrile to form colloidal dispersions. In particular, NPs dispersions in DMF were stable for several years without any detectable aggregation or precipitation. Attempts to directly redisperse NPs in water resulted in partial aggregation, while the addition of a small amount of DMF in water yielded stable colloidal dispersions. Significantly, this surface modification strategy was applicable to NPs having a variety of compositions, sizes, and shapes, including metal oxides such as Fe_3O_4, TiO_2 nanorods, metal alloys such as FePt and $CoPt_3$ NPs, semiconductors such as Bi_2S_3 nanorods, and dielectrics such as rare-earth doped upconversion $NaYF_4$ nanorods and nanoplates.

It should be mentioned that the attempts to apply this approach to other semiconductor quantum dots such as PbS and PbTe result in unstable colloidal DMF dispersions with significant NP agglomeration or fusion. The BF_4^- modified NPs are highly soluble in DMF due to the dual stabilization from BF_4^- and DMF. The NP solubility is even higher than the initial solubility in hexane, especially for larger NPs. In contrast, the solubility was significantly improved when $NaYF_4$ nanoplates were dispersed in DMF after $NOBF_4$ treatment as well as the high stability of the colloidal dispersion. Even highly concentrated colloidal dispersions were stable for several months without sedimentation or precipitation despite the large NP size. It was speculated that the improved NP solubility reduces the scattering of the incident laser light, which increased the excitation efficiency and therefore effectively enhanced the emission intensity. It should be mentioned that when smaller upconversion $NaYF_4$:Yb/Er nanorods, which can form transparent and stable dispersions in hexane, were subjected to a similar $NOBF_4$ treatment, the emission intensity remained virtually unchanged, implying that the observed emission enhancement for the upconversion $NaYF_4$ nanoplates indeed resulted from the improved NP solubility rather than surface capping or solvent effect.

One of the major benefits of $NOBF_4$ treatment is that the hydrophilic, BF_4^--modified NPs obtained can be further functionalized by a variety of capping molecules regardless of the NP composition and morphology. This allows for fully reversible phase transfer and surface functionalization of NPs. Upon the addition of organic ligands such as OA, OAm, TDPA, or HAm to the NP dispersion in DMF combined with hexane, NPs were found to transfer immediately from the DMF layer to the hexane layer, suggesting that NPs were successfully capped by organic ligands. TEM results revealed the preserved NP size and shape after the secondary surface modification, and the formation of ordered NP superlattices provided further evidence of the complete recovery of surface functionalization by organic ligands. This facile reversed surface modification can be attributed to the weak binding affinity of BF_4^- anions to the NP surface. In addition to hydrophobic capping molecules, the BF_4^--modified NPs can also be functionalized by water-soluble polymers such as PVP, stabilizing NPs in aqueous media without aggregation or precipitation for several months. It was demonstrated that $NOBF_4$ treatment can serve as an intermediate step toward the systematic surface property engineering of NPs by introducing delicately designed functional capping molecules.

Water soluble organic ligand-free UCNPs could be obtained efficiently using both of the above two processes. These as-prepared organic ligand-free UCNPs with inorganic ions are suitable for further bio-conjunctions in water solutions with electronegative groups such as −SH, −COOH, $-NH_2$, −OH of hydrophilic and biocompatible molecules and may be used for imaging and detection applications in biological assays.

Shi et al. reported the organic ligand-free UCNPs-based nanotheranostics by means of multistep chemical processes as well as the successful application as nanoprobes for simultaneous dual magnetic resonance (MR)/optical imaging and direct nuclear drug delivery [33]. To obtain physiologically stable UCNPs as contrast agents, the oleate-coated UCNPs were treated with HCl at pH 4. Then, the pegylation of oleate-free UCNPs with amine-poly (ethylene glycol)-thiol (NH_2–PEG–SH 5000) was conducted to increase particle stability in culture media, because this resulted in a covalent PEG overcoating on the nanoparticle (UCNPs-PEG) through the strong thiole–metal interactions [34]. PEG component could provide high water solubility and greatly enhanced biocompatibility, while the amino group attached to PEG was conjugated to carboxyl group on the nuclear localizing signal peptide (TAT). DOX was loaded into the as-prepared materials by a simple physical adsorption approach used in supramolecular chemistry. Such nuclear-targeted theranostic systems, which combined multiple imaging modalities and therapeutic functions, may ultimately enable the drug delivery into the cell nuclei more efficiently under the real-time monitoring of the therapy process.

Surface modification with absorbed proteins and specific binding peptides could significantly alter the nanomaterial-cell interaction and nanomaterial toxicity. The study of Wen group showed that a short synthetic peptide, RE-1, identified by means of phage display, binds to organic ligand-free UCNPs, forms a stable coating layer on the NPs' surface, and effectively abrogates their autophagy-inducing activity [35]. During this process, the amino acid sequence, rather than the composition, played a key role in determining the peptide binding to the organic ligand-free NPs. The results revealed that the feasibility of surface engineering with a high-affinity binding peptide and its variants to "tune" the cell interaction and autophagy-inducing activity for Ln-based NPs. Whereas other surface-modifying polymers such as PEG were widely employed to alter cell interaction and biological response for nanomaterials, peptides like RE-1 work through specific high-affinity interactions rather than through non-specific absorption or covalent modification. This lead to several advantages: high consistency and ease in scaling up as a result of the simple add-and-mix coating procedure; ability to "tune" the cell interaction to a desired level with the use of a series of peptide variants; material selectivity, with the possibility of selective coating in a mixed environment or on a composite material containing both Ln-based and non-Ln-based material. The stable coating formed by the bound RE-1 peptide abrogates the autophagy-inducing activity and toxicity by preventing NP sedimentation and reducing NP-cell interaction, with the latter as the predominant mechanism for smaller NPs. The retained ability of the RE-1 motif to coat the nanoparticles and reduce cell interactions in general would clear out non-specific cell interactions, while the targeting motifs would enable the nanoparticles to reach the target cells and perform the intended diagnostic or therapeutic tasks.

5.2.1.2 Bifunctional Molecules Ligand Exchange

Bifunctional molecules ligand exchange is currently the most popular method to modify the surface of hydrophobic UCNPs without obviously affecting their shapes, compositions, and morphologies. In a typical bifunctional molecules ligand exchange procedure, the original capping hydrophobic ligand such as OA and OM bound on the UCNP surface can be replaced with new bifunctional organic molecules or polymers. Such bifunctional organic molecules or polymers should have at least one functional group capable of anchoring to the NP surface tightly and another hydrophilic group at the other end to guarantee the well dispersion in water or can be further functionalized. Hitherto, a variety of organic molecules or polymers meeting this structural requirement have been investigated for the decoration of the NP surface with hydrophilic moieties, which including PEG diacid [36], poly (ethylene glycol) (PEG)-phosphate [37–39], polyacrylic acid (PAA) [40–44], 3-mercaptopropionic acid, hexanedioic acid, 6-aminohexanoic acid (AA) [45], and 2-aminoethyl dihydrogen phosphate (AEP) [46, 47].

In a pioneering work, Yin and coworkers first demonstrated an exchange procedure by using PAA as new capping ligand (Fig. 5.2c) [48]. In this paper, they described a robust ligand exchange method that uses short-chain hydrophilic polyelectrolyte molecules to replace the original hydrophobic ligands at an elevated temperature in a glycol solvent. The polyelectrolyte molecules were chosen on the basis of strong coordination of their functional groups to the nanoparticle surface, with typical examples including poly(acrylic acid) (PAA), poly(allylamine) (PAAm), and poly(sodium styrene sulfonate) (PSS). Diethylene glycol (DEG) was selected as the polar solvent because of its high boiling point (~245 °C), high dissolving capacity for polyelectrolytes, and high miscibility with both water and typical organic solvents. Compared to previously reported ligand exchange processes, their approach has several advantages. Firstly, the exchange reaction is performed at relatively high temperatures so that the dynamic solvation of the ligands favors the exchange of original surfactants with new ones through mass action. Secondly, each polyelectrolyte chain binds to the nanoparticle surface through multiple anchoring points, providing more robust surface adhesion than that achievable with a small molecule that typically has only one binding group [49]. The strong binding also prevents desorption of ligands from the particle surface, which has been one of the sources of toxicity in practical applications. Thirdly, the abundant uncoordinated groups on each polymer chain extend into water, making the particles highly soluble. In addition, these remaining functional groups such as –COOH and $-NH_2$ are available for further surface modification. Fifthly, the process described here uses commercial polyelectrolyte materials, which are much more economical than some surfactants such as phospholipids or specially synthesized polymers used in other procedures [50].

Another example of PAA surface modification of UCNPs was demonstrated by Hilderbrand who first coated Y_2O_3 UCNPs with PAA, and then linked amine-modified PEG molecules and NIR dyes with the free carboxyl groups of PAA via conventional amide coupling techniques. As a general approach to preparation of water-soluble UCNPs, they developed a simple procedure for the surface coating of unmodified Y_2O_3 particles based on carboxylic acid coordination to the particles (Fig. 5.2d) [51]. This is accomplished by sonication of the uncoated NPs in ethylene glycol solution containing PAA. However, nanoparticles coated with carboxylic acid groups are not ideal for vascular imaging as they may show pronounced non-specific tissue binding or shortened circulation half-lives in the blood. To solve this problem, the carboxylic acid coated Y_2O_3 NPs were further coupled to 2,000 MW mPEG-NH_2 polyethylene glycol (PEG) polymers using standard amide coupling chemistry. Preliminary data with these materials show they have low cytotoxicity. Practical utility for imaging the vasculature in live mice is demonstrated with the NPs. These initial studies indicate that this functionalized Y_2O_3-based upconverting nanomaterial is a promising platform for in vivo optical-based diagnostic imaging.

Besides, there are also several works of ligand exchange modifying UCNPs involving the use of poly(ethylene glycol) (PEG)-based polymers. The use of PEG as a stabilizing ligand is well documented and has several advantages [52]. It renders the NPs dispersible in aqueous solvents as well as provides improved stability in pH buffers. PEG is known to prevent undesired protein adhesion while at the same time being low toxic and non-immunogenic, thus improving biocompatibility. Yi and Chow [36] first discussed the hydrophilic surface modification of oleylamine stabilized β-$NaYF_4$:Yb/Er nanoparticles by (PEG)-based acid ligand exchange method. The as-synthesized hydrophobic $NaYF_4$:Yb,Er NPs were modified to have hydrophilic surfaces by replacing oleylamine on the particles' surfaces with a bipolar regent, polyethylene glycol 600 diacid (HOOC–PEG–COOH). After surfactant exchange, the NPs remained as a clear colloid in deionized water, without any noticeable settling or precipitation after 2 weeks. No noticeable change in luminescence properties was observed after the surfactant exchange. These NPs and their dispersions in various media are potentially useful in optical nanodevices and bioprobes.

Another attractive research of those works was monitored by van Veggel group (Fig. 5.2e) [52]. They presented a technique for the replacement of oleate with a PEG-phosphate ligand as an efficient method for the generation of water-dispersible $NaYF_4$ UCNPs. The PEG-phosphate ligands were proved to exchange with the original oleate ligands anchoring on the surface of the NPs, obtaining water-dispersible UCNPs. Compared to the original UCNPs in organic solvents, the UCL intensity of the NPs in aqueous environments was found to be severely quenched. This could be attributed to an increase in the multiphonon relaxations of the lanthanide excited state in aqueous environments due to high energy vibrational modes of water molecules. This problem could be overcome partially by the synthesis of core/shell UCNPs which was aimed at improving photophysical properties in water over the original core NPs. The PEG-phosphate coated upconverting NPs were then employed to image a line ovarian cancer cell line (CaOV3) to demonstrate their potential in biological application.

The ligand exchange strategy could also be applied to DNA modification of lanthanide-based UCNPs. Lu reported a simple and versatile one-step approach to prepare DNA-functionalized UCNPs based on ligand exchange at the liquid−liquid interface [53]. DNA molecules could replace the original oleic acid ligands on UCNPs just through vigorously stirring of the oil (chloroform)/water phases. The UCNPs were observed to transfer into the water layer from the chloroform layer due to DNA attachment through the interaction between the negatively charged phosphates of the DNA with surface lanthanide ions. The approach does not require preparation of functionalizable UCNPs or chemical modification of oligonucleotides. The as-prepared DNA-UCNPs could cross cell membranes without the need of transfection agents, and visualized under NIR light excitation with high photostability. It is suitable for the DNA-UCNPs to be used for DNA delivery and targeted imaging of cancer cells. This novel approach and the resulting outstanding combination of properties will advance both the research fields of UCNPs and DNA-based bionanotechnology, and will find wide application in the areas including bioinspired nanoassembly, biomedicine, and biological assays.

5.2.2 Oxidation of Oleic Acid

Direct ligand oxidation method is primarily based on the oxidation of the unsaturated carbon–carbon double bonds of the capping organic ligands such as OA into carboxylic acid groups to generate water-soluble UCNPs.

Using this technique, Li and coworkers first demonstrated the conversion of OA-capped $NaYF_4$:Yb/Er NPs into water-soluble ones with preserved morphology, dispersity, and luminescence intensity via the Lemieux–von Rudloff reagent [54] (Fig. 5.3a). The Lemieux-von Rudloff oxidation method was well-known to oxidize selectively a carbon–carbon double bond (R–CH=CH–R′) to give two carboxylic acids. As a result of the ligand evolution from oleic acid to azelaic acid on the surface of UCNPs, all of the oxidized samples could be readily dispersed in water and in some polar organic solvents such as DMF, or DMSO due to the presence of free carboxylic acid groups. Colloidal aqueous solutions of UCNPs exhibited similar luminescent spectra as the pH of the solution is adjusted over a wide range, indicating excellent pH-independent luminescent properties of UCNPs as compared to organic dyes and semiconductor quantum dots. Covalent coupling of streptavidin to the surface of UCNPs was facilitated by cross-linkers (EDC and sulfo-NHS) which activated the surface carboxyl groups on UCNPs and led to the formation of amide bonds with streptavidin. The successful fabrication of a DNA sensor indicated that streptavidin molecules on the surface of $NaYF_4$:Yb,Er NPs remain highly active. Therefore, it is anticipated that this streptavidin coating surface chemistry provides a novel approach for the production of biocompatible photoluminescent

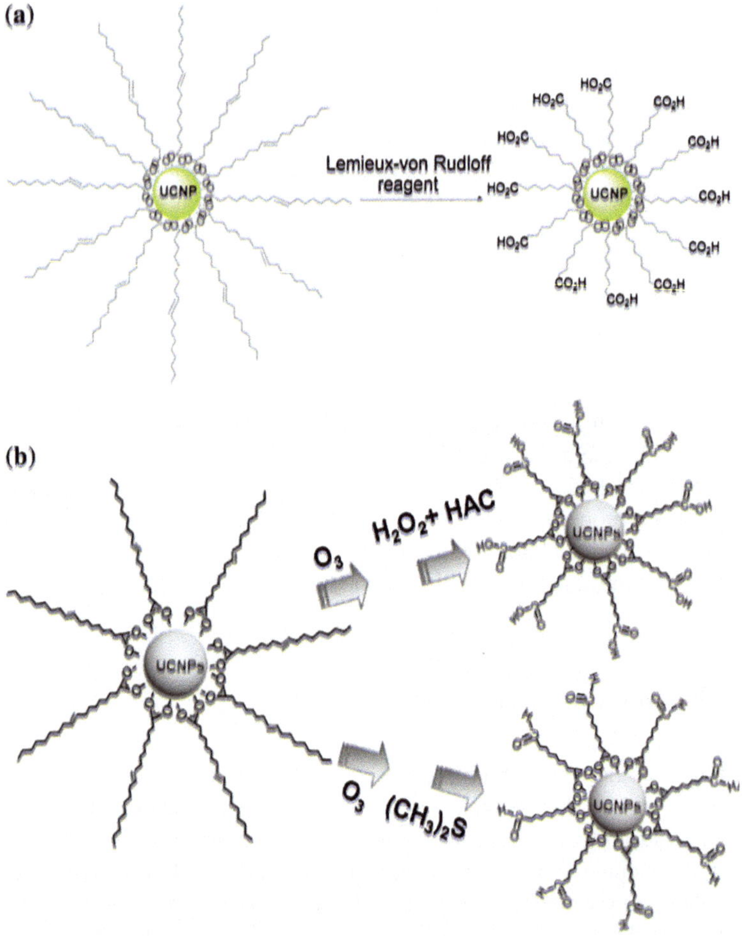

Fig. 5.3 Schematic illustration of direct oxidation of oleic acid. **a** The conversion of OA-capped $NaYF_4$:Yb/Er NPs into water-soluble ones via the Lemieux–von Rudloff reagent. **b** The oxidation of OA into the azelaic aldehyde or azelaic acid through ozonolysis on the surface of $NaYF_4$:Yb/Er NPs. **a** Reprinted with the permission from Ref. [54]. Copyright 2008 American Chemical Society. **b** Reprinted with the permission from Ref. [55]. Copyright 2009 Wiley-VCH Verlag GmbH & Co. KGaA

UCNP probes, giving UCNPs great potential as luminescent labeling materials for biological applications.

By contrast, Yan and coworkers reported the oxidation of OA into the azelaic aldehyde or azelaic acid through ozonolysis on the surface of $NaYF_4$:Yb/Er NPs [55]. In this work, the hydrophobic OA-capped UCNPs were converted into amphiphilic ones by ozonolysis, combined with a subsequent oxidation or reduction treatment (Fig. 5.3b). The whole process consisted of two main steps: The first one was the formation of a three-membered-ring epoxide, and the second one was to obtain the final products bearing –COOH or –CHO ending groups, whereby a mixture of H_2O_2 and CH_3COOH was used as the oxidant, and CH_3SCH_3 as the reductant, respectively. In the product, the oleic acid ligands on the UCNP surface were transformed into azelaic acid or azelaic aldehyde. After surface modification, the carboxyl-modified UCNPs were no longer dispersible in non-polar solvents, but they had good solubility in polar solvents (such as ethanol or water). Also, the NPs maintained their original spherical shapes without any damage, indicating that the modification takes place only on the surface, while the intrinsic upconversion luminescence property of the NPs was not essentially altered after the above surface-modification process. The procedure could be applied to various other hydrophobic NPs, such as rare-earth, semiconductor, and metal NPs with double-bond-containing surface ligands. The active sites on the surface of the UCNPs not only afford an enhanced solubility in polar solvents, but also give the capability of further conjugation with functional receptors. The presence of aldehyde groups on the surface of the UCNPs allowed further conjugation with amino-containing molecules, and the MPTEA-functionalized UCNPs derived from azelaic aldehyde-capped precursors afforded a novel approach for the detection of Ce^{4+}. On the other hand, HEMA-functionalized UCNPs open up the possibility of constructing hybrid structures between UCNPs and other functional NPs, which might exhibit interesting multiplex properties. The successful fabrication of carboxylic acid functionalized UCNPs suggested that it is possible to expand the application of these UCNPs as luminescent labels into other biological fields such as bioimaging.

The ligand oxidation approach turned out to be very simple and easily handled for surface modification of hydrophobic NPs. However, it was only applicable to a limited number of capping ligands which contained unsaturated carbon–carbon double bond.

5.3 Ligand Attraction

The ligand attraction approach is usually based on hydrophobic–hydrophobic van der Waals interactions between the well-chosen amphiphilic polymers and original hydrophobic ligand to realize the successful conversion from hydrophobic UCNPs to water-soluble ones. In this process, the hydrophobic portion of the amphiphilic polymers interacts with the organic capping layer of the NPs via hydrophobic attraction, while the hydrophilic portion faces outwards, interacting with the aqueous solvent and rendering the particle water soluble. During this process, the coating amphiphilic molecules do not directly interact with the NP surface. This method is expected to produce stable water-soluble UCNPs with preserved luminescent properties. Up to now, the most widely used amphiphilic polymers for ligand attraction include OA–PAA–PEG [56], poly(maleic anhydride-alt-1-octadecene) [57], poly (ethylene glycol)-block-poly(carprolactone) [58], poly((ethylene glycol)-block-lactic acid) [58], 6-aminohexanoic acid [59], PAA [60], and poly(L-lysine) [61].

Chow and Yi successfully modified the surface of $NaYF_4$:Yb/Er@$NaYF_4$ core-shell NPs with a desirable carboxylic functional group via the ligand attraction method by using the octylamine and isopropylamine modified PAA as coating polymers [60]. In this work, the PAA polymer coated core/shell NPs were synthesized by coating the core/shell NPs with an amphiphilic layer, 25 % octylamine and 40 % isopropylamine modified PAA (Fig. 5.4a). The coating of PAA arose from the hydrophobic interactions of the octyl and isopropyl groups of PAA with the octadecyl group of oleylamine on the NPs surface. After surface modification, the hydrophilic carboxyl group of PAA extended outward, rendering NPs water soluble and allowing attachment of biomolecules.

Liu and coworkers utilized an amphiphilic polymer of OA–PAA–PEG to render OA-capped $NaYF_4$:Yb/Er NPs water soluble [56] (Fig. 5.4b). A number of $NaYF_4$ UCNPs doped with different lanthanide elements were synthesized following a literature procedure with slight modifications. As-synthesized UCNPs coated by oleic acid were not soluble in water. A PEG and octylamine modified poly(acrylic acid) polymer (OA–PAA–PEG) was used to transfer the hydrophobic UCNPs into the aqueous phase. The coating of OA–PAA–PEG on UCNPs was formed by hydrophobic interactions between hydrocarbon chains of the octylamine and the oleic acid. Multicolor in vivo UCL imaging has been realized by imaging subcutaneously injected UCNPs and further demonstrated in multiplexed lymph node mapping and multicolor in vivo cancer cell tracking.

In particular, van Veggel and coworkers proposed a simple and effective cross-linking methodology to achieve stable dispersion of $NaYF_4$ core/shell UCNPs in water [57]. They reported a method in which the polymer poly(maleic anhydride-alt-1-octadecene) (PMAO) (modified with and without PEG) was coated over UCNPs using the hydrophobic interaction between the octadecene chain of the polymer and the oleate ligand on the surface of the NPs, followed by cross-linking the anhydride rings of the PMAO with bis(hexamethylene)triamine (BHMT) (Fig. 5.4c). These hydrophilic shell cross-linked UCNPs exhibit very high stability at different pH values, physiological buffers, and biological growth media. Finally, we show that these cross-linked UCNPs can potentially be used in biolabeling applications, by using them to label human prostate cancer cells.

Recently, Zhang and coworkers developed a methodology to modify UCNPs with a new family of compact zwitterionic ligand systems composed with functional phospholipids [62] (Fig. 5.4d). Efficient upconverting $NaGdF_4$:20%Yb^{3+}/2% Er^{3+} UCNPs were synthesized using the successive layer-by-layer strategy reported previously with oleic acid as the stabilizing agent. The compact lipids system containing 1,2-disteraroyl-sn-glycero-3-phosphoethanolamine-*N*-[carboxyl (polyethylene glycol)-2000] (PEG2000-PE) and 1,2-dioleoyl-sn-glycero-3-phosphocholine (DOPC) were designed and used to coat the surface of UCNPs to form the zwitterionic UCMC. PEG2000-PE acts as the stabilizer to ensure the monolayer cladding of the phospholipids onto the UCNPs, preventing the formation of lipid vesicles. The DOPC is used as the basic component to realize the zwitterionic surface. The hydrophilic region of DOPC contains two functional groups: a phosphate group acting as a negative-charged group, and a quaternary amine group

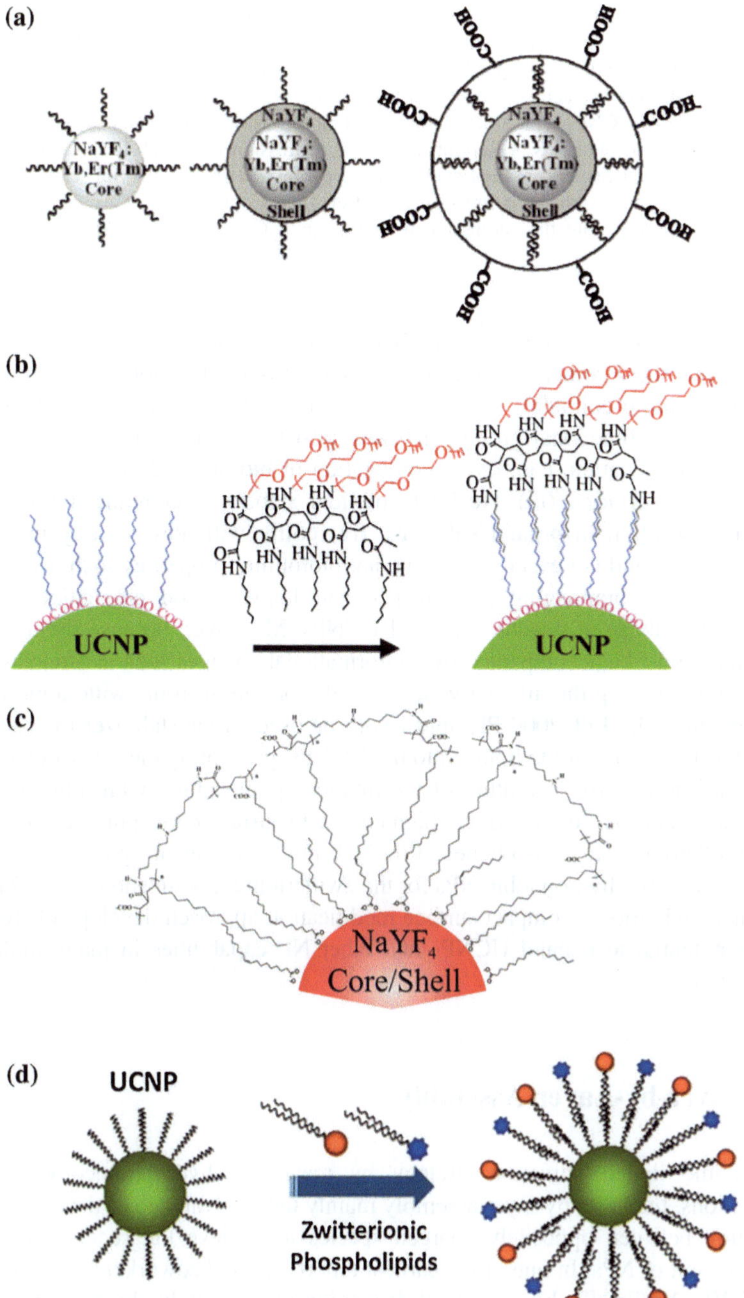
(a)
NaYF4: Yb,Er(Tm) Core
NaYF4
NaYF4: Yb,Er(Tm) Core
Shell
COOH
(b)
UCNP
UCNP
(c)
NaYF4 Core/Shell
(d)
UCNP
Zwitterionic Phospholipids

◀ **Fig. 5.4** Schematic illustration of ligand attraction. **a** Modification the surface of $NaYF_4$:Yb/Er@$NaYF_4$ core–shell NPs with PAA. **b** Utilization of an amphiphilic polymer of OA–PAA–PEG to render OA-capped $NaYF_4$:Yb/Er NPs water-soluble. **c** Cross-linking methodology with PMAO to achieve stable dispersion of NaYF4 core–shell UCNPs in water. **d** Modification of UCNPs with a new family of compact zwitterionic ligand systems composed with functional phospholipids. **a** Reprinted with the permission from Ref. [60]. Copyright 2006 American Chemical Society. **b** Reprinted with the permission from Ref. [56]. Copyright 2010 Springer. **c** Reprinted with the permission from Ref. [57]. Copyright 2012 American Chemical Society. **d** Reprinted with the permission from Ref. [62]. Copyright 2014 American Chemical Society

acting as a positive-charged group. Because of the zwitterionic nature, inherent benefits that these compact surface modification provide for biological applications include: (i) more colloidal stability under a wide pH range and high salt concentration, (ii) the functional groups on ZwitLipo-UCNPs are easily used for further chemical modification, e.g., direct conjugation of biotinylated DNA and antibody onto UCNP surfaces, (iii) excellent colloidal stability and reduced nonspecific protein adsorption in protein solutions, (iv) utility with low toxicity in cellular environments, and (v) excellent in vivo antibiofouling properties such as the more efficient tumor bioimaging performance and longer blood circulation half-life compared to those of negative-charged UCNPs. Moreover, because the hydrocarbon chain must be accompanied by conformational changes such as stretching and compression during the high curvature micelle formation from zwitterionic DOPC phospholipid, the PEG2000-PE lipid has to be used as the stabilizer to ensure the monolayer cladding of the lipids onto the UCNPs, preventing the formation of lipid vesicles. The intrinsic curvature of the multicomponent lipid with different headgroup size is proper for covering a high curvature surface corresponding to the NPs size. Furthermore, it is also believed that the multicomponent lipid system can be extended to other hydrophobic NPs for the zwitterionic modification. Therefore, the designed zwitterionic compact surface modification approach developed here has a strong potential to expand UCNPs and other NP capabilities in many biological applications.

5.4 Layer-by-Layer Assembly

Unlike the ligand attraction strategy by means of hydrophobic–hydrophobic interactions, the layer-by-layer assembly mainly takes advantage of the electrostatic attraction between oppositely charged species to achieve the successful surface modification of NPs. In one of the earliest reports, Li and coworkers functionalized the $NaYF_4$:Yb/Er NPs by sequential deposition of oppositely charged poly(allylamine hydrochloride) (PAH), poly(styrene sulfonate) (PSS), and PAH to generate hydrophilic PAH–PSS–PAH-coated NPs with stable amino-rich surfaces [63]. The layer-by-layer approach is based on the electrostatic attraction between the

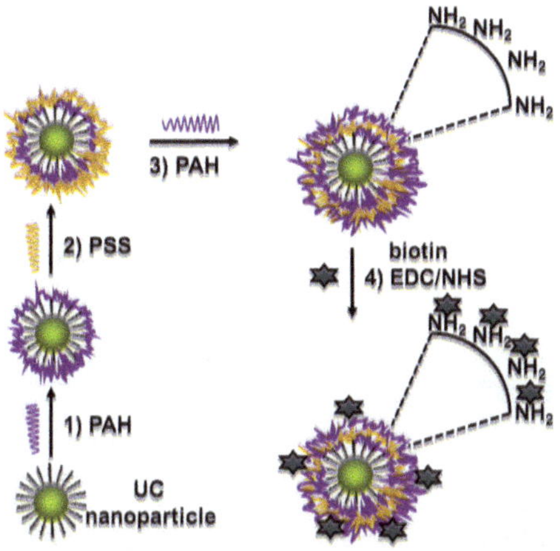

Fig. 5.5 Schematic illustration of layer-by-layer assembly. Sequential deposition of oppositely charged poly(allylamine hydrochloride) (PAH), poly(styrene sulfonate) (PSS), and PAH to generate hydrophilic PAH–PSS–PAH-coated NPs with stable amino-rich shells. Reprinted with the permission from Ref. [63]. Copyright 2005 Wiley-VCH Verlag GmbH & Co. KGaA

oppositely charged species deposited, and its major advantage is that it permits the preparation of coated colloids of different shapes and sizes, with uniform layers of diverse composition as well as controllable thickness. The synthetic methodology also can be used to prepare other single-emission upconversion-luminescent NPs. The fabrication process slightly affected the luminescence of the phosphors. The functionalized upconversion-luminescent NPs were then conjugated with biotin. A FRET system with upconversion-luminescent NPs as energy donors and gold NPs as energy acceptors has been developed and applied to detect trace amounts of avidin. The results indicated that such a FRET system is sensitive and simple for use in biological analyses. Although this simple method permits the preparation of water-soluble NPs with uniform layers of diverse compositions and controllable thickness, the time-consuming post-treatment procedure along with the significant increase in the hydrodynamic size of the NPs is somewhat unfavorable for their further bioapplications (Fig. 5.5).

5.5 Surface Modifications Using Silane Reagents

Surface silanization is a powerful technique for covalent surface modification of various kinds of materials including silica, metal oxides, semiconductors, and so on [64].

Silanization on UCNP surface could be performed in either of the following ways: in one way, the silane reagents directly react with the particles to form a silane-modified surface (end-capping); in the other way, coating the surface with tetraethoxysilane (TEOS) to create a layer of silica thereby yielding particles of

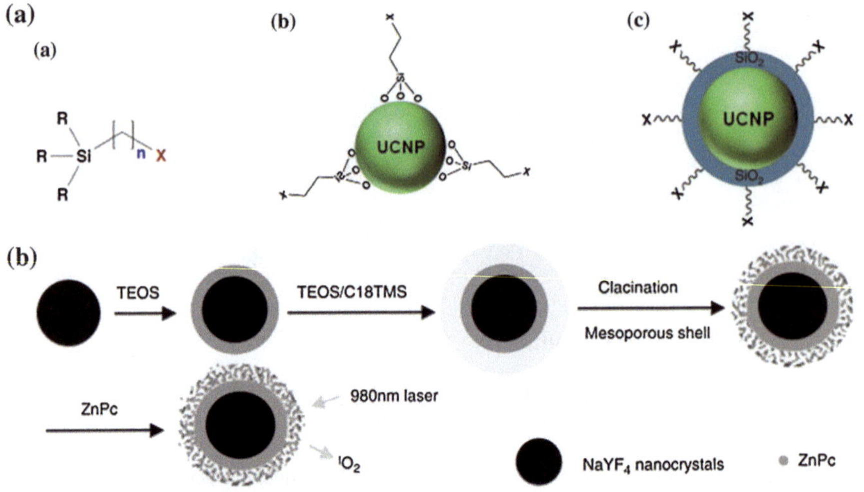

Fig. 5.6 **a** Schematic illustration of modifications using silane reagents. UCNPs coated with a thin layer of silica (SiO_2) by controlled hydrolysis and polycondensation of precursors. **b** Mesoporous silica-coated UCNPs prepared by the microemulsion method. Reprinted with the permission from Ref. [25]. Copyright 2005 Elsevier Ltd.

UCNP@SiO_2, and then using silanizing reagents to coat on this layer. The surface of the UCNPs might also be coated with a mixture of tetraalkoxysilane and a reagent; this would result in simultaneous coating of the UCNPs with silica and the UCNP functionalization, only required one separation step [65]. The resulting UCNPs were of low toxicity, mono-dispersion which could be well dispersed in aqueous solution. The silica layer in particles could also protect the UCNP from undesired external effects such as quenching, penetrating, or leaching.

In order to obtain particles of the type shown in Fig. 5.6a, a thin layer of silica (SiO_2) was coated on UCNPs by controlled hydrolysis and polycondensation of precursors such as TEOS or other tetraalkoxysilanes of the type $(RO)_4Si$, in which R was an alkyl substituent. Aminosilanes (e.g., aminopropyl triethoxysilane) were used most in direct modifications of the UCNP surface. Amino-modified UCNPs could be further biofunctionalized by covalent attachment of biomolecules such as biotin [66], folic acid (FA) [67], peptides [68], proteins [69], antibodies [70, 71], and DNA [72, 73].

Hu et al. [74] used particles modified with FA for the detection of tumor cells. In this work, micro-emulsion method was applied to obtain silica-coated and amino-modified UCNPs. To construct a multimodal luminescent probe with the potential for both down- and upconverting luminescence, a SiO_2-coated upconversion Yb^{3+}/Er^{3+}-codoped $NaYF_4$ nanophosphor capped by the downconversion luminescence dye fluorescein isothiocyanate was designed and synthesized. Fluorescein isothiocyanate (FITC) was incorporated into the silica shell as the secondary fluorophore. These nanocomposites are abbreviated as UCNP@SiO_2(FITC). To demonstrate the targeting capability of the as-prepared UCNP@SiO_2(FITC) nanocomposites, they used FR as a ligand for conjugation with the amine-functionalized nanocomposites (UCNP@SiO_2(FITC)−NH_2) to synthesize FR-functionalized NPs (UCNP@-SiO_2(FITC)−FA). Cytotoxicity tests revealed a good biocompatibility of the resulting particles. Further targeted imaging of tumor cells was performed by confocal luminescence imaging of a human nasopharyngeal epidermal carcinoma KB cell line with overexpressed surface folate receptors. The results clearly indicated that the silica-coated UCNP nanocomposites would be promising candidates as multimodal luminescent probes in bioimaging.

In order to immobilize antibodies, Wang et al. [75] first reported the preparation of silica-coated UCNPs using the Stöber method. They modified the surface with aminopropyl triethoxysilane, and subsequently conjugated rabbit anti-goat IgG to the amino-functionalized surface by glutaraldehyde. In the sandwich-like luminescence resonance energy transfer (LRET) system, the IgG-modified UCNPs acted as energy donors, and rabbit anti-goat IgG-modified gold NPs acted as energy acceptors, in which system, the green emission of the UCNPs following excitation at 980 nm around 540 nm could be absorbed maximally by the gold NPs.

In combination with TEOS, carboxyethyl triethoxysilane enables the coating of the surface of the NPs and introducing anionic carboxyl groups [76]. The surface charge of these modified NPs is about −30 mV. To stabilize the silica shell and control its thickness, some neutral polymers such as poly (vinyl pyrrolidone) took part in the modification [77]. The silica coating technique could be applied to not only hydrophilic but also hydrophobic UCNPs and enables the entrapment of secondary reporters such as magnetic Fe_3O_4 NPs [65] or organic dyes [78]. The microemulsion technique was also applied to coat UCNPs with TEOS and aminopropyl triethoxysilane, simultaneously incorporating fluorescein or rhodamine derivatives, or even QDs into the silica shell. This paved the way for the use of UCNPs in multiplexing or encoding [78].

The silica-coated UCNPs also showed potential for application in photodynamic therapy. Mesoporous silica-coated UCNPs were prepared by the microemulsion method [79] by first coating the UCNPs with TEOS (Fig. 5.6b). Subsequently, the mesoporous layer was formed by treating the silica layer with a mixture of TEOS and octadecyl trimethoxysilane (C18TMS). The coated particles were calcinated at 500 °C to remove excess C18TMS to obtain a mesoporous layer. The photosensitizer zinc phthalocyanine was then incorporated into this layer. The upconversion emission of the UCNPs activated the photosensitizer to release active singlet oxygen which possibly kills cancer cells.

5.6 Surface Polymerization

Surface polymerization can be defined as a process of growing an inorganic shell on the NP core. Qian et al. polymerized styrene on the surface of UCNPs successfully. Monodisperse polystyrene nanospheres with an average size of 400 nm in diameter, tagged with different color β-$NaYF_4$ NPs, were prepared using a miniemulsion polymerization method [80]. In this work, polystyrene nanospheres, encapsulated with β-$NaYF_4$:Yb/Er and β-$NaYF_4$:Yb/Tm NPs, were prepared using the mini-emulsion method. More than 20 β-$NaYF_4$ NPs were encapsulated in a single polystyrene nanosphere. The nanospheres emit multicolor NIR-to-visible up-conversion fluorescence upon excitation at a wavelength of 980 nm. The ratio between β-$NaYF_4$:Yb/Er and β-$NaYF_4$:Yb/Tm NPs was changed to prepare the polystyrene nanospheres with different up-conversion emission colors. The nanospheres could be used for a variety of biological applications which require high-sensitivity detection of biomolecules.

5.7 Surface Control During Synthesis

For UCNPs to be successfully used as luminescent labels, they must exhibit good aqueous solubility and suitable surface functionality [i.e., coating surface groups with carboxyl groups (–COOH), amino groups (–NH_2), or mercapto groups (–SH)] to allow further conjugation with biologically active molecules [81]. Typically, two-step controllable synthetic strategies are used to convert the hydrophobic UCNPs into hydrophilic ones. Such two-step synthetic strategies are usually associated with some intrinsic limitations, such as relatively complicated synthesis processes and post-treatment procedures. It is meaningful to develop a versatile surface-modification method for the preparation of functional surface-tunable UCNPs.

Under the guidance of this concept, quite a few works have been developed. The groups of Zhang and Liu have reported a one-step approach for preparing water-soluble and compatible NPs using polyethylenimine (PEI) as a surfactant [82, 83]. The synthesis of PEI/$NaYF_4$ NPs doped with upconverting lanthanide ions (Er^{3+}, Tm^{3+}) was carried out using a hydrothermal method. PEI is a hydrophilic polymer with primary, secondary, and tertiary amino groups and an overall positive charge. The positively charged amino groups stabilize the NPs in solution and can also be used for covalently bonding to biomolecules. PEI is a highly branched polymer with an appreciable thermal stability, which makes it the polymer of choice for this hydrothermal synthesis reaction. Free PEI has been reported to be toxic to cells, but when it is in particulate form, the detrimental effects are greatly lessened [84]. High molecular weight PEI is more efficient for controlling the particle growth and stabilizing the particles against aggregation. The as-prepared UCNPs have a spherical shape with an average size of about 50 nm in diameter. Furthermore, the NPs are soluble in water and biocompatible with human cells. The UCNPs doped with Yb^{3+}, Er^{3+} or Yb^{3+}, Tm^{3+} showed strong green or blue fluorescence in aqueous solutions when excited at 980 nm with a NIR laser. The method developed may also be used for one-pot synthesis of other water-soluble and biocompatible UCNPs.

The hydrothermal method is not limited to synthesizing only amine-functionalized UCNPs. Yang et al. [85] reported a simple one-step hydrothermal method to

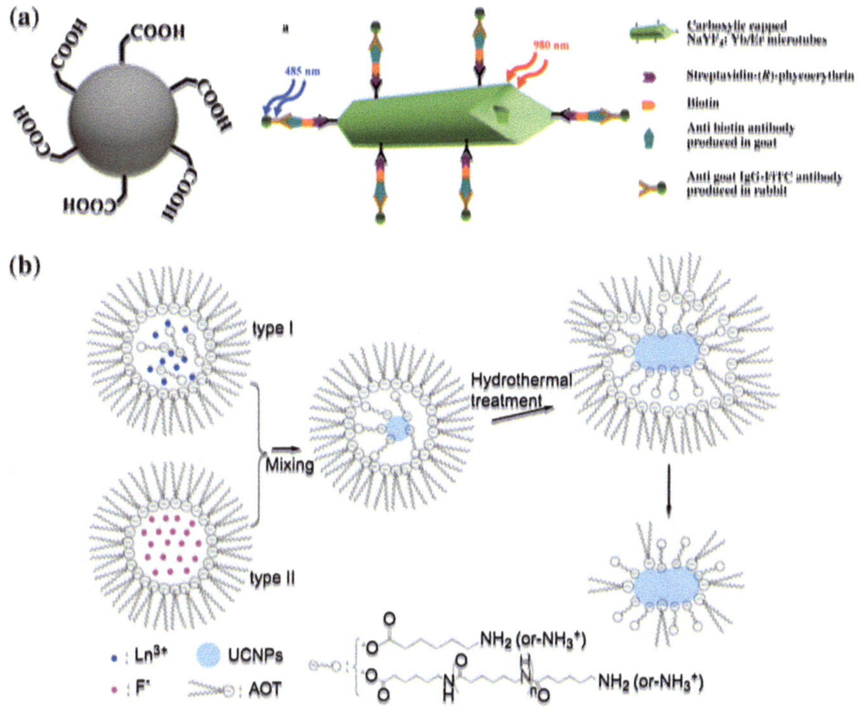

Fig. 5.7 **a** Schematic illustration of the surface control during synthesis. Carboxyl-functionalized $NaYF_4$:Yb/Er NP and the conjugation of antibodies with the carboxyl-functionalized microtubes. **b** Synthesis mechanism of biocompatible UCNPs by a modified hydrothermal microemulsion route. **a** Reprinted with the permission from Ref. [85]. Copyright 2012 Wiley-VCH Verlag GmbH & Co. KGaA. **b** Reprinted with the permission from Ref. [86]. Copyright 2009 Elsevier Ltd.

synthesize shape-controlled, water-dispersible, and carboxyl-functionalized $NaYF_4$ upconversion phosphors by using small-molecule binary acids as capping agents. The diverse phases, sizes, and shapes can be successfully controlled under different conditions such as reaction temperature, reaction time, molar ratio of capping agent to sodium hydroxide, and the varied structure of binary acids. Simultaneously, the carboxyl-functionalized $NaYF_4$:Yb/Er materials can be directly grafted onto the antibody for immunoassays, which indicates the potential bioapplications of these kinds of materials (Fig. 5.7a).

In addition to the above method, a modified hydrothermal microemulsion route to prepare water-soluble and amine-functionalized UCNPs assisted with bifunctional ligand 6-aminohexanoic acid has been developed by Li group [86] (Fig. 5.7b). The as-prepared samples after hydrothermal treatment for 6 h are egg-like or cocoon-like nanoparticles with diameter of $\sim$20 nm and length of 20–40 nm, and have an excellent dispersibility in water. Due to the presence of free amine moieties on their surfaces, the tumor-targeting molecule FA can be covalently linked to the UCNPs.

By utilizing near-infrared to visual (NIR-to-Vis) upconversion transition in FA-conjugated UCNPs which was injected into a subcutaneous HeLa tumor-bearing athymic nude mice, targeted up-conversion luminescence imaging of tumors in vivo and ex vivo has been demonstrated.

5.8 Summary

The understanding of surface modification of UCNPs is vital for improving the upconversion efficiency and aqueous solubility as well as providing potential capabilities for different biomedical applications. Ideally modified UCNPs generally have the following characteristics like stability over a wide pH range and at high salt concentrations, low nonspecific interaction with other species in biological environments, and most importantly, easy functionalization for further bioapplications. So far, a variety of surface modification strategies have been developed to transfer UCNPs with hydrophobic surfaces into water soluble, such as ligand exchange, ligand oxidation, ligand-free synthesis, ligand attraction, electrostatic layer-by-layer assembly, and surface silanization. Some of the existing strategies achieved good application efficiency, it still remains a matter for us to explore better methods for the surface modification and bioconjugation of UCNPs.

References

1. Lapshin, R.V., Alekhin, A.P., Kirilenko, A.G., Odintsov, S.L., Krotkov, V.A.: Vacuum ultraviolet smoothing of nanometer-scale asperities of Poly(methyl methacrylate) surface. J. Surf. Inv. **4**, 1 (2010)
2. Alekhin, A.P., Boleiko, G.M., Gudkova, S.A., Markeev, A.M., Sigarev, A.A., Toknova, V.F., Kirilenko, A.G., Lapshin, R.V., Kozlov, E.N., Tetyukhin, D.V.: Synthesis of biocompatible surfaces by nanotechnology methods. Nanotechnol. Russ. **5**, 696–708 (2010)
3. Bertazzo, S., Rezwan, K.: Control of α-alumina surface charge with carboxylic acids. Langmuir **26**, 3364–3371 (2009)
4. Bertazzo, S., Zambuzzi, W.F., Da Silva, H.A., Ferreira, C.V., Bertran, C.A.: Bioactivation of alumina by surface modification: a possibility for improving the applicability of alumina in bone and oral repair. Clin. Oral Implant Res. **20**, 288–293 (2009)
5. London, G., Chen, K.-Y., Carroll, G.T., Feringa, B.L.: Towards dynamic control of wettability by using functionalized altitudinal molecular motors on solid surfaces. Chem. A Eur. J. **19**, 10690–10697 (2013)
6. Wei, Q.F., Gao, W.D., Hou, D.Y., Wang, X.Q.: Surface modification of polymer nanofibres by plasma treatment. Appl. Surf. Sci. **245**, 16–20 (2005)
7. Yuan, Y., Zhang, J., An, L., Cao, Q., Deng, Y., Liang, G.: Oligomeric nanoparticles functionalized with NIR-emitting CdTe/CdS QDs and folate for tumor-targeted imaging. Biomaterials **35**, 7881–7886 (2014)
8. Jaque, D., Martinez Maestro, L., del Rosal, B., Haro-Gonzalez, P., Benayas, A., Plaza, J.L., Martin Rodriguez, E., Garcia Sole, J.: Nanoparticles for photothermal therapies. Nanoscale **6**, 9494–9530 (2014)

9. Höbel, S., Vornicescu, D., Bauer, M., Fischer, D., Keusgen, M., Aigner, A.: A novel method for the assessment of targeted PEI-based nanoparticle binding based on a static surface plasmon resonance system. Anal. Chem. **86**, 6827–6835 (2013)
10. Wu, M., Meng, Q., Chen, Y., Xu, P., Zhang, S., Li, Y., Zhang, L., Wang, M., Yao, H., Shi, J.: Ultrasmall confined iron oxide nanoparticle MSNs as a pH-responsive theranostic platform. Adv. Funct. Mater. **24**, 4273–4283 (2014)
11. Venditti, I., Fontana, L., Fratoddi, I., Battocchio, C., Cametti, C., Sennato, S., Mura, F., Sciubba, F., Delfini, M., Russo, M.V.: Direct interaction of hydrophilic gold nanoparticles with dexamethasone drug: loading and release study. J. Colloid Interface Sci. **418**, 52–60 (2014)
12. Bao, B., Ma, M., Chen, J., Yuwen, L., Weng, L., Fan, Q., Huang, W., Wang, L.: Facile preparation of multicolor polymer nanoparticle bioconjugates with specific biorecognition. ACS Appl. Mater. Interfaces **6**, 11129–11135 (2014)
13. Biju, V.: Chemical modifications and bioconjugate reactions of nanomaterials for sensing, imaging, drug delivery and therapy. Chem. Soc. Rev. **43**, 744–764 (2014)
14. Mura, S., Zouhiri, F., Lerondel, S., Maksimenko, A., Mougin, J., Gueutin, C., Brambilla, D., Caron, J., Sliwinski, E., LePape, A., Desmaele, D., Couvreur, P.: Novel isoprenoyl nanoassembled prodrug for paclitaxel delivery. Bioconjug. Chem. **24**, 1840–1849 (2013)
15. Lu, Y., Wang, L., Chen, D., Wang, G.: Determination of the concentration and the average number of gold atoms in a gold nanoparticle by osmotic pressure. Langmuir **28**, 9282–9287 (2012)
16. Gamrad, L., Rehbock, C., Krawinkel, J., Tumursukh, B., Heisterkamp, A., Barcikowski, S.: Charge balancing of model gold-nanoparticle-peptide conjugates controlled by the peptide's net charge and the ligand to nanoparticle ratio. J. Phys. Chem. C **118**, 10302–10313 (2014)
17. Ma, X., Cheng, Z., Jin, Y., Liang, X., Yang, X., Dai, Z., Tian, J.: SM5-1-conjugated PLA nanoparticles loaded with 5-fluorouracil for targeted hepatocellular carcinoma imaging and therapy. Biomaterials **35**, 2878–2889 (2014)
18. Algar, W.R., Prasuhn, D.E., Stewart, M.H., Jennings, T.L., Blanco-Canosa, J.B., Dawson, P. E., Medintz, I.L.: The controlled display of biomolecules on nanoparticles: a challenge suited to bioorthogonal chemistry. Bioconjug. Chem. **22**, 825–858 (2011)
19. Liu, H.Y., Gao, X.: Engineering monovalent quantum dot—antibody bioconjugates with a hybrid gel system. Bioconjugate Chem. **22**, 510–517 (2011)
20. Su, H., Yuan, R., Chai, Y., Zhuo, Y.: Enzyme-nanoparticle conjugates at oil–water interface for amplification of electrochemical immunosensing. Biosens. Bioelectron. **33**, 288–292 (2012)
21. Zhu, Y., Chandra, P., Shim, Y.-B.: Ultrasensitive and selective electrochemical diagnosis of breast cancer based on a hydrazine–Au nanoparticle—aptamer bioconjugate. Anal. Chem. **85**, 1058–1064 (2012)
22. Longmuir, K., Robertson, R., Haynes, S., Baratta, J., Waring, A.: Effective targeting of liposomes to liver and hepatocytes in vivo by incorporation of a plasmodium amino acid sequence. Pharmaceut. Res. **23**, 759–769 (2006)
23. Rossi, L., Quach, A., Rosenzweig, Z.: Glucose oxidase–magnetite nanoparticle bioconjugate for glucose sensing. Anal. Bioanal. Chem. **380**, 606–613 (2004)
24. Lin, Q., Jin, C.S., Huang, H., Ding, L., Zhang, Z., Chen, J., Zheng, G.: Nanoparticle-enabled, image-guided treatment planning of target specific RNAi therapeutics in an orthotopic prostate cancer model. Small **10**, 3072–3082 (2014)
25. Mader, H.S., Kele, P., Saleh, S.M., Wolfbeis, O.S.: Upconverting luminescent nanoparticles for use in bioconjugation and bioimaging. Curr. Opin. Chem. Biol. **14**, 582–596 (2010)
26. Sperling, R.A., Parak, W.J.: Surface modification, functionalization and bioconjugation of colloidal inorganic nanoparticles. Philos. Trans. Math. Phys. Eng. Sci. **368**, 1333–1383 (2010)
27. Lin, S.-Y., Tsai, Y.-T., Chen, C.-C., Lin, C.-M., Chen, C.-H: Two-step functionalization of neutral and positively charged thiols onto citrate-stabilized Au nanoparticles. J. Phys. Chem. B **108**, 2134–2139 (2004)

28. Ji, X., Copenhaver, D., Sichmeller, C., Peng, X.: Ligand bonding and dynamics on colloidal nanocrystals at room temperature: the case of alkylamines on CdSe nanocrystals. J. Am. Chem. Soc. **130**, 5726–5735 (2008)
29. Kim, M., Chen, Y., Liu, Y., Peng, X.: Super-stable, high-quality Fe_3O_4 dendron-nanocrystals dispersible in both organic and aqueous solutions. Adv. Mater. **17**, 1429–1432 (2005)
30. Gittins, D.I., Caruso, F.: Biological and physical applications of water-based metal nanoparticles synthesised in organic solution. ChemPhysChem **3**, 110–113 (2002)
31. Bogdan, N., Vetrone, F., Ozin, G.A., Capobianco, J.A.: Synthesis of ligand-free colloidally stable water dispersible brightly luminescent lanthanide-doped upconverting nanoparticles. Nano Lett. **11**, 835–840 (2011)
32. Dong, A., Ye, X., Chen, J., Kang, Y., Gordon, T., Kikkawa, J.M., Murray, C.B.: A generalized ligand-exchange strategy enabling sequential surface functionalization of colloidal nanocrystals. J. Am. Chem. Soc. **133**, 998–1006 (2010)
33. Liu, J.-N., Bu, W., Pan, L.-M., Zhang, S., Chen, F., Zhou, L., Zhao, K.-L., Peng, W., Shi, J.: Simultaneous nuclear imaging and intranuclear drug delivery by nuclear-targeted multifunctional upconversion nanoprobes. Biomaterials **33**, 7282–7290 (2012)
34. Lu, Z., Gao, C., Zhang, Q., Chi, M., Howe, J.Y., Yin, Y.: Direct assembly of hydrophobic nanoparticles to multifunctional structures. Nano Lett. **11**, 3404–3412 (2011)
35. Zhang, Y., Zheng, F., Yang, T., Zhou, W., Liu, Y., Man, N., Zhang, L., Jin, N., Dou, Q., Zhang, Y., Li, Z., Wen, L.-P.: Tuning the autophagy-inducing activity of lanthanide-based nanocrystals through specific surface-coating peptides. Nat. Mater. **11**, 817–826 (2012)
36. Yi, G.S., Chow, G.M.: Synthesis of hexagonal-phase $NaYF_4$:Yb,Er and $NaYF_4$:Yb,Tm nanocrystals with efficient up-conversion fluorescence. Adv. Funct. Mater. **16**, 2324–2329 (2006)
37. Wong, H.-T., Vetrone, F., Naccache, R., Chan, H.L.W., Hao, J., Capobianco, J.A.: Water dispersible ultra-small multifunctional $KGdF_4$:Tm^{3+}, Yb^{3+} nanoparticles with near-infrared to near-infrared upconversion. J. Mater. Chem. **21**, 16589–16596 (2011)
38. Boyer, J.-C., Manseau, M.-P., Murray, J.I., van Veggel, F.C.J.M.: Surface modification of upconverting $NaYF_4$ nanoparticles with PEG-phosphate ligands for NIR (800 nm) biolabeling within the biological window. Langmuir **26**, 1157–1164 (2009)
39. Hou, Y., Qiao, R., Fang, F., Wang, X., Dong, C., Liu, K., Liu, C., Liu, Z., Lei, H., Wang, F., Gao, M.: $NaGdF_4$ nanoparticle-based molecular probes for magnetic resonance imaging of intraperitoneal tumor xenografts in vivo. ACS Nano **7**, 330–338 (2012)
40. Zhang, C., Yuan, Y., Zhang, S., Wang, Y., Liu, Z.: Biosensing platform based on fluorescence resonance energy transfer from upconverting nanocrystals to graphene oxide. Angew. Chem. Int. Ed. **50**, 6851–6854 (2011)
41. Liu, C., Wang, Z., Jia, H., Li, Z.: Efficient fluorescence resonance energy transfer between upconversion nanophosphors and graphene oxide: a highly sensitive biosensing platform. Chem. Commun. **47**, 4661–4663 (2011)
42. Liu, Y., Tu, D., Zhu, H., Li, R., Luo, W., Chen, X.: A strategy to achieve efficient dual-mode luminescence of Eu^{3+} in lanthanides doped multifunctional $NaGdF_4$ nanocrystals. Adv. Mater. **22**, 3266–3271 (2010)
43. Chen, G., Ohulchanskyy, T.Y., Law, W.C., Agren, H., Prasad, P.N.: Monodisperse $NaYbF_4$: Tm^{3+}/$NaGdF_4$ core/shell nanocrystals with near-infrared to near-infrared upconversion photoluminescence and magnetic resonance properties. Nanoscale **3**, 2003–2008 (2011)
44. Liu, C., Wang, H., Li, X., Chen, D.: Monodisperse, size-tunable and highly efficient [small beta]-$NaYF_4$:Yb,Er(Tm) up-conversion luminescent nanospheres: controllable synthesis and their surface modifications. J. Mater. Chem. **19**, 3546–3553 (2009)
45. Meiser, F., Cortez, C., Caruso, F.: Biofunctionalization of fluorescent rare-earth-doped lanthanum phosphate colloidal nanoparticles. Angew. Chem. Int. Ed. **43**, 5954–5957 (2004)
46. Liu, R., Tu, D., Liu, Y., Zhu, H., Li, R., Zheng, W., Ma, E., Chen, X.: Controlled synthesis and optical spectroscopy of lanthanide-doped $KLaF_4$ nanocrystals. Nanoscale **4**, 4485–4491 (2012)

47. Liu, K., Liu, X., Zeng, Q., Zhang, Y., Tu, L., Liu, T., Kong, X., Wang, Y., Cao, F., Lambrechts, S.A.G., Aalders, M.C.G., Zhang, H.: Covalently assembled NIR nanoplatform for simultaneous fluorescence imaging and photodynamic therapy of cancer cells. ACS Nano **6**, 4054–4062 (2012)
48. Zhang, T., Ge, J., Hu, Y., Yin, Y.: A general approach for transferring hydrophobic nanocrystals into water. Nano Lett. **7**, 3203–3207 (2007)
49. Willis, A.L., Turro, N.J., O'Brien, S.: Spectroscopic characterization of the surface of iron oxide nanocrystals. Chem. Mater. **17**, 5970–5975 (2005)
50. Nikolic, M.S., Krack, M., Aleksandrovic, V., Kornowski, A., Förster, S., Weller, H.: Tailor-made ligands for biocompatible nanoparticles. Angew. Chem. Int. Ed. **45**, 6577–6580 (2006)
51. Hilderbrand, S.A., Shao, F., Salthouse, C., Mahmood, U., Weissleder, R.: Upconverting luminescent nanomaterials: application to in vivo bioimaging. Chem. Commun. 4188–4190 (2009)
52. Boyer, J.-C., Manseau, M.-P., Murray, J.I., van Veggel, F.C.J.M.: Surface modification of upconverting $NaYF_4$ nanoparticles with PEG-phosphate ligands for NIR (800 nm) biolabeling within the biological window. Langmuir **26**, 1157–1164 (2009)
53. Li, L.-L., Wu, P., Hwang, K., Lu, Y.: An exceptionally simple strategy for DNA-functionalized up-conversion nanoparticles as biocompatible agents for nanoassembly, DNA delivery, and imaging. J. Am. Chem. Soc. **135**, 2411–2414 (2013)
54. Chen, Z., Chen, H., Hu, H., Yu, M., Li, F., Zhang, Q., Zhou, Z., Yi, T., Huang, C.: Versatile synthesis strategy for carboxylic acid-functionalized upconverting nanophosphors as biological labels. J. Am. Chem. Soc. **130**, 3023–3029 (2008)
55. Zhou, H.-P., Xu, C.-H., Sun, W., Yan, C.-H.: Clean and flexible modification strategy for carboxyl/aldehyde-functionalized upconversion nanoparticles and their optical applications. Adv. Funct. Mater. **19**, 3892–3900 (2009)
56. Cheng, L., Yang, K., Zhang, S., Shao, M., Lee, S., Liu, Z.: Highly-sensitive multiplexed in vivo imaging using pegylated upconversion nanoparticles. Nano Res. **3**, 722–732 (2010)
57. Jiang, G., Pichaandi, J., Johnson, N.J.J., Burke, R.D., van Veggel, F.C.J.M.: An effective polymer cross-linking strategy to obtain stable dispersions of upconverting $NaYF_4$ nanoparticles in buffers and biological growth media for biolabeling applications. Langmuir **28**, 3239–3247 (2012)
58. Budijono, S.J., Shan, J., Yao, N., Miura, Y., Hoye, T., Austin, R.H., Ju, Y., Prud'homme, R. K.: Synthesis of stable block-copolymer-protected $NaYF_4$:Yb^{3+}, Er^{3+} up-converting phosphor nanoparticles. Chem. Mater. **22**, 311–318 (2009)
59. Cao, T., Yang, Y., Gao, Y., Zhou, J., Li, Z., Li, F.: High-quality water-soluble and surface-functionalized upconversion nanocrystals as luminescent probes for bioimaging. Biomaterials **32**, 2959–2968 (2011)
60. Yi, G.-S., Chow, G.-M.: Water-soluble $NaYF_4$:Yb,Er(Tm)/$NaYF_4$/polymer core/shell/shell nanoparticles with significant enhancement of upconversion fluorescence. Chem. Mater. **19**, 341–343 (2006)
61. Nichkova, M., Dosev, D., Gee, S.J., Hammock, B.D., Kennedy, I.M.: Microarray immunoassay for phenoxybenzoic acid using polymer encapsulated Eu:Gd_2O_3 nanoparticles as fluorescent labels. Anal. Chem. **77**, 6864–6873 (2005)
62. Yao, C., Wang, P., Zhou, L., Wang, R., Li, X., Zhao, D., Zhang, F.: Highly biocompatible zwitterionic phospholipids coated upconversion nanoparticles for efficient bioimaging. Anal. Chem. (2014)
63. Wang, L., Yan, R., Huo, Z., Wang, L., Zeng, J., Bao, J., Wang, X., Peng, Q., Li, Y.: Fluorescence resonant energy transfer biosensor based on upconversion-luminescent nanoparticles. Angew. Chem. Int. Ed. **44**, 6054–6057 (2005)
64. Sivakumar, S., Diamente, P.R., van Veggel, F.C.J.M.: Silica-coated Ln^{3+}-doped LaF_3 nanoparticles as robust down- and upconverting biolabels. Chem. A Eur. J. **12**, 5878–5884 (2006)

65. Liu, Z., Yi, G., Zhang, H., Ding, J., Zhang, Y., Xue, J.: Monodisperse silica nanoparticles encapsulating upconversion fluorescent and superparamagnetic nanocrystals. Chem. Commun. 694–696 (2008)
66. Sivakumar, S., Diamente, P.R., van Veggel, F.C.J.M.: Silica-coated Ln^{3+}-doped LaF_3 nanoparticles as robust down- and upconverting biolabels. Chem. A Eur. J. **12**, 5878–5884 (2006)
67. Hu, H., Xiong, L., Zhou, J., Li, F., Cao, T., Huang, C.: Multimodal-luminescence core-shell nanocomposites for targeted imaging of tumor cells. Chem. A Eur. J. **15**, 3577–3584 (2009)
68. Zako, T., Nagata, H., Terada, N., Utsumi, A., Sakono, M., Yohda, M., Ueda, H., Soga, K., Maeda, M.: Cyclic RGD peptide-labeled upconversion nanophosphors for tumor cell-targeted imaging. Biochem. Biophys. Res. Commun. **381**, 54–59 (2009)
69. Kong, D., Quan, Z., Yang, J., Yang, P., Li, C., Lin, J.: Avidin conjugation to up-conversion phosphor $NaYF_4$:Yb^{3+}, Er^{3+} by the oxidation of the oligosaccharide chains. J. Nanopart. Res. **11**, 821–829 (2009)
70. Wang, M., Hou, W., Mi, C.-C., Wang, W.-X., Xu, Z.-R., Teng, H.-H., Mao, C.-B., Xu, S.-K.: Immunoassay of goat antihuman immunoglobulin G antibody based on luminescence resonance energy transfer between near-infrared responsive $NaYF_4$:Yb,Er upconversion fluorescent nanoparticles and gold nanoparticles. Anal. Chem. **81**, 8783–8789 (2009)
71. Wang, M., Mi, C.-C., Wang, W.-X., Liu, C.-H., Wu, Y.-F., Xu, Z.-R., Mao, C.-B., Xu, S.-K.: Immunolabeling and NIR-excited fluorescent imaging of HeLa cells by using $NaYF_4$:Yb,Er upconversion nanoparticles. ACS Nano **3**, 1580–1586 (2009)
72. Wang, L., Li, Y.: Green upconversion nanocrystals for DNA detection. Chem. Commun. 2557–2559 (2006)
73. Jiang, S., Zhang, Y.: Upconversion nanoparticle-based FRET system for study of siRNA in live cells. Langmuir **26**, 6689–6694 (2010)
74. Hu, H., Xiong, L., Zhou, J., Li, F., Cao, T., Huang, C.: Multimodal-luminescence core-shell nanocomposites for targeted imaging of tumor cells. Chem. A Eur. J. **15**, 3577–3584 (2009)
75. Wang, M., Hou, W., Mi, C.-C., Wang, W.-X., Xu, Z.-R., Teng, H.-H., Mao, C.-B., Xu, S.-K.: Immunoassay of goat antihuman immunoglobulin G antibody based on luminescence resonance energy transfer between near-infrared responsive $NaYF_4$:Yb,Er upconversion fluorescent nanoparticles and gold nanoparticles. Anal. Chem. **81**, 8783–8789 (2009)
76. Guo, H., Li, Z., Qian, H., Hu, Y., Muhammad, I.N.: Seed-mediated synthesis of $NaYF_4$:Yb,Er/$NaGdF_4$ nanocrystals with improved upconversion fluorescence and MR relaxivity. Nanotechnology **21**, 125602 (2010)
77. Li, Z., Zhang, Y.: Monodisperse silica-coated polyvinylpyrrolidone/$NaYF_4$ nanocrystals with multicolor upconversion fluorescence emission. Angew. Chem. **118**, 7896–7899 (2006)
78. Li, Z., Zhang, Y., Jiang, S.: Multicolor core/shell-structured upconversion fluorescent nanoparticles. Adv. Mater. **20**, 4765–4769 (2008)
79. Qian, H.S., Guo, H.C., Ho, P.C.-L., Mahendran, R., Zhang, Y.: Mesoporous-silica-coated up-conversion fluorescent nanoparticles for photodynamic therapy. Small **5**, 2285–5590 (2009)
80. HaiSheng, Q., ZhengQuan, L., Yong, Z.: Multicolor polystyrene nanospheres tagged with up-conversion fluorescent nanocrystals. Nanotechnology **19**, 255601 (2008)
81. Zhou, J., Yao, L., Li, C., Li, F.: A versatile fabrication of upconversion nanophosphors with functional-surface tunable ligands. J. Mater. Chem. **20**, 8078–8085 (2010)
82. Feng, W., Dev, K.C., Zhengquan, L., Yong, Z., Xianping, F., Minquan, W.: Synthesis of polyethylenimine/$NaYF_4$ nanoparticles with upconversion fluorescence. Nanotechnology **17**, 5786–5789 (2006)
83. Wang, F., Liu, X.: Upconversion multicolor fine-tuning: visible to near-infrared emission from lanthanide-doped $NaYF_4$ nanoparticles. J. Am. Chem. Soc. **130**, 5642–5643 (2008)
84. Godbey, W.T., Wu, K.K., Mikos, A.G.: Poly(ethylenimine) and its role in gene delivery. J. Controlled Release **60**, 149–260 (1999)

85. Yang, J., Shen, D., Li, X., Li, W., Fang, Y., Wei, Y., Yao, C., Tu, B., Zhang, F., Zhao, D.: One-step hydrothermal synthesis of carboxyl-functionalized upconversion phosphors for bioapplications. Chem. A Eur. J. **18**, 13642–13650 (2012)
86. Xiong, L.-Q., Chen, Z.-G., Yu, M.-X., Li, F.-Y., Liu, C., Huang, C.-H.: Synthesis, characterization, and in vivo targeted imaging of amine-functionalized rare-earth up-converting nanophosphors. Biomaterials **30**, 5592–5600 (2009)
87. Liu, Y., Tu, D., Zhu, H., Ma, E., Chen, X.: Lanthanide-doped luminescent nano-bioprobes: from fundamentals to biodetection. Nanoscale **5**, 1369–1384 (2013)

Chapter 6
Upconversion Nanoparticles for Biomedical Imaging

Abstract Rare-earth upconversion (UC) nanoparticles, when excited by continuous-wave near-infrared (NIR) light, exhibit a unique narrow photoluminescence with higher energy. Such special UC luminescence makes UCNPs promising as bio-imaging probes with attractive features, such as no auto-fluorescence from biological samples and deep penetration depth. As a result, UCNPs have emerged as novel imaging agents for small animals. In this chapter, recent reports regarding the biomedical imaging investigation using UCNPs are summarized. The applications of UCNPs for small-animal imaging, including optical imaging, optical tomography, and multimodal imaging are reviewed in detail. The recent techniques for upconversion imaging and toxicity assessment of UCNPs are also presented. Finally, we discuss the challenges and opportunities in the development of UCNPs for biomedical imaging.

Keywords Rare-earth upconversion nanoparticles · UCNPs · Toxicity assessment · NIR probe · Biomedical imaging · Luminescence diffuse optical tomography · Multimodal imaging · Magnetic resonance imaging · Positron emission tomography · Single-photon emission computed tomography

6.1 Introduction

Recently, the rapid development of life science and medicine has stimulated the investigation of biomedical imaging. This rising research area includes plenty of bioassay methods such as computed tomography (CT) [1–5], positron emission tomography (PET) [2, 3, 5–8], single-photon emission computed tomography (SPECT) [4, 5, 9], and magnetic resonance imaging (MRI) [5, 8, 10–12]. These tomographic imaging techniques play an important role on biomedical imaging due to their superiority of unlimited penetration depth. However, they also suffer from

Rui Wang and Fan Zhang contributed together to this chapter.

F. Zhang, *Photon Upconversion Nanomaterials*,
Nanostructure Science and Technology, DOI 10.1007/978-3-662-45597-5_6

some inevitable disadvantages such as limited spatial resolution and inability to visualize real-time dynamics owing to their long acquisition time [13].

Moreover, these tomographic imaging techniques always require radioactive sources and radioactive contrast agents which may be harmful to both patients and medical personnel. Compared to tomographic imaging techniques, optical biomedical imaging has the advantage of fast feedback, high sensitivity, and high resolution. Recent advances in this field have relied on the development and use of fluorescent-based probes such as organic dyes and quantum dots (QDs). However, current fluorescent probes all have drawbacks that limit their use for high-contrast in vivo imaging. For example, synthetic organic dyes are rapidly photobleached, making them unsuitable for longitudinal tracking experiments. Common binary QDs (e.g., CdTe, PbS) contain highly toxic elements, which significantly hampers their clinical application. To address this problem, Cd-free QDs (e.g., CuInS) were recently developed. However, the emission efficacy of these QDs decreases considerably when they are excited at longer wavelengths (>700 nm). To understand the wavelength dependence of optical imaging performance, we must consider the optical properties of biological tissue [14].

When biological tissue is irradiated with a beam of light, the main factors that lead to attenuation of signal proportional to the depth of the feature of interest are absorption and scattering of the photons induced by the biological tissue and water. This kind of attenuation is especially severe in the traditional visible imaging region (400–750 nm) [15]. As a result, the penetration depth in visible imaging region is limited to 1–3 mm. To overcome this problem, much effort has been made to develop and implement imaging agents outside visible imaging region and within the "biological transparency window" or the NIR window (750–900 nm) (Fig. 6.1a) [16–19].

The tissue optical penetration depth is generally determined by the absorption and scattering of the excitation and emission light, Bashkatov etc. have proposed a formula to calculate it: $\delta = (3\mu_a(\mu_a + \mu_s'))^{-1/2}$, where μ_a is the optical absorption extinction coefficient which depends on the wavelength (Fig. 6.1b), $\mu_s'(\sim\lambda^{-w})$ is the reduced scattering coefficient, and δ is the resulting penetration depth. The exponent (w) depends on the size and concentration of scatterers in the tissue and ranges from 0.22 to 1.68 for different tissues (Fig. 6.1c) [20, 21]. From the formula and above-mentioned figures, we can know that both water absorption and tissue scattering are much stronger in the visible region than in the NIR region, demonstrating the superiority of NIR light in theoretical analysis. Since the 1990s, NIR biomedical imaging has obtained extensive attention because of its deeper penetration in biological tissue than visible light [22–26]. To date, numerous luminescent materials, including NIR organic dyes [27], NIR metal complexes [28, 29], NIR QDs [30], and single-walled carbon nanotubes (SWNTs) [21, 31], have been developed as NIR fluorescent probes for fluorescence biomedical imaging. However, the NIR QDs always contain toxic elements such as mercury, lead, and arsenic, which are undesirable in biomedical applications [32, 33]. NIR Ag_2S QDs, in spite of low toxicity, suffer from bleaching from both air and irradiation light.

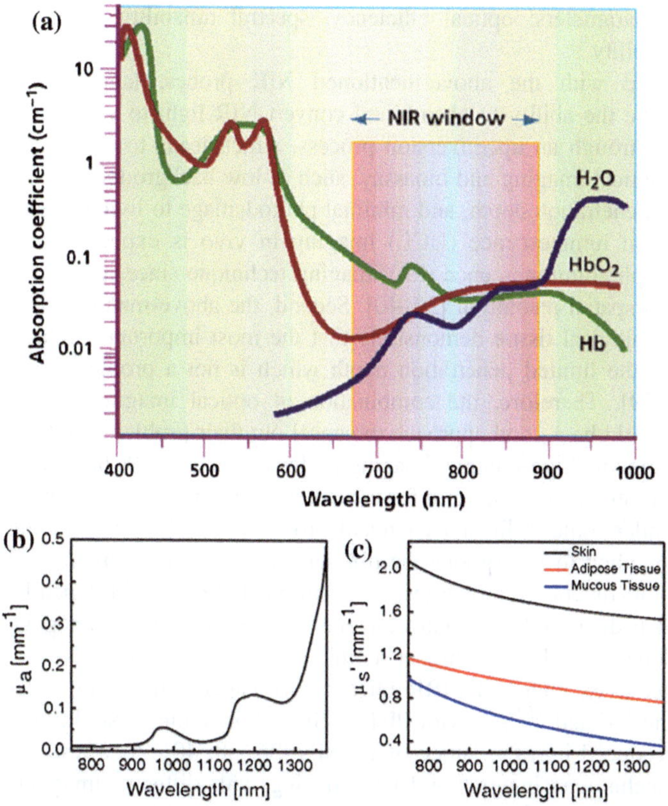

Fig. 6.1 **a** The absorption coefficient of hemoglobin and water in 400–1,000 nm. **b** Absorption coefficient, μ_a, of water. **c** Reduced scattering coefficient, μ_s', for skin, adipose tissue, and mucous tissue in 800–1,400 nm. **a** Reprinted with the permission from Ref. [16]. Copyright 2001 Nature Publishing Group. **b**, **c** Reprinted with the permission from Ref. [21]. Copyright 2011 National Academy of Sciences (USA)

Although The SWNTs have revealed nice performance in short-wavelength infrared (SWIR) biomedical imaging [21], there are several drawbacks which cannot be ignored before they are widely applied as a SWIR imaging agent, such as broad full width at half maximum (FWHM) and high excitation energy. Typically, the emission peaks of the SWNTs span are ~400 nm, hindering their tunability in spectral wavelength which puts them at a disadvantage in multi-label imaging because of the overlapping signals. SWNTs also require high-powered, pulsed excitation sources (>2 W) to achieve sufficient signal-to-background ratios (SBRs). Furthermore, additional nanotoxicology studies are still required to prove their biocompatibility [34]. Hence, the development of efficient and biocompatible NIR probes with excitation or emission within this NIR window is of great interest for high-contrast optical imaging of deep tissues, with careful consideration of the following parameters: optical efficiency, spectral tunability, photostability, and biocompatibility.

Compared with the above-mentioned NIR probes, lanthanide (Ln)-doped UCNPs have the ability to absorb and convert NIR light to visible/UV/NIR light efficiently through an upconversion process, which leads to many unique advantages for optical imaging and bioassay, such as low background auto-fluorescence, deep light penetration depth, and minimal photodamage to living organisms. Thus, upconversion luminescence (UCL) imaging in vivo is expected to be the next generation photoluminescence (PL) imaging technique since it provides high sensitivity and spatial resolution [35–40]. Second, the above-mentioned optical properties of biological tissue demonstrate that the most important obstacle to optical imaging is the limited penetration depth which is not a problem in tomographic imaging [29]. Therefore, the combination of optical imaging and tomographic imaging would be a good approach to cancel out their problems and combine their merits, achieving high-contrast imaging performance. UCNPs have moved into the spotlight as an efficient platform for constructing multi-functional nanoprobes. Due to the similar ionic radii and chemical properties of Ln ions, doping is a very common method to incorporate the different functional Ln ions in the UCNPs for desired multi-functional properties. For example, Gd-based/Gd-doped UCNPs have been used as dual-modality contrast agents for MRI and optical imaging. Gd-, Yb-, and Lu-doped UCNPs with higher atomic number and stronger X-ray attenuation have served as excellent CT/MRI/optical contrast agents. In addition, radionuclides, such as the ^{18}F and ^{153}Sm, with PET or SPECT imaging capability could also be easily introduced into the lattice of UCNPs [36, 38, 40]. As a result, multimodal imaging techniques integrating UCL imaging with different imaging modalities have been rapidly developed, which may allow medical imaging to be more sensitive and accurate [38].

6.2 In Vitro and In Vivo Toxicity Assessment

The potential toxicity and biocompatibility of UCNPs are of great importance for their biomedical applications. Similar to many other inorganic nanocrystals, UCNPs cannot be easily degraded in biological environment. Thus, understanding the behaviors of UCNPs in biological systems as well as their toxicology profiles is a critical fundamental question to be addressed in order to use this class of nanomaterials in medical applications in the future. Toxicity considerations should be integrated into materials selection since elements with non-toxicity profiles are most likely accepted for biomedical applications. Therefore, $NaYF_4$-based UCNPs are most investigated since yttrium has an innocuous toxicity property. Manganese is a kind of essential trace element in human bodies, but the tolerable limit of manganese is lower than that of yttrium. Other elements such as Gd^{3+} ions are toxic, which need further investigation when they are used in vivo [38].

6.2.1 Recent Progress on Toxicity Assessment of UCNPs

To date, various cell lines have been used to evaluate the cytotoxicity of different UCNPs through the CCK-8, MTT, and MTS assays [41–55]. Most of the results suggested that UCNPs were non-cytotoxic to a broad range of tested cell lines within a certain range of concentrations and a limited incubation period. The distribution of UCNPs in living HeLa cells was also investigated by the luminescence imaging technique [56]. The particles were internalized through endocytosis, transported by microtubule-dependent motor proteins (dyneins), accumulated at the perinuclear region, transported by another type of motor proteins (kinesins), and finally released out of the cells. As one of the simplest multi-cellular eukaryotic organisms with a nervous system, *Caenorhabditis elegans* (*C. elegans*) have also been used as a model animal for evaluating the toxicity of UCNPs. Lim's group [57], Yan's group [58], and Xu's group [59] reported in vivo toxicity studies with the feeding of UCNPs in the *C. elegans* through ingestion and excretion of UCNPs, observing life span, egg viability, the growth rate of UCNPs-treated worms, and so on. Their results indicated that UCNPs were not significantly toxic except at high concentrations of 10 mg/mL or higher [38].

The in vivo toxicity of UCNPs was also assessed through animal studies. For further applications of UCNPs in bioscience and medicine, the most important issue is still the potential long-term toxicity of UCNPs and their nanocomposites. It is well known that the long-term toxicity of UCNPs in small animals can be assessed by body weight measurement, histology analysis, serum biochemistry assays, etc. Many groups have reported that the UCNPs-treated mice could survive for more than 1 month without any apparent adverse effects to their health [45, 60–62]. For example, Li et al. [60] examined the biocompatibility of UCNPs coated with a silica shell or a layer of azelaic acid molecules. They found that these UCNPs had essentially no effect on the cell viability of human nasopharyngeal epidermal carcinoma KB cells after incubation for 20 h at a concentration of 800 mg/mL. Shan et al. [63] have also reported that there is limited or no toxicity of carboxyl- and amino-functionalized UCNPs after incubation for 9 days with human osteosarcoma cells. Zhang et al. [64] reported the ex vivo bio-distribution of PEI-modified and SiO_2-coated $NaYF_4$:Yb,Er/Tm UCNPs by measuring the concentration of Y^{3+} using coupled plasma atomic emission spectroscopy. They observed that the SiO_2-coated UCNPs first accumulated in the lung and heart, were gradually excreted by urine or feces, and finally cleared from a normal Wistar rat's body in 7 days. To investigate their biocompatibility and tissue distribution for longer times, healthy rats were injected intravenously with PAA-UCNPs. Biodistribution results showed that the uptake and retention of PAA-UCNPs took place primarily in the liver and the spleen. Most of the PAA-UCNPs were excreted from the body of mice in a very slow manner. In addition, histological, hematological, and biochemical analysis results indicated that there was no apparent toxicity of PAA-UCNPs in mice at long exposure times (up to 115 days) [60]. Zhao's group [65, 66] evaluated in vivo toxic effects of Gd/Mn-contained UCNPs using male and female mice.

Thirty days after injection, the survival rate of the mice was 100 %. And the weights of the mice did not noticeably decrease. These results indicated that Gd/Mn-contained UCNPs had low toxicity even though Gd and Mn are usually considered as toxic elements. The reason for its low toxicity may be that the Gd^{3+}/Mn^{2+} ions were confined in a rigid matrix of host NCs and not easily released to the environment. This has been confirmed by inductively coupled plasma mass spectrometry (ICP-MS) analysis. However, to ensure the suitability of the UCNPs for in vivo biological applications, further long-term (up to several months) toxicity studies concerning particle size, shape, and surface chemistry are necessary [38].

6.2.2 Evaluation Methods for the Long-Term Toxicity of UCNPs in Small Animals

6.2.2.1 Behavior Observation

Behavior observation is a straightforward method to evaluate the toxicity of UCNPs in small animals. Many research groups have described the behaviors of small animals after injection with various UCNPs. For example, Li et al. [60] reported that no changes in eating, drinking, fur color, exploratory behavior, activity, and neurological status were observed for the mice injected with PAA-UCNPs and for the control groups. They [62] also reported that $NaGdF_4$:Yb/Er/Tm nanocrystals-treated mice could survive for more than 1 month without showing abnormal behavior [54].

6.2.2.2 Body Weight Measurement

The fluctuation in body weight is a useful indicator for studying the toxicity effects of the UCNPs. The body weight of two groups of mice with and without tail vein injection of 15 mg/kg PAA-UCNPs was recorded [60]. Only small weight differences between the mice of the two groups were observed, indicating the low toxicity of PAA-UCNPs in mice. Furthermore, mice injected with the PAA-UCNPs underwent observational evaluations for 115 days. In addition, Zhang et al. [44] have reported that the silica-coated $NaYF_4$:Yb/Er nanocrystals showed no toxic effect on the body weight of rats at a dose of 10 mg/kg [54].

6.2.2.3 Histology Analysis

Usually, histological assessment of tissues has been conducted to determine whether or not the nanocrystals cause tissue damage, inflammation, or lesions from toxic exposure. At a dose of 15 mg/kg PAA-UCNPs [60], the structures of the organs (heart, lung, liver, spleen, and kidney) of the exposed mice were normal and hardly different from those of the control group. Cardiac muscle tissue in the heart samples showed no hydropic degeneration. Hepatocytes in the liver samples appeared normal, and there were no inflammatory infiltrates. No pulmonary fibrosis was observed in the lung samples. The glomerulus structure could be distinguished easily in the kidney samples. No necrosis was found in any of the groups. However, the spleen was slightly affected by the injection of PAA-UCNPs, since there was slight hyperplasia in the periarteriolar lymphoid sheath (PALS) of white pulp. This might be caused by the nanotoxicity of PAA-UCNPs [54].

6.2.2.4 Hematology Analysis

Nanocrystals are special particulate materials within the size regime of viruses and large proteins, and consequently, they may induce an inflammatory response, increase or decrease the activity of the immune system, and alter related hematological factors such as white blood cell count. Therefore, an assessment of standard hematological and biochemical markers was used to further quantify the potential toxicity of UCNPs [60].

For mice injected with 15 mg/kg PAA-UCNPs, blood smears indicated that the number and shape of red blood cells, platelet, and white blood cells were normal, suggesting that there was no significant toxicity in the test mice compared to the control mice. Furthermore, established serum biochemistry assays indicated that the three important hepatic indicators (alanine aminotransferase, aspartate aminotransferase, and total bilirubin) and two indicators for kidney functions (creatinine and urea) were at similar levels for the mice exposed to PAA-UCNPs and for the control mice. These results suggested no toxicity of PAA-UCNPs in mice over long exposure times (up to 115 days) [54].

6.3 Biomedical Imaging Using UCNPs Under 980-nm Excitation

UC materials were first used in tissue imaging in 1999, where Zijlmans et al. [67] reported the first UC bio-imaging based on submicron-sized Y_2O_2S:Yb/Tm particles (0.2–0.4 μm). Upon excitation at 980 nm, they observed a low auto-fluorescence signal and no bleaching even after continuous exposure to high excitation energy levels. After that, the idea of UC bio-imaging was realized by employing some other oxysulfides or oxides nanomaterials [66, 68]. In 2009, Hilderbrand et al. [68] reported the synthesis and surface modification of Y_2O_3:Yb/Er UCNPs. These UCNPs could be excited with non-harmful doses of light and had luminescence emission centered at 660 nm. Then, the utility of these particles for in vivo blood pool imaging was demonstrated by visualization of the vasculature in a mouse model after intravenous injection of these nanoparticles. Late in 2012, Zhao et al. [66] demonstrated the successful formation of size tunable Ln ion-doped Gd_2O_3 nanoparticles with novel multi-color UCL. When the Gd_2O_3:Yb/Er nanoparticles (50 nm in size) were injected into superficial skin sole of the foot, the thicker muscle of inner thigh (about 5 mm deep) and back, luminescence from the injected area could be observed in a darkroom by exciting with a 980-nm laser and recorded using CCD-based digital camera. However, the submicron size of the applied oxysulfides or oxides particles limited the applications. The necessary requirements for material selection in practical bio-imaging are small size, bright luminescence, and biological safety.

Recently, with the rapid development of synthesis techniques, the fluoride-based UCNPs with smaller size but high-quality luminescence has been explored extensively and widely used in cell, tissue, and animal imaging. In contrast to the oxysulfide or oxide host materials, fluorides are considered as a better kind of host material for the doping of Ln ions to achieve intense UC emissions, owning to their low phonon energies and the resulting minimization of the quenching of the excited state of the Ln ions. One of the first reports of in vivo imaging of UCNPs in small animals reported spot measurements of subcutaneously injected UCNPs in rats by Zhang et al. [64], which showed much deeper penetration depth under 980-nm excitation compared to commercial green emitting QDs excited by UV. In a later work, Liu et al. [42] compared the in vivo imaging sensitivity of UCNPs and QDs (QD545 and QD625) side by side and demonstrated the in vivo detection limit of UCNPs to be at least one order of magnitude lower than that of QDs. Although the NIR excitation light for UC materials has strong penetration ability, the UV/VIS UC emissions are still easily absorbed by biological samples, which definitely limit their further applications in the observation of deep biological tissues. To further increase the penetration depth of UCL imaging, tuning the upconversion emission into NIR spectral range (750–900 nm) and the red light region (600–750 nm) is essential for the deep-tissue imaging since the light scatting, absorbance, and auto-fluorescence of tissue in those range are minimal.

For small-animal in vivo imaging, highly sensitive NIR–NIR systems have attracted increasing attention, because both the excitation and emission lights are located in the "optical transmission window" of tissues (750–900 nm) (Fig. 6.1a). For this purpose, Tm^{3+} ions are frequently chosen as dopants since they can exhibit strong NIR emission between 750 and 850 nm due to the transition from $^3H_4$ to $^3H_6$ under CW laser excitation at 980 nm. The feature of NIR–NIR UCNPs allows high-contrast in vivo optical bio-imaging, as both light attenuation and scattering are significantly reduced in the NIR spectral range and the auto-fluorescence of tissues is absent under the conditions of UC excitation and emission. Prasad et al. [46] first reported high-contrast in vivo bio-imaging using NIR–NIR UC PL $NaYF_4:Yb^{3+}/Tm^{3+}$ UCNPs (Fig. 6.2a, b). However, the weak PL that can only be detected by highly sensitive electron multiplied charge-coupled devices (EMCCD) is still a substantial limitation for improving the SBR and the imaging depth. Li et al. [69] reported on sub-10-nm citric acid-capped hexagonal $NaLuF_4:Gd^{3+}/Yb^{3+}/Tm^{3+}$ UCNPs for whole-body PL imaging of a mouse. The UC quantum yield (800 nm)

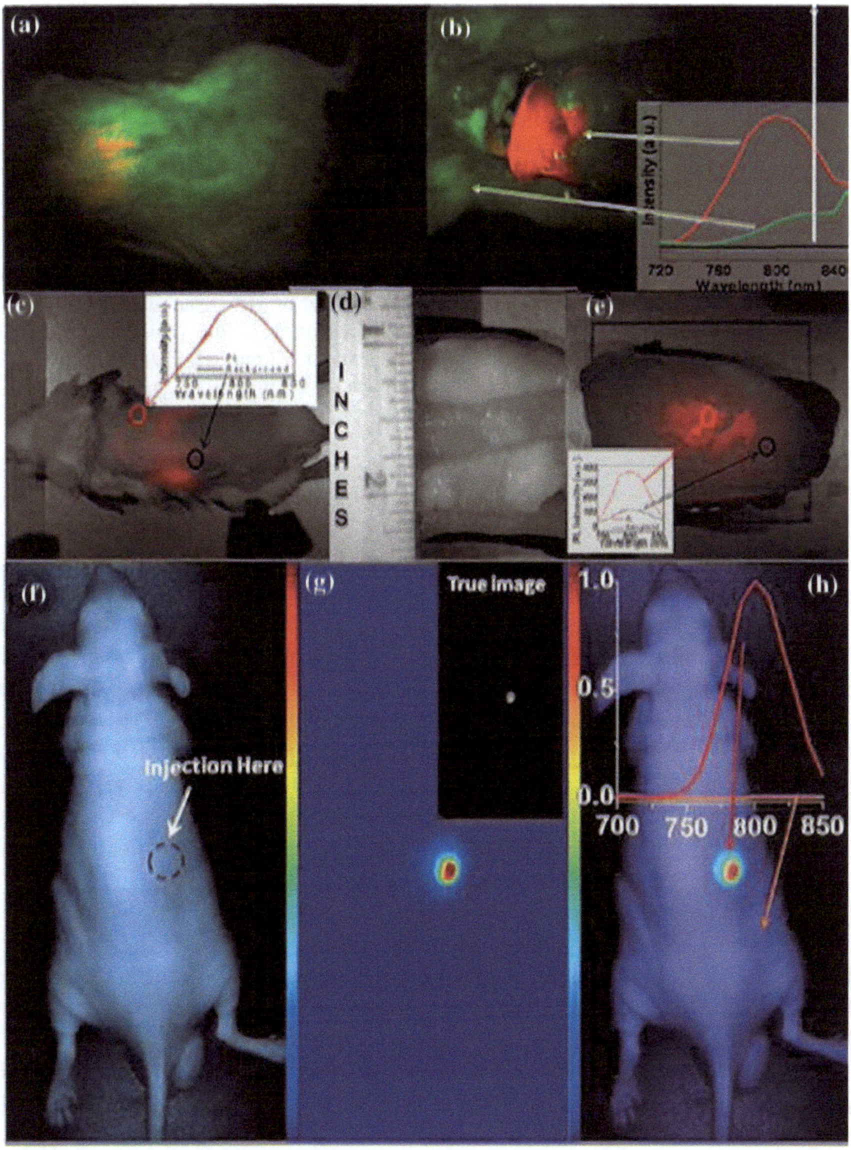
(a)
(b)
Intensity (a.u.)
720
760
800
840
(c)
(d)
INCHES
(e)
(f)
Injection Here
(g)
True image
1.0
0.5
0.0
700
750
800
850
(h)

◀ **Fig. 6.2** Whole-body images of mouse injected with $NaYF_4$:Yb^{3+}/Tm^{3+} UCNPs; intact mouse (**a**), same mouse after dissection (**b**). The *red color* indicates emission from UCNPs; *green* and *black* show background as indicated by the *arrows*. The *inset* presents the PL spectra corresponding to the spectrally unmixed components of the multispectral image obtained with the Maestro system. **c** Whole-body images of a BALB/c mouse injected via tail vein with the hyaluronic acid-coated α-($NaYbF_4$:Tm^{3+})/CaF_2. **d** Bright-field image of the pork tissue (*side view*), displaying the imaging depth; **e** UC PL image of the cuvette containing α-($NaYbF_4$:Tm^{3+})/CaF_2 covered with a pork tissue. The *insets* in **c** and **e** show the spectra of the NIR UC PL and background taken from the *circled area*. **f–h** In vivo whole-body image of a $NaYbF_4$:Yb^{3+}/Tm^{3+}-injected nude mouse: **f** bright-field image, **g** pseudo-color image obtained from true image (the *inset black/white image*), and **h** superimposed image (bright-field image and pseudo-color image) with the unmixed spectra of in vivo image (the *inset chart*) of UC signal and background as indicated by the *arrows*. **a**, **b** Reprinted with the permission from Ref. [46]. Copyright 2008 American Chemical Society. **c–e** Reprinted with the permission from Ref. [70]. Copyright 2012 American Chemical Society. **f–h** Reprinted with the permission from Ref. [76]. Copyright 2011 American Chemical Society

for these nanoparticles was measured to be about 0.47 % under excitation density of 17.5 W/cm^2 at 980 nm. A high-contrast UC PL imaging of the whole body of a mouse with a penetration depth of about 2 cm and an excellent detection limit of 50 cells was achieved. More recently, Prasad et al. [70] reported on the development of novel efficient core/shell NIR–NIR $NaYbF_4$:Tm^{3+}/CaF_2 UCNPs, which were intravenously injected in BALB/c mice to perform whole-body imaging; a high SBR of 310 was achieved (Fig. 6.2c). The quantum yield could reach as high as 0.6 % under imaging excitation conditions of 0.3 W/cm^2 at 980 nm. It was shown that the UC PL signal could be readily detected and imaged, with a low background, through a 3.2-cm-thick pork tissue (Fig. 6.2d, e). This study provides a foundation for the development of 3-D imaging systems for advanced whole-body optical imaging. Up to now, in vivo whole-body imaging of small animals has been successfully realized based on Tm^{3+}-doped $NaYF_4$ [71], $NaGdF_4$ [62], $NaLuF_4$ [72], and $NaYbF_4$ [73] nanocrystals [35].

Research on biomedical imaging using red-emitting UCNPs is rising up recently. In 2011, Liu et al. [47] reported an oil-based synthetic method for the preparation of $KMnF_3$ nanocrystals with Ln dopants homogeneously incorporated in the host lattice. As a result of efficient energy transfer between the dopant ion and host Mn^{2+} ion, remarkably pure single-band UC emissions were generated in the red region, showing its promise in deep penetration biomedical imaging. In 2012, Zhao et al. [65] reported the single-band red-emitting Mn-doped $NaYF_4$:Yb/Er UCNPs for in vivo imaging and extended the imaging depth to 15 mm. Dominant red UCL makes them attractive bio-probes for in vitro and in vivo bio-imaging. Moreover, the results from in vitro cell imaging, localized UCL spectra, and in vivo bio-imaging reveal that these UCNPs can have applications in optical bio-imaging with many advantages, such as high contrast, high signal-to-noise ratio (SNR), deeper tissue penetration, and absence of auto-fluorescence.

6.4 Biomedical Imaging Using UCNPs Excited at Different Wavelength

Traditionally, highly efficient UCNPs require a NIR laser at wavelength of about 980 nm as the excitation source since the sensitizer ion Yb^{3+} has a high absorption cross section in its absorption band. However, 980-nm light suffers from strong absorption by water and biological specimens (Fig. 6.1a). Thus, 980-nm excitation is unattractive due to problems such as limited penetration depth and tissue damages because of sample overheating. Therefore, the excitation band tuning of UCNPs into an appreciate range is also an important issue for improving the performance of UCNPs. Zou et al. [74] reported the concept of an UCNPs where an organic NIR dye is used as an antenna to harvest the NIR photons with a broad band (740–850 nm) for the β-$NaYF_4$:Yb/Er nanoparticles in which the upconversion occurs. The overall upconversion by the dye-sensitized nanoparticles is dramatically enhanced (by a factor of ~3,300) as a result of increased absorptivity and overall broadening of the absorption spectrum of the upconverter. Prasad group [75] also reported the intense UC PL in colloidal $LiYF_4$:Er^{3+} nanocrystals under excitation with telecom wavelength at 1,490 nm. The intensities of two- and three-photon anti-Stokes UC PL bands are higher than or comparable to that of the Stokes emission under excitation with low power density in the range 5–120 W/cm^2. The QY of the UC PL was measured to be as high as $\sim 1.2 \pm 0.1$ %, which is almost four times higher than the highest UC PL efficiency (0.3 ± 0.1 %) reported to date for Ln-doped nanocrystals in 100-nm-sized hexagonal $NaYF_4$:Yb^{3+}/Er^{3+} using excitation at ~980 nm. All of the above-mentioned works are very important to the investigation of a novel excitation source for traditional UCNPs. But they also suffer from their own troubles: organic dyes are unstable, and the efficiency of antenna is very sensitive to the loading capacity of the organic dyes, even a small addition of dyes will result in extremely serious quenching. Meanwhile, as an excitation source, 1,490-nm light suffers from serious water absorption, and this kind of upconversion is restricted to the crystal parameters of host matrix, hindering their further application in bio-imaging.

To avoid the drawbacks which are brought by the excitation source, such as heat generation, water absorption, and tissue damage, further optimization on the excitation source is required for UCNPs-based bio-imaging. Zhan et al. [76] demonstrated a promising excitation approach for better NIR-to-NIR UC PL for in vitro or in vivo imaging employing a cost-effective 915-nm laser. This novel laser excitation method provides drastically less heating of the biological specimen and larger imaging depth in the animals or tissues due to low water absorption (Fig. 6.2f–h). However, it is still far from applications with flexibility due to the sacrifice of UC efficiency. Recently, researchers have moved their attention to 800-nm excitation UCNPs based on the large absorption cross section of Nd^{3+} at 800 nm (around one order of magnitude higher than Yb^{3+} at 980 nm). Han et al. [77] found that when Nd^{3+}, Yb^{3+}, and Er^{3+}/Tm^{3+} were codoped in $NaYF_4$, UC signals could be distinguished by the naked eye upon 800-nm excitation. In this novel UCNPs, Nd^{3+} serves as an 800-nm photon sensitizer and Yb^{3+} as a bridging ion, with an energy transfer path of $Nd^{3+} \rightarrow Yb^{3+} \rightarrow Er^{3+}/Tm^{3+}$. However, the doping limit of the Nd^{3+} was only 1 % because of concentration quenching. Novel structures need to be designed to raise the Nd^{3+} concentration limit. In 2013, Liu et al. [78] reported a new type of core/shell UCNPs. Through spatially confined doping of Nd^{3+} which can be effectively excited at 795 nm, he claimed that the active $NaYF_4:Nd^{3+}$ shell layer can effectively prevent surface quenching of Yb^{3+} emission and simultaneously promote the transfer of excitation energy to Yb^{3+} ions, which significantly enhances the UC efficiency. Furthermore, 800 nm generates less heat than 980-nm light, confirmed by cell irradiation experiment. Almost at the same time, Yan and his coworker [79] reported a similar core/shell structure with the similar design idea of doping Nd^{3+} in the shell to ensure successive $Nd^{3+} \rightarrow Yb^{3+} \rightarrow$ activator energy transfer. In vivo imaging of a nude mouse subcutaneously injected with UCNPs demonstrated that comparable photon numbers can be measured when irradiated with 980-nm laser and 808-nm laser, respectively. In addition to the UC process, the authors also explored downconversion (DC) PL of Ln-doped nanocrystals. In vivo NIR imaging of a nude mouse injected with "UCNPs" showed high SNR under 808-nm laser excitation. Later in 2013, Li et al. [80] designed and synthesized a novel structure: $NaGdF_4$:Nd/$NaYF_4$/$NaGdF_4$:Nd/Yb/Er/$NaYF_4$C/S1/S2/S3 NCs, each of the parts assume a respective role and work together to achieve dual-mode UC and DC luminescence upon the low heat effect 800-nm excitation. Nd^{3+}, Yb^{3+}, Er^{3+} tri-doped $NaGdF_4$:Nd/Yb/Er UC layer [NIR (800 nm)-to-visible (540 nm)] with a constitutional efficient 800-nm excitable property was achieved for in vitro bio-imaging with low auto-fluorescence and photodamage effects. Moreover, typical NIR (800 nm)-to-NIR (860–895 nm) DC luminescence of Nd^{3+} has also been realized with this designed nanostructure. With the low heat effect, high penetration depth of the excitation, and the high efficiency of the DC luminescence, the in vivo high-contrast DC imaging of a whole-body nude mouse was achieved. These above results indicate that UCNPs which can be excited under 800-nm nanocrystals have a bright future in biomedical imaging development.

6.5 Ultrasmall UCNPs for Biomedical Imaging

Photoluminescent nanomaterials with large surface areas, tunable optical properties, and strong signal output can potentially serve as a new generation of biomedical imaging probes. These nanomaterials often passively accumulate at disease sites such as tumors in much higher concentrations and for longer times than small drug molecules through the well-known enhanced permeability and retention (EPR) effect [81, 82], making nanomaterials even more promising in addressing many challenges that small drug molecules encounter in early cancer diagnosis and therapy. For example, $Y_2O_3:Yb^{3+}/Er^{3+}$ nanoparticles coated with PEG polymer were used for in vivo vascular imaging of nude mice. The polymer coating minimizes non-specific tissue binding and prolongs the circulation half-lives of the particles in the blood [83].

This unique strength of nanomaterials in tumor targeting is fundamentally due to the ability of nanomaterials to escape kidney filtration and be retained in the blood plasma for longer times than small molecules. However, unlike small, renal-clearable, clinically used molecular probes such as ultrasmall noble metal nanoparticles, quantum dots, and organic dyes, Ln-based UCNPs often severely accumulate in reticuloendothelial system (RES) organs (liver, spleen, etc.) [84], resulting in low targeting specificity (defined as the amount of probe in tumor vs. that in liver) and potential long-term toxicity, hampering their clinical use. To minimize non-specific accumulation in RES organs and potential toxicity of UCNPs, several different kinds of renal-clearable Ln-based UCNPs have recently been developed as we have discussed in Chap. 3. In this chapter, we mainly focus on the in vitro and in vivo imaging using these ultrasmall UCNPs [85, 86].

In 2011, Li et al. [69] reported the synthesis of sub-10-nm Gd^{3+}-doped β-$NaLuF_4$ nanocrystals material with bright UCL by thermal decomposition in the presence of oleylamine. Such sub-10-nm β-$NaLuF_4$:Gd,Yb,Tm nanocrystals display bright UCL emission with a quantum yield of 0.47 ± 0.06 %, indicating that sub-10-nm hexagonal $NaLuF_4$:Gd^{3+}/Yb^{3+}/Tm^{3+} (7.8 nm) is an excellent nanomaterial for UC-based biomedical imaging. After synthesis, the hydrophobic $NaLuF_4$:Gd^{3+}/Yb^{3+}/Tm^{3+} and $NaYF_4$:Yb^{3+}/Tm^{3+} nanocrystals were converted into hydrophilic ones by replacing the surface oleylamine with citric acid. The obtained hydrophilic nanocrystals were denoted as cit-Lu6-Tm and cit-Y1-Tm, respectively.

To confirm further the penetration depth of cit-Lu6-Tm for UC imaging in vivo, a black mouse was selected as a model animal. As far as we know, a black mouse has never been used to investigate the penetration depth of conventional fluorescent materials because of the strong absorption of excitation and emission. The SNR of a black mouse injected subcutaneously with cit-Lu6-Tm was 12-fold larger than that with cit-Y1-Tm. The UC signal from cit-Lu6-Tm also could be detected even from the back side, suggesting that high-contrast UC imaging of a whole-body black mouse with a penetration depth of ~2 cm was achieved. This showed that ultrasmall $NaLuF_4$:Gd^{3+}/Yb^{3+}/Tm^{3+} has great penetration capability. Li et al. also investigated the detection limits of cells which were labeled by ultrasmall $NaLuF_4$:Gd^{3+}/Yb^{3+}/Tm^{3+}. By subcutaneous injection of cit-Lu6-Tm-labeled KB cells into a nude mouse, the detection threshold of cit-Lu6-Tm-labeled cells was measured. They found that UC signal can be detected with SNR > 3 for a nude mouse injected subcutaneously with 50 cit-Lu6-Tm-labeled KB cells (incubation with 200 μg/mL cit-Lu6-Tm in pH 7 phosphate-buffered saline for 2 h at 37 °C) under 980-nm excitation at 120 mW/cm^2, indicating that the detection limit of UCNP-labeled cells injected subcutaneously is <50.

Up to now, few works involving renal-clearable UC imaging have been reported, although several groups have investigated the synthesis and luminescence optimization of ultrasmall UCNPs. The biomedical imaging using ultrasmall UCNPs is limited to the in situ imaging (subcutaneous imaging or Sentinel lymph node imaging), rather than targeted imaging. Thus, although the synthesis of ultrasmall

UCNPs has made a great progress in this field, the clearance property of ultrasmall UCNPs is still unclear by now.

6.6 Optical Tomography

Luminescence diffuse optical tomography (LDOT) is a form of CT that creates a digital volumetric model of a fluorophore-sustained object by reconstructing images made from light scattered through an object. In a typical experiment, a highly scattering tissue medium is illuminated by a narrow collimated beam, and the luminescence from the sustained target that propagates through the medium is collected by an array of detectors attached to the tissue surface. The shape of the target can be reconstructed on the basis of the recorded optical data, which discerns the target from the surrounding environment. LDOT is a compact, fast, and highly sensitive technique for three-dimensional deep-tissue imaging of fluorescent targets. These properties and the noninvasive nature of LDOT have made it very attractive for longitudinal studies of small animals, where it has been applied to follow the development in time of, for example, cancer tumors, proteases, Alzheimer's disease, and different drug effects [35].

The contrast agents used in LDOT are based on endogenous fluorophores such as hemoglobin, collagen, and elastin. The utilization of exogenous Stokes-shifted contrast agents such as molecular dyes or QDs can improve the detection sensitivity. However, auto-fluorescence and noise are still nuisances for high-contrast LDOT. Although much of the noise can be suppressed by employing low-noise equipment, the tissue auto-fluorescence remains to plague the measurements with traditional Stokes-shifted fluorophores. In addition, the linear power density dependence of the emission from these fluorophores results in poor spatial resolution in the reconstructed images, as sharp spatial features are inevitably smeared out during diffuse-light propagation [35].

The use of UCNPs can completely eliminate the tissue auto-fluorescence because the emission from endogenous fluorophores is Stokes-shifted [87]. Hence, the UCNPs have been suggested as an alternative to fluorophores in LDOT. Xu et al. [88] have recently demonstrated the use of $NaYF_4{:}Yb^{3+}/Tm^{3+}$ nanocrystals for DOT scanning in a controlled environment by using a gelatin-based tissue phantom. The reconstructed optical data obtained from UCNPs showed a uniform and confined phosphor distribution (Fig. 6.3a). In contrast, the reconstructed optical data obtained from the use of an organic fluorophore (rhodamine 6G) showed significant artifacts at two ends of the fluorescent target (Fig. 6.3b). In addition to the anti-Stokes-shifted emission, it has been shown that the nonlinear power dependence of the UC emission can be further exploited to enhance both the quality and the contrast of the resulting images [89]. The experiments show that the spatial resolution of the LDOT reconstruction images can be significantly improved by the use of the synthesized $NaYF_4{:}Yb^{3+}/Tm^{3+}@NaYF_4$ UCNPs and break the current spatial resolution limits of LDOT images obtained from using conventional linear

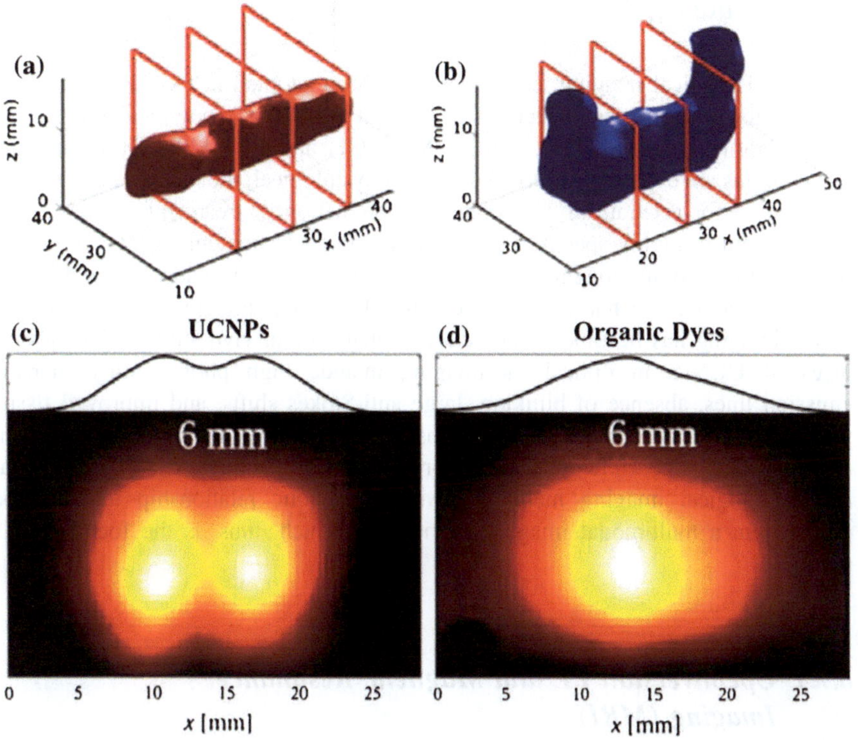

Fig. 6.3 A comparison of $NaYF_4:Yb^{3+}/Tm^{3+}$ UCNPs (*left column*) and organic dyes (*right column*) in LDOT. **a**, **b** Three-dimensional rendering of the reconstructed fluorophores; the *boxes* indicate the position of the cross-sectional slices. **a** Reconstruction using UCNPs shows a smooth and uniform rendering. **b** Reconstruction using rhodamine 6G shows several artifacts at the two ends of the fluorescent target. **c**, **d** Two-dimensional plots of the LDOT reconstructions with **c** the quadratically power-dependent $NaYF_4:Yb^{3+}/Tm^{3+}@NaYF_4$ UCNPs as contrast agents and **d** the linearly power-dependent DY-781 fluorophores, and their corresponding intensity profiles (*line plots*). The true depth was $z = 7$ mm, while the separation distance between the fluorescent tubes was set to 6 mm. The use of quadratic UCNPs clearly leads to reconstructions with higher spatial resolutions and qualities. **a**, **b** Reprinted with the permission from Ref. [88]. Copyright 2009 AIP Publishing LLC. **c**, **d** Reprinted with the permission from Ref. [90]. Copyright 2012 American Chemical Society

fluorophores as contrast agents (Fig. 6.3c, d) [90]. Although the advantages of UCNPs for LDOT imaging have been shown, the application of UCNPs for in vivo model or human tissue model remains unassessed. Further works of UCNPs on LDOT in either of these models will be essential to evaluate their potential for clinical practices [35].

6.7 Multimodal Imaging Using UCNPs

Current biomedical imaging techniques have been adopted in the development of imaging the structures and functions of biological systems. They include all of the major radiological modalities, namely MRI, CT, and PET [29]. These imaging technologies are expected to furnish atomic, physiological, molecular, and genomic information for accurate disease diagnosis, prediction of treatment response, and development of highly specific and sensitive drugs and imaging agents. However, none of the current imaging methods used in humans provide comprehensive medical imaging. To harness the strengths of different imaging methods, multimodal imaging has become an attractive strategy for in vivo studies. The advantages of UCNPs in optical bio-imaging include high photostability, narrow emission lines, absence of blinking, large anti-Stokes shifts, and improved tissue penetration with NIR excitation. The considerable attention and rapid development of UCNPs offers possible opportunity for multimodal imaging [35]. UC PL optical and radiological imaging methods have a symbiotic relationship that can be explored in a multimodal imaging framework, which, thus, is the focus of this section.

6.7.1 Upconversion PL and Magnetic Resonance Imaging (MRI)

UCL imaging displays high sensitivity, but is limited by its relatively low penetration depth in biological samples. In contrast, MRI is a powerful noninvasive imaging technique providing high-resolution, three-dimensional images of atomic structures as well as functional and physiological information about tissues in vivo. To bridge the gaps in resolution and sensitivity, constructing nanoparticles carrying both UCL emissions and magnetic properties within one system is highly desirable for dual-modal UCL/MRI imaging.

MRI is a biomedical diagnostic technique based on nuclear magnetic resonance (NMR). NMR has been used in component analysis and molecule structure analysis by chemists and physicists for 20 years since its discovery by Block and Purcenll in 1946. In 1967, when Jasper Johns et al. detected the signal of H, P, and N element distributions in a small animal in their in vivo experiment, human beings first realized that NMR may possess wonderful performance in life science. In 1971, Damadlan et al. used NMR in a clinical application and found that the longitudinal relaxation time ($T1$) and transverse relaxation time ($T2$) of diseased human tissues are longer than those of normal tissues. Later in 1973, Lauterbur took the first image using NMR, and since then the NMR imaging technology was defined as MRI. In 1980, for the first time, Hawkes et al. reported the detection of diseased tissues in the human brain by MRI; in the same year, the first commercial MRI machine was successfully invented. Today, MRI is one of the most efficient biomedical imaging technologies in the world.

NMR is a physical phenomenon in which nuclei in a magnetic field absorb and re-emit electromagnetic radiation. This energy is at a specific resonance frequency which depends on the strength of the magnetic field and the magnetic properties of the isotope of the atoms. If the high energy nuclei return to the low energy state without energy radiation, we call it relaxation. In other words, the term relaxation describes several processes by which nuclear magnetization prepared in a non-equilibrium state returns to the equilibrium distribution. Typically, there are two principal relaxation processes which are termed as *T*1 and *T*2 relaxation, respectively. The longitudinal (or spin-lattice) relaxation *T*1 refers to the transition of the non-equilibrium state to the equilibrium distribution state through transferring its energy into heat; The transverse (or spin–spin) relaxation time *T*2 refers to the transition of the non-equilibrium state to the equilibrium distribution state by transferring its energy to nearby equilibrium distribution state nuclei. For both *T*1 and *T*2 relaxation, the half-life period of relaxation is denoted as relaxation time *T*1 and *T*2, respectively. Generally speaking, the nucleus "imaged" in the MRI is hydrogen nucleus. Excluding fat, water comprises approximately 70 % of the human body by mass. Water content and the relaxation time of the hydrogen nucleus in the water molecule are the keys to efficient MRI. Thus, the high water content in the human body is the foundation of MRI applied in biomedical area. The difference between diseased tissues and normal tissues can be seen in different *T*1 or *T*2 relaxation times.

MRI contrast agents are a group of contrast media used to improve the visibility of internal body structures in MRI. In the early stages of MRI application, scientists and doctors both thought that the use of these agents was insignificant. However, they found that MRI can hardly detect small or early diseased tissue due to its low resolution. Clinical doctors complained that signals from small disease tissues and normal tissues cannot be distinguished easily, and hoped the contrast ratio could be further improved. MRI contrast agents certainly provided a good option for them. They can alter the relaxation times of atoms within body tissues where they are present after oral or intravenous administration, improving the SNR in the diseased area. The MRI contrast agents can be divided into two classes according to their pharmacological mechanism: *T*1-weighted agent and *T*2-weighted agent. Paramagnetic substances such as Gd^{3+} and Mn^{2+} can dramatically shorten the *T*1 relaxation time due to their all spins parallel outer-layer electronic configuration, resulting in significant enhancement in MRI signals. These are called *T*1-weighted agents. Superparamagnetic iron oxide and ultrasmall superparamagnetic iron oxide consist of suspended colloids of iron oxide nanoparticles and when injected during imaging, they can reduce the *T*2 signals of absorbing tissues. These are called *T*2-weighted agents. Unlike *T*1 weighted agents, *T*2-weighted agents will decrease the MRI signals in the contacted area, which means the clinical image is a black field image.

6.7.1.1 *T*1-Weighted Magnetic Imaging

Due to seven unpaired inner 4*f* electrons, Gd^{3+} has become the most useful *T*1-weighted MRI agent. Currently, there are some strategies for constructing dual-modal *T*1-MRI/UCL imaging agents. One attractive strategy is codoping Yb^{3+}, Er^{3+} and/or Tm^{3+} into the Gd^{3+}-based host. Codoping Yb^{3+}, Er^{3+}, and/or Tm^{3+} within one nanophosphor is a straightforward route to construct a UCL and MR dual-modality imaging agent for bimodal imaging. Tan et al. [91] reported the Gd_2O_3: Tb/Yb,Er nanorods as multimodal contrast agents with optical (DC and UC luminescence) and paramagnetic properties (relaxation r_1 = 1.5 s^{-1} mM^{-1}). Hao et al. [92] reported $NaGdF_4$:Yb/Er submicron disks showing upconversion luminescent and paramagnetic properties with mass magnetic susceptibility (Ms) values of 9.82×10^{-5} emu g^{-1}. Song et al. [93] reported NiNTA moiety-functionalized PAA-$NaGdF_4$:Yb/Er/Tm nanocrystals of 10 and 40 nm with UCL emissions and magnetic properties. The r_1 values are 0.99 and 0.47 s^{-1} mM^{-1} for the case of 10 and 40 nm nanocrystals, respectively. In addition to the aforementioned application of upconversion/magnetic nanophosphors in cell imaging, we have recently explored the use of upconversion/magnetic nanophosphors in small-animal imaging in vivo. For example, Li et al. [62] reported that Yb, Er, Tm codoped $NaGdF_4$ nanoparticles with NIR-to-NIR UCL emission and high magnetic properties (r_1 = 5.6 s^{-1} mM^{-1}) were successfully applied in UCL and MRI dual-modal bio-imaging of small animals. By the UCL and MR imaging in vivo, the 25–60-nm nanoparticles were demonstrated to accumulate in the liver and spleen of the mouse (Fig. 6.4a–c) [54].

Another strategy for constructing *T*1-MRI/UCL agents is doping Gd^{3+} into the hosts or coating Gd^{3+} onto the surface layer of the nanoparticles. For example, Prasad et al. [94] reported that Gd^{3+}-doped $NaYF_4$:Gd/Yb/Er nanocrystals displayed UCL emission and magnetic properties (r_1 = 0.14 s^{-1} mM^{-1}). Wang et al. [95] reported $NaYb_{1-x}Gd_xF_4$:Tm/Er/Ho nanoparticles with tunable UCL and super-paramagnetic properties (1.5 emu g^{-1}). Li et al. [96] reported Gd, Yb, Er codoped $NaYF_4$ nanophosphors by doping Gd^{3+} up to 60 % with r_1 value of 0.41 s^{-1} mM^{-1} for UCL and MR small-animal imaging. Due to the high affinity of Gd^{3+} with the fluoride ion, loading Gd^{3+} on the surface of fluoride-based UCNPs is achievable and provides high MR relaxation constants. Recently, Li et al. [97] developed a cation-assisted ligand assembly method which easily provided UCNPs with *T*1-magnetic properties by loading Gd^{3+} onto the surface, and the r_1 value of the $NaYF_4$:Yb,Er nanoparticles could be as high as 28.39 s^{-1} mg^{-1} (Fig. 6.4d, e) [54].

In addition, the core/shell structure with separate core and nanoshell of Gd-based compound becomes a useful method to construct *T*1 magnetic/luminescent difunctional probes. For example, using a $NaGdF_4$ nanostructure as the shell, a series of $NaGdF_4$:Er/Yb@$NaGdF_4$ (r_1 = 1.05–1.40 s^{-1} mM^{-1}), $NaYF_4$:Yb, Er@$NaGdF_4$ (r_1 = 0.48 s^{-1} mM^{-1}), and $NaYbF_4$:Tm@$NaGdF_4$ (r_1 = 2.6 s^{-1} mM^{-1}) nanocomposites [54] were synthesized to show good UCL emission and magnetic properties. Zhang et al. [98] have reported that a core/shell hybrid nanocomposite consisting of a $NaYF_4$:Yb,Er/Tm core and a paramagnetic silanized complex

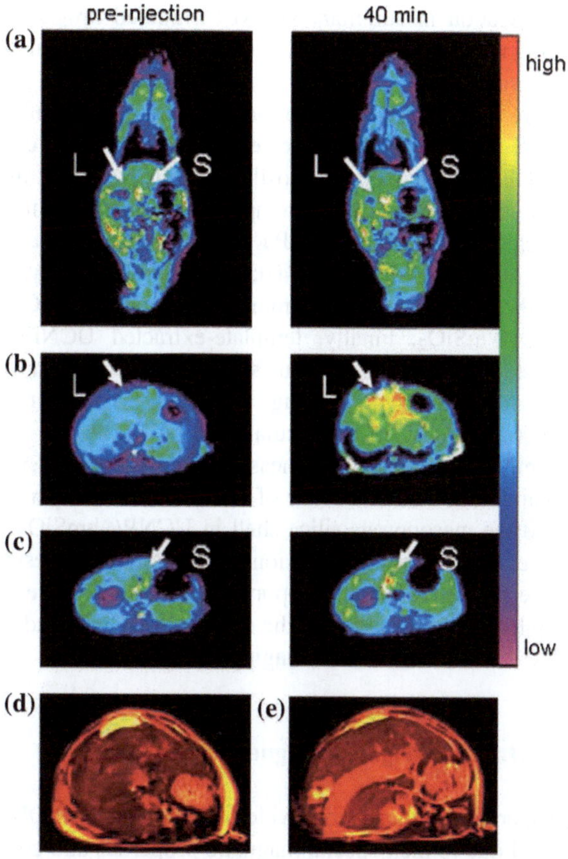

Fig. 6.4 Mapped **a** coronal images of the whole body and **b**, **c** transversal cross-sectional images of the liver (*L*) and spleen (*S*) of mice at pre-injection and at 40 min after intravenous injection of nanocrystals at 1.5 mg/kg. **d**, **e** MRI imaging of small animals using $NaYF_4$:Yb,Er. **a–c** Reprinted with the permission from Ref. [62]. Copyright 2010 Elsevier. **d**, **e** Reprinted with the permission from Ref. [96]. Copyright 2011 American Chemical Society

3-amin opropyl(trimethoxysilyl)-diethylenetriamine tetraacetic acid (Si-DTTA)-Gd^{3+} shell showed both high MR relaxation constant ($r_1 = 6.8\ s^{-1}\ mM^{-1}$) and good UCL emission. Generally speaking, the MRI performance of *T*1 is highly related to the feasibility of water molecules bonding to the Gd^{3+} ions: the relaxation constant increases with ease of bonding, leading to better performance. The above-mentioned UCNPs composites suffer from several drawbacks: (1) the Gd^{3+} content in the outmost layer is too low; (2) the hydrophilicity is not enough; (3) the bonding chance between water molecules in the bio-tissues and the surface of the UCNPs is very low. Hence, to enhance the *T*1-MR contrast signal while maintaining the UC PL intensity, the following issues should be resolved in the design and synthesis of the UCNPs composites: (1) make sure all (most) of the Gd^{3+} ions are located in the outmost layer of the nanostructure; (2) increase the hydrophilicity, while keeping the connection between water and the surface of the UCNPs; and (3) make sure that the size of the UCNPs is small (to get a high surface ratio). Recently, Shi et al. [99] reported the synthesis of multi-functional $NaYF_4$:Yb/Tm@$NaGdF_4$ core/hollow mesoporous silica shell nanoparticles (designated as UCNP@hmSiO_2) and their application in simultaneous NIR-triggered drug delivery and in situ quantitative drug release monitoring by both upconverted luminescence (UCL) and longitudinal relaxation time (*T*1) MR imaging in real time. The oleic acid-stabilized UCNPs were synthesized by a thermal decomposition method. Subsequently, the UCNPs were incorporated into reverse micelles containing cyclohexane and Igepal CO-520. Addition of tetraethylorthosilicate (TEOS) to the mixture initiated the controlled sol-gel reaction at room temperature, which led to the formation of dense silica shells (designated as UCNP@dSiO_2). A mesoporous silica shell was deposited onto the surface of the UCNP@dSiO_2 through hydrolysis of TEOS and self-assembly with the cetyltrimethylammonium chloride (CTAC) template, to form UCNP@d-SiO_2@mSiO_2. Finally, template-extracted UCNP@dSiO_2@mSiO_2 was directly tuned to hollow mesoporous silica by incubation in water at 95 °C for 2 h through a "surface-protected etching" process. The longitudinal relaxivity (r_1) of UCNP@hmSiO_2 was calculated to be 12.29 mM^{-1} s^{-1}. This relaxivity is significantly higher than that measured for UCNP@dSiO_2@mSiO_2 (3.58 mM^{-1} s^{-1}), mainly due to the absence of the dense silica layer in between the Gd-UCNP core and the mesoporous silica shell in UCNP@hmSiO_2, which facilitates water molecules bonding to the Gd^{3+} ions. This work is successful in both synthesis and MRI. The hollow structure supports an excellent space for the coordination of water molecules to the Gd^{3+} on the surface of UCNPs and also gives us inspiration on the UC-MRI (*T*1) bimodal imaging [54].

6.7.1.2 *T*2-Weighted Magnetic Imaging

Superparamagnetic iron oxide nanoparticles (SPION) have been widely applied in MRI due to their special magnetic properties and good biocompatibility. Currently, some prototypes of UCNPs coupled with SPION as bifunctional probes have been demonstrated. Early synthesis of magnetic/upconversion nanoparticles was focused on SiO_2-coating both UCNPs and Fe_3O_4 nanoparticles into a big silica layer with uncontrollable size distribution [100]. Yan et al. [101] have developed a cross-linker anchoring process to synthesize Fe_3O_4/$NaYF_4$:Yb,Er heteronanoparticles, using DDA or MUA as a cross-linker. The obtained heteronanoparticles adequately preserved both UCL emission and room temperature superparamagnetic properties. The magnetic saturation was 9.25 and 7.05 emu g^{-1} for the cases of DDA and MUA, respectively. Lin et al. [102] reported core/shell Fe_3O_4@nSiO_2@mSiO_2@$NaYF_4$:Yb,Er/Tm nanostructures with UCL and magnetic properties (38.0 emu g^{-1}), simultaneously. Very recently, a Fe_3O_4@SiO_2@Y_2O_3:Yb,Er nanocomposite was synthesized by the Chen group [103]. The inner Fe_3O_4 core of 18.4 nm and outer Y_2O_3:Yb,Er layer of 3.3 nm endowed the nanocomposites with saturation magnetization of 7.22 emu g^{-1}, r_2 value of 186.2 s^{-1} mM^{-1} at 1.5 *T* and NIR-light-excited UCL emission. Shi et al. [104] have developed a "neck-formation" strategy to combine SPION and $NaYF_4$:Yb,Er nanophosphors into heteronanoparticles by silica-shielding. The silica-shielded heteronanoparticles (<250 nm) displayed saturation magnetization of 89.8 emu g^{-1}, r_2 value of 211.76 s^{-1} mM^{-1} at

3.0 *T* and UCL emission at 515, 537, and 651 nm. Wu et al. [105] reported a facile method to synthesize $NaYF_4$:Yb,Er/Tm@SiO_2@Fe_3O_4 nanocomposites at room temperature without complicated surface modification. The prepared nanocomposite had good superparamagnetic (5.6 emu g^{-1}) and luminescent properties for cell imaging. Among these magnetic/luminescent nanoparticles, only silica-shielded heteronanoparticles have been applied in bimodal *T*2-MRI and UC luminescence small-animal imaging by intra-tumoral injection [104]. It should be noted that these difunctional probes are based on SiO_2-coated nanocomposites, resulting in the relatively large size of 460 nm [54].

Recently, Li et al. [71] have developed non-silica core/shell $NaYF_4$:Yb, Tm@Fe_xO_y nanocrystals, with a $NaYF_4$:Yb,Tm core of 20 nm and a Fe_xO_y shell of 5 nm. These core/shell nanocrystals exhibit excellent NIR UCL emission at 800 nm under CW 980-nm excitation and superparamagnetic properties with M_s value of ~12 emu g^{-1}. Furthermore, these obtained core/shell nanocrystals have been applied in dual-modality *T*2-enhanced MR and UCL imaging in vivo of the lymphatic systems of small animals. Using UCL emission at 800 nm as the detected signal, in vivo tracking from the injection point to the auxillary lymph nodes along the right antebrachium of the mouse could be observed. In addition, they also developed a novel class of UCNP-based multi-functional nanoparticles (MF nanocrystals) which integrated a hexagonal $NaYF_4$-based UCNP core and a layer of ultrasmall superparamagnetic Fe_3O_4 nanoparticles (IO nanocrystals) by electrostatic attraction [106]. The MF nanocrystals with diameter of ~230 nm showed combined optical and magnetic properties (~50 emu g^{-1}, r_2 = 352.8 s^{-1} mM^{-1} at 7.0 *T*). In vivo UCL/MR dual-model tumor-targeted and lymphatic imaging of MF nanocrystals were further demonstrated in animal experiments [54].

6.7.2 Upconversion PL and X-Ray Computed Tomography (CT)

X-ray CT is a technology that uses computer-processed X-rays to produce tomographic images (virtual slices) of specific areas of the scanned object, allowing the user to see what is inside it without cutting it open. These cross-sectional images are used for diagnostic and therapeutic purposes in various medical disciplines. In order to improve the contrast ratio, radiocontrast agent is always desired in the operation. Typically, radiocontrast agents are compounds which contain heavy elements, such as iodine and barium. Compared with them, Ln elements also have higher atomic numbers and K-edge values within the X-ray spectrum in medical use, thus providing an opportunity to create strong X-ray attenuation for enhanced CT. The combined use of Ln-doped UCNPs as CT and UC PL imaging contrast agents has recently attracted widespread interest. For example, Cui et al. [107] reported on Ln-doped $NaGdF_4$ UCNPs as effective contrast agents for dual-mode UC PL imaging and CT imaging in vivo, as can be seen in Fig. 6.5. Notably, by optimizing the CT

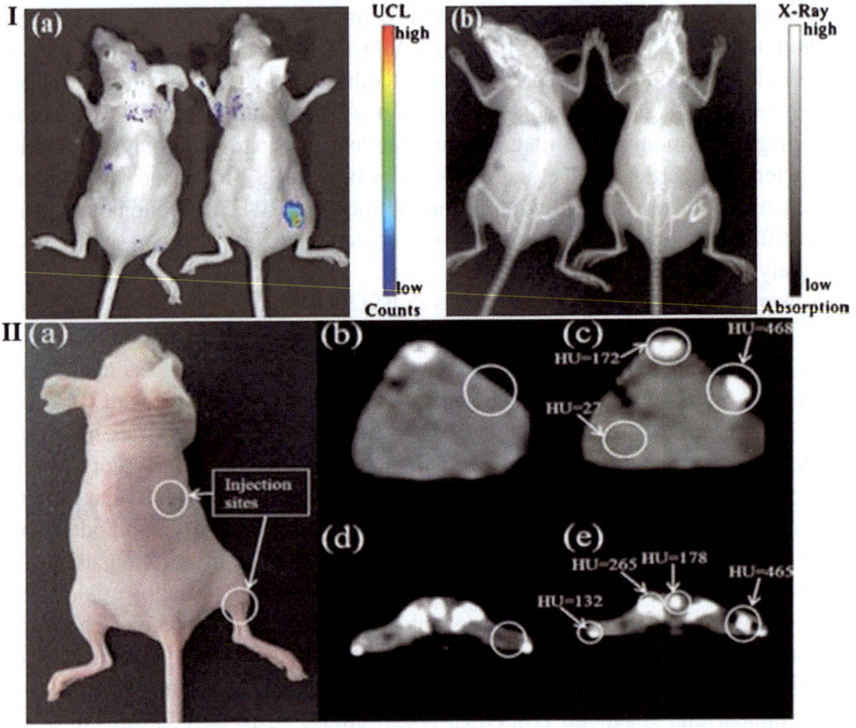

Fig. 6.5 In vivo UC PL imaging (**I***a*) and X-ray imaging (**I***b*) of mice after subcutaneous injection (*left*) without and (*right*) with $NaGdF_4:Yb^{3+}/Er^{3+}$. In vivo CT images of mice after subcutaneous injection with UCNPs. **II***a* The photograph of the injected mouse; **II***b*, *c* the transverse image of the back, the HU value of the injection site is 468; **II***d*, *e* the transverse image of the buttock, the HU value of the injection site is 465. **I**, **II** Reprinted with the permission from Ref. [107]. Copyright 2011 Wiley-VCH Verlag GmbH & Co. KGaA

contrast elements in the UCNPs according to the fundamental principles of X-ray imaging, Lu et al. [108] have synthesized several nanoparticulate CT contrast agents with excellent imaging performance. For example, an Yb^{3+}-based $NaYbF_4$ agent provides enhanced contrast as compared to currently available CT contrast agents under normal operating conditions. They also integrated both Ba^{2+} and Yb^{3+}, which have great differential in *K*-edge value, into a single NP to yield the first example of binary contrast agents, which display much higher contrast than iodinated agents at different voltages and are highly suited to diagnostic imaging of various patients. However, the nanocrystals developed by Lu et al. are incapable of implementing simultaneous CT and UC PL imaging due to the lack of luminescent activators in the matrix. This drawback can be overcome by introducing Tm^{3+}- or Er^{3+}-dopants into $NaYbF_4$ and $BaYbF_5$ nanoparticles [109]. Interestingly, Shi et al. [109] found that, in addition to CT and UC PL bio-imaging, $BaYbF_5:Er^{3+}$ nanocubes can also work as irradiation dose enhancers in tumors during radiotherapy.

This provides a theranostic paradigm for UCNPs, wherein the precise positioning of tumor is entailed by their role in multimodal imaging, and tumor eradiating is implemented via radiotherapy through their role as irradiation enhancer. Various heavy metal elements in the host lattice of UCNPs to have a prominent radiotherapy enhancement effect will be an interesting area to explore [35].

6.7.3 Upconversion PL and Positron Emission Tomography (PET)

The concept of emission and transmission tomography was introduced by David E. Kuhl, Luke Chapman, and Roy Edwards in the late 1950s. Their work later led to the design and construction of several tomographic instruments at the University of Pennsylvania. Tomographic imaging techniques were further developed by Michel Ter-Pogossian, Michael E. Phelps, and others at Washington University School of Medicine.

Work by Gordon Brownell, Charles Burnham, and their associates at the Massachusetts General Hospital beginning in the 1950s contributed significantly to the development of PET technology and included the first demonstration of annihilation radiation for medical imaging. Their innovations, including the use of light pipes and volumetric analysis, have been important in the deployment of PET imaging. In 1961, James Robertson and his associates at Brookhaven National Laboratory built the first single-plane PET scan, nicknamed the "head-shrinker."

One of the factors most responsible for the prominence of positron imaging was the development of radiopharmaceuticals. In particular, the development of labeled 2-fluorodeoxy-D-glucose (2FDG) by the Brookhaven group under the direction of Al Wolf and Joanna Fowler was a major factor in expanding the scope of PET imaging. The compound was first administered to two normal human volunteers by Abass Alavi in August 1976 at the University of Pennsylvania. Brain images obtained with an ordinary (non-PET) nuclear scanner demonstrated the concentration of FDG in that organ. Later, the substance was used in dedicated positron tomographic scanners, to yield the modern procedure.

The logical extension of positron instrumentation was a design using two 2-dimensional arrays. PC-I was the first instrument using this concept and was designed in 1968, completed in 1969 and reported in 1972. The first applications of PC-I in tomographic mode as distinguished from the computed tomographic mode were reported in 1970. It soon became clear to many of those involved in PET development that a circular or cylindrical array of detectors was the logical next step in PET instrumentation. Although many investigators took this approach, James Robertson and Zang-Hee Cho were the first to propose a ring system that has become the prototype of the current shape of PET [35].

PET is a nuclear medicine, functional imaging technique that produces a three-dimensional image of functional processes in the body. The system detects pairs of gamma rays emitted indirectly by a positron-emitting radionuclide (tracer), which is introduced into the body on a biologically active molecule. Three-dimensional images of tracer concentration within the body are then constructed by computer analysis. In modern PET-CT scanners, three-dimensional imaging is often accomplished with the aid of a CT X-ray scan performed on the patient during the same session, in the same machine.

Nuclear imaging, especially PET, is an established clinical tool for whole-body imaging with detection sensitivity reaching below the picomolar range for functional imaging, albeit with a low (~mm) spatial resolution. ^{18}F is the most widespread radionuclide used in PET imaging for clinical diagnosis. In the light of the short half-life of ^{18}F, the quick conjugation of ^{18}F into the probe with a high reaction yield is necessary to improve the efficiency of the labeling and to reduce the cost. Unlike the conventional ^{18}F-labeled methods based on organic reactions, Li et al. [110] has recently developed a simple synthetic strategy for ^{18}F-labeled rare-earth nanoparticles through a straightforward inorganic reaction between rare-earth cations and fluoride ions. The ^{18}F-labeling process based on rare-earth elements was achieved efficiently in water at room temperature with an ^{18}F-labeling yield of >90 % and completed within 5 min, with only simple purification by aqueous washing and centrifugation, and without the use of organic agents. For example, radioactivity measurements indicated that 90.13, 97.34, 96.44, and 88.66 % of $^{18}F^-$ were incorporated onto $NaYF_4$, $Y(OH)_3$, $Gd(OH)_3$, and $NaYF_4$:Yb,Tm nanoparticles, respectively. In particular, ^{18}F-labeled $NaYF_4$:Yb,Tm UCNPs could be further evaluated by PET imaging of their in vivo distribution giving applications in lymph monitoring. Real-time PET quantification detection of lymph node signals within 60 min indicated the rapid accumulation of materials from the paw into lymph nodes. Combining the application of UCL imaging in cell and tissue, ^{18}F-labeled UCNPs become a useful strategy for the creation of a dual-modal bio-imaging technique, combining UCL and PET imaging. Moreover, Li et al. [111] has successfully exploited ^{18}F-labeled $NaYF_4$:Yb,Tm nanoparticles for UCL/PET whole-body imaging of small animals [54].

6.7.4 Upconversion PL and Single-Photon Emission Computed Tomography (SPECT)

SPECT, or less commonly, SPET is a nuclear medicine tomographic imaging technique using gamma rays. It is very similar to conventional nuclear medicine planar imaging using a gamma camera. However, it is able to provide true 3D information. This information is typically presented as cross-sectional slices through the patient, but can be freely reformatted or manipulated as required [38].

The technique also requires delivery of a gamma-emitting radioisotope (a radionuclide) into the patient, normally through injection into the bloodstream. On occasion, the radioisotope is a simple soluble dissolved ion, such as a radioisotope of gallium (III). Ln ions used in UCNPs contain many radioisotopes which provide an opportunity to use UCNPs as a SPECT contrast agent [38].

As mentioned in, Sect. 5.7.3, the UCNPs could be tracked in real time with high sensitivity in vivo from mice to large animals by PET imaging. However, the half-life of ^{18}F is too short (1.829 h), greatly limiting its biological applications in long-term imaging. To overcome this problem, Li et al. [112] prepared ^{153}Sm-labeled $NaLuF_4$:Yb/Tm nanocrystals for UCL and SPECT dual-modality imaging. The half-life of ^{153}Sm (46.3 h) is much longer compared to ^{18}F. In addition, it also emits medium-energy beta rays, making it suitable for long-term SPECT imaging. These radioactive/UCL $NaLuF_4$:^{153}Sm/Yb/Tm nanocrystals have been applied for SPECT and UCL dual-modality bio-imaging in vivo (Fig. 6.6). The bio-distribution of the NaLuF4:^{153}Sm/Yb/Tm nanocrystals has been quantified in vivo using SPECT bio-imaging [38].

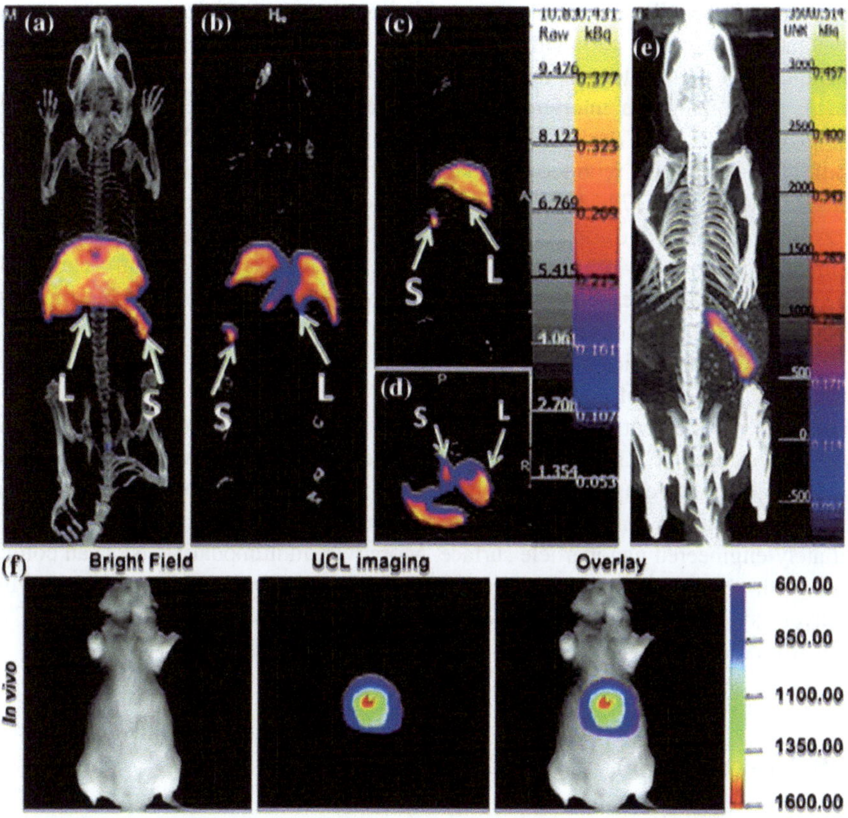

Fig. 6.6 In vivo SPECT and UCL imaging study after intravenous injection of ^{153}Sm-UCNPs. **a** whole-body three-dimensional projection, **b** coronal, **c** sagittal and **d** transversal images acquired at 1 h and **e** whole-body three-dimensional projection images acquired at 24 h are shown, respectively. The *arrows inset* point to the liver (*L*) and spleen (*S*). **f** In vivo UCL imaging of the Kunming mouse 1 h after tail vein injection of the ^{153}Sm-UCNPs (20 mg/kg). **a–f** Reprinted with the permission from Ref. [112]. Copyright 2013 Elsevier

6.7.5 *Upconversion PL and Multimodal Imaging (MRI/PET/SPECT/CT)*

The combination of UCNPs with different properties can offer synergistic multifunctional nanomedical platforms with optical, magnetic, and radioactive properties, which can make accomplishing multimodal imaging possible. The incorporation of multimodal capability in UCNPs generally relies on the selection of appropriate Ln ions in the host lattice or the choice of proper surface treatment with desired imaging capability. For example, Li and coworkers developed a radioactive, magnetic UCNPs based on ^{18}F-labeled $NaYF_4:Gd^{3+}/Yb^{3+}/Er^{3+}$ or $NaYF_4:Yb^{3+}/Tm^{3+}@NaGdF_4$, in which the ^{18}F radionuclide on the treated nanoparticle surface enables the PET imaging, the Gd^{3+} ions enable the MRI imaging, and the Yb^{3+}/Er^{3+} provides the ability for UC PL imaging [97]. Another effective way to incorporate a set of imaging modalities in UCNPs is to employ a core/shell structure in which separate layers and nanoparticle surface parts are intended for different imaging modalities. The advantage of this strategy is that the different imaging modalities are spatially isolated, minimizing the interference between each other. For example, core/shell $Fe_3O_4@NaLuF_4:Yb^{3+}/Er^{3+}/Tm^{3+}$ nanostructure and $NaLuF_4:Yb^{3+}/Tm^{3+}@SiO_2$–GdDTPA have been reported as imaging probes for trimodality CT/MRI/UCPL imaging (Fig. 6.7) [72, 113]. TaO_x-decorated $NaYF_4:Yb^{3+}/Er^{3+}/Tm^{3+}@NaGdF_4$ core/shell UCNPs have also been reported by Shi et al. [114] for in vivo CT/MR/UCL trimodal imaging. More recently, $NaLuF_4$:Yb,Tm@$NaGdF_4$(^{153}Sm) core/shell UCNPs have been reported for SPECT/CT/MR/UC PL four-modal bio-imaging [115]. The essence to construct multimodal imaging nanoprobes is to utilize the paramagnetic properties of Gd^{3+}, Fe_3O_4, TaO_x to realize MRI imaging, the high attenuation of X-ray of Ln ions (e.g., Lu^{3+} or Yb^{3+}) to realize CT imaging, the Yb^{3+}, Er^{3+}/Tm^{3+} Ln elements to realize the UC PL, and the radionuclide ^{153}Sm for SPECT imaging or the ^{18}F radionuclide for PET imaging. Multimodal imaging contrast agents can be accomplished in particulate formulations by carefully selecting a combination of Ln ions in well-defined hierarchical nanostructures with an appropriately engineered nanoparticle surface. Moreover, multimodal UCNPs can conjugate with photoactivated platinum (IV) pro-drug to realize a multi-functional drug delivery system for imaging-guided cancer therapy, which incorporates UCPL/CT/MR trimodal imaging and NIR-activated platinum pro-drug delivery [35, 116].

6.8 Upconversion Imaging Techniques

6.8.1 *Imaging System*

UC-based biomedical imaging belongs to fluorescent-based imaging, so the components of an upconversion imaging system are very similar to those of traditional fluorescent imaging systems, including detector, irradiation source (different

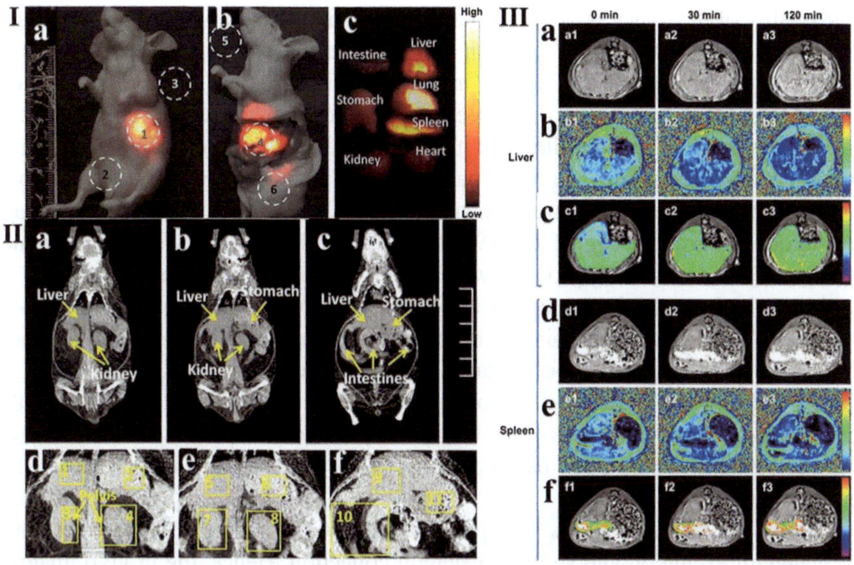

Fig. 6.7 **I***a* In vivo UC PL imaging of a nude mouse after injection with UC/SiO_2-GdDTPA for 10 min. I*b* Ex vivo UC PL imaging of the same but sacrificed nude mouse. **I***c* Ex vivo UC PL image of main organs of the sacrificed nude mouse. **II***a–c* Serials of coronal CT images of Kunming mouse at different layers after injection of UC/SiO_2-GdDTPA. **II***e–g* Enlarged CT views of the abdomen. **III***a*, *d* *T*1-weighted MRI images of liver (**III***a*) and spleen (**III***d*) at 0, 30, and 120 min after injection of UC/SiO_2-GdDTPA. **III***b*, *e* *T*1 distribution images of liver (**III***b*) and spleen (**III***e*) at 0, 30, and 120 min after injection of UC/SiO_2-GdDTPA. **III***c*, f Local colorized *T*1-weighted MR images of liver (**III***c*) and spleen (**III***f*) at 0, 30, and 120 min after injection of UC/SiO_2-GdDTPA. **I–III** Reprinted with the permission from Ref. [72]. Copyright 2012 Elsevier

wavelength), and filters. Hence, we would introduce the photographic methods before talking about the UC imaging system.

Typically, the biomedical imaging can be divided into in vitro and in vivo imaging. Confocal microscopy is always used for in vitro imaging, and in vivo imaging system is required if we need to perform an in vivo imaging experiment.

6.8.1.1 In Vitro Imaging System

Confocal laser scanning microscopy (CLSM or LSCM) is a technique for obtaining high-resolution optical images with depth selectivity. The key feature of confocal microscopy is its ability to acquire in-focus images from selected depths, a process known as optical sectioning. Images are acquired point-by-point and reconstructed with a computer, allowing three-dimensional reconstructions of topologically complex objects. For opaque specimens, this is useful for surface profiling, while for non-opaque specimens, interior structures can be imaged. For interior imaging, the quality of the image is greatly enhanced over simple microscopy because image information from multiple depths in the specimen is not superimposed. A conventional microscope "sees" as far into the specimen as the light can penetrate, while a confocal microscope only "sees" images one depth level at a time. In effect, the CLSM achieves a controlled and highly limited depth of focus. The principle of confocal microscopy was originally patented by Marvin Minsky in 1957, but it took another 30 years and the development of lasers for CLSM to become a standard technique toward the end of the 1980s. In 1978, Thomas and Christoph Cremer designed a laser scanning process, which scans the three-dimensional surface of an object point-by-point by means of a focused laser beam, and creates the over-all picture by electronic means similar to those used in scanning electron microscopes. This CLSM design combined the laser scanning method with the 3D detection of biological objects labeled with fluorescent markers for the first time. During the next decade, confocal fluorescence microscopy was developed into a fully mature technology which has been used to image many luminescent nanoprobe-labeled samples including UCNPs.

Based on the anti-Stokes UC process, the UCL imaging system uses long-wavelength excitation and short-wavelength collection. Hence, the in vivo imaging system of UCL-based imaging is slightly different from the traditional reflectance imaging system. For Yb^{3+}-sensitized UCNPs as luminescent labels, an external CW 980-nm laser as excitation source is introduced by an optical fiber. Scheme 6.1 shows the setup for UCL laser scanning confocal microscopy [117].

In the laser scanning UCL microscope (LSUCLM) [118], the laser beam is expanded to parallel light and fills the objective lens to provide excitation irradiation onto the specimen. As shown in Scheme 6.1a, unlike with a traditional confocal microscope, a selected short-pass excitation dichroic mirror was adapted to separate the UCL signal from the excitation source. Laser scanning UCL confocal microscope is utilized for cell imaging studies [117].

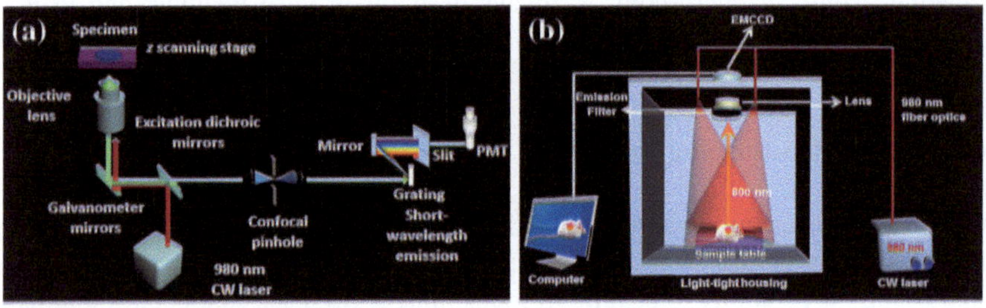

Scheme 6.1 Schematic illustration of **a** UCL laser scanning confocal microscope and **b** UCL in vivo imaging system. Reprinted with the permission from Ref. [117]. Copyright 2014 Elsevier

6.8.1.2 In Vivo Imaging System

Simple "photographic methods," in which the light source and the detector reside on the same side of the animal imaged, are generally referred to as "reflectance imaging." Reflectance imaging is currently the typical method of choice for accessing the distribution of fluorescent probes in vivo, and this method was applied more generally to imaging fluorescent proteins or even bioluminescence even if in the latter case no excitation light is used.

NIR luminescence reflectance imaging in particular operates on light with a defined bandwidth as a source of photons that encounters a fluorescent molecule ("optical contrast agent or molecular probe"), which emits a signal with different spectral characteristics, that can be resolved with an emission filter and captured by a high-sensitivity CCD camera.

A typical reflectance imaging system is shown in Scheme 6.1b. The light source can be either a laser at an appropriate wavelength for the fluorochrome targeted or white light sources using appropriate short-pass filters. Generally, laser sources are preferable because they offer higher power delivery at narrower and better-defined spectral windows (typically ±3 nm for laser diodes vs. ±10 nm or more for filtered white light sources). The laser beam is expanded on the animal surface with an optical system of lenses (not shown). Narrow wavelength selection is important, especially in the NIR, whereas the excitation and emission spectra overlap and it is likely that excitation photons can propagate into the fluorescent images. The CCD camera is usually a high-sensitivity camera since fluorescent signals are of low intensity. On the other hand, since the targeted measurement is a diffuse-light measurement, emanating from a virtually flat surface, CCD chip resolution and dynamic range are not crucial factors in these types of systems. The noise requirements of a CCD camera for luminescence imaging are not as stringent since in the general case the detection sensitivity limit is set by the background tissue fluorescence (background probe distribution and auto-fluorescence), and luminescence intensity can be adjusted above the camera noise floor by optimizing illumination intensity and acquisition times. These features are in contrast to bioluminescence measurements, where the detection sensitivity is usually set by the CCD camera noise floor. Detection systems that scan a single spot over a field of view and detect light using photomultiplier tube detectors have also been reported. Such systems may offer greater dynamic range than CCD-based systems but compromise resolution, acquisition frame rate, and overall noise characteristics, and are not widely used in in vivo luminescence reflectance measurements.

Considering the fact that Yb^{3+}, Tm^{3+}-codoped UCNPs with excitation at 980 nm and emission at 800 nm are usually used for UCL imaging in vivo of small animals, [119] adjustable CW 980-nm lasers are used as the excitation sources and UCL signals peaked at 800 nm are collected by using a band-pass filter set (Scheme 6.1b). Due to the relatively low UCL efficiency of the UCNPs, a cooled electron-multiplying charge-coupled device (EMCCD) is used as a highly sensitive detector to collect the weak signal [117].

To obtain a high-contrast UC image, there are several parameters should be learned first.

Power Density

As we discussed in the UC mechanism (Chap. 1), the UC process is a nonlinear process that means the UC intensity would exponentially grow along with increasing power density. High power density can lead to high-contrast imaging. However, the NIR laser also suffers from one of its unique disadvantages: thermal effects on biological tissues. Generally speaking, the maximum power density of 980 nm is $\sim$3 W/cm^2, while the maximum power density would be $\sim$6 W/cm^2 for 800-nm excitation. This data were obtained through the heat generation experiments using a nude mouse irradiated by 980/800-nm laser.

Exposure Time

Exposure time also has a great impact on the resulting signals. Typically, the resulting signal is proportional to the exposure time. In traditional luminescence biomedical imaging system, the exposure time cannot be very high, because the background and auto-fluorescence signal would increase along with the signal from the region of interest. The final SNR would not be improved after prolonged exposure in the traditional luminescence biomedical imaging system, limiting the exposure time to several microseconds (ms). In the case of UC-based biomedical imaging, the auto-fluorescence signal is very low because the auto-fluorescence signal can be easily blocked along with the excitation light by using a filter. However, the exposure time is still restricted due to the limitation of the filter. No commercial filter completely blocks the designated wavelength, indicating that this wavelength can slightly penetrate the filter. Therefore, if the exposure time is relatively short, the irradiation light cannot be detected due to insufficient energy accumulation of the irradiation light; it can also become noise if the exposure time is long enough for too much energy accumulation, in which case the SNR also cannot be improved. Another obstacle to the increase of exposure time is the detector. If the signal is in saturation for the detector, it will result in overexposure. The overexposure mechanism is the equivalent to overexposure in photography. In photography, overexposure is caused by exposing the film (detector/CCD) to too much light or for too long a time. To sum up the reported papers which involve biomedical imaging, the exposure time of UC-based biomedical imaging is determined to be hundreds microseconds to several seconds.

The State of Mice

Both exposure time and power density are quantifiable parameters. By comparison, the state of mice is the experience of the experimenter, this experience, like certain skills, should be obtained from experiments. Therefore, we only suggest some key points, while the details should be learned in practice. First, the anesthesia of mice should be appropriate. During imaging, the mice should not be struggling, because this struggling would affect the precise localization of UCNPs; but the dose of anesthetic cannot be very high, in case it affects the health of the mice. Pentobarbital, chloral hydrate, and urethane are commonly used anesthetics, whose usage methods can be found in medical papers and books. In recent years, some imaging companies have developed novel anesthetic coupled imaging systems, involving not only the basic components required for an in vivo imaging experiment, but also an anesthetic flow system. This anesthetic flow system uses gaseous anesthetics, such as isoflurane, enflurane, and desflurane. The depth of anesthesia can be controlled by the flow rate and the anesthetic concentration. Second, the region of interest should be as flat as possible. It is a common knowledge that light will diffuse reflect if it is irradiate upon a rough surface. However, the diffuse reflection light is not always detected by the detector, resulting in errors in signal intensity in different regions, which is undesirable in biomedical imaging because it would skew the judgment of the diseased position. In some circumstances, the signal would not be detected at all. Third, the irradiation laser should be expanded before it reaches the surface of the mouse due to the heat generation and possibility of overexposure. Moreover, the laser beam may cause signals that are too high in a localized region. Fourth, time interval between imaging and injection/gavage should be appropriate. Generally, if the UCNPs were subcutaneously injected, imaging should be performed as soon as possible, because the UCNPs will slowly be transferred to the whole body by blood circulation through the nearby blood vessels. If the UCNPs were forced into the stomach of the mice by oral gavage, imaging of the stomach should be performed 15 min after gavage due to the 0.5–1-h dwell time of the mouse stomach. The imaging intensity would decrease slowly because of the excretion of the UCNPs. The case of intravenous injection is more complicated, but in general, intravenous injection imaging can be divided into passive targeting imaging and active targeting imaging. Passive targeting imaging is based on EPR effect. The UCNPs often severely accumulate in RES organs (liver, spleen, etc.), resulting in low targeting specificity (defined as the amount of probe in tumor vs. that in liver) and potential long-term toxicity. After injection, some of the UCNPs will accumulate in the liver quickly and cannot be eliminated within several days because these inorganic nanocrystals are difficult to decompose. Thus, the imaging of the liver can be seen in 2 h; by comparison, passive targeting imaging of diseased tissues, such as tumors, takes relative longer (6–12 h). Active targeting imaging is based on the specific recognition between diseased tissues and specific ligands/molecules/characters on the surface of UCNPs. In other words, the UCNPs are modified in order to avoid the accumulation in RES organs and better recognize the diseased tissues. In this circumstance, the imaging of diseased tissues can be performed in 2 h and may last several days due to the specific recognition between diseased tissues and UCNPs.

6.8.2 Penetration Depth Measurement

As we discussed in the introduction, absorption and scattering of the photons induced by biological tissue and water are the main factors that lead to attenuation of signal proportional to the depth of the feature of interest. However, both absorption and scatting processes are highly related to the wavelength of light; that is, penetration depth depends on the wavelength of the light. Up to now, there are many papers involving the penetration depth measurement of the UCNPs, some through theoretical models, some using in vitro model (such as beef slices, agarose and mice tissue), and others directly observing living mice (the penetration depth is determined by the injection site, specific organs such as the stomach and the whole thickness of the mouse). The penetration depth may be totally different depending on measurement methods. So we will introduce them separately in the following contents.

6.8.2.1 Theoretical Model

Up to now, several groups have established theoretical models which can be used for easy but accurate assessment of the depth of luminescence from UCNPs. Actually, the biological tissues are very complicated; the different components (such as muscle, skin, and fat) possess different absorption and scatting properties. The above-mentioned researchers also used some tissues for more precise results. It should be noted that this is totally different from "in vitro model," which determined the penetration depth by experiment directly. Theory-based researchers merely utilized the tissues model to correct their formula, so their results benefit from precision, repeatability, and good predictability.

In 2014, Zhang et al. [120] established a theoretical model which can be used for easy but accurate assessment of the depth of luminescence probes embedded in tissue based on multispectral luminescence of UCNPs. The parameters in the deduced quantitative relation between the light propagation depth and the UCL spectrum were fixed from tissue-mimicking liquid phantoms, and the setup is depicted in Fig. 6.8Ia, where UCNPs were encapsulated into a capillary tube and embedded in the tissue-mimicking liquid phantoms. The optical path length on excitation and emission could be separately adjusted and the corresponding UCL spectra were recorded by PMT. The integrated intensity ratio of the green and red emission was used for sensing the depth. To simulate the UC signal attenuation in tissue, a special sample chamber equipped with a two-dimensional (2D) translation stage was prepared in this study, as shown in Fig. 6.8Ia. The propagation distance

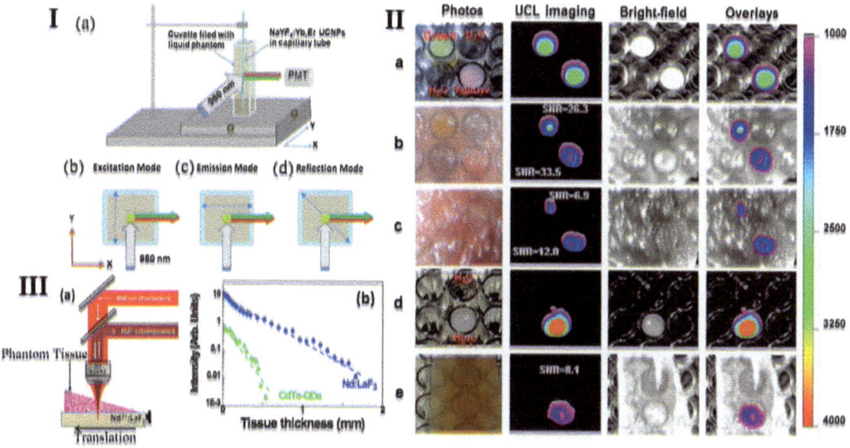

Fig. 6.8 **I***a* Schematic of the setup used for light penetration depth dependent UCL spectrum study. **I***b–d* are the three different working modes. The liquid phantom cuvette moves along the directions shown by the *dashed lines*. **II***a–c* Penetration depth experiment of UCNC-G (*green*) and UCNC-Y (*yellow*) without beef slice (**II***a*), with 0.1-cm beef slice (**II***b*) and with 0.2-cm beef slice (**II***c*) (λex = 635 nm, λUCL = 530 ± 25 nm, excitation power density of 12.5 mW/cm^2). **II***d*, *e* Penetration depth experiments of UCNC-B (*blue*) without beef slice (**II***d*) and with 0.1-cm beef slice (**II***e*) (λex = 532 nm, λUCL = 452 ± 25 nm, excitation power density of 13 mW/cm^2). Water was used as the control sample in this penetration depth experiment. **III***a* Schematic diagram showing the experimental procedure to determine the penetration depth of Nd^{3+}:LaF_3 NPs. **III***b* Intensity of the one-photon excited luminescence generated by Nd^{3+}:LaF_3 nanocrystals as a function of the tissue thickness. **I***a–d* Reprinted with the permission from Ref. [120]. Copyright 2014 Royal Society of Chemistry. **II***a–e* Reprinted with the permission from Ref. [121]. Copyright 2013 American Chemical Society. **III***a*, *b* Reprinted with the permission from Ref. [122]. Copyright 2013 American Chemical Society

of the excitation light and emission light could be separately controlled. UCNPs, encapsulated in a small glass capillary (1 mm outer diameter) at a concentration of 10 mg mL^{-1}, were dipped into the liquid phantom vertically. The tissue-equivalent liquid phantom was used as a simulation model and poured into a 10 mm × 10 mm silica cuvette, which was fixed on the 2-D translation platform. The optical properties were adjusted by the relative concentration of India ink (absorption component) and Intralipid (scattering component). The spectra at different depths were recorded by PMT in the spectrophotometer with a 980-nm laser excitation of 700 mW cm^2. In excitation mode (Ex mode, Fig. 6.8Ib), the liquid phantom cuvette moves along the excitation direction, i.e., *Y*-axis, in steps of 1 mm, the UC luminescence spectra were recorded at each step with an SPEX spectrophotometer. In emission mode (Em mode, Fig. 6.8Ic), the cuvette moves along the emission direction, i.e., *X*-axis, in steps of 1 mm. In reflection mode (Ref mode, Fig. 6.8Id), the cuvette moves along the excitation direction and the emission direction simultaneously.

The UCL was thus treated as a line light source, and the energy fluence attenuated isotropically in the tissue. Based on the optical diffusion theory, the distribution of the excitation light and the emission light along their propagation direction (z) inside tissue could be written as:

$$D_x \frac{\mathrm{d}^2\phi_x(z)}{\mathrm{d}z^2} - \mu_{a,x}\phi_x(z) = -\mu'_{s,x}\mathrm{P}_{xo}\mathrm{e}^{-\mu'_{t,x}z} \quad (6.1)$$

$$\frac{\mathrm{d}^2\phi_m(z)}{\mathrm{d}z^2} + \frac{2}{z}\frac{\mathrm{d}\phi_m(z)}{\mathrm{d}z} - \mu^2_{\mathrm{eff},m}\phi_m(z) = -\frac{P_{m0}}{D_m}\delta(z) \quad (6.2)$$

where ϕ_x and ϕ_m are the influent intensities of excitation light and emission light inside the tissue; P_{x0} and P_{m0} are the initial intensities of the incident excitation light (e.g., 980 nm) and the emission light (e.g., 540 or 650 nm); $\mu_{a,x}$, $\mu'_{s,x}$, and $\mu'_{t,x}$ are the absorption coefficient, reduced scattering coefficient, and the total attenuation coefficient for the excitation light; $\mu_{\mathrm{eff},m}$ is the effective attenuation coefficient for the emission light; and D_x and D_m are the diffusion coefficients of excitation and emission, respectively. The solution for the emission diffusion equations is

$$\phi_m(z) = \frac{P_{m0}(z)\mathrm{e}^{-\mu_{\mathrm{eff},m}z}}{4\pi D_m z} \quad (6.3)$$

From Eq. (6.3), we can see that the luminescence energy fluence ϕ_m is affected not only by the initial luminescence intensity P_{m0} but also by the tissue optical property $\mu_{\mathrm{eff},m}$. Regarding $NaYF_4$:Yb/Er UC luminescence, we can divide the luminescence energy fluence into two parts ϕ_{540} and ϕ_{650}, corresponding to the two emission bands around 540 and 650 nm, respectively. The intensity ratio R detected is therefore

$$R = \frac{\phi_{540}(z)}{\phi_{650}(z)} = \frac{\frac{P_{0,540}(z)\mathrm{e}^{-\mu_{\mathrm{eff},540}z}}{4\pi D_{540}z}}{\frac{P_{0,650}(z)\mathrm{e}^{-\mu_{\mathrm{eff},650}z}}{4\pi D_{650}z}} = \frac{P_0^{540}(z)}{P_0^{650}(z)}\frac{D_{650}}{D_{540}}\mathrm{e}^{-(\mu_{\mathrm{eff},540}+\mu_{\mathrm{eff},650})z} \quad (6.4)$$

$$\mu^2_{\mathrm{eff}} = 3\,\mu_\alpha\mu'_t = 3\,\mu_\alpha(\mu_\alpha + \mu'_s) \quad (6.5)$$

The first item at the right side of Eq. (6.4) is constant that is determined by the intrinsic optical properties of UCNPs. The diffusion coefficients D_{540} and D_{650} in the second part are also constant for a homogeneous tissue. Thus, from this equation, we can deduce that the *G*/*R* ratio detected at the surface follows an exponential decay pattern with increasing tissue depth, and the attenuation slope can be calculated from the difference of effective attenuation coefficients at these two wavelengths.

6.8.2.2 In Vitro Model

The in vitro model for UC penetration depth measurement includes pork/beef tissues and simulated tissue.

Pork/Beef tissues

Using pork/beef tissues as in vitro model is the most simple, convenient, and intuitive method to measure the penetration depth of UCNPs. In a typical procedure, the researchers need to place a certain amount of UCNPs beneath the pork/beef slices, the NCs are excited at NIR, and the UC signals are imaged using a CCD detector through increasing thickness of tissue (pork/beef slices). The penetration depth is determined to be the slice thickness at which the UC signals cannot be detected.

In 2013, Li et al. [121] compared the abilities of blue-emissive upconversion nanocapsule (UCNC-B), green-emissive upconversion nanocapsule (UCNC-G), and yellow-emissive upconversion nanocapsule (UCNC-Y) to penetrate tissue by covering UCNCs with beef slices. In the UCL imaging without beef slices, the UCNC-B displays much brighter UC signal than UCNC-G and UCNC-Y, as shown in Fig. 6.8II. However, when covered with a 0.1-cm beef slice, the UC signal from UCNC-B (Fig. 6.8IIe) decreased remarkably compared with the cases of UCNC-G and UCNC-Y (Fig. 6.8IIb), and the SNR = 8.1 of UCNC-B is significantly lower than those of UCNC-G and UCNC-Y, which is attributed to the low penetration depth of the blue emission. The penetration abilities of UCNC-G and UCNC-Y were further investigated when covered with a 0.2-cm beef slice. Clear UC signals could be observed for UCNC-G and UCNC-Y with SNRs of 6.9 and 12.0 (Fig. 6.8IIc), respectively. These results indicated that UCNC-G and UCNC-Y provide better penetration ability than UCNC-B. This is a classic example of penetration depth measurement based on beef tissue slices. The authors have demonstrated that the green light and yellow light (essentially a mixture of green and red light) can penetrate deeper than the blue light.

As opposed to placing a certain amount of UCNPs directly beneath the pork/beef slices, many researchers prefer to use agarose to disperse the as-synthesized nanocrystals. An agarose is a polysaccharide polymer material, generally extracted from seaweed. Agarose is a linear polymer made up of the repeating unit of agarobiose, which is a disaccharide made up of D-galactose and 3,6-anhydro-L-galactopyranose. Agarose is a preferred matrix for work with proteins and nucleic acids as it has widespread physical, chemical, and thermal stability, and its low degree of chemical complexity also makes it less likely to interact with biomolecules. Moreover, agarose may be modified by hydroxyethylation, and this substitution reduces the number of intrastrand hydrogen bonds, resulting in lower melting and setting temperature than standard agaroses. The exact temperature is determined by the degree of substitution, and many available low-melting point (LMP) agarose can remain as a fluid in the 30–35 °C (86–95 °F) range. This property allows the mixing of UCNPs with agarose to take place in a homogeneous system. After cooling to the room temperature, the nanocrystals are well distributed in the solidified agarose gel. Then, the gel is placed into pork muscle tissues at different depths for imaging experiments.

In 2014, Zhao et al. [80] reported the penetration depth measurement of agarose-loaded UCNPs. Typically, water-soluble Tm^{3+}-based UCNPs were dissolved in warm 1 % agarose solution (5 mg/mL). After cooling to room temperature, the solidified agarose gel containing the NCs was placed into pork muscle tissues at different depths for imaging experiments. They found that the signals can be detected even in pork tissues of 15 mm thickness. The advantage of agarose-loaded UCNPs is that the UCNPs are uniformly dispersed, rather than randomly dispersed, eliminating the disadvantages of high local concentration.

Although the pork/beef tissues are favored by the researchers due to their low cost, convenience, simplicity, and intuitive properties, there are still drawbacks when pork/beef tissues are used as an in vitro model. Firstly, the fat and protein contents in the pork/beef tissues are different from those in mouse or human tissues. As we discussed in the introduction, the absorption and scatting coefficients are different in various tissues, so the penetration depth results obtained from pork/beef tissues may have large deviations compared with the real values. Secondly, the pork/beef slices are not easy to store, fresh meat must be bought for every experiment, resulting in unscientific comparison due to different tissues being used in different experiments.

Simulated Tissue

Simulated tissue, also known as phantom tissue, is a mixture containing a certain amount of absorber and scattering medium to emulate tissues. Simulated tissue benefits from tunable absorption and scatting coefficients (simply modify the amount of absorber and scattering medium), ease of shaping (it can be cut into slices or molded to specific templates) ease of storage (preservative can be added during the fabrication procedure), and properties comparable to pork/beef slices and real tissues. Thus, simulated tissue-based penetration depth measurement is gaining popularity among researchers.

The typical simulated tissue fabrication method: the phantom tissue is fabricated by mixing a scattering medium (agar) with several absorbing components (blue India ink, olive oil, and water) so as to mimic the optical properties of in vitro human skin. The use of agarose (one of the main components of agar) as a scattering medium in phantom tissues was first reported by Cubeddu et al. Hemoglobin and melanin are the main components responsible for the absorption in the visible and the ultraviolet regions, as reported by Kobayashi et al. A certain amount (0.36 g) of microbiological agar powder is first dissolved in distilled water to get a solution that is around 1 % by volume. Then, appropriate amounts of olive oil and blue India ink are added to the mixture. The optimal percentages to mimic the absorption of in vitro human skin are 0.50 % (200 μL) of blue India ink and 0.32 % (128 μL) of olive oil as added directly to the mentioned agar/water solution. Then, the solution is cooled down to 4 °C. Note that if long-stem storage is required, a small amount of sodium azide (explosive) can be added to the solution at first step [122].

In 2013, Veggel et al. [122] reported the penetration depth measurement of luminescence from $LaF_3:Nd^{3+}$ nanocrystals using a phantom tissue. Figure 6.8IIIa schematically shows the experimental setup used for the determination of the tissue penetration depth of $LaF_3:Nd^{3+}$ nanocrystals. Briefly, 15 atom % Nd^{3+}-doped nanocrystals were dispersed in distilled water with a resulting nanocrystals concentration of 0.3 % in mass. The colloidal solution was placed into a 200-μm-deep microchannel, and it was optically excited by a continuous-wave 808-nm laser beam. This beam was focused into the microchannel by a 50 × microscope objective (0.55 NA, leading to a spot size close to 2.5 μm). The same microscope objective was used for collection of the Nd^{3+} luminescence that, after passing through different filters and apertures, was analyzed by a high-resolution spectrometer (not shown in the figure for the sake of simplicity). A phantom tissue with a variable thickness was placed between the microchannel and the focusing objective. This setup allowed them to measure the luminescence as a function of the tissue thickness by simply translating the colloidal suspension. The phantom tissue used in this work consists of a scattering medium (such as agar) containing two main absorbing components (India ink and olive oil). The ratio between the different components was adjusted to reproduce the main features of the extinction spectrum of an in vitro human skin tissue. This is mainly given by the visible and infrared absorption bands of hemoglobin and water, respectively. Figure 6.8IIIb shows the $LaF_3:Nd^{3+}$ nanocrystals emission intensity as a function of the phantom tissue thickness. From data included in Fig. 6.8IIIb, it is evident that the use of $LaF_3:Nd^{3+}$ nanocrystals allows for penetration depths into in vitro human skin tissues close to 2 mm. Of course, this penetration depth could be improved by increasing the excitation power as well as by improving the sensitivity of the detection systems. In any case, the penetration depth reported here is on the same order of magnitude as those achieved by using other rare-earth-doped nanocrystals also emitting within the biological window. In order to get a reliable comparison of the penetration depth reached by $LaF_3:Nd^{3+}$ nanocrystals with that of the commercial quantum dots, they measured, in the same experimental conditions, the penetration depth of 8-nm CdTe QDs. These luminescent nanoprobes have recently been proposed as promising nanoprobes for subtissue imaging. In this case, the $LaF_3:Nd^{3+}$ nanocrystals solution within the microchannel was replaced by an 8-nm CdTe QDs solution at the same concentration (0.3 % in mass). Excitation of CdTe QDs was achieved by using a 100 fs Ti:sapphire laser tuned to 920 nm so that the CdTe QDs emission at 780 nm was generated after a two-photon excitation process. The obtained CdTe QDs emitted intensity as a function of the tissue thickness is included in Fig. 6.8IIIb. As can be observed, much larger penetration depths were obtained by using $LaF_3:Nd^{3+}$ nanocrystals instead of CdTe QDs.

Simulated tissue has attracted much interest among researchers due to their absorption and scatting coefficients being very close to those of real tissues. However, further optimization is still required for more precise measurement of the penetration depth.

6.8.2.3 In Vivo Model

There is no doubt that in vivo measurement is the most credible method among various penetration depth measurement methods. Typically, the mouse tissue-based penetration depth measuring can be divided into two types: subcutaneous injection method and organs-based method.

In subcutaneous injection method, UCNPs are subcutaneously injected at a certain depth position and then UC images are taken with NIR irradiation. The penetration depth is determined to be the subtissue depth at which the UC signals cannot be detected. This method is very rough and only a few papers used this method to determine the penetration depth.

In organs-based method, a specific organ is always chosen as model because it has a fixed depth. For example, the depth of the stomach of a nude mouse is 3–5 mm from the front side; the depth of liver is a little shallower. By gavage procedure, UCNPs can be easily imbued into the stomach of a nude mouse which has a specific depth. Then, UC images are taken to determine whether or not the UC luminescence can penetrate. If the liver is chosen as the model organ, intravenous injection can be performed because the UCNPs can accumulate in the liver according to the RES recognition.

In 2014, Zhang et al. [80] reported the penetration depth measurement using the mouse stomach as model organ. They imbued 0.2 ml different concentration (1, 5, 20, 50, 100 nM) of water-soluble UCNPs into mice stomach by using a gastric syringe. After 20-min digestion, UC images were taken upon 800-nm irradiation (0.2 W/cm^2) with appropriate filters equipped. Bright UC signals were detected, indicating a penetration depth deeper than 3–5 mm.

Although the in vivo model is the most credible model among various penetration depth measuring model, it also has some limitations, such as the limited choice of organs. The size of the stomach is small, and its position is relatively fixed compared to the liver. The size of the liver is too big to determine its depth. It is difficult to transport to and maintain UCNPs in other organs.

6.9 Conclusion and Future Opportunity for Upconversion-Based Biomedical Imaging

UCNPs have been successfully used in in vitro and in vivo biomedical imaging with high SNR. The optical imaging depth can reach 1.0–3.2 cm in biological tissues using NIR-to-NIR UCNPs. However, there are still several problems and obstacles need to be investigated and solved. Thus, we present several obstacles for the development of UC-based biomedical imaging and outlook the future opportunity for it.

1. The penetration depth in biological tissues is very important to a luminescent probe. However, the penetration depth of the same UCNPs reported by different groups differs a lot. We consider that this difference is resulted from different measuring model used in different reports. Therefore, we propose that a standard penetration depth measurement method should be established. This method must include tissue model, measuring environment condition, and operating procedure.
2. Although UCNPs have been proved to be low toxic in in vitro and in vivo experiment. However, these results were obtained from small animals such as mice and zebra fish, which are probably different from human body. Therefore, further toxic investigation should be carried on before their clinical application.
3. Despite a few works on active target bio-imaging, most investigations on PL or multimodal imaging are limited to passive bio-imaging. However, the passive route would lead to undesired accumulation in RES organs, which is harmful to the specific organs such as liver, spleen, and kidney. Next step for UCNPs-based biomedical imaging is to explore high efficient targeted imaging UC probes.
4. Although several groups have investigated the synthesis and luminescence optimization of ultrasmall UCNPs, the biomedical imaging using ultrasmall UCNPs is limited to the in situ imaging (subcutaneous imaging or Sentinel lymph node imaging), rather than targeted imaging. Thus, the clearance property of ultrasmall UCNPs is still unclear by now. We predict that efficient UCNPs with clearable capability will be an appealing future direction.
5. Limited penetration depth is a permanent obstacle to whole luminescent biomedical imaging techniques including UC-based biomedical imaging. But UCNPs have its advantage in multimodal imaging because UCNPs can be easily redesigned to multimodal probes. Development of a multi-functional nanoprobe for simultaneous bio-imaging covering optical imaging, MRI, PET, SPECT, and CT monitoring would be an interesting topic.
6. Photoacoustic imaging is an emerging high-resolution imaging modality, which can work perfectly based on plasmonic nanostructures. UCNPs have been reconstructed into photoacoustic-UC bimodal nanoprobes due to the large absorption coefficient of Ln ions (such as Yb^{3+}) in NIR region. The combination of photoacoustic and UC imaging not only enhances the SNR of the imaging results, but also keeps the biosafety of the UC imaging because photoacoustic imaging is also harmless to the biological tissues, unlike above-mentioned tomographic imaging.

References

1. Galperin, A., Margel, D., Baniel, J., Dank, G., Biton, H., Margel, S.: Radiopaque iodinated polymeric nanoparticles for X-ray imaging applications. Biomaterials **28**, 4461–4468 (2007)
2. Bar-Shalom, R., Yefremov, N., Guralnik, L., Gaitini, D., Frenkel, A., Kuten, A., Altman, H., Keidar, Z., Israel, O.: Clinical performance of PET/CT in evaluation of cancer: additional value for diagnostic Imaging and patient management. J. Nucl. Med. **44**, 1200–1209 (2003)
3. Beyer, T., Townsend, D.W., Brun, T., Kinahan, P.E., Charron, M., Roddy, R., Jerin, J., Young, J., Byars, L., Nutt, R.: A combined PET/CT scanner for clinical oncology. J. Nucl. Med. **41**, 1369–1379 (2000)
4. Buck, A.K., Nekolla, S., Ziegler, S., Beer, A., Krause, B.J., Herrmann, K., Scheidhauer, K., Wester, H.J., Rummeny, E.J., Schwaiger, M., Drzezga, A.: SPECT/CT. J. Nucl. Med. **49**, 1305–1319 (2008)
5. Pichler, B.J., Wehrl, H.F., Judenhofer, M.S.: Latest advances in molecular imaging instrumentation. J. Nucl. Med. **49**, 5S–23S (2008)
6. Strauss, L.G., Conti, P.S.: The applications of pet in clinical oncology. J. Nucl. Med. **32**, 623–648 (1991)
7. Degrado, T.R., Turkington, T.G., Williams, J.J., Stearns, C.W., Hoffman, J.M., Coleman, R. E.: Performance-characteristics of a whole-body pet scanner. J. Nucl. Med. **35**, 1398–1406 (1994)
8. Pichler, B.J., Kolb, A., Nagele, T., Schlemmer, H.P.: PET/MRI: paving the way for the next generation of clinical multimodality imaging applications. J. Nucl. Med. **51**, 333–336 (2010)
9. Merkel, O.M., Librizzi, D., Pfestroff, A., Schurrat, T., Behe, M., Kissel, T.: In vivo SPECT and real-time gamma camera imaging of biodistribution and pharmacokinetics of siRNA delivery using an optimized radiolabeling and purification procedure. Bioconjug. Chem. **20**, 174–182 (2009)
10. Dirksen, A., Langereis, S., de Waal, B.F.M., van Genderen, M.H.P., Hackeng, T.M., Meijer, E.W.: A supramolecular approach to multivalent target-specific MRI contrast agents for angiogenesis. Chem. Commun. **22**, 2811–2813 (2005)
11. Bridot, J.L., Faure, A.C., Laurent, S., Riviere, C., Billotey, C., Hiba, B., Janier, M., Josserand, V., Coll, J.L., Vander Elst, L., Muller, R., Roux, S., Perriat, P., Tillement, O.: Hybrid gadolinium oxide nanoparticles: multimodal contrast agents for in vivo imaging. J. Am. Chem. Soc. **129**, 5076–5084 (2007)
12. Liu, J.A., Bu, W.B., Zhang, S.J., Chen, F., Xing, H.Y., Pan, L.M., Zhou, L.P., Peng, W.J., Shi, J.L.: Controlled synthesis of uniform and monodisperse upconversion core/mesoporous silica shell nanocomposites for bimodal imaging. Chem. Eur. J. **18**, 2335–2341 (2012)
13. Adams, S., Baum, R.P., Stuckensen, T., Bitter, K., Hor, G.: Prospective comparison of F-18-FDG PET with conventional imaging modalities (CT, MRI, US) in lymph node staging of head and neck cancer. Eur. J. Nucl. Med. **25**, 1255–1260 (1998)
14. Ntziachristos, V., Ripoll, J., Wang, L.H.V., Weissleder, R.: Looking and listening to light: the evolution of whole-body photonic imaging. Nat. Biotechnol. **23**, 313–320 (2005)
15. Frangioni, J.V.: In vivo near-infrared fluorescence imaging. Curr. Opin. Chem. Biol. **7**, 626–634 (2003)
16. Weissleder, R.: A clearer vision for in vivo imaging. Nat. Biotechnol. **19**, 316–317 (2001)
17. Pansare, V.J., Hejazi, S., Faenza, W.J., Prud'homme, R.K.: Review of long-wavelength optical and NIR imaging materials: contrast agents, fluorophores, and multifunctional nano carriers. Chem. Mater. **24**, 812–827 (2012)
18. Chance, B.: Near-infrared images using continuous, phase-modulated, and pulsed light with quantitation of blood and blood oxygenation. Adv. Opt. Biopsy Opt. Mammogr. **838**, 29–45 (1998)
19. Ntziachristos, V., Bremer, C., Weissleder, R.: Fluorescence imaging with near-infrared light: new technological advances that enable in vivo molecular imaging. Eur. Radiol. **13**, 195–208 (2003)

20. Bashkatov, A.N., Genina, E.A., Kochubey, V.I., Tuchin, V.V.: Optical properties of human skin, subcutaneous and mucous tissues in the wavelength range from 400 to 2000 nm. J. Phys. D Appl. Phys. **38**, 2543–2555 (2005)
21. Welsher, K., Sherlock, S.P., Dai, H.J.: Deep-tissue anatomical imaging of mice using carbon nanotube fluorophores in the second near-infrared window. Proc. Natl. Acad. Sci. **108**, 8943–8948 (2011)
22. Weissleder, R., Tung, C.H., Mahmood, U., Bogdanov, A.: In vivo imaging of tumors with protease-activated near-infrared fluorescent probes. Nat. Biotechnol. **17**, 375–378 (1999)
23. Gao, X.H., Cui, Y.Y., Levenson, R.M., Chung, L.W.K., Nie, S.M.: In vivo cancer targeting and imaging with semiconductor quantum dots. Nat. Biotechnol. **22**, 969–976 (2004)
24. Lim, Y.T., Kim, S., Nakayama, A., Stott, N.E., Bawendi, M.G., Frangioni, J.V.: Selection of quantum dot wavelengths for biomedical assays and imaging. Molecular imaging **2**, 50–64 (2003)
25. Chen, R.J., Bangsaruntip, S., Drouvalakis, K.A., Kam, N.W.S., Shim, M., Li, Y.M., Kim, W., Utz, P.J., Dai, H.J.: Noncovalent functionalization of carbon nanotubes for highly specific electronic biosensors. Proc. Natl. Acad. Sci. **100**, 4984–4989 (2003)
26. Smith, A.M., Mancini, M.C., Nie, S.M.: Bioimaging second window for in vivo imaging. Nat. Nanotechnol. **4**, 710–711 (2009)
27. Yuan, A., Wu, J.H., Tang, X.L., Zhao, L.L., Xu, F., Hu, Y.Q.: Application of near-infrared dyes for tumor imaging, photothermal, and photodynamic therapies. J. Pharm. Sci. **102**, 6–28 (2013)
28. Khlebtsov, N.G., Dykman, L.A.: Optical properties and biomedical applications of plasmonic nanoparticles. J. Quant. Spectrosc. Radiat. Transfer **111**, 1–35 (2010)
29. Wang, R., Zhang, F.: NIR luminescent nanomaterials for biomedical imaging. J. Mater. Chem. B **2**, 2422–2443 (2014)
30. Du, Y.P., Xu, B., Fu, T., Cai, M., Li, F., Zhang, Y., Wang, Q.B.: Near-infrared photoluminescent Ag_2S quantum dots from a single source precursor. J. Am. Chem. Soc. **132**, 1470–1471 (2010)
31. Liu, Z., Cai, W., He, L., Nakayama, N., Chen, K., Sun, X., Chen, X., Dai, H.: In vivo biodistribution and highly efficient tumour targeting of carbon nanotubes in mice. Nat. Nanotechnol. **2**, 47–52 (2006)
32. Rogach, A.L., Eychmüller, A., Hickey, S.G., Kershaw, S.V.: Infrared-emitting colloidal nanocrystals: synthesis, assembly, spectroscopy, and applications. Small **3**, 536–557 (2007)
33. Hardman, R.: A toxicologic review of quantum dots: Toxicity depends on physicochemical and environmental factors. Environ. Health Persp. **114**, 165–172 (2006)
34. Naczynski, D.J., Tan, M.C., Zevon, M., Wall, B., Kohl, J., Kulesa, A., Chen, S., Roth, C.M., Riman, R.E., Moghe, P.V.: Rare-earth-doped biological composites as in vivo shortwave infrared reporters. Nat. Commun. **4**, 2199 (2013)
35. Chen, G.Y., Qju, H.L., Prasad, P.N., Chen, X.Y.: Upconversion nanoparticles: design, nanochemistry, and applications in theranostics. Chem. Rev. **114**, 5161–5214 (2014)
36. Feng, W., Zhu, X.J., Li, F.Y.: Recent advances in the optimization and functionalization of upconversion nanomaterials for in vivo bioapplications. NPG Asia Mater. **5**, e75 (2013)
37. Gorris, H.H., Wolfbeis, O.S.: Photon-upconverting nanoparticles for optical encoding and multiplexing of cells, biomolecules, and microspheres. Angew. Chem. Int. Ed. **52**, 3584–3600 (2013)
38. Gu, Z.J., Yan, L., Tian, G., Li, S.J., Chai, Z.F., Zhao, Y.L.: Recent advances in design and fabrication of upconversion nanoparticles and their safe theranostic applications. Adv. Mater. **25**, 3758–3779 (2013)
39. Haase, M., Schafer, H.: Upconverting nanoparticles. Angew. Chem. Int. Ed. **50**, 5808–5829 (2011)
40. Wang, F., Liu, X.G.: Recent advances in the chemistry of lanthanide-doped upconversion nanocrystals. Chem. Soc. Rev. **38**, 976–989 (2009)

41. Boyer, J.C., Vetrone, F., Cuccia, L.A., Capobianco, J.A.: Synthesis of colloidal upconverting $NaYF_4$ nanocrystals doped with Er^{3+}, Yb^{3+} and Tm^{3+}, Yb^{3+} via thermal decomposition of lanthanide trifluoroacetate precursors. J. Am. Chem. Soc. **128**, 7444–7445 (2006)
42. Cheng, L.A., Yang, K., Zhang, S.A., Shao, M.W., Lee, S.T., Liu, Z.A.: Highly-sensitive multiplexed in vivo Imaging using PEGylated upconversion nanoparticles. Nano Res. **3**, 722–732 (2010)
43. Dong, N.N., Pedroni, M., Piccinelli, F., Conti, G., Sbarbati, A., Ramirez-Hernandez, J.E., Maestro, L.M., Iglesias-de la Cruz, M.C., Sanz-Rodriguez, F., Juarranz, A., Chen, F., Vetrone, F., Capobianco, J.A., Sole, J.G., Bettinelli, M., Jaque, D., Speghini, A.: NIR-to-NIR two-photon excited CaF^2: Tm^{3+}, Yb^{3+} nanoparticles: multifunctional nanoprobes for highly penetrating fluorescence bio-imaging. ACS Nano **5**, 8665–8671 (2011)
44. Jalil, R.A., Zhang, Y.: Biocompatibility of silica coated $NaYF_4$ upconversion fluorescent nanocrystals. Biomaterials **29**, 4122–4128 (2008)
45. Liu, Z., Pu, F., Huang, S., Yuan, Q.H., Ren, J.S., Qu, X.G.: Long-circulating Gd_2O_3:Yb^{3+}, Er^{3+} up-conversion nanoprobes as high-performance contrast agents for multi-modality imaging. Biomaterials **34**, 1712–1721 (2013)
46. Nyk, M., Kumar, R., Ohulchanskyy, T.Y., Bergey, E.J., Prasad, P.N.: High contrast in vitro and in vivo photoluminescence bioimaging using near infrared to near infrared up-conversion in TM^{3+} and Yb^{3+} doped fluoride nanophosphors. Nano Lett. **8**, 3834–3838 (2008)
47. Wang, J., Wang, F., Wang, C., Liu, Z., Liu, X.G.: Single-band upconversion emission in lanthanide-doped $KMnF_3$ nanocrystals. Angew. Chem. Int. Ed. **50**, 10369–10372 (2011)
48. Wei, X.J., Wang, W., Chen, K.Z.: Preparation and characterization of ZnS:Tb, Gd and ZnS: Er, Yb, Gd nanoparticles for bimodal magnetic-fluorescent imaging. Dalton Trans. **42**, 1752–1759 (2013)
49. Yan, L., Chang, Y.N., Zhao, L.N., Gu, Z.J., Liu, X.X., Tian, G., Zhou, L.J., Ren, W.L., Jin, S., Yin, W.Y., Chang, H.Q., Xing, G.M., Gao, X.F., Zhao, Y.L.: The use of polyethylenimine-modified graphene oxide as a nanocarrier for transferring hydrophobic nanocrystals into water to produce water-dispersible hybrids for use in drug delivery. Carbon **57**, 120–129 (2013)
50. Yang, T.S., Sun, Y., Liu, Q., Feng, W., Yang, P.Y., Li, F.Y.: Cubic sub-20 nm $NaLuF_4$-based upconversion nanophosphors for high-contrast bioimaging in different animal species. Biomaterials **33**, 3733–3742 (2012)
51. Yin, W.Y., Zhao, L.N., Zhou, L.J., Gu, Z.J., Liu, X.X., Tian, G., Jin, S., Yan, L., Ren, W.L., Xing, G.M., Zhao, Y.L.: Enhanced red emission from GdF_3:Yb^{3+}, Er^{3+} upconversion nanocrystals by Li^+ doping and their application for bioimaging. Chem. Eur. J. **18**, 9239–9245 (2012)
52. Yin, W.Y., Zhou, L.J., Gu, Z.J., Tian, G., Jin, S., Yan, L., Liu, X.X., Xing, G.M., Ren, W.L., Liu, F., Pan, Z.W., Zhao, Y.L.: Lanthanide-doped $GdVO_4$ upconversion nanophosphors with tunable emissions and their applications for biomedical imaging. J. Mater. Chem. **22**, 6974–6981 (2012)
53. Zeng, S.J., Tsang, M.K., Chan, C.F., Wong, K.L., Hao, J.H.: PEG modified $BaGdF_5$:Yb/Er nanoprobes for multi-modal upconversion fluorescent, in vivo X-ray computed tomography and biomagnetic imaging. Biomaterials **33**, 9232–9238 (2012)
54. Zhou, J., Liu, Z., Li, F.Y.: Upconversion nanophosphors for small-animal imaging. Chem. Soc. Rev. **41**, 1323–1349 (2012)
55. Zhou, J., Zhu, X.J., Chen, M., Sun, Y., Li, F.Y.: Water-stable $NaLuF_4$-based upconversion nanophosphors with long-term validity for multimodal lymphatic imaging. Biomaterials **33**, 6201–6210 (2012)
56. Bae, Y.M., Park, Y.I., Nam, S.H., Kim, J.H., Lee, K., Kim, H.M., Yoo, B., Choi, J.S., Lee, K. T., Hyeon, T., Suh, Y.D.: Endocytosis, intracellular transport, and exocytosis of lanthanide-doped upconverting nanoparticles in single living cells. Biomaterials **33**, 9080–9086 (2012)
57. Lim, S.F., Riehn, R., Ryu, W.S., Khanarian, N., Tung, C.K., Tank, D., Austin, R.H.: In vivo and scanning electron microscopy imaging of upconverting nanophosphors in Caenorhabditis elegans. Nano Lett. **6**, 169–174 (2006)

58. Zhou, J.C., Yang, Z.L., Dong, W., Tang, R.J., Sun, L.D., Yan, C.H.: Bioimaging and toxicity assessments of near-infrared upconversion luminescent $NaYF_4$:Yb, Tm nanocrystals. Biomaterials **32**, 9059–9067 (2011)
59. Chen, J., Guo, C.R., Wang, M., Huang, L., Wang, L.P., Mi, C.C., Li, J., Fang, X.X., Mao, C. B., Xu, S.K.: Controllable synthesis of $NaYF_4$:Yb, Er upconversion nanophosphors and their application to in vivo imaging of Caenorhabditis elegans. J. Mater. Chem. **21**, 2632–2638 (2011)
60. Xiong, L.Q., Yang, T.S., Yang, Y., Xu, C.J., Li, F.Y.: Long-term in vivo biodistribution imaging and toxicity of polyacrylic acid-coated upconversion nanophosphors. Biomaterials **31**, 7078–7085 (2010)
61. Yang, Y., Sun, Y., Liu, Y., Peng, J.J., Wu, Y.Q., Zhang, Y.J., Feng, W., Li, F.Y.: Long-term in vivo biodistribution and toxicity of $Gd(OH)_3$ nanorods. Biomaterials **34**, 508–515 (2013)
62. Zhou, J., Sun, Y., Du, X.X., Xiong, L.Q., Hu, H., Li, F.Y.: Dual-modality in vivo imaging using rare-earth nanocrystals with near-infrared to near-infrared (NIR-to-NIR) upconversion luminescence and magnetic resonance properties. Biomaterials **31**, 3287–3295 (2010)
63. Shan, J.N., Chen, J.B., Meng, J., Collins, J., Soboyejo, W., Friedberg, J.S., Ju, Y.G.: Biofunctionalization, cytotoxicity, and cell uptake of lanthanide doped hydrophobically ligated $NaYF_4$ upconversion nanophosphors. J. Appl. Phys. **104**, 094308 (2008)
64. Chatteriee, D.K., Rufalhah, A.J., Zhang, Y.: Upconversion fluorescence imaging of cells and small animals using lanthanide doped nanocrystals. Biomaterials **29**, 937–943 (2008)
65. Tian, G., Gu, Z.J., Zhou, L.J., Yin, W.Y., Liu, X.X., Yan, L., Jin, S., Ren, W.L., Xing, G.M., Li, S.J., Zhao, Y.L.: Mn^{2+} dopant-controlled synthesis of $NaYF_4$:Yb/Er upconversion nanoparticles for in vivo imaging and drug delivery. Adv. Mater. **24**, 1226–1231 (2012)
66. Zhou, L.J., Gu, Z.J., Liu, X.X., Yin, W.Y., Tian, G., Yan, L., Jin, S., Ren, W.L., Xing, G.M., Li, W., Chang, X.L., Hu, Z.B., Zhao, Y.L.: Size-tunable synthesis of lanthanide-doped Gd_2O_3 nanoparticles and their applications for optical and magnetic resonance imaging. J. Mater. Chem. **22**, 966–974 (2012)
67. Zijlmans, H., Bonnet, J., Burton, J., Kardos, K., Vail, T., Niedbala, R.S., Tanke, H.J.: Detection of cell and tissue surface antigens using up-converting phosphors: a new reporter technology. Anal. Biochem. **267**, 30–36 (1999)
68. Hilderbrand, S.A., Shao, F.W., Salthouse, C., Mahmood, U., Weissleder, R.: Upconverting luminescent nanomaterials: application to in vivo bioimaging. Chem. Commun. **28**, 4188–4190 (2009)
69. Liu, Q., Sun, Y., Yang, T.S., Feng, W., Li, C.G., Li, F.Y.: Sub-10 nm hexagonal lanthanide-doped $NaLuF_4$ upconversion nanocrystals for sensitive bioimaging in vivo. J. Am. Chem. Soc. **133**, 17122–17125 (2011)
70. Chen, G.Y., Shen, J., Ohulchanskyy, T.Y., Patel, N.J., Kutikov, A., Li, Z.P., Song, J., Pandey, R.K., Agren, H., Prasad, P.N., Han, G.: (alpha-$NaYbF_4$:Tm^{3+})/CaF_2 core/shell nanoparticles with efficient near-infrared to near-infrared upconversion for high-contrast deep tissue bioimaging. ACS Nano **6**, 8280–8287 (2012)
71. Xia, A., Gao, Y., Zhou, J., Li, C.Y., Yang, T.S., Wu, D.M., Wu, L.M., Li, F.Y.: Core-shell NaYF4:Yb^{3+},Tm^{3+}@Fe_xO_y nanocrystals for dual-modality T-2-enhanced magnetic resonance and NIR-to-NIR upconversion luminescent imaging of small-animal lymphatic node. Biomaterials **32**, 7200–7208 (2011)
72. Xia, A., Chen, M., Gao, Y., Wu, D.M., Feng, W., Li, F.Y.: Gd^{3+} complex-modified $NaLuF_4$-based upconversion nanophosphors for trimodality imaging of NIR-to-NIR upconversion luminescence, X-ray computed tomography and magnetic resonance. Biomaterials **33**, 5394–5405 (2012)
73. Xing, H.Y., Bu, W.B., Ren, Q.G., Zheng, X.P., Li, M., Zhang, S.J., Qu, H.Y., Wang, Z., Hua, Y.Q., Zhao, K.L., Zhou, L.P., Peng, W.J., Shi, J.L.: A $NaYbF_4$:Tm^{3+} nanoprobe for CT and NIR-to-NIR fluorescent bimodal imaging. Biomaterials **33**, 5384–5393 (2012)
74. Zou, W.Q., Visser, C., Maduro, J.A., Pshenichnikov, M.S., Hummelen, J.C.: Broadband dye-sensitized upconversion of near-infrared light. Nat. Photonics **6**, 560–564 (2012)

75. Chen, G.Y., Ohulchanskyy, T.Y., Kachynski, A., Agren, H., Prasad, P.N.: Intense visible and near-infrared upconversion photoluminescence in colloidal $LiYF_4$:Er^{3+} nanocrystals under excitation at 1490 nm. ACS Nano **5**, 4981–4986 (2011)
76. Zhan, Q.Q., Qian, J., Liang, H.J., Somesfalean, G., Wang, D., He, S.L., Zhang, Z.G., Andersson-Engels, S.: Using 915 nm laser excited Tm^{3+}/Er^{3+}/Ho^{3+}-doped $NaYbF_4$ upconversion nanoparticles for in vitro and deeper in vivo bioimaging without overheating irradiation. ACS Nano **5**, 3744–3757 (2011)
77. Shen, J., Chen, G., Vu, A.-M., Fan, W., Bilsel, O.S., Chang, C.-C., Han, G.: Engineering the upconversion nanoparticle excitation wavelength: cascade sensitization of tri-doped upconversion colloidal nanoparticles at 800 nm. Adv. Opt. Mater. **1**, 644–650 (2013)
78. Xie, X.J., Gao, N.Y., Deng, R.R., Sun, Q., Xu, Q.H., Liu, X.G.: Mechanistic investigation of photon upconversion in Nd^{3+}-sensitized core-shell nanoparticles. J. Am. Chem. Soc. **135**, 12608–12611 (2013)
79. Wang, Y.F., Liu, G.Y., Sun, L.D., Xiao, J.W., Zhou, J.C., Yan, C.H.: Nd^{3+}-sensitized upconversion nanophosphors: efficient in vivo bioimaging probes with minimized heating effect. ACS Nano **7**, 7200–7206 (2013)
80. Li, X., Wang, R., Zhang, F., Zhou, L., Shen, D., Yao, C., Zhao, D.: Nd^{3+} sensitized up/down converting dual-mode nanomaterials for efficient in-vitro and in-vivo bioimaging excited at 800 nm. Sci. Rep. **3**, 3536 (2013)
81. Matsumura, Y., Maeda, H.: A new concept for macromolecular therapeutics in cancer-chemotherapy: mechanism of tumoritropic accumulation of proteins and the antitumor agent smancs. Cancer Res. **46**, 6387–6392 (1986)
82. Iyer, A.K., Khaled, G., Fang, J., Maeda, H.: Exploiting the enhanced permeability and retention effect for tumor targeting. Drug Discov. Today **11**, 812–818 (2006)
83. Kamimura, M., Kanayama, N., Tokuzen, K., Soga, K., Nagasaki, Y.: Near-infrared (1550 nm) in vivo bioimaging based on rare-earth doped ceramic nanophosphors modified with PEG-b-poly(4-vinylbenzylphosphonate). Nanoscale **3**, 3705–3713 (2011)
84. Zamboni, W.C.: Liposomal, nanoparticle, and conjugated formulations of anticancer agents. Clin. Cancer Res. **11**, 8230–8234 (2005)
85. Dobrovolskaia, M.A., McNeil, S.E.: Immunological properties of engineered nanomaterials. Nat. Nanotechnol. **2**, 469–478 (2007)
86. Kareem, H., Sandstrom, K., Elia, R., Gedda, L., Anniko, M., Lundqvist, H., Nestor, M.: Blocking EGFR in the liver improves the tumor-to-liver uptake ratio of radiolabeled EGF. Tumor Biol. **31**, 79–87 (2010)
87. Xu, C.T., Svensson, N., Axelsson, J., Svenmarker, P., Somesfalean, G., Chen, G.Y., Liang, H.J., Liu, H.C., Zhang, Z.G., Andersson-Engels, S.: Autofluorescence insensitive imaging using upconverting nanocrystals in scattering media. Appl. Phys. Lett. **93**, 171103 (2008)
88. Xu, C.T., Axelsson, J., Andersson-Engels, S.: Fluorescence diffuse optical tomography using upconverting nanoparticles. Appl. Phys. Lett. **94**, 251107 (2009)
89. Liu, H.C., Xu, C.T., Andersson-Engels, S.: Multibeam fluorescence diffuse optical tomography using upconverting nanoparticles. Opt. Lett. **35**, 718–720 (2010)
90. Xu, C.T., Svenmarker, P., Liu, H.C., Wu, X., Messing, M.E., Wallenberg, L.R., Andersson-Engels, S.: High-resolution fluorescence diffuse optical tomography developed with nonlinear upconverting nanoparticles. ACS Nano **6**, 4788–4795 (2012)
91. Das, G.K., Heng, B.C., Ng, S.C., White, T., Loo, J.S.C., D'Silva, L., Padmanabhan, P., Bhakoo, K.K., Selvan, S.T., Tan, T.T.Y.: Gadolinium oxide ultranarrow nanorods as multimodal contrast agents for optical and magnetic resonance imaging. Langmuir **26**, 8959–8965 (2010)
92. Wang, Z.L., Hao, J.H., Chan, H.L.W.: Down- and up-conversion photoluminescence, cathodoluminescence and paramagnetic properties of $NaGdF_4$:Yb^{3+}, Er^{3+} submicron disks assembled from primary nanocrystals. J. Mater. Chem. **20**, 3178–3185 (2010)
93. Ryu, J., Park, H.Y., Kim, K., Kim, H., Yoo, J.H., Kang, M., Im, K., Grailhe, R., Song, R.: Facile synthesis of ultrasmall and hexagonal $NaGdF_4$: Yb^{3+}, Er^{3+} nanoparticles with magnetic and upconversion imaging properties. J. Phys. Chem. C **114**, 21077–21082 (2010)

94. Kumar, R., Nyk, M., Ohulchanskyy, T.Y., Flask, C.A., Prasad, P.N.: Combined optical and MR bioimaging using rare earth ion doped NaYF4 nanocrystals. Adv. Funct. Mater. **19**, 853–859 (2009)
95. Chen, D.Q., Yu, Y.L., Huang, F., Yang, A.P., Wang, Y.S.: Lanthanide activator doped $NaYb_{1-x}Gd_xF_4$ nanocrystals with tunable down-, up-conversion luminescence and paramagnetic properties. J. Mater. Chem. **21**, 6186–6192 (2011)
96. Zhou, J., Yu, M.X., Sun, Y., Zhang, X.Z., Zhu, X.J., Wu, Z.H., Wu, D.M., Li, F.Y.: Fluorine-18-labeled $Gd^{3+}/Yb^{3+}/Er^{3+}$ co-doped $NaYF_4$ nanophosphors for multimodality PET/MR/UCL imaging. Biomaterials **32**, 1148–1156 (2011)
97. Liu, Q., Sun, Y., Li, C.G., Zhou, J., Li, C.Y., Yang, T.S., Zhang, X.Z., Yi, T., Wu, D.M., Li, F.Y.: F-18-labeled magnetic-upconversion nanophosphors via rare-earth cation-assisted ligand assembly. ACS Nano **5**, 3146–3157 (2011)
98. Li, Z.Q., Zhang, Y., Shuter, B., Idris, N.M.: Hybrid lanthanide nanoparticles with paramagnetic shell coated on upconversion fluorescent nanocrystals. Langmuir **25**, 12015–12018 (2009)
99. Liu, J.N., Bu, J.W., Bu, W.B., Zhang, S.J., Pan, L.M., Fan, W.P., Chen, F., Zhou, L.P., Peng, W.J., Zhao, K.L., Du, J.L., Shi, J.L.: Real-time in vivo quantitative monitoring of drug release by dual-mode magnetic resonance and upconverted luminescence imaging. Angew. Chem. Int. Ed. **53**, 4551–4555 (2014)
100. Lu, H.C., Yi, G.S., Zhao, S.Y., Chen, D.P., Guo, L.H., Cheng, J.: Synthesis and characterization of multi-functional nanoparticles possessing magnetic, up-conversion fluorescence and bio-affinity properties. J. Mater. Chem. **14**, 1336–1341 (2004)
101. Shen, J., Sun, L.D., Zhang, Y.W., Yan, C.H.: Superparamagnetic and upconversion emitting $Fe_3O_4/NaYF_4$:Yb, Er hetero-nanoparticles via a crosslinker anchoring strategy. Chem. Commun. **46**, 5731–5733 (2010)
102. Gai, S.L., Yang, P.P., Li, C.X., Wang, W.X., Dai, Y.L., Niu, N., Lin, J.: Synthesis of magnetic, up-conversion luminescent, and mesoporous core-shell-structured nanocomposites as drug carriers. Adv. Funct. Mater. **20**, 1166–1172 (2010)
103. Yu, X.G., Shan, Y., Li, G.C., Chen, K.Z.: Synthesis and characterization of bifunctional magnetic-optical $Fe_3O_4@SiO_2@Y_2O_3:Yb^{3+}$, Er^{3+} near-infrared-to-visible up-conversion nanoparticles. J. Mater. Chem. **21**, 8104–8109 (2011)
104. Chen, F., Zhang, S.J., Bu, W.B., Liu, X.H., Chen, Y., He, Q.J., Zhu, M., Zhang, L.X., Zhou, L.P., Peng, W.J., Shi, J.L.: A "neck-formation" strategy for an antiquenching magnetic/upconversion fluorescent bimodal cancer probe. Chem. Eur. J. **16**, 11254–11260 (2010)
105. Gao, Q., Xiu, Y., Li, G.D., Chen, J.S.: Sensor material based on occluded trisulfur anionic radicals for convenient detection of trace amounts of water molecules. J. Mater. Chem. **20**, 3307–3312 (2010)
106. Cheng, L., Yang, K., Li, Y.G., Chen, J.H., Wang, C., Shao, M.W., Lee, S.T., Liu, Z.: Facile preparation of multifunctional upconversion nanoprobes for multimodal imaging and dual-targeted photothermal therapy. Angew. Chem. Int. Ed. **50**, 7385–7390 (2011)
107. He, M., Huang, P., Zhang, C.L., Hu, H.Y., Bao, C.C., Gao, G., He, R., Cui, D.X.: Dual phase-controlled synthesis of uniform lanthanide-doped $NaGdF_4$ upconversion nanocrystals via an OA/ionic liquid two-phase system for in vivo dual-modality imaging. Adv. Funct. Mater. **21**, 4470–4477 (2011)
108. Liu, Y.L., Ai, K.L., Liu, J.H., Yuan, Q.H., He, Y.Y., Lu, L.H.: A high-performance ytterbium-based nanoparticulate contrast agent for in vivo X-ray computed tomography imaging. Angew. Chem. Int. Ed. **51**, 1437–1442 (2012)
109. Xing, H.Y., Zheng, X.P., Ren, Q.G., Bu, W.B., Ge, W.Q., Xiao, Q.F., Zhang, S.J., Wei, C. Y., Qu, H.Y., Wang, Z., Hua, Y.Q., Zhou, L.P., Peng, W.J., Zhao, K.L., Shi, J.L.: Computed tomography imaging-guided radiotherapy by targeting upconversion nanocubes with significant imaging and radiosensitization enhancements. Sci. Rep. **3**, 1751 (2013)
110. Sun, Y., Yu, M.X., Liang, S., Zhang, Y.J., Li, C.G., Mou, T.T., Yang, W.J., Zhang, X.Z., Li, B. A., Huang, C.H., Li, F.Y.: Fluorine-18 labeled rare-earth nanoparticles for positron emission tomography (PET) imaging of sentinel lymph node. Biomaterials **32**, 2999–3007 (2011)

111. Liu, Q., Chen, M., Sun, Y., Chen, G.Y., Yang, T.S., Gao, Y., Zhang, X.Z., Li, F.Y.: Multifunctional rare-earth self-assembled nanosystem for tri-modal upconversion luminescence/fluorescence/positron emission tomography imaging. Biomaterials **32**, 8243–8253 (2011)
112. Yang, Y., Sun, Y., Cao, T.Y., Peng, J.J., Liu, Y., Wu, Y.Q., Feng, W., Zhang, Y.J., Li, F.Y.: Hydrothermal synthesis of $NaLuF_4$:Sm153,Yb,Tm nanoparticles and their application in dual-modality upconversion luminescence and SPECT bioimaging. Biomaterials **34**, 774–783 (2013)
113. Zhu, X.J., Zhou, J., Chen, M., Shi, M., Feng, W., Li, F.Y.: Core-shell Fe_3O_4@$NaLuF_4$:Yb, Er/Tm nanostructure for MRI, CT and upconversion luminescence tri-modality imaging. Biomaterials **33**, 4618–4627 (2012)
114. Xiao, Q.F., Bu, W.B., Ren, Q.G., Zhang, S.J., Xing, H.Y., Chen, F., Li, M., Zheng, X.P., Hua, Y.Q., Zhou, L.P., Peng, W.J., Qu, H.Y., Wang, Z., Zhao, K.L., Shi, J.L.: Radiopaque fluorescence-transparent TaO_x decorated upconversion nanophosphors for in vivo CT/MR/UCL trimodal imaging. Biomaterials **33**, 7530–7539 (2012)
115. Sun, Y., Zhu, X.J., Peng, J.J., Li, F.Y.: Core-shell lanthanide upconversion nanophosphors as four-modal probes for tumor angiogenesis imaging. ACS Nano **7**, 11290–11300 (2013)
116. Dai, Y.L., Xiao, H.H., Liu, J.H., Yuan, Q.H., Ma, P.A., Yang, D.M., Li, C.X., Cheng, Z.Y., Hou, Z.Y., Yang, P.P., Lin, J.: In vivo multimodality imaging and cancer therapy by near-infrared light-triggered trans-platinum pro-drug-conjugated upconverison nanoparticles. J. Am. Chem. Soc. **135**, 18920–18929 (2013)
117. Liu, Q., Feng, W., Li, F.Y.: Water-soluble lanthanide upconversion nanophosphors: synthesis and bioimaging applications in vivo. Coordin. Chem. Rev. **273**, 100–110 (2014)
118. Yu, M.X., Li, F.Y., Chen, Z.G., Hu, H., Zhan, C., Yang, H., Huang, C.H.: Laser scanning up-conversion luminescence microscopy for imaging cells labeled with rare-earth nanophosphors. Anal. Chem. **81**, 930–935 (2009)
119. Xiong, L.Q., Chen, Z.G., Tian, Q.W., Cao, T.Y., Xu, C.J., Li, F.Y.: High contrast upconversion luminescence targeted imaging in vivo using peptide-labeled nanophosphors. Anal. Chem. **81**, 8687–8694 (2009)
120. Liu, K., Wang, Y., Kong, X., Liu, X., Zhang, Y., Tu, L., Ding, Y., Aalders, M.C.G., Buma, W.J., Zhang, H.: Multispectral upconversion luminescence intensity ratios for ascertaining the tissue imaging depth. Nanoscale **6**, 9257–9263 (2014)
121. Liu, Q., Yin, B.R., Yang, T.S., Yang, Y.C., Shen, Z., Yao, P., Li, F.Y.: A general strategy for biocompatible, high-effective upconversion nanocapsules based on triplet-triplet annihilation. J. Am. Chem. Soc. **135**, 5029–5037 (2013)
122. Rocha, U., Jacinto, C., Silva, W.F., Guedes, I., Benayas, A., Maestro, L.M., Elias, M.A., Bovero, E., van Veggel, F., Sole, J.A.G., Jaque, D.: Subtissue thermal sensing based on neodymium-doped LaF_3 nanoparticles. ACS Nano **7**, 1188–1199 (2013)

Chapter 7
The Applications of Upconversion Nanoparticles in Bioassay

Abstract Highly sensitive detection and accurate analysis of diverse biomolecules are critical in biology, medicine, and life science. Due to the unique nature of upconversion nanoparticles (UCNPs), there is no background luminescence significantly decreasing the limit of detection (LOD). Upconversion (UC) materials feature large anti-Stokes shifts and long lifetimes, leading to largely reduced background noise. Their optical properties are independent of the environment. Thereby, efficient bioassay methods have been developed based on UCNPs in the past decade. In this chapter, we provide a comprehensive survey on the advances of UCNPs' applications in homogeneous detection, heterogeneous detection, and encoding for multiplexed detection of biomolecules and tumor markers.

Keywords Bioassay · Luminescence upconversion · Homogeneous detection · Heterogeneous detection · Multiplexed detection · Barcode encoding · Tumor marker

7.1 Introduction

Bioassay (shorthand for biological assay), or biological standardization, is a type of scientific experiment which determines the relative strength of a substance by comparing its effect on a test organism with that of a standard preparation. Bioassays are procedures that can determine the concentration, purity, or biological activity of a substance such as a vitamin, hormone, or plant growth factor by measuring its effect on an organism, tissue, cells, enzyme, or receptor. Bioassays may be qualitative or quantitative. Qualitative bioassays are used for assessing the unquantifiable physical effects of a substance, whereas quantitative bioassays involve estimation of the concentration or functional activity of a substance by measurement of the biological response that it produces. Quantitative bioassays are typically analyzed using biostatistical methods.

Lei Zhou and Fan Zhang contributed together to this chapter.

F. Zhang, *Photon Upconversion Nanomaterials*,
Nanostructure Science and Technology, DOI 10.1007/978-3-662-45597-5_7

In short, a bioassay is a biochemical test that measures the presence or concentration or functional activity of a biomolecule (antibody, enzyme, immunoglobulin, etc.), often referred to as an "analyte", in a solution. Highly sensitive detection and accurate analysis of diverse biomolecules, particularly cancerous biomarkers in human fluid samples, are critical for early cancer theranostics which can consequently increase patient survival. To achieve this goal, efficient bioassay methods have been developed based on lanthanide-doped luminescent nanobioprobes in the past decade. Based on their signal detection formats, the presently established biological assays can be generally classified into two major categories, namely heterogeneous and homogeneous bioassays. A heterogeneous assay is performed in multiple steps with analytes being added, washed, and separated at different points in the assay, while a homogeneous assay is carried out simply by mixing the analytes and the sample and then making an optical measurement. The use of upconversion nanoparticles (UCNPs) in assays provides several advantages: (i) Because of the unique nature of upconversion (UC), there is no background luminescence from the carrier fluid or the assay biochemistry, significantly decreasing the limit of detection (LOD) to a degree unreachable by conventional assays. (ii) The optical properties of UCNPs are independent of the environment (e.g., buffer chemistry and assay temperature). These properties make them an ideal choice for using in diverse biological assays without the adverse effect of the undesired short-lived background noises, such as auto-fluorescence from biological tissues or scattered light in biological labeling. As a result, the detection process is unaffected by the sample fluid and is robust with respect to a large variety of environmental sampling conditions. In this chapter, we will provide a comprehensive survey on the advances in the development of lanthanide-doped luminescent nanobioprobes and their applications in homogeneous assay, heterogeneous assay, and encoding for multiplexed detection of biomolecules and tumor markers [1, 2].

7.2 Heterogeneous Assay

Heterogeneous assays have the advantages of specific recognition and high binding affinity between target analytes and capture molecules anchored on a solid substrate which allows the detection of trace amounts of analyte. The heterogeneous assay platform utilizes bio-functionalized molecules (capture molecules) that are immobilized on a solid substrate to detect target analytes with the assistance of UCNPs. As illustrated in Fig. 7.1, in a typical procedure for a sandwich assay, the capture molecules are first immobilized on a solid substrate like a 96-well microplate via incubation and then conjugated with the analyte solution. Washing is performed in order to eliminate other biological species present in the sample which might interfere with the last detection step. A solution of capture molecules labeled with NPs is further added in order to mark the analyte. As such, the analyte is trapped between the capture molecule immobilized on the solid phase and the NP-labeled capture molecule. After thoroughly washing off the excess unbound NP-labeled

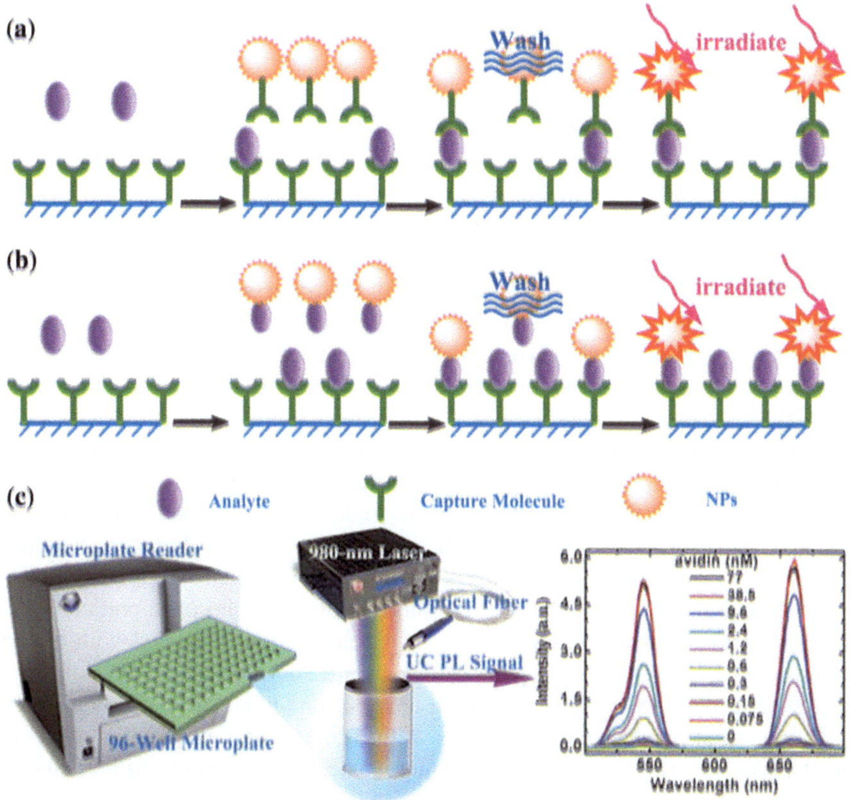

Fig. 7.1 Schematic illustration of typical heterogeneous assays based on Ln^{3+}-doped nanoprobes: **a** sandwich-type assay, **b** competitive assay, and **c** the diagram depicts the experimental setup for heterogeneous UC PL detection of analyte by employing Yb^{3+}, Er^{3+} codoped UCNPs as probes (*left*), and UC emission spectra of the bioassays as a function of the analyte concentration (*right*). In this detection process, an external 0–4 W adjustable CW980 nm laser was used as the excitation light source, and the synergy 4 multi-mode microplate reader (BioTek) was used as the signal collector. Reprinted with the permission from Ref. [1]. Copyright 2012 Royal Society of Chemistry

capture molecules, the microplate is finally subjected to a photoluminescence (PL) measurement. As a consequence, the concentration of the analyte can be quantified by directly measuring the optical signal of the NP label, as indicated in Fig. 7.1a. On the other hand, in a competitive assay a solution containing NP-labeled analyte instead of NP-labeled capture molecule is added in the second step in order to compete with the free analyte to bind to the capture molecule anchored on the substrate, causing the optical signal to vary inversely with the concentration of the free analyte (Fig. 7.1b). Conversely, in a non-competitive heterogeneous assay (i.e., sandwich assay), the optical signal detected is directly proportional to the concentration of the analyte. In the non-competitive heterogeneous assay, with the increase of the analyte concentration, an increased UC PL is observed resulting from the less analyte-labeled UCNPs on the substrate. A calibration between the concentration of the analyte and the UC PL intensity is needed to quantify the analyte in the test sample. It is worth mentioning that a sandwich heterogeneous assay is often applied to detect large biomolecules that have multiple binding sites, while a competitive assay is generally applied to detect small biomolecules [1].

The first heterogeneous bioassay based on UC particles was reported by Zijlmans and coworkers in 1999. Two upconverting compositions including a green-emitting ytterbium/erbium $(Y.Yb.Er)O_2S$ and a blue-emitting ytterbium/thulium $(Y.Yb.Tm)_2O_2S$ were used for the detection of a prostate-specific antigen (PSA) in human prostate tissue. In their work, avidin-modified UC phosphors were bound to biotin-conjugated antibodies, which were then targeted to the antigens of interest. The formalin-fixed, paraffin-embedded 3-mm sections of human prostate mounted on APTS-coated slides were used as the detecting samples [3].

Since then, numerous efforts have been made to use UC particles for diverse heterogeneous bioassays to detect analytes. Tanke et al. used the green-emitting phosphor $(Y_2O_2S$:Yb,Er) particles to detect nucleic acid in 2001. The 400-nm diameter monodispersed spherical particles were prepared by homogeneous precipitation of $(YYbEr)OHCO_3$ in the presence of urea at elevated temperatures. The precursors were then converted to oxysulfide at 850°C. Phosphor particles were then encapsulated with a uniform layer of silica and functionalized with amino-groups, and the aminated phosphor particles were coupled with biotin-polyethyleneglycol. Probe DNAs were spotted on amino-silane-modified glass slides [4].

Cooper et al. used phosphor particles with a composition of $(YYbEr)_2O_2S$ and $(YYbTm)_2O_2S$. The phosphors were silica coated and silanized with triethoxy-amino-propylsilane to put amino-functional groups on the surface of the particles to serve as attachment sites for the antibodies. Then these particles were used as reporters in lateral flow (immunochromatographic) assays, combining the advantages of this assay format with those of this reporter system. Using human chorionic gonadotropin (hCG) as a model analyte to describe the properties of phosphors in immunochromatographic assays, a detection limit of 10 pg hCG in a 100-ml sample has been achieved. UC lateral flow assays are considerably more sensitive than traditional lateral flow assays using colored beads and a visual readout. Sensitive assays can be performed more sensitively than traditional lateral flow assays using colored beads and a visual readout. Sensitive assays can be performed in a variety of aqueous samples and analyzed in 15 min using a hand-held detector. Multiple assays can be performed simultaneously on the same membrane strip, since UC of different compositions with unique emission spectra are available [5].

Therefore, Niedbala and coworkers developed a lateral flow-based strip assay for the simultaneous detection of amphetamine, methamphetamine, phencyclidine, and opiates in saliva by employing multi-color $NaYF_4$:Yb/Er UCNPs. In their work, upconverting phosphors are used as labels in rapid lateral flow assays. The lateral flow assay strip is designed similarly to commercial lateral flow test strips that utilize colloidal gold or latex particles. The upconverting phosphor labels utilized are spherical, approximately 400 nm in diameter and are covalently conjugated to antibodies. The phosphor conjugates are placed into the test by premixing with the sample or dried into a sample pad [6].

Most heterogeneous assays reported before are based on submicrometer-sized UC particles that are not stable in colloidal solution. Smaller UCNPs with improved stability and a decreased number of probe molecules per particle should result in better target-to-reporter ratios and improved assay kinetics. Li and Wang recently proposed a sandwich-hybridization assay for the ultrasensitive detection of DNA using sub-50 nm $NaYF_4$:Yb,Er UCNPs. In their study, a three-component sandwich assay was employed, in which Fe_3O_4 NPs modified with capture DNA were conjugated to probe DNA-modified $NaYF_4$:Yb/Er NPs through the hybridization of the target DNA. When combined with magnetic techniques for facile separation, this method produced a LOD of 10 nM without polymerase chain reaction (PCR) amplification [7].

In 2012, Henna Päkkilä utilizing UCNPs for quantitative multianalyte microarray immunoassay. The single-crystal $NaYF_4$:Yb^{3+},Er^{3+} nanomaterials (crystallite size 25–28 nm) were synthesized. The author demonstrated the possibility of printing multiple antibodies in the array format onto the bottom of microtiter well plate, and the performance of the array-based immunoassay was evaluated using a mix of secondary antibody-coated UCNPs, each specific to one of the measured analytes. To demonstrate the performance of this platform, PSA, thyroid-stimulating hormone (TSH), and luteinizing hormone (LH) were selected as model analytes. The LODs for PSA, LH, and TSH (0.05 ng/mL, 0.20 U/L, and 1.7 m U/L, respectively) [8].

More recently, Chen group developed a novel heterogeneous UC bioassay approach, which makes use of the surface-functionalized UCNPs as bio-probes to detect trace amounts of analyte in solutions, ranging from a commonly used model biomolecules like avidin to important tumor biomarkers like urokinase plasminogen activator receptor (uPAR), carcinoembryonic antigen (CEA), and a-fetoprotein (AFP). The concentration of analyte can be quantified by measuring the optical signals of UC nanoprobes bound to a 96-well microplate upon 980 nm near-infrared (NIR) excitation (Fig. 7.1c). The LOD for avidin, defined as the concentration that corresponds to three times the standard deviation above the signal measured in the control experiment, was determined to be $\sim$90 pM. The results obtained for the detection of uPAR, CEA, and AFP with their LODs range from 40 to 100 pM [1]. What's more, they synthesized monodispersed and size-controllable $NaScF_4$:Ln^{3+}NPs via thermal decomposition method. The biotinylated $NaScF_4$:Yb/Er NPs were employed for the detection of avidin by means of our self-customized microplate reader equipped with a fiber laser at 980 nm. The UC luminescence signal gradually increased with the avidin concentration ranging from 0.15 to 76.9 nM. The LOD of avidin was estimated to be 180 pM [9].

7.3 Homogeneous Assay

Homogeneous UC assays generally utilize a sandwich structure to detect an analyte in which a UCNP and an acceptor molecule are both labeled with the capture molecule that can specifically seize the analyte. Generally, the capture molecule-labeled UCNPs and the capture molecule-labeled acceptor molecules are far from each other in the solution, producing no detectable signal (either the quenching of UC PL or the fluorescence of acceptor molecule). Only the presence of analytes can bring the UCNPs and the acceptor molecules into close proximity to establish a Förster resonance energy transfer (FRET) process that gives birth to the binding-modulated signal. Compared with the heterogeneous bioassay which is labor intensive and time consuming due to the tedious separation and washing steps before signal measurement, homogeneous assays are convenient for fast bio-detection. Since the homogeneous bioassay, commonly based on the technique of FRET, can be performed in one step without the laborious separation and washing typical in heterogeneous bioassays, it is particularly easy to handle and thereby has attracted growing interest in the past decades [1, 10–18]. A variety of homogeneous assay methods, such as conventional FRET and UC-FRET, have been proposed in an effort to develop the most convenient bio-detection method by employing UCNPs as labels.

In a typical FRET process, excitation of the donor by an energy source (e.g., flashlamp or laser) triggers an energy transfer to the acceptor; if they are within a given proximity to each other and there is spectral overlap between the absorption of the acceptor and the emission of the donor, the acceptor emits light at its given wavelength. To achieve improved FRET efficiency, the donor and acceptor pair used in the FRET bioassay process should be considered carefully. UCNPs, characterized by long-lived luminescence lifetimes and large apparent Stokes or anti-Stokes shifts, are regarded as ideal energy donors in FRET bioassay. In particular, the long-lived luminescence of Ln^{3+} provides a means to effectively inhibit short-lived background noise (in the nanosecond range) from biological samples, thereby yielding a higher detection sensitivity and enhanced ratio of signal to noise as compared to conventional FRET bioassays [19, 20]. Moreover, such unwanted background noise in bioassays can also be easily overcome by using the unique anti-Stokes luminescence of Ln^{3+} denoted as UC, which is usually achieved upon a 980-nm laser excitation [21–23]. As a new type of luminescent bio-probes, the main advantage of UCNPs is that the interference of background noise arising from biological substances can be thoroughly circumvented in the bio-detection process, since the UC phosphors are excited in the NIR region to achieve visible emissions. For this reason, various types of UCNPs have recently been synthesized and then used as a background-free label for various biological applications like UC-FRET bioassay and UC bio-imaging [24–28]. Consequently, a number of homogeneous bioassay methods based on UC-FRET have recently emerged in order to develop the most convenient bioassay method for fast detection in some practical applications, as briefly exemplified in the following (Fig. 7.2).

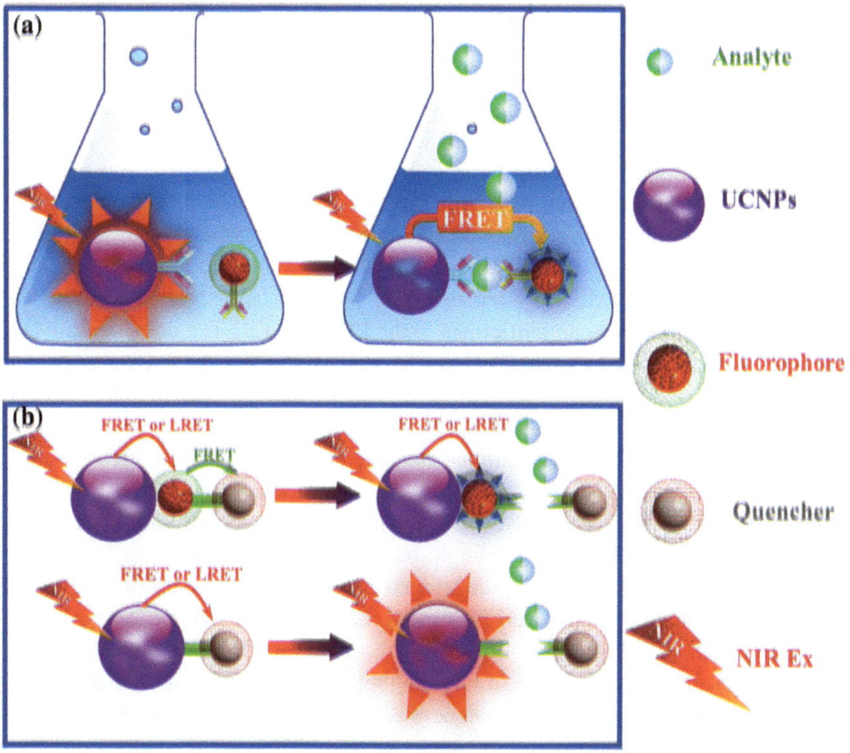

Fig. 7.2 Schematic illustration of typical homogeneous assays based on UCNPs. **a** Assays utilizing a sandwich-type structure. These systems involve the use of UCNPs as energy donors and strong-absorbing materials as energy acceptors. The donors and acceptors are brought into close proximity by analytes. **b** Assays utilizing the inhibition of energy transfer process (**b**, *upper*). The fluorophore utilized as the energy donor of UCNPs is initially deactivated by a linked quencher through a FRET process. The linkage between the fluorophore and the quencher can be cut off by an analyte (generally, an enzyme), restoring the PL of the fluorophore (**b**, *lower*). The PL of UCNP, quenched by FRET or LRET to a quencher, can be recovered by analyte-induced separation of the quencher from the UCNP or elimination of quenchers' strong absorption. Reprinted with the permission from Ref. [2]. Copyright 2014 American Chemical Society

The application of UC particles in homogeneous assay dates back to the work by Mitchell and Morgan in 2007. Upconverting waveguides of "Yaglass" were prepared in their work. "Yaglass" is an optically transparent glass-ceramic material composed primarily of lead fluoride doped with erbium and ytterbium. The paper revealed that upconverting lanthanide-based labels have much promise both as potential energy donors in FRET assays and also for evanescent wave-coupled detection of surface binding [18].

Subsequently, Kuningas and coworkers utilized the streptavidin-conjugated UC particles as donors and biotinylated phycobiliprotein (biotin-BPE) as the acceptor to experimentally corroborate the UC-FRET. A competitive model assay for biotin was constructed using Y_2O_2S:Yb/Er phosphors. UC phosphor was excited at NIR (980 nm) and sensitized acceptor emission was measured at red wavelength (600 nm) by using a microtitration plate fluorometer equipped with an infrared laser diode and suitable excitation and emission filters. The competitive detection of biotin could reach a LOD of 0.7–9 nM [29].

Simultaneous detection of multiple analytes by using multi-peak emission profiles of UC particle donors has also been reported by Rantanen et al. By making use of the multi-peak emission profiles of UC particle donors, there is no interference from autofluorescence [10]. However, these works involve the use of large-sized UC particles as energy donors. Because only emitting lanthanide dopants located near the particle surfaces can participate in the LRET or FRET, a considerably large fraction of emissive ions embedded in the core structures produce signals only through non-proximity-based reabsorptive energy transfer, limiting the detection sensitivity [2].

Zhang and coworkers demonstrated that the small-sized oligonucleotide-modified UCNPs can be used as a sensitive bio-sensor to detect target DNA reaching a LOD down to 1.3 nM. They also revealed that oligonucleotide-modified UCNPs can distinguish target DNAs with single-base mismatches [30].

In addition, Li and coworkers developed a sensitive DNA sensor using streptavidin-functionalized small-sized $NaYF_4$:Yb^{3+}/Er^{3+} UCNPs as the energy donors and reporter-DNA-TAMRA as the energy acceptor to detect a trace amount of target DNA with a non-optimized LOD ranging from 10 to 60 nM. Which is on the basis of the highly specific interaction between streptavidin and the capture between two short oligonucleotides (capture- and reporter-DNA) and a longer one (target DNA) [31]. Meanwhile, Li et al. [32] demonstrated the use of the chromophoric iridium (III) complex-coated $NaYF_4$:Yb/Er/Tm NPs as a ratiometric UC probe for highly selective detection of the cyanide anion in aqueous solution with a LOD of 0.18 mM, through the inhibition of the energy transfer from the UCNPs to the chromophoric complex.

In addition to organic dyes, metallic NPs like gold NPs and graphene oxide have also been frequently investigated as superior quenchers of UC donors inhomogeneous assays. For instance, highly sensitive UC homogeneous bio-sensors employing $NaYF_4$:Yb^{3+}/Er^{3+} UCNPs as the energy donors and gold NPs as the energy acceptors have been accomplished for detection of avidin and rabbit anti-goat IgG, yielding a very low LOD of 0.5 nM for avidin and 0.88 μg/mL for rabbit anti-goat IgG [33, 34]. The efficient energy transfer from UCNPs to graphene oxide has been shown in GO-$NaYF_4$:Yb/Er nanocomposite prepared by a one-step facile method [35]. Liu and coworkers then constructed a novel bio-sensor for glucose determination based on FRET from UCNPs to GO, which was capable of homogeneously detecting glucose in human serum samples without any background interference [36].

An intriguing fluorescence-quenching-based enzyme activity assay has been developed, in which an Alexa Fluor 680 (AF680) fluorophore is utilized as the energy acceptor to receive the upconverted energy from UCNPs and is quenched by the Blackberry Quencher 650 (BBQ650) that is linked to AF680 through a single-stranded oligonucleotide sequence. The oligonucleotide can be cut into shorter fragments upon an enzymatic reaction catalyzed by benzonase endonuclease, which leads to the fluorescence recovery of AF680. It was found that this assay offers an eightfold increase in signal-to-background ratio as compared to the conventional method relying on direct excitation into AF680 at 655 nm. (ii) Separation of the UCNPs and the acceptor dye made FRET or LRET process deactivated, then the PL of UCNPs recovered. More recently, Liu and coworkers proposed a novel method for the rapid detection of glutathione in aqueous solutions and living cells. Their method was established based on the observation that the UC luminescence of hexagonal $NaYF_4$:Yb/Tm core–shell UCNPs was quenched by the manganese dioxide (MnO_2) nanosheets formed on UCNP surfaces and such a MnO_2-induced UC quenching effect would be inhibited when a small amount of GSH was added to the particle solutions, which reduces MnO_2 into Mn^{2+} [2, 37].

Similarly, an assay for detection of metalloproteinase-2 (MMP-2, important biomarker in blood) has been developed by utilizing PEI-modified $NaYF_4$:Yb^{3+}/Er^{3+}UCNPs (energy donors) linked through a polypeptide chain to carbon NPs (energy acceptors). The polypeptide chain contains a specific MMP-2 substrate domain and p-electron-rich region. Upon NIR excitation, the UC-FRET process was initiated by virtue of the p–p stacking interaction between the probe peptide and CNPs, there by leading to the quenching of emissions from UC donors. By contrast, such a UC-FRET process was inhibited due to the cleavage of the substrate peptide between UC NPs and CNPs induced by the addition of MMP-2 in solution. The addition of MMP-2 in solution can scissor the substrate peptide between the UCNPs and the carbon NPs and recover the UC PL of UCNPs, creating a LOD as low as 10 pg/mL for this homogeneous assay [13].

7.4 UCNPs for Encoding and Multiplexed Detection

The need to decipher many biological events simultaneously has driven the development of multiplexed fluorescent tags. And the development of new encoding strategies is a driving force for multiplexed analyte detection in complex mixtures with minimized sample volume, assay time, and cost. Multiplexing in biotechnology and life sciences refers to high throughput technologies capable of simultaneous identification and quantification of multiple distinctive things [38, 39]. Multiplexing molecular diagnostics impacts broadly on the bioinformatics areas of genomics [40], proteomics [41], metabolomics [42], cytomics [43], and personalized medicine [44], where personal gene expression profilingholds promise for individually targeted therapies. Multiplexing typically requires a matrix of optical codes [45].

Barcoding materials have been widely used as luminescent probes for multiplexed assays in biological species because of their distinct optical characteristics [46–48]. The use of these materials for anti-counterfeiting applications has also attracted much attention in recent years [49, 50]. However, their application in practical settings has been largely hindered by high-cost materials fabrication, low reaction yields, complex instrumentation setup, and high background noise when excited by common ultraviolet (UV) or visible light sources [51, 52]. Besides, for barcoding, the number of available dyes that have single wavelength excitation and distinguishable emission maxima is limited. In addition, the fluorescent signal from dye molecules is weak, and the dye molecules are susceptible to irreversible photodestruction. To address these issues, barcoding NPs have been developed to serve as alternative substrates for multiplexed bioassays [53–55]. Barcoding NPs have exhibited advantages over energy transfer fluorescent dyes in that the NPs are not vulnerable to photobleaching or degradation in the complex environment, their optical signal is higher than single dye molecules, and a large barcoding library can be built. Quantum dots (QDs) have attracted great interest in this area because of their broad excitation spectra and tunable emission wavelengths. In a proof-of-concept investigation, it is estimated that by using QDs with just 6 colors in 10 different intensities, a library of approximately one million entries could be encoded [53]. But the "blinking" characteristic is a limiting factor for raster scanning systems such as confocal microscopy and flow cytometry. QD's cytotoxicity is of concern for in vivo applications [56–58].

UCNPs may provide a much needed solution for the above-mentioned problems [23, 59, 60]. Firstly, their fabrication methods are typically based on low-cost bottom-up processes and can be easily scaled up for massive production [61–63]. Secondly, the morphology of the UC materials is highly tunable with features ranging from nanosized particles to microscale rods [64, 65], allowing for direct decoding without the concern of the diffraction limit using conventional optical microscopes. The third aspect is that UC materials feature large anti-Stokes shifts and long lifetimes, leading to largely reduced background noise [14]. Furthermore, the ability to tune the emission wavelength from UV to visible light to NIR spectral regions offered by photon UC enables the generation of a large, diverse library of optical barcodes. Despite their promise, it has been challenging to prepare multicolored UC barcodes using a single spherical nanoparticle due to the resolution limit of conventional optical microscopes [52].

Two multiplexed assay formats can be broadly distinguished as follows: In the first, a multitude of probe elements is identified by spatial position on planar arrays [66]. In the second, the probes are attached to suspended microspheres that are labeled with an identifier code or "barcode" [67]. Microspheres in suspension have a large surface area and enable fast binding kinetics. Whereas planar arrays have to be spotted individually, encoded microspheres can be generated faster and more homogeneously in large batch syntheses [68]. Thus, encoded microspheres have the potential to greatly reduce the costs per array, provided they can be detected using simple and inexpensive instrumentation [69]. Both modes of detection allow for screening a large number of different probe elements in parallel. In addition, each type of encoded microsphere is present in multiple copies, which allows the observation of a statistically relevant number of probe elements in a single experiment.

The maximal degree of multiplexing based on the use of (suspended) microspheres depends on the number of identifier codes that can be detected separately. The labeling of microparticles with fluorophores is probably the most established method for spectral encoding [70, 71]. Microspheres, usually made from polystyrene, can be swollen in an organic solvent containing defined concentrations of fluorophores. Following the transfer of the microparticles to an aqueous solution, they shrink and permanently enclose the fluorophores.

Spectral encoding is compatible with standard optical detection schemes, such as flow cytometry and is thus widely applicable. Equation (7.1) gives the maximal number of spectral codes (*C*)

$$C = i^w - 1 \tag{7.1}$$

where i is the number of intensity levels that can be differentiated and w is the number of wavelengths. One intensity level always has to be subtracted as it is indistinguishable from the background. C can be increased exponentially by using multiple intensity levels and multiple wavelengths at the same time. For example, four intensity levels and two wavelengths result in a coding capacity of 15 ($=4^2 - 1$; Fig. 7.3a), while ten intensity levels and four wavelengths in principle yield 9999 codes. The actual coding capacity, however, is limited because of (a) the width and overlap of emission spectra, (b) variations in intensity levels, and (c) interfering background signals [69].

Several approaches have been investigated to extend the limited number of codes that can be generated from conventional fluorophores. For example, the emission of fluorophore-doped silica NPs can be modulated by tuning the efficiency of FRET to various acceptors [58]. Luminescence lifetime data could provide additional information [72]. A combinatorial code can be generated by using a biomolecule, such as DNA as the coding element. In this case, a specific DNA barcode is attached to each microsphere. The DNA barcode can then be decoded through sequential hybridization and dehybridization steps using complementary oligonucleotides that carry different fluorophores [73, 74].

Efficient spectral encoding strategies for multiplexed analysis of biological samples depend on both low background signals from biomolecules and good separation of emission signals to distinguish as many codes as possible. UCNPs emit visible light under NIR excitation (anti-Stokes emission). This unique optical property precludes background fluorescence and light scattering from biological materials. The emission of multiple and narrow emission lines is an additional hallmark of UCNPs that opens up new avenues for optical encoding. Distinct emission signatures can be obtained if the multiple emission of UCNPs is tuned by their dopant composition or by surface modification with dyes. Tuning the intensity of only one of the multiple emission lines and using another one as a constant reference signal enables the design of ratiometric codes that are resistant to fluctuations in absolute signal intensities. Combining several UCNPs each displaying a distinct set of emission lines expands the coding capacity exponentially and lays the foundation for highly multiplexed analyte detection. Manipulating the luminescence decay lifetimes can create

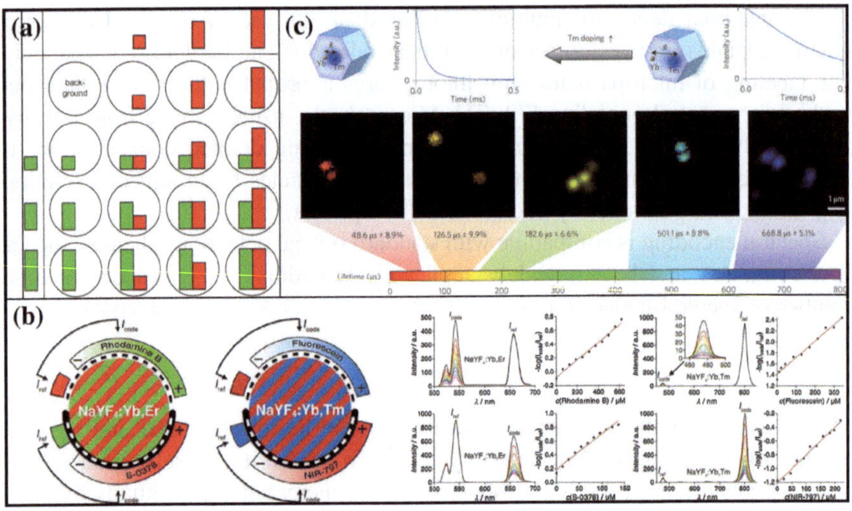

Fig. 7.3 **a** Schematic representation of an encoding scheme. Microspheres are labeled with combinations of two fluorescent dyes that emit either *green* or *red light*. The emission intensity of each dye can be modulated individually by adjusting their concentrations. **b** Tuning the emission intensity of UCNPs differentially by an organic dye. UCNPs of type $NaYF_4$:Yb,Er are covered with various concentrations of either rhodamine B or the dye S-0378. The concentration of rhodamine B is adjusted in ten increments to selectively reabsorb the *green* (l_{max} = 543 nm) emission while the *red* (l_{max} = 657 nm) emission remains constant. UCNPs of type NaYF4:Yb,Tm display dual emission lines that can be adjusted individually either by fluorescein for the *blue emission line* (l_{max} = 475 nm) or by the dye NIR-797 for the NIR emission line (l_{max} = 802 nm). **c** Lifetime tuning scheme and time-resolved confocal images for $NaYF_4$:Yb,Tm upconversion nanocrystals. **a** Reprinted with the permission from Ref. [69]. Copyright 2013 John Wiley and Sons. **b** Reprinted with the permission from Ref. [55]. Copyright 2011 John Wiley and Sons. **c** Reprinted with the permission from Ref. [45]. Copyright 2013 Nature Publishing Group

a temporal coding dimension in a wide microsecond-to-millisecond range. These advantages of lanthanides make it beneficial for designing lanthanide-doped micro-barcodes [45, 69].

7.4.1 Tuning UCNPs' Material Composition for Encoding and Multiplexed Detection

The modular assembly of sensitizers and activators in a common host lattice allows the generation of a combinatorial code of different colors. The dopant Yb^{3+} which has a single dominant energy transfer from $^2F_{7/2}$ to $^2F_{5/2}$ can be used to specifically absorb NIR light and to sensitize activator ions that display distinct emission lines. The kind of activator and its concentration in the $NaYF_4$ lattice allow the color to be tuned. In the case of ratio metric encoding, the absolute signal intensity *i* in Eq. (7.1) is replaced by the ratio of I_{code} to I_{ref}. The number of analytical wavelengths (w) can be defined by choosing different activator ions (ai). Considering that one activator ion is required for the constant reference signal that has to be distinguishable from the background, the coding capacity of UCNPs can be expressed by Eq. (7.2).

$$C_{UCNP} = \left(\frac{I_{code}}{I_{ref}}\right)^{ai-1} \tag{7.2}$$

The narrow emission lines of UCNPs allow many more codes to be accommodated in the visible region than is possible with the broad emission bands of conventional organic fluorophores. In practice, the number of emission lines of UCNPs suitable for encoding is limited because the emission arises from transitions between discrete energy levels. However, the commonly used emission lines of Er^{3+}, Tm^{3+}, and Ho^{3+} are sufficient to obtain a large set of codes [75]. Recently, a new energy migration mechanism in core–shell UCNPs was described that incorporates Tb^{3+}, Eu^{3+}, Dy^{3+}, and Sm^{3+} as activator ions. Their additional emission lines further increase the encoding capacity of UCNPs [76].

Instead of incrementally increasing the intensity of the encoding signal, the complex signature of the multiple UCNP emission may be used as a "barcode". Each dopant composition would thus define a unique barcode. This encoding strategy was first developed for multiplexed bioanalysis by the group of Nann. Silica-coated UCNPs of 20 nm diameter were doped with different activator ions to obtain the following set of four barcodes: (a) $NaYF_4$:Yb for blue, green, red, and IR light, (b) $NaYbF_4$:Er for green and red light, (c) $NaYbF_4$:Tm for blue and NIR light, and (d) $NaYbF_4$:Ho for green light [77]. This barcoding system is robust but the number of codes is ultimately restricted by the number of available activator ions. More barcodes can be generated if different activator ions, each with a unique signature of multiple emission lines, are varied relative to each other in a way similar to ratiometric encoding. In this case, the relative intensities of the signatures define the code [69].

Later, Fan Zhang and coworkers present to separate the coding and target signals. UCNPs were employed as the coding part and organic dyes as the reporters (Fig. 7.4a). The codes of UCNP-tagged beads depend on the composition of UCNPs. In particular, it is extremely important to note that there is no optical cross talk between the UC optical code and any reporter dyes. The single wavelength 980-nm continuous wave (CW) laser used to excite the UC materials does not excite dyes which absorb in the visible and UV region and, conversely, the UC materials are not excited by the visible lasers used to excite the organic dyes. All these favorable properties contribute to the promise of a great potential of UPNCs in the multiplexed detection. In order to demonstrate the use of UC optically encoded microbeads for biological multiplexed assays, a model DNA hybridization system was designed using oligo probes and triple-color UPNC encoded beads [79, 80]. Target DNA molecules #2–#4 are directly labeled with three different fluorescent dyes (Fig. 7.4a). Optical spectroscopy at the single-bead level yields both the coding and the target

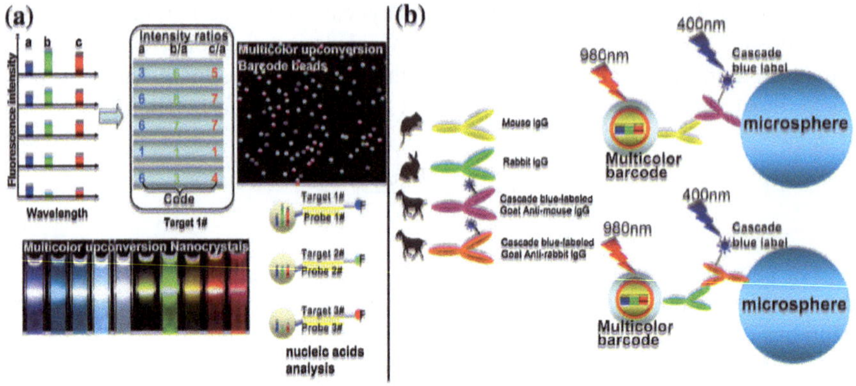

Fig. 7.4 **a** Schematic illustration and detection results of DNA hybridization assays using multicolor $NaYF_4$:Yb/Ho/Tm upconversion nanocrystals encoded beads. **b** Scheme for the conjugation of antibody-UCNPs with secondary antibody microspheres. **a** Reprinted with the permission from Ref. [78]. Copyright 2011 John Wiley and Sons. **b** Reprinted with the permission from Ref. [59]. Copyright 2011 John Wiley and Sons

signals. The coding signals identify the DNA sequence, whereas the target signal indicates the presence and the abundance of that sequence. From the encoding detection results, it is extremely important to note that there is no optical cross talk between the UC optical code and any reporter dyes. The midrange IR radiation used to excite the UC materials does not excite dyes which absorb in the visible region (Fig. 7.4a) and, conversely, the UC materials are not excited by the visible lasers used to excite the organic dyes [78]. Furthermore, they showed the use of UC nanobarcodes for biological multiplexed assays, a model multiplexed immunoassay system was designed using antibody-conjugated UCNPs and secondary antibody-coated microspheres (Fig. 7.4b). Mouse lgG and rabbit lgG were conjugated to different UPNPs using a carbodiimide-based approach. Cascade blue-labeled goat anti-mouse lgG and cascade blue-labeled goat anti-rabbit lgG were immobilized onto microsphere in a similar method. The UC nanobarcodes/microsphere interaction simulates the recognition process between conjugated UC nanobarcodes and potential receptors/antigens on the surface of a cell or bacterium. Two sets of antibody-modified UPNPs with different UC coding signal and two sets of secondary antibody-coated microspheres were cross-reacted [59].

Recently, Liu group reported the rational design and synthesis of multi-color-banded UC barcodes based on tip modified hexagonal-phase $NaYF_4$ microrods with different activators doped at the tips. With varying sets of activators, they prepared a library of single-crystal-based UC barcodes comprising different combinations of three primary colors (red, green, and blue) that are easily readable with conventional optical microscopes [45].

7.4.2 Tuning UCNPs' Optical Surface Layer for Encoding and Multiplexed Detection

Rather than changing the composition of the upconverting material, the surface of the UCNPs can also be modulated with an optical-screening layer capable of filtering off certain emission lines. This strategy is particularly effective for UCNPs displaying two main emissions, such as $NaYF_4$:Yb,Er (green and red) or $NaYF_4$:Yb,Tm (blue and NIR). One of the dual emission lines can be reabsorbed selectively by adjusting the amount of an organic dye [55] or of gold NPs [81] contained in the screen layer. The second emission is not reabsorbed and can serve as a reference signal. For example, the green emission line of UCNPs of type $NaYF_4$:Yb,Er was adjusted in ten increments by coating the NPs with varying quantities of the purple dye rhodamine B. Each intensity level was then used to define a code (I_{code}; Fig. 7.3b). As the absorption spectrum of rhodamine B does not display any overlap with the red emission line, the intensity of the red emission served as a constant reference signal. UCNPs of type $NaYF_4$:Yb,Tm display mainly blue and NIR emission lines that are well separated and whose intensities can be tuned individually by a screening layer possessing overlapping absorption spectra.

Therefore, Gorris et al. exploit the unique luminescent properties of UCNPs as background-free codes for multiplexed analyses. The surface of UCNPs is covered with a screen layer containing different amounts of an organic dye such that one of the two emission bands is reabsorbed to various degrees. The second emission band, by contrast, is not absorbed by the dye on the surface and serves as a ratiometric reference to the emission of the first band. The red (657 nm) emission of $NaYF_4$:Yb,Er was filtered by dye S-0378 and the NIR (802 nm) emission of $NaYF_4$:Yb,Tm by dye NIR-797. Ratiometric codes were generated by using three intensity levels (I_{code}) and one reference intensity (I_{ref}) for each type of UCNP. By using this barcoding method, it expands the number of codes in the same range [55].

7.4.3 Tuning UCNPs' Luminescence Lifetimes for Encoding and Multiplexed Detection

Fluorescence lifetimes are typically less than tens of nanoseconds; however, materials containing trivalent lanthanide ions usually exhibit longer lifetimes from microseconds to even milliseconds [82–84]. Recently it was demonstrated that their microsecond region lifetimes can be fine-tuned to create multiplexing optical codes and assigned to single UCNPs [45, 85]. These tunable lifetime values constitute a new temporal dimension of optical codes for data storage, document security and multiplexing assays, which can simultaneously assess a large number of biomolecular species [86].

Fluorescence color codes used in a broad range of fields have been one of the most popular methods for multiplexing. For example, cytometry combined with fluorescence staining has been at the frontier of rapid detection of single cells for the last several decades. However, the crowded spectral domain limits the modern multi-color flow cytometry to fewer than 20 channels, requiring three to five lasers, tens of filters and up to 20 light detectors [87]. The unavoidable spectral overlap requires complicated and tedious color compensations. Thus, the outstanding challenge in multiplexing is to find ways of creating additional distinguishable coding dimensions, such as mass spectra [43], fluorescence lifetime [88] and Raman spectra [89]. These new dimensions become highly useful when the new identities can be accurately decoded (detected) by a simple, rapid and low-cost device.

Jin' group reported a new multiplexing concept by manipulating the luminescence decay lifetimes to create a temporal coding dimension in a wide microsecond-to-millisecond range. Among typical materials that emit such long-lived luminescence, rare-earth-doped luminescent nanocrystals are attracting significant attention with their applications in sensing, bio-imaging, nanomedicine, solar cells and three-dimensional displays [84]. Recent advances in the synthesis of rare-earth-doped UCNPs, in particular, have led to accurate control of composite, crystal phase, morphology, and emission colors [65]. They have developed an approach to precisely tune the micro-/millisecond luminescence decays and produce individual populations of rare-earth-doped nanocrystals with distinct lifetimes. The results constitute a new optical coding and decoding dimension, which broadens the application of UC materials in nanoscale photonics. Their lifetime multiplexing concept based on the tunability of lifetimes has been implemented following a simple scheme of codoping sensitizer Yb^{3+} ions and blue-emitting Tm^{3+} ions at stepwise varied concentrations into the $NaYF_4$ nanocrystals. In this case, energy transfer from the sensitizer to the emitter ion at varying sensitizer–emitter distances provides tunability. By using this approach they have been able to tune the lifetime of 40 nm $NaYF_4$:Yb,Tm nanocrystals in the blue emission band from 48 ms (4 mol % Tm) to 668 ms (0.2 mol % Tm). Thanks to the photostability of the nanocrystals and the background-free advantage of time-resolved detection, the purpose-built time-resolved confocal scanning system achieved a sufficiently high signal-to-background ratio (over 102) for lifetime measurement of a single nanocrystal's emission at 475 nm. Figure 7.3c shows confocal images of five typical groups of lifetime-encoded Tm-doped "t-dots", using pseudo-color to map the luminescence lifetime for each pixel.

References

1. Liu, Y., Tu, D., Zhu, H., Ma, E., Chen, X.: Lanthanide-doped luminescent nano-bioprobes: from fundamentals to biodetection. Nanoscale **5**, 1369–1384 (2013)
2. Chen, G., Qju, H., Prasad, P.N., Chen, X.: Upconversion nanoparticles: design, nanochemistry, and applications in theranostics. Chem. Rev. **114**, 5161–5214 (2014)

3. Zijlmans, H., Bonnet, J., Burton, J., Kardos, K., Vail, T., Niedbala, R.S., Tanke, H.J.: Detection of cell and tissue surface antigens using up-converting phosphors: a new reporter technology. Anal. Biochem. **267**, 30–36 (1999)
4. van de Rijke, F., Zijlmans, H., Li, S., Vail, T., Raap, A.K., Niedbala, R.S., Tanke, H.J.: Up-converting phosphor reporters for nucleic acid microarrays. Nat. Biotechnol. **19**, 273–276 (2001)
5. Hampl, J., Hall, M., Mufti, N.A., Yao, Y-mM, MacQueen, D.B., Wright, W.H., Cooper, D.E.: Upconverting phosphor reporters in immunochromatographic assays. Anal. Biochem. **288**, 176–187 (2001)
6. Niedbala, R.S., Feindt, H., Kardos, K., Vail, T., Burton, J., Bielska, B., Li, S., Milunic, D., Bourdelle, P., Vallejo, R.: Detection of analytes by immunoassay using up-converting phosphor technology. Anal. Biochem. **293**, 22–30 (2001)
7. Wang, L., Li, Y.: Green upconversion nanocrystals for DNA detection. Chem. Commun. 2557–2559 (2006)
8. Pakkila, H., Yliharsila, M., Lahtinen, S., Hattara, L., Salminen, N., Arppe, R., Lastusaari, M., Saviranta, P., Soukka, T.: Quantitative multianalyte microarray immunoassay utilizing upconverting phosphor technology. Anal. Chem. **84**, 8628–8634 (2012)
9. Ai, Y., Tu, D., Zheng, W., Liu, Y., Kong, J., Hu, P., Chen, Z., Huang, M., Chen, X.: Lanthanide-doped $NaScF_4$ nanoprobes: crystal structure, optical spectroscopy and biodetection. Nanoscale **5**, 6430–6438 (2013)
10. Rantanen, T., Jarvenpaa, M.-L., Vuojola, J., Arppe, R., Kuningas, K., Soukka, T.: Upconverting phosphors in a dual-parameter LRET-based hybridization assay. Analyst **134**, 1713–1716 (2009)
11. Wang, F., Banerjee, D., Liu, Y., Chen, X., Liu, X.: Upconversion nanoparticles in biological labeling, imaging, and therapy. Analyst **135**, 1839–1854 (2010)
12. Wang, G., Peng, Q., Li, Y.: Lanthanide-doped nanocrystals: synthesis, optical-magnetic properties, and applications. Acc. Chem. Res. **44**, 322–332 (2011)
13. Wang, Y., Shen, P., Li, C., Wang, Y., Liu, Z.: Upconversion fluorescence resonance energy transfer based biosensor for ultrasensitive detection of matrix metalloproteinase-2 in blood. Anal. Chem. **84**, 1466–1473 (2012)
14. Wang, F., Liu, X.: Recent advances in the chemistry of lanthanide-doped upconversion nanocrystals. Chem. Soc. Rev. **38**, 976–989 (2009)
15. Wang, M., Abbineni, G., Clevenger, A., Mao, C., Xu, S.: Upconversion nanoparticles: synthesis, surface modification and biological applications. Nanomed. Nanotechnol. Biol. Med. **7**, 710–729 (2011)
16. Zhang, C., Yuan, Y., Zhang, S., Wang, Y., Liu, Z.: Biosensing platform based on fluorescence resonance energy transfer from upconverting nanocrystals to graphene oxide. Angew. Chem. Int. Ed. **50**, 6851–6854 (2011)
17. Wang, L., Yan, R., Huo, Z., Wang, L., Zeng, J., Bao, J., Wang, X., Peng, Q., Li, Y.: Fluorescence resonant energy transfer biosensor based on upconversion-luminescent nanoparticles. Angew. Chem. Int. Ed. **44**, 6054–6057 (2005)
18. Morgan, C.G., Mitchell, A.C.: Prospects for applications of lanthanide-based upconverting surfaces to bioassay and detection. Biosens. Bioelectron. **22**, 1769–1775 (2007)
19. Casanova, D., Giaume, D., Gacoin, T., Boilot, J.-P., Alexandrou, A.: Single lanthanide-doped oxide nanoparticles as donors in fluorescence resonance energy transfer experiments. J. Phys. Chem. B **110**, 19264–19270 (2006)
20. Wang, L., Li, Y.: Luminescent nanocrystals for nonenzymatic glucose concentration determination. Chem. Eur. J. **13**, 4203–4207 (2007)
21. Ju, Q., Liu, Y., Tu, D., Zhu, H., Li, R., Chen, X.: Lanthanide-doped multicolor GdF_3 nanocrystals for time-resolved photoluminescent biodetection. Chem. Eur. J. **17**, 8549–8554 (2011)

22. Liu, Y., Zhou, S., Tu, D., Chen, Z., Huang, M., Zhu, H., Ma, E., Chen, X.: Amine-functionalized lanthanide-doped zirconia nanoparticles: optical spectroscopy, time-resolved fluorescence resonance energy transfer biodetection, and targeted imaging. J. Am. Chem. Soc. **134**, 15083–15090 (2012)
23. Liu, Y., Tu, D., Zhu, H., Chen, X.: Lanthanide-doped luminescent nanoprobes: controlled synthesis, optical spectroscopy, and bioapplications. Chem. Soc. Rev. **42**, 6924–6958 (2013)
24. Kumar, M., Zhang, P.: Highly sensitive and selective label-free optical detection of DNA hybridization based on photon upconverting nanoparticles. Langmuir **25**, 6024–6027 (2009)
25. Kuningas, K., Ukonaho, T., Pakkila, H., Rantanen, T., Rosenberg, J., Lovgren, T., Soukka, T.: Upconversion fluorescence resonance energy transfer in a homogeneous immunoassay for estradiol. Anal. Chem. **78**, 4690–4696 (2006)
26. Rantanen, T., Jaervenpaeae, M.-L., Vuojola, J., Kuningas, K., Soukka, T.: Fluorescence-quenching-based enzyme-activity assay by using photon upconversion. Angew. Chem. Int. Ed. **47**, 3811–3813 (2008)
27. Kumar, M., Guo, Y., Zhang, P.: Highly sensitive and selective oligonucleotide sensor for sickle cell disease gene using photon upconverting nanoparticles. Biosens. Bioelectron. **24**, 1522–1526 (2009)
28. Rantanen, T., Pakkila, H., Jamsen, L., Kuningas, K., Ukonaho, T., Lovgren, T., Soukka, T.: Tandem dye acceptor used to enhance upconversion fluorescence resonance energy transfer in homogeneous assays. Anal. Chem. **79**, 6312–6318 (2007)
29. Kuningas, K., Rantanen, T., Ukonaho, T., Lovgren, T., Soukka, T.: Homogeneous assay technology based on upconverting phosphors. Anal. Chem. **77**, 7348–7355 (2005)
30. Zhang, P., Rogelj, S., Nguyen, K., Wheeler, D.: Design of a highly sensitive and specific nucleotide sensor based on photon upconverting particles. J. Am. Chem. Soc. **128**, 12410–12411 (2006)
31. Chen, Z., Chen, H., Hu, H., Yu, M., Li, F., Zhang, Q., Zhou, Z., Yi, T., Huang, C.: Versatile synthesis strategy for carboxylic acid-functionalized upconverting nanophosphors as biological labels. J. Am. Chem. Soc. **130**, 3023–3029 (2008)
32. Liu, J., Liu, Y., Liu, Q., Li, C., Sun, L., Li, F.: Iridium (III) complex-coated nanosystem for ratiometric upconversion luminescence bioimaging of cyanide anions. J. Am. Chem. Soc. **133**, 15276–15279 (2011)
33. Wang, L.Y., Yan, R.X., Hao, Z.Y., Wang, L., Zeng, J.H., Bao, J., Wang, X., Peng, Q., Li, Y. D.: Fluorescence resonant energy transfer biosensor based on upconversion-luminescent nanoparticles. Angew. Chem. Int. Ed. **44**, 6054–6057 (2005)
34. Wang, M., Hou, W., Mi, C.-C., Wang, W.-X., Xu, Z.-R., Teng, H.-H., Mao, C.-B., Xu, S.-K.: Immunoassay of goat antihuman immunoglobulin G antibody based on luminescence resonance energy transfer between near-infrared responsive $NaYF_4$:Yb,Er upconversion fluorescent nanoparticles and gold nanoparticles. Anal. Chem. **81**, 8783–8789 (2009)
35. Wei, W., He, T., Teng, X., Wu, S., Ma, L., Zhang, H., Ma, J., Yang, Y., Chen, H., Han, Y., Sun, H., Huang, L.: Nanocomposites of graphene oxide and upconversion rare-earth nanocrystals with superior optical limiting performance. Small **8**, 2271–2276 (2012)
36. Zhang, C., Yuan, Y., Zhang, S., Wang, Y., Liu, Z.: Biosensing platform based on fluorescence resonance energy transfer from upconverting nanocrystals to graphene oxide. Angew. Chem. Int. Ed. **50**, 6851–6854 (2011)
37. Deng, R., Xie, X., Vendrell, M., Chang, Y.-T., Liu, X.: Intracellular glutathione detection using MnO_2-nanosheet-modified upconversion nanoparticles. J. Am. Chem. Soc. **133**, 20168–20171 (2011)
38. Schena, M., Shalon, D., Davis, R.W., Brown, P.O.: Quantitative monitoring of gene-expression patterns with a complementary-dna microarray. Science **270**, 467–470 (1995)
39. Pregibon, D.C., Toner, M., Doyle, P.S.: Multifunctional encoded particles for high-throughput biomolecule analysis. Science **315**, 1393–1396 (2007)
40. Bartel, D.P.: MicroRNAs: genomics, biogenesis, mechanism, and function. Cell **116**, 281–297 (2004)

41. Pawson, T., Nash, P.: Assembly of cell regulatory systems through protein interaction domains. Science **300**, 445–452 (2003)
42. Nicholson, J.K., Lindon, J.C.: Systems biology: metabonomics. Nature **455**, 1054–1056 (2008)
43. Bendall, S.C., Simonds, E.F., Qiu, P., Amir, El-ad D., Krutzik, P.O., Finck, R., Bruggner, R. V., Melamed, R., Trejo, A., Ornatsky, O.I., Balderas, R.S., Plevritis, S.K., Sachs, K., Pe'er, D., Tanner, S.D., Nolan, G.P.: Single-cell mass cytometry of differential immune and drug responses across a human hematopoietic continuum. Science **332**, 687–696 (2011)
44. van 't Veer, L.J., Bernards, R.: Enabling personalized cancer medicine through analysis of gene-expression patterns. Nature **452**, 564–570 (2008)
45. Lu, Y.Q., Zhao, J.B., Zhang, R., Liu, Y.J., Liu, D.M., Goldys, E.M., Yang, X.S., Xi, P., Sunna, A., Lu, J., Shi, Y., Leif, R.C., Huo, Y.J., Shen, J., Piper, J.A., Robinson, J.P., Jin, D.Y.: Tunable lifetime multiplexing using luminescent nanocrystals. Nat. Photonics **8**, 33–37 (2014)
46. Nam, J.M., Thaxton, C.S., Mirkin, C.A.: Nanoparticle-based bio-bar codes for the ultrasensitive detection of proteins. Science **301**, 1884–1886 (2003)
47. Agasti, S.S., Liong, M., Peterson, V.M., Lee, H., Weissleder, R.: Photocleavable DNA barcode-antibody conjugates allow sensitive and multiplexed protein analysis in single cells. J. Am. Chem. Soc. **134**, 18499–18502 (2012)
48. Kim, S.-H., Shim, J.W., Yang, S.-M.: Microfluidic multicolor encoding of microspheres with nanoscopic surface complexity for multiplex immunoassays. Angew. Chem. Int. Ed. **50**, 1171–1174 (2011)
49. Osberg, K.D., Rycenga, M., Bourret, G.R., Brown, K.A., Mirkin, C.A.: Dispersible surface-enhanced raman scattering nanosheets. Adv. Mater. **24**, 6065–6070 (2012)
50. Creran, B., Yan, B., Moyano, D.F., Gilbert, M.M., Vachet, R.W., Rotello, V.M.: Laser desorption ionization mass spectrometric imaging of mass barcoded gold nanoparticles for security applications. Chem. Commun. **48**, 4543–4545 (2012)
51. Zhou, J., Liu, Z., Li, F.: Upconversion nanophosphors for small-animal imaging. Chem. Soc. Rev. **41**, 1323–1349 (2012)
52. Zhang, Y., Zhang, L., Deng, R., Tian, J., Zong, Y., Jin, D., Liu, X.: Multicolor barcoding in a single upconversion crystal. J. Am. Chem. Soc. **136**, 4893–4896 (2014)
53. Han, M.Y., Gao, X.H., Su, J.Z., Nie, S.: Quantum-dot-tagged microbeads for multiplexed optical coding of biomolecules. Nat. Biotechnol. **19**, 631–635 (2001)
54. Mirkin, C.A., Park, S.J., Nam, J.M.: Bio-barcodes based on oligonucleotide-modified nanoparticles. US 06974669 (2005)
55. Gorris, H.H., Ali, R., Saleh, S.M., Wolfbeis, O.S.: Tuning the dual emission of photon-upconverting nanoparticles for ratiometric multiplexed encoding. Adv. Mater. **23**, 1652–1655 (2011)
56. Medintz, I.L., Uyeda, H.T., Goldman, E.R., Mattoussi, H.: Quantum dot bioconjugates for imaging, labelling and sensing. Nat. Mater. **4**, 435–446 (2005)
57. Michalet, X., Pinaud, F.F., Bentolila, L.A., Tsay, J.M., Doose, S., Li, J.J., Sundaresan, G., Wu, A.M., Gambhir, S.S., Weiss, S.: Quantum dots for live cells, in vivo imaging, and diagnostics. Science **307**, 538–544 (2005)
58. Wang, L., Tan, W.H.: Multicolor FRET silica nanoparticles by single wavelength excitation. Nano Lett. **6**, 84–88 (2006)
59. Zhang, F., Haushalter, R.C., Haushalter, R.W., Shi, Y., Zhang, Y., Ding, K., Zhao, D., Stucky, G.D.: Rare-earth upconverting nanobarcodes for multiplexed biological detection. Small **7**, 1972–1976 (2011)
60. Podhorodecki, A., Banski, M., Noculak, A., Sojka, B., Pawlik, G., Misiewicz, J.: On the nature of carrier relaxation and ion-ion interactions in ultrasmall [small beta]-$NaYF_4$:Eu^{3+} nanocrystals–effect of the surface. Nanoscale **5**, 429–436 (2013)
61. Dai, Y.L., Xiao, H.H., Liu, J.H., Yuan, Q.H., Ma, P.A., Yang, D.M., Li, C.X., Cheng, Z.Y., Hou, Z.Y., Yang, P.P., Lin, J.: In vivo multimodality imaging and cancer therapy by near-infrared light-triggered trans-platinum pro-drug-conjugated upconversion nanoparticles. J. Am. Chem. Soc. **135**, 18920–18929 (2013)

62. Liu, Q., Sun, Y., Yang, T., Feng, W., Li, C., Li, F.: Sub-10 nm hexagonal lanthanide-doped $NaLuF_4$ upconversion nanocrystals for sensitive bioimaging in vivo. J. Am. Chem. Soc. **133**, 17122–17125 (2011)
63. Anderson, R.B., Smith, S.J., May, P.S., Berry, M.T.: Revisiting the NIR-to-visible upconversion mechanism in beta-$NaYF_4$:Yb^{3+},Er^{3+}. J. Phys. Chem. L. **5**, 36–42 (2014)
64. Zhang, F., Wan, Y., Yu, T., Zhang, F., Shi, Y., Xie, S., Li, Y., Xu, L., Tu, B., Zhao, D.: Uniform nanostructured arrays of sodium rare-earth fluorides for highly efficient multicolor upconversion luminescence. Angew. Chem. Int. Ed. **46**, 7976–7979 (2007)
65. Wang, F., Han, Y., Lim, C.S., Lu, Y., Wang, J., Xu, J., Chen, H., Zhang, C., Hong, M., Liu, X.: Simultaneous phase and size control of upconversion nanocrystals through lanthanide doping. Nature **463**, 1061–1065 (2010)
66. Fodor, S.P.A., Read, J.L., Pirrung, M.C., Stryer, L., Lu, A.T., Solas, D.: Light-directed, spatially addressable parallel chemical synthesis. Science **251**, 767–773 (1991)
67. Finkel, N.H., Lou, X., Wang, C., He, L.: Barcoding the microworld. Anal. Chem. **76**, 352A–359A (2004)
68. Wilson, R., Cossins, A.R., Spiller, D.G.: Encoded microcarriers for high-throughput multiplexed detection. Angew. Chem. Int. Ed. **45**, 6104–6117 (2006)
69. Gorris, H.H., Wolfbeis, O.S.: Photon-upconverting nanoparticles for optical encoding and multiplexing of cells, biomolecules, and microspheres. Angew. Chem. Int. Ed. **52**, 3584–3600 (2013)
70. Fulton, R.J., McDade, R.L., Smith, P.L., Kienker, L.J., Kettman, J.R.: Advanced multiplexed analysis with the FlowMetrix(TM) system. Clin. Chem. **43**, 1749–1756 (1997)
71. Steemers, F.J., Ferguson, J.A., Walt, D.R.: Screening unlabeled DNA targets with randomly ordered fiber-optic gene arrays. Nat Biotech **18**, 91–94 (2000)
72. Grabolle, M., Kapusta, P., Nann, T., Shu, X., Ziegler, J., Resch-Genger, U.: Fluorescence lifetime multiplexing with nanocrystals and organic labels. Anal. Chem. **81**, 7807–7813 (2009)
73. Epstein, J.R., Ferguson, J.A., Lee, K.H., Walt, D.R.: Combinatorial decoding: an approach for universal DNA array fabrication. J. Am. Chem. Soc. **125**, 13753–13759 (2003)
74. Oh, B.K., Nam, J.M., Lee, S.W., Mirkin, C.A.: A fluorophore-based bio-barcode amplification assay for proteins. Small **2**, 103–108 (2006)
75. Wang, F., Liu, X.G.: Upconversion multicolor fine-tuning: visible to near-infrared emission from lanthanide-doped $NaYF_4$ nanoparticles. J. Am. Chem. Soc. **130**, 5642–5643 (2008)
76. Wang, F., Deng, R., Wang, J., Wang, Q., Han, Y., Zhu, H., Chen, X., Liu, X.: Tuning upconversion through energy migration in core–shell nanoparticles. Nat. Mater. **10**, 968–973 (2011)
77. Ehlert, O., Thomann, R., Darbandi, M., Nann, T.: A four-color colloidal multiplexing nanoparticle system. ACS Nano **2**, 120–124 (2008)
78. Zhang, F., Shi, Q., Zhang, Y., Shi, Y., Ding, K., Zhao, D., Stucky, G.D.: Fluorescence upconversion microbarcodes for multiplexed biological detection: nucleic acid encoding. Adv. Mater. **23**, 3775–3779 (2011)
79. Krutzik, P.O., Nolan, G.P.: Fluorescent cell barcoding in flow cytometry allows high-throughput drug screening and signaling profiling. Nat. Methods **3**, 361–368 (2006)
80. Chan, W.C.W., Maxwell, D.J., Gao, X.H., Bailey, R.E., Han, M.Y., Nie, S.M.: Luminescent quantum dots for multiplexed biological detection and imaging. Curr. Opin. Biotechnol. **13**, 40–46 (2002)
81. Li, Z.Q., Wang, L.M., Wang, Z.Y., Liu, X.H., Xiong, Y.J.: Modification of $NaYF_4$:Yb, Er@SiO_2 nanoparticles with gold nanocrystals for tunable green-to-red upconversion emissions. J. Phys. Chem. C **115**, 3291–3296 (2011)
82. Eliseeva, S.V., Buenzli, J.-C.G.: Lanthanide luminescence for functional materials and bio-sciences. Chem. Soc. Rev. **39**, 189–227 (2010)
83. Leif, R.C., Vallarino, L.M., Becker, M.C., Yang, S.: Increasing the luminescence of lanthanide complexes. Cytometry Part A **69A**, 767–778 (2006)

84. Gnach, A., Bednarkiewicz, A.: Lanthanide-doped up-converting nanoparticles: merits and challenges. Nano Today **7**, 532–563 (2012)
85. Zhao, J., Lu, Z., Yin, Y., McRae, C., Piper, J.A., Dawes, J.M., Jin, D., Goldys, E.M.: Upconversion luminescence with tunable lifetime in $NaYF_4$:Yb,Er nanocrystals: role of nanocrystal size. Nanoscale **5**, 944–952 (2013)
86. Lu, Y., Lu, J., Zhao, J., Cusido, J., Raymo, F.M., Yuan, J., Yang, S., Leif, R.C., Huo, Y., Piper, J.A., Robinson, J.P., Goldys, E.M., Jin, D.: On-the-fly decoding luminescence lifetimes in the microsecond region for lanthanide-encoded suspension arrays. Nat. Commun. **5**, 3741 (2014)
87. Perfetto, S.P., Chattopadhyay, P.K., Roederer, M.: Innovation—Seventeen-colour flow cytometry: unravelling the immune system. Nat. Rev. Immunol. **4**, 645–655 (2004)
88. Cui, H.H., Valdez, J.G., Steinkamp, J.A., Crissman, H.A.: Fluorescence lifetime-based discrimination and quantification of cellular DNA and RNA with phase-sensitive flow cytometry. Cytometry Part A **52A**, 46–55 (2003)
89. Watson, D.A., Brown, L.O., Gaskill, D.R., Naivar, M., Graves, S.W., Doorn, S.K., Nolan, J. P.: A flow cytometer for the measurement of raman spectra. Cytometry Part A **73A**, 119–128 (2008)

Chapter 8
Upconversion Nanoparticles for Biosensing

Abstract The last few decades have witnessed the emergence and rapid development of nanoscience and nanotechnology including nanochemistry, nanophysics, nanomedicine, and nanobiomaterials, which allows for the continuous exploration of facile, inexpensive, sensitive, and specific biological analytical techniques by employing nanobioprobes. Recently, upconversion nanoparticles (UCNPs) demonstrated their great potential in many fields of biological science including cells and tissue labeling for bio-imaging, bio-detection, therapy, and multiplexed analysis. In this chapter, we mainly summarize recent advanced UCNPs-based nanosensors with different energy acceptors, including organic dyes, gold nanoparticles, carbon nanomaterials, and MnO_2 nanosheets for various analytes (ions, gas molecules, biomolecules, etc.).

Keywords Upconversion nanoparticles · Biosensing · Spectrum overlap · UCNPs-based nanosensor · Acceptor · Föster resonance energy transfer (FRET) · Luminescence resonance energy transfer (LRET) · Magnetic separation

8.1 Introduction

The last few decades have witnessed the emergence and rapid development of nanoscience and nanotechnology including nanochemistry, nanophysics, nanomedicine, and nanobiomaterials, which allows for the continuous exploration of facile, inexpensive, sensitive, and specific biological analytical techniques by employing nanobioprobes [1–11]. Previously, the most common types of luminescent bio-probes were organic dyes, and semiconductor quantum dots (QDs), which were well studied [12–15]. However, the use of these conventional bio-probes in bio-detection always suffered from several disadvantages.

Lei Chen and Fan Zhang contributed together to this chapter.

F. Zhang, *Photon Upconversion Nanomaterials*,
Nanostructure Science and Technology, DOI 10.1007/978-3-662-45597-5_8

For instance, traditional organic dyes, such as fluorescent dyes or proteins, and QDs, demonstrate many useful features, but their luminescence relies on energy downconversion; thus, they have to be excited either by UV, which is potentially harmful and carcinogenic, or blue/green visible radiation. As is widely known, strong scattering and absorption in tissues and in many biomolecules intrinsically limits the penetration depth of this short-wavelength excitation light [16]. Possible damage to biomolecules or even cell death may be also induced due to long-term irradiation with photons of relatively high energy. Moreover, under UV–Vis photoexcitation, most of endogenous bio-components, such as flavins, porphyrins, NADH, and FAD+, exhibit auto-fluorescence (AF), which decrease the signal-to-noise detection ratio (SNR) [17]. Due to their nature, different organic fluorescent molecules have broad absorption and emission spectra which overlap between each other and the AF as well. This feature makes them not very suitable for multiplexed bio-labeling [6]; thus, complicated and cumbersome spectral deconvolution methods need to be applied. Moreover, organic dyes exhibit short detection times because of their high photobleaching rate and resulting chemical degradation [6]. Their luminescence yield and spectral properties may also be strongly affected by local chemical environments, which often complicate the analysis. Concerning QDs, which possess relatively narrow size-tunable emission, bright photoluminescence, good photostability, broad ultraviolet (UV) excitation, and relatively narrow emission, their use in bio-field is restrained because of their toxicity [18], chemically instability [19], hydrophobic nature, and cost [20]. A promising alternative to traditional fluorophores and QDs may be UCNPs, not only because of their advantageous luminescent properties such as non-toxicity, photostability, and chemical stability, but also mostly because lanthanide ions have the ability to efficiently upconvert NIR radiation into shorter wavelengths [9, 21]. It must also be pointed out that, compared with dyes and QDs, UCNPs demonstrate high and spectrally narrow absorption and emission cross sections, together with sufficiently large Stokes shift in order to selectively photoexcite and detect its luminescence; in addition, they are resistant to photobleaching or biochemical degradation to allow for prolonged (i.e., hours and days) luminescence detection in vivo. Also, they enable the synthesis of many probes with unique spectral codes and distinguish between them in a single sample for multiplexing detection. Lastly, the availability of bio-functionalization and bio-conjugation with specific biomolecules on the surface of nanoparticles through flexible and simple methods also adds to the superiority of UCNPs being used in biosensing [2, 3, 22].

Recently, UCNPs demonstrated their great potential in many fields of biological science including cells and tissue labeling for bio-imaging, bio-detection, therapy, and multiplexed analysis [17, 23–25]. UCNPs, being excited by near-infrared (NIR) light, enable detection of signal in AF-free conditions [23] and provide deeper, yet noninvasive penetration of the radiation into bio-samples [26].

Detection applications are one of the most important research fields for UCNPs. Because the luminescent properties of rare earth ions are mainly decided by the host lattice, and the quenching centers exist inside or on the surface of nanocrystals, UCNPs usually respond little to changes in the surroundings except temperature [27].

This makes UCNPs difficult to use as direct sensors for detection applications. Fortunately, energy transfer processes provide a chance to introduce an analyte-responsive chromophore as the optical detection component, and the wavelength conversion ability of UCNPs can extend the working wavelength of this chromophore from the original short-wavelength region to the NIR range.

These are several basic concepts and principles in energy transfer. Usually, the energy transfer from an excited energy donor (denoted as D) to the acceptor (denoted as A) may occur in two different manners, radiatively and non-radiatively. The radiative energy transfer involves the emission of a photon from the donor, and the subsequent absorption of that photon by the acceptor [29]. In contrast, the non-radiative energy transfer does not involve the emission or re-absorption of photons, but instead, the deactivation of the donor and the excitation of the acceptor occur simultaneously, while the energy transfers through coulombic interactions [22]. As a result, non-radiative energy transfer is also termed as "resonance energy transfer" (RET, generally including föster resonance energy transfer (FRET) or luminescence resonance energy transfer (LRET)), indicating the energy is transferred due to the "resonance" between the donor and the acceptor [22]. Although it is required that the emission spectrum of the donor overlaps the absorption spectrum of the acceptor in both cases, an RET process possibly occurs only when the distance r between donor and acceptor is rather small (typically <10 nm) (Fig. 8.1). Although energy transfer takes place simply on condition of the spectral overlap of the D–A pair, yet in related studies, it is also necessary that the excitation wavelength of the donor resides out of the absorption band of the acceptor, so as to eliminate the possibility of direct excitation of acceptor. This ensures that the acceptor is only activated by the energy transferred from the donor [22].

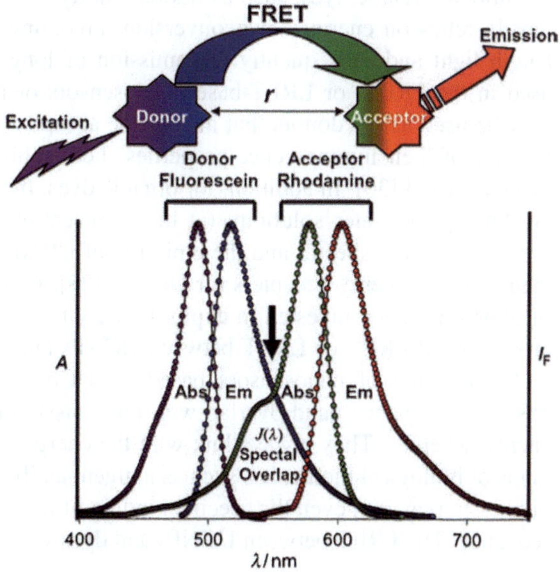

Fig. 8.1 Schematic of the FRET process: upon excitation, the excited-state donor molecule transfers energy non-radiatively to a proximal acceptor molecule located at distance r from the donor. The acceptor releases the energy either through luminescence or non-radiative channels. Reprinted with the permission from Ref. [28]. Copyright 2006 Wiley-VCH Verlag GmbH & Co. KGaA

However, this requirement is usually not easy to meet for conventional D–A pairs, because classic donors are commonly excited in the UV and visible range, while the acceptors (e.g., dye molecules, semiconductors, noble metal nanoparticles) generally show broad absorption bands in the same range. But it is not the case when UCNPs serve as the donor. Specifically, UCNPs are excited in the NIR region, where the acceptors are barely absorptive. Therefore, UCNPs can be employed as ideal donors for different acceptors in the studies of energy transfer.

8.2 The Construction of UCNPs-Based FRET or LRET Nanosensors with Different Energy Acceptors

Biosensing based on UCNPs is generally implemented by FRET or LRET mechanism. The photoluminescence of UCNPs is not directly related to any biochemical property of a system except for temperature. Therefore, to be useful in a biochemical recognition process, UCNPs have to be combined with suitable recognition elements such as indicator dyes. The recognition element of a biosensor may consist of an enzyme, an antibody, a polynucleotide, or even living cells. Several related systems based on FRET process, which are constructed by strategies with UCNPs and different acceptors, will be summarized in the following section.

8.2.1 Organic Dyes as Energy Acceptors

Traditional organic dyes, such as fluorescent dyes or proteins, whose luminescence mostly relies on energy downconversion, involving the excitation of short-wavelength light and subsequently the emission of long-wavelength light usually, are used in the FRET- or LRET-based nanosensors or nanoprobes [5]. They can not only be used as the donors, but also as the acceptors in nanobio-detection systems because of their luminescence properties such as high luminescence intensity and quantum yield [30]. In addition, for organic dyes, their small size, low cost, ease of labeling, good water solubility (at high concentrations), and their overlap in the broad absorption spectra and the emission of UCNPs are all advantages that make them suitable energy accepters for UCNPs [28]. In the past few years, a great deal of efforts has been invested in exploring easy-to-use colorimetric and fluorometric probes-based FRET or LRET between UCNPs and organic dyes.

In one kind of nanosensors based on UCNPs, organic dyes, which usually possess absorbance band overlap with the emission of UCNPs, can be chosen for energy acceptor. They always link with the energy donor UCNPs through specific kinds of biological molecules, such as antigen–antibody, two complementary DNA, biotin–streptavidin, even the specific binding of metal ions and basic group of DNA sequence. The FRET between UCNPs and dyes will never happen with the absence of the analyte. Only the introduction of analyte, leading to the close proximity, can trigger the FRET from energy donor to energy acceptor. This type of nanosensor is always used for the detection of DNA, biotin, metal ions, etc.

Previously, Zhang and coworkers designed a versatile scheme of developing a highly sensitive and specific single-stranded nucleotide sensor based on the UCNPs and organic dyes with two shorter designed oligonucleotides to capture the longer target oligonucleotide [31]. One of the two short oligonucleotides was covalently bound to the UCNPs, while the other one was labeled with a fluorophore, which was chosen because that its excitation spectrum overlaps with the emission spectrum of the UCNPs (Fig. 8.2a). Upon illumination by an infrared laser, the photon UCNPs would emit visible light. When the target DNA sequence was introduced in solution, the DNA oligonucleotides modified on the surface of UCNPs assembled with dye-labeled DNA oligonucleotides to form a double-strand DNA. It was at this point that LRET occurred between the UCNPs and the dyes, allowing a decrease in the luminescence of the UCNPs and an increase in the emission of the fluorophore. Essentially, the quantification of the target oligonucleotide was performed by monitoring the fluorophore emission upon an infrared excitation. Many similar sensing systems which are listed in Table 8.1 have been reported before.

To allow tuning of the spectral overlap between UCNPs and dyes to generate the detection signal, it is important to select a chemodosimeter with an absorption band matched to the upconversion emission of the UCNPs [27]. For instance, Li and

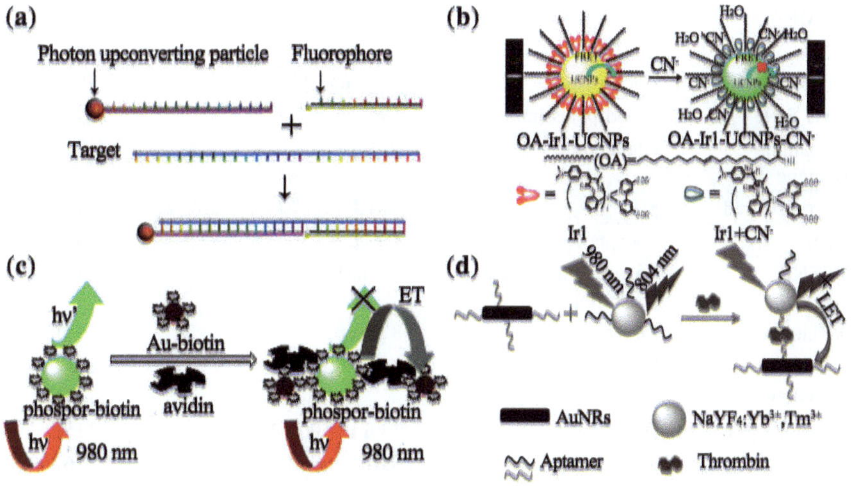

Fig. 8.2 FRET- or LRET-based nanosensors integrated by UCNPs as energy donor and different quenching materials as energy acceptors: **a** and **b** Organic dyes. **c** Gold nanoparticles. **d** Gold nanorods. **a** Reprinted with the permission from Ref. [31]. Copyright 2010 American Chemical Society. **b** Reprinted with the permission from Ref. [32]. Copyright 2011 American Chemical Society. **c** Reprinted with the permission from Ref. [33]. Copyright 2005 Wiley-VCH Verlag GmbH & Co. KGaA. **d** Reprinted with the permission from Ref. [34]. Copyright 2014 Wiley-VCH Verlag GmbH & Co. KGaA

Table 8.1 Examples of LRET-based biosensing using UCNPs as donors and various acceptors for detection

Energy accepter	Target	Linear range or LOD	References
Organic dye			
TAMRA	DNA	1.3 nM	[31]
TAMRA	DNA	10–50 nM	[92]
TAMRA	DNA	0.6 nM	[93]
SYBR green I	DNA	0.1 nM	[95]
SYBR green	DNA	0.06 nM	[72]
I[Ir(CS)2(acac)]	O_2		[87]
Iridium(III) complex	CN^-	0.18 μM	[32]
N719	Hg^{2+}	1.95 ppb	[73]
Chelate complex of Cr(III)	Cr^{3+}	24 nM	[82]
ANDBS	NO^{2-}	4.6 ng mL^{-1}	[109]
RhB hydrazide	Cu^{2+}	4.6 ppb	[74]
Iridium complex	CN^-	62.6 μM	[35]
RhB hydrazide	Cu^{2+}	1.0 μM	[75]
R-phycoerythrin	R-phycoerythrin	0.5 μg/ml	[98]
RhB derivative	Fe^{3+}	0–500 μM	[84]
BHQ1	ATP	1.70 μM	[110]
RBT	Hg^{2+}	3.7 nM	[78]
TAMRA	MRSA DNA sequence	0.18 nM	[96]
hCy7	$MeHg^+$	0.18 ppb	[76]
Rh derivative	Hg^{2+}	0–12 μM	[79]
ETH 5294	Blood potassium	10^{-4} M	[85]
Dopamine-quinone	Glutathione	0.29 μM	[102]
pHrodo™ red	pH	3–6.7	[111]
RhB	Phospholipase D	4 U L^{-1}	[94]
PETNR	Enzyme turn over and flavin reduction	2.5 nM	[112]
Melamine	Melamine	8.2 ppb	[113]
Ru1	Hg^{2+}	0.000 05 U/μL	[80]
TAMRA	Protein kinase activity		[114]
$[Ru(dpp)_3]^{2+}Cl_2$	Hypoxia	15.6 to 500 μg/mL	[115]
TNT	TNT	0.019 mg/mL	[116]
ANP	Cysteine/CN^-	28.5 μM 6.7 μM	[117]
RBH1	HOCl	0.32 μM	[118]

(continued)

Table 8.1 (continued)

Energy accepter	Target	Linear range or LOD	References
AuNPs			
	Avidin	0.5–370 nm	[33]
	Goat anti-human IgG	0.88 μg mL^{-1}	[97]
	pH	3–11	[119]
	Glucose	0.043 μM	[101]
	Biothiols	12.4 μg mL^{-1} for Cys; 15.1 μg mL^{-1} for Hcy	[120]
	Platelet-derived growth factor	10 nM	[121]
	Cr^{3+}	0.8 nM	[83]
	TNT	0–8.0 mg mL^{-1}	[122]
	Cartap	1 ppm	[123]
	Fumonisin B_1	0.01 ng mL^{-1}	[124]
AuNRs			
	Thrombin	1.5 nM	[99]
	Hg^{2+}	2 nM	[77]
AuNPs and AuNRs			
	Thrombin	0.118 nM	[34]
	Streptavidin	0.2 μg/mL	[41]
	Pb^{2+}, Hg^{2+}	50, 150 pM	[81]
Graphene oxide			
	ATP	80 nM	[63]
	Glucose	0.025 μM	[54]
	OTA	0.02 ng mL^{-1}	[62]
	FB1	0.1 ng mL^{-1}	[62]
	Kanamycin	9 pM	[64]
	HIV antibodies in human	2 nM	[65]
	Serum		
CNPs			
	Thrombin	0.18 nM	[67]
	MMP-2	10 pg/mL	[59]
CNP oxide			
	ATP	0.5–200 μM	[125]
	Insulin in human plasma	1 pM to 2.0 nM	[126]
Polymers			
	DNA	0.036 nM	[60]
MnO_2 nanosheets	Glutathione	0.9 μM	[61]
	OTA	0.02 ng mL^{-1}	[69]
	Cat D	1–100 ng mL^{-1}	[69]

coworkers presented the first example of UCNPs-based FRET detection of CN^- in a semi-aqueous system (DMF/H_2O = 9:1, v/v) through modifying a CN^- responsive chromophoric iridium complex (**Ir1**) on the surface of UCNPs [32]. Iridium complexes showed a significant absorbance with the maximum at 505 nm (full width of half-maximum is ~70 nm) and a molar extinction coefficient, corresponding to the significant spectral overlap between the green emission of UCNPs and the absorption of **Ir1**. The design strategy was based on CN^-modulating the FRET process in OA-**Ir1**-UCNPs, with ratiometric UCL emission as the detection signal. As a result, FRET occurred, and a ratio (I_{540}/I_{800}) of UCL intensity at 540 and 800 nm became weak. In the presence of CN^-, there was a reaction of the α, β-unsaturated carbonyl ligand in the **Ir1** seen with the CN^- induced weak absorption in the visible region, corresponding to low spectral overlap between the absorption of **Ir1** and the green UC luminescence (Fig. 8.2b). Therefore, FRET was restrained, and the ratio I_{540}/I_{800} was enhanced. Unfortunately, CN^- usually exists in the aqueous media, which makes the nanoprobe they designed fail to work in many cases. To solve this problem, they introduced a general self-assembly strategy based on hydrophobic–hydrophobic interaction to fabricate hydrophilic hybrid nanoparticles with CN^--responsive chemodosimeters in the hydrophobic layers [35]. Firstly, they coated amphiphilic block-polymer poly(maleic anhydride-alt-1-octadecene)-polyethylene glycol (C_{18}PMH-PEG) onto the surface of hydrophobic oleic acid (OA)-capped $NaYF_4$:20%Yb, 2%Ho UCNPs to fabricate a hydrophilic nanostructure (denoted as PEG-UCNPs). In PEG-UCNPs, a hydrophobic layer composed of the long carbon chains from OA molecules and C_{18}PMH-PEG was formed. Then, the CN^--sensitive iridium(III) complex ([$(ppy)_2Ir(dmpp)$]PF_6, **Ir1**), which had been reported to show high selectivity and sensitivity for CN^- in acetonitrile with significant color change from pink to colorless, was loaded into the hydrophobic layer [35]. This nanosized hybrid structure was utilized as a colorimetric and upconversion luminescence ratiometric probe for cyanide detection in pure water with high selectivity, based on a highly efficient LRET with UCNPs and **Ir1** as energy donor and energy accepter, respectively. On the basis of similar FRET processes, from upconversion luminescence of UCNPs (donors) to the absorption of organic dyes (acceptors), some other upconversion nanoprobes are also constructed for the detection of different species, by selecting the appropriate chemodosimeter as the acceptor. It is listed in Table 8.1.

8.2.2 Noble Metal Nanoparticles as Energy Acceptors

In addition to organic dyes, noble metal colloidal nanoparticles, as a unique class of chromophores, have also received great research interests. When the organic quencher is replaced with a gold NP in a fluorophore-quencher system, the energy transfer following the FRET mechanism is dependent on the plasmonic nature of the Au NP. The surface plasmon resonance (SPR) of gold and silver nanoparticles endows them with high molar extinction coefficients (typically 105–106 M^{-1} cm^{-1}, while 103–105 M^{-1} cm^{-1} for organic chromophores). Also, the SPR band across the visible and NIR region can be well adjusted by tuning the size and shape of the nanoparticles, so as to cater to different spectral demands. More importantly, as FRET acceptors, noble metal nanoparticles do not lay stringent requirements on their geometrical orientation with respect to the donor, because unlike organic dyes with specific and anisotropic transition dipoles, these nanospheres show an isotropic distribution of the transition dipoles. In all, Au NP quenchers have several appealing features [36–39]. (i) Au NPs have no defined dipole moment as organic molecules. Hence, the energy transfers in any orientation relative to the Au NPs. (ii) Au NPs have large absorption cross sections near the plasmon resonance. (iii) Au NPs have no luminescence emission, avoiding the cross-excitation that usually occurs in organic dyes. (iv) Au NPs have low toxicity. Because of these reasons, noble metal nanoparticles have recently also been adopted as energy acceptors pairing with a variety of donors, including both conventional UV-excited ones [40] and NIR-excited ones.

To the best of our knowledge, the first demonstration of a FRET-based biological detection using $NaYF_4$:Yb/Er UCNPs, and gold nanoparticles (Au NPs) was reported by Wang et al. [33]. In the FRET system, both colloidal nanoparticles were functionalized with amino groups using the layer-by-layer method based on the electrostatic attraction between the oppositely charged species deposited and then conjugated with biotin following established protocols. The biotin-functionalized $NaYF_4$:Yb/Er UCNPs served as the donor and the biotin-modified gold nanoparticles acted as the acceptor. In a mixed solution of both nanoparticles, when avidin was added, a reaction between the biotin and avidin occurred where the distance between the donor and the acceptor got shorter and energy emitted from the donor was transferred to the acceptor effectively. The more avidin that was added, the more energy could be transferred to the Au NPs. Consequently, the photoluminescence intensity of UCNPs was largely suppressed, and a linear correlation between the relative luminescence intensity and the concentration of avidin was established (Fig. 8.2c).

Besides gold nanoparticles, gold nanorods (Au NRs) also can exhibit strong surface plasma absorption in the NIR-to-IR region depending on their aspect ratios [34]. Gold nanorods have several advantages for application as the acceptor in an LRET system. The stronger optical absorption of gold NRs can enhance quenching efficiency. Compared with common quenchers, such as Au NPs, Au NRs exhibit favorable absorption properties in the NIR region. The surface of the gold nanorods may be chemically functionalized by facile procedures [41]. Besides such characteristics, Au NRs also offer the advantages of good biocompatibility and low toxicity. Therefore, Au NRs, which do not cause photoinduced damage and light interference, are promising nanomaterials for applications in biological and medical studies such as biosensing, bio-imaging, and photothermal therapy [42, 43]. These properties point to the great potential of gold nanorods in LRET systems for use as the acceptor for NIR range biosensing.

Wang's group recently developed a new LRET system for the detection of thrombin in the NIR region by utilizing NIR-to-NIR upconversion lanthanide nanophosphors (UCNPs) as the donor and gold nanorods (Au NRs) as the acceptor [34]. The emission of upconverting $NaYF_4$:Yb^{3+}/Tm^{3+} nanoparticles with sharp NIR emission peaks upon NIR excitation synthesized by an inexpensive infrared continuous wave laser diode showed spectral overlap with the absorption of the Au NRs. Both UCNPs and Au NRs were conjugated with thrombin aptamers (TBAs). These two single-stranded DNA (ssDNA) sequences were non-complementary in design, but they were also bio-recognition elements for thrombin [34]. In solution, the two probes were not in close proximity with the absence of thrombin because of the electrostatic repulsion from the negatively charged ssDNA attached on the surfaces of the nanoparticles. In this context, UCNPs fluoresced at 804 nm under the excitation of 980 nm lasers. When thrombin was present, specific recognition would occur between the two TBAs and thrombin (Fig. 8.2d). As a result, UCNPs and Au NRs would be brought into close proximity, which induced the occurrence of LET between the UCNPs (donor) and Au NRs (acceptor). Consequently, the detection was obtained by monitoring the quenching efficiency of the NIR emission of UCNPs.

8.2.3 Carbon Nanomaterials as Energy Acceptors

Graphene is a single layer of carbon atoms in a closely packed two-dimensional lattice [44]. Ever since the discovery of graphene in 2004 [45], it has been a popular nanomaterial due to its remarkable electronic, mechanical, and thermal properties. Particularly, graphene, along with its water-soluble derivative, graphene oxide (GO), has found important applications in nanoelectronics due to the fast electron mobility [46]. In addition to nanoelectronics, potential applications of graphene in biology have attracted much interest recently [47–51]. The molecular dynamics (MD) simulation revealed that this strong adsorption of ssDNA was attributed to the stacking interaction between the ring structure bases and the hexagonal cells of graphene [51]. In the previous work of Swathi et al. [52, 53] it was proposed through theoretical calculations that graphene could act as a super quencher of organic dyes, as a result of non-radiative transfer of electronic excitation energy from dye excited states to the π-system of graphene. The rate of this long-range RET was suggested to have an r^{-4} dependence on distance r and presented a striking contrast to traditional FRET, for which the rate has an r^{-6} dependence. Simultaneously, the abundance of carboxy, hydroxy, and epoxy groups on the surface of GO nanosheets made the material more water-soluble and also enabled covalent conjugation with other molecules [54]. Given this property, graphene and GO have been used as FRET acceptors with organic dyes and QDs as energy donors [47, 49, 55–58], in which they exhibited high efficiency in quenching the donor emission and thus provide good sensitivity.

Due to their unique physical and chemical properties, GO have also been frequently employed as superior quenchers of UC donors, as a result of non-radiative energy transfer from UCNPs to GO [54, 62–65]. For example, Huang and coworkers observed the UCNPs-to-GO energy transfer in a novel GO-$NaYF_4$:Yb/Er nanocomposite prepared by a one-step facile method [66]. Li's group discovered the construction of a biosensing platform using ssDNA-functionalized UCNPs and water-soluble GO, which could efficiently quench the luminescence of UCNPs [63]. First, they modified the surface of UCNPs with carboxylic acid groups by using poly(acrylic acid) (PAA). And then, amine-containing ssDNA which was designed as adenosine triphosphate (ATP)-specific aptamer, was covalently bound to the carboxylic groups on the UCNPs surface with EDC/NHS-assisted standard procedures. The strong upconversion luminescence of ssDNA-functionalized UCNPs could be quenched effectively with the addition of GO. Because it had been documented that ssDNA could be strongly adsorbed on the surface of GO based on the strong π–π stacking effect between nucleobases of ssDNA and sp^2 atoms of GO. Therefore, the ssDNA–UCNPs could be brought in close proximity to the GO surface through the strong on-covalent π–π stacking interaction between ssDNA and GO. Since there was a good overlap between luminescence emission of UCNPs and absorption spectrum of GO, the highly efficient luminescence quenching of ssDNA-UCNPs could be considered as the direct consequence of efficient FRET between UCNPs and GO. When ATP was added into the system, the high affinity between ATP and its specific aptamer led to the formation of aptamer–ATP complexes accompanied by rigid structure conformation of aptamer sequences and, therefore, decrease of the surface charge of the DNA molecules and the exposure of nucleobases [63]. As a result, the π–π stacking was weakened and UCNPs were separated from the GO surface, resulting in the decreasing quenching efficiency and increasing upconversion luminescence. The ultrahigh quenching efficiency led to a high signal-to-background ratio and thus high sensitivity for target detection. Therefore, a sensitive platform for detection of biomolecules could be rationally constructed by using UCNPs and GO as the FRET pair.

However, Liu and coworkers reported a biosensing platform using UCNPs as the energy donor and GO as the energy acceptor [54]. Unlike commonly used biosensing methods, they utilized a different model in which donor and acceptor are brought into FRET proximity through specific molecular recognition [Concanavalin A (conA)] and chitosan (CS), instead of π–π stacking effect between ssDNA and GO. The conjugation of conA and CS molecules on the surface of UCNPs and GO, respectively, was successfully implemented by an EDC/NHS coupling protocol. The known tight binding of conA with CS brought UCNP and GO into close proximity and, thus, induced energy transfer from UCNPs to GO. Thereafter, the resulting FRET process was anticipated to be inhibited when glucose was introduced because of the competition between glucose and CS for binding to conA (Fig. 8.3a). The resulting FRET was inhibited, which forms the foundation of glucose sensing. This sensor was capable of detecting glucose in serum samples without background interference and extended to detect other biomolecules [54].

In comparison with the two-dimensional graphene (or GO) with a relatively large plane, the size of zero-dimensional carbon nanoparticles (CNPs) is more comparable with biomolecules. Lately, CNPs have inspired intensive research

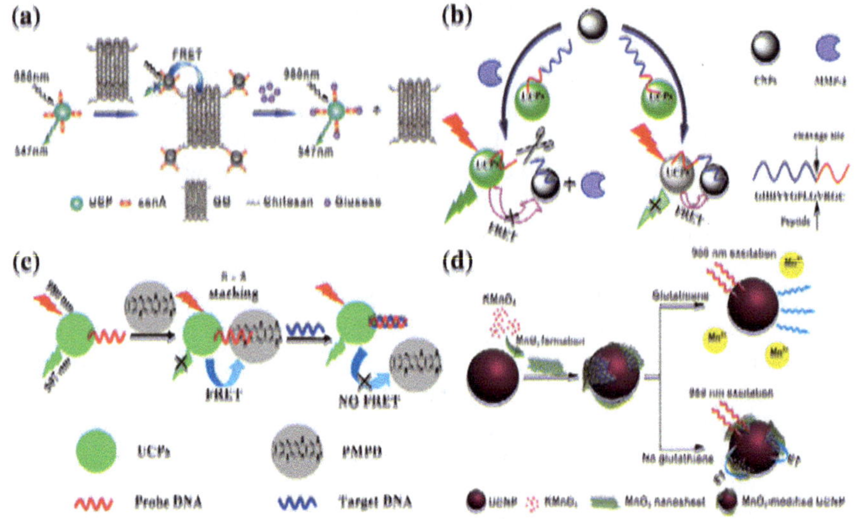

Fig. 8.3 FRET- or LRET-based nanosensors integrated by UCNPs as energy donor and different quenching materials as energy acceptors: **a** Graphene oxide. **b** Carbon nanoparticles. **c** PMPD nanospheres. **d** MnO_2 nanosheets. **a** Reprinted with the permission from Ref. [54]. Copyright 2011 Wiley-VCH Verlag GmbH & Co. KGaA. **b** Reprinted with the permission from Ref. [59]. Copyright 2012 American Chemical Society. **c** Reprinted with the permission from Ref. [60]. Copyright 2013 American Chemical Society. **d** Reprinted with the permission from Ref. [61]. Copyright 2011 American Chemical Society

efforts in various fields. Since CNPs hold the similar sp^2 electronic structure with graphene, it is reasonable to expect for their luminescence quenching ability [67]. Liu's group developed a new FRET sensor using UCNPs and CNPs as the energy donor acceptor pair for direct determination of thrombin for the first time. In particular, they designed a novel UC-FRET biosensor for the detection of matrix metalloproteinase-2 (MMP-2), a very important biomarker in blood, by using PEI-modified $NaYF_4$:Yb/Er UCNPs coupled with a polypeptide chain as energy donors, and CNPs as energy acceptors for the first time [59, 68]. The polypeptide chain (GHHYYGPLGVRGC) composed of both the MMP-2 specific substrate domain (PLGVR) and a π-rich motif (HHYY, which consisted of four aromatic residues) was designed and linked to the surface of UCNPs. Unlike the previous CNP-involved FRET systems that all relied on ssDNA, a peptide chain was revealed to self-assemble on CNPs. Upon NIR excitation, the UC-FRET process was initiated by virtue of the π–π stacking interaction between the probe peptide and CNPs, thereby leading to the quenching of emissions from UC donors. By contrast, such a FRET process was inhibited due to the cleavage of the substrate peptide separated UCPs from CNPs induced by the addition of MMP-2 in solution and thereby the emission of UCNPs recovered (Fig. 8.3b). Owing to the NIR excitation nature of UCPs, this FRET sensor, which provided a quite simple and straightforward approach to quantify MMP-2, was successfully used to directly detect MMP-2 in human plasma and whole blood samples [59].

Afterward, they discovered a new energy acceptor poly-m-phenylenediamine (PMPD). Since aromatic polymers possess a large conjugate plane similar to the aforementioned nanosized graphitic carbon materials, it was reasonable to expect their good luminescence quenching capability toward upconversion donors. Besides, the large conjugate system with a delocalized π–electron structure also enabled label-free assembly of ssDNAs and peptides on the surface of APs through π–π stacking [60]. More attractively, it was very convenient to fuse hydrophilic motifs such as carboxyl or amine groups onto the aromatic rings, and thus, good water solubility could be easily obtained with minimum influences on the conjugate structure [60]. Inspired by these prospects, they designed the FRET-based sensing platform using UCNPs and PMPD nanospheres as the energy donor–acceptor pair and ssDNA as the model analyte [60]. Taking the advantage of the large conjugate structure of PMPD, the sensor was set up by assembling the probe ssDNA on the surface of UCNPs, followed by simply mixing the UCNPs-probe DNA complex with PMPD (Fig. 8.3c) without any chemical assembling or packing. Then, the FRET from UCNPs to PMPD was initiated and led to the quenching of upconversion luminescence. Upon the hybridization between the target DNA and the probe DNA, double-stranded DNA formed with sharply reduced π–π stacking interaction on PMPD surface. Accordingly, the luminescence of UCPs was recovered in response to the release of the quencher, which made the foundation of the DNA sensor and showed high sensitivity for the target DNA with the sequence complementary to the probe DNA. Moreover, further extensions to other sensing targets could be performed by coupling specific probes (such as aptamers and peptide substrates) with UCNPs, because of the very flexible configuration of the FRET system [60].

8.2.4 MnO_2 Nanosheets as Energy Acceptors

Another efficient FRET quencher is manganese dioxide (MnO_2) nanosheets. A MnO_2 nanosheet consisting of an edge-shared MnO_6 octahedral crystal lattice has the thickness on the order of nanometers with a lateral size ranging from submicrometers to micrometers [69]. As a result of the *d–d* transition of Mn ions in the ligand field of MnO_6 octahedra, MnO_2 has intense and broad light absorption [70]. During the course of their investigation on energy migration upconversion, Liu and coworkers discovered that the upconverted luminescence could be effectively quenched by MnO_2 nanosheets formed in a solution of UCNPs [61]. The quenching was highly efficient as 20 mol% of MnO_2 nanosheets almost completely quenched the upconverted emissions. Interestingly, the quenched luminescence was turned on by introducing glutathione (GSH) to reduce MnO_2 to Mn^{2+} (Fig. 8.3d). On the basis of these findings, they proposed a hybrid biosensing system, based on MnO_2-modified UCNPs, for rapid screening and quantification of GSH levels occurring in cancer cells [61]. When liver hepatocellular cells (HepG2) were incubated with MnO_2-modified UCNPs, they exhibited a strong luminescence. A significant decrease in the emission intensity of UCNPs was observed with adding buthionine sulfoximine (a glutathione synthesis inhibitor) or N-methylmaleimide (a glutathione scavenger). This indicated that the sensing system can be used to monitor the concentration of glutathione in cells [4].

Inspired by the above knowledge and findings, Liu and coworkers expected that MnO_2 nanosheets could be another potential quencher for homogeneous biosensing with wide applications. Hence, they proposed a homogeneous FRET-sensing protocol, using MnO_2 nanosheets as the energy acceptor for the first time, which made full use of the light absorption of MnO_2 [69]. At an initial trial, they used an ssDNA chain as the probe and ochratoxin A (OTA) as the sensing target. The OTA aptamers tagged to fluorophores were spontaneously assembled on the flat surface, resulting in the energy transfer from the fluorophore to MnO_2 and the quenching of the luminescence. The interaction of the aptamers with OTA molecules changed the conformation of the aptamers, which reduced the exposure of the nucleobases. As a consequence, the physisorption of the aptamers on MnO_2 surface was weakened. In this situation, the emission of the fluorophore was expected to be recovered, which enabled the recognition and quantification of the target. In order to validate the universality of the sensing protocol, they then used a peptide chain as the probe and constructed another FRET sensor for cathepsin D (Cat D), which was based on the specific cleavage of the peptide by Cat D [69]. In this case, they employed polyethylenimine (PEI)-modified $NaYF_4$:Yb/Tm nanoparticles as the energy donor. A polypeptide chain containing six consecutive aromatic amino acids, and the Cat D-specific substrate was designed as the probe and linked to the PEI–UCPs (at the C terminus) following the sulfo-SMCC coupling protocol. Similar to the ssDNA probe, the peptide was assembled on the surface of MnO_2 naosheets due to the van der Waals interaction between the aromatic residues and the nanosheet, resulting in the quenching of the UCNPs luminescence. With the introduction of Cat D, which specifically cleaved the substrate, leading to the detachment of UCNPs from the MnO_2 nanosheet [69], the luminescence intensity of UCNPs is gradually restored. The relative luminescence intensity at 480 nm was linearly dependent on Cat D concentration within the range of 1–100 ng mL^{-1}.

8.3 Biosensing of Different Species of Analytes

In the above section, the energy transfer mechanism between UCNPs and different quenchers was briefly summarized, together with the construction strategies of sensing systems. Different sensors can be used to detect various analytes. In the following section, the usage of these sensors for different analytes will be discussed.

8.3.1 Biosensing of Ions

Human bodies contain many different kinds of ions. Some of them are necessary for life, while some are extremely toxic to mammals; detecting and evaluation of the concentration of these chemicals, therefore, are of particular interest and importance for the lives [71]. Sensing ions using UCNPs have been shown to be highly sensitive in living cells [32, 35, 72–75].

As mentioned in the above section, Li's group developed a OA-Ir1-UCNPs nanosystem to serve as a ratiometric UCL probe for highly selective detection of CN^- over other anions in a semi-aqueous system (dimethylformamide (DMF)/H_2O = 9:1, v/v) [32]. The low detection limit of 0.18 μM CN^- made OA-Ir1-UCNPs applicable for detecting CN^- in drinking water. Furthermore, by means of laser-scanning upconversion luminescence microscopy (LSUCLM) experiments, the OA-Ir1-UCNPs nanosystem could be used as a ratiometric UCL probe for monitoring CN^- in living cells. Using a self-assembly method based on hydrophobic–hydrophobic interaction, they combined UCNPs, an amphiphilic polymer, and a CN^- responsive **Ir**(III) complex in one nanosized hybrid structure to detect cyanide in pure water with high selectivity, based on a highly efficient LRET process [35].

Heavy metal (such as Hg^{2+}, Pb^{2+}, Ag^+, Cr^{3+}) pollution is one of the most serious environmental problems, which undermines global sustainability [5]. Mercury, as one of the most toxic heavy metals, causes serious environmental and health problems, and it exists in a variety of different forms (metallic, ionic, and as a part of organic salts and complexes). As such, sensitive detection of Hg^{2+} with simple procedure is highly desirable in disease prevention [5]. Lately, Li and coworkers developed a highly selective water-soluble nanoprobe based on dye-assembled nanoparticles for upconversion luminescence sensing and bio-imaging of mercury ions [73]. $NaYF_4$:20mol% of Yb, 1.6mol% of Er, 0.4mol% Tm UCNPs and the Hg^{2+}-responsive ruthenium complex N719 was chosen as the FRET donor–acceptor pair. The overlap between the green upconversion luminescence emission of UCNPs and the absorbance of N719 indicated the possibility of the energy transfer process from UCNPs to N719. The FRET process occurred in the absence of Hg^{2+}. The addition of Hg^{2+} led to decreasing spectral overlap between the visible upconversion luminescence emission of UCNPs and absorption of adduct N719–Hg^{2+}; thereby, the FRET Efficiency was reduced, corresponding to an increasing green upconversion luminescence. This dye-assembled nanosystem of ratiometric upconversion luminescence for detection of Hg^{2+} in aqueous solution possessed high selectivity, a rapid response time (<10 s), and high sensitivity (detection limit of 1.95 ppb). The limit of detection they achieved for Hg^{2+} was lower than the maximum level (2 ppb) of Hg^{2+} in drinking water set by the US EPA.

It should be noted that Hg^{2+} can be converted into methylmercury ($MeHg^+$) through microbial methylation. $MeHg^+$ is a strong liposoluble ion. The accumulation of $MeHg^+$ in the organs of animals will cause prenatal nervous system and visceral damage. Recently, Li and coworkers fabricated a highly sensitive water-soluble probe for UCL monitoring and bio-imaging of $MeHg^+$ [76]. The LRET nanosystem (hCy7-UCNPs) was composed of $NaYF_4$:20%Yb, 1.6%Er, 0.4%Tm UCNPs and $MeHg^+$-responsible NIR cyanine dye hCy7, and the emission spectrum of UCNPs and the absorption band of a $MeHg^+$-sensitive chromophore were overlap (Fig. 8.4a). The addition of $MeHg^+$ induced a significant red shift of the maximum absorption peaks of the NIR chemodosimeter Cy7 from 670 to 845 nm (Fig. 8.4b). These two absorption band peaks at 845 and 670 nm matched the UCL emissions of $^3H_4 \rightarrow {}^3H_6$ transition ($\lambda_{UCL} = 800$ nm) of Tm^{3+} and $^4F_{9/2} \rightarrow {}^4I_{15/2}$ transition ($\lambda_{UCL} = 660$ nm) of Er^{3+}, respectively. Thus, it was reasonable that the degree of energy transfer could be modulated by the presence or absence of $MeHg^+$ when Er^{3+}- and Tm^{3+}-codoped

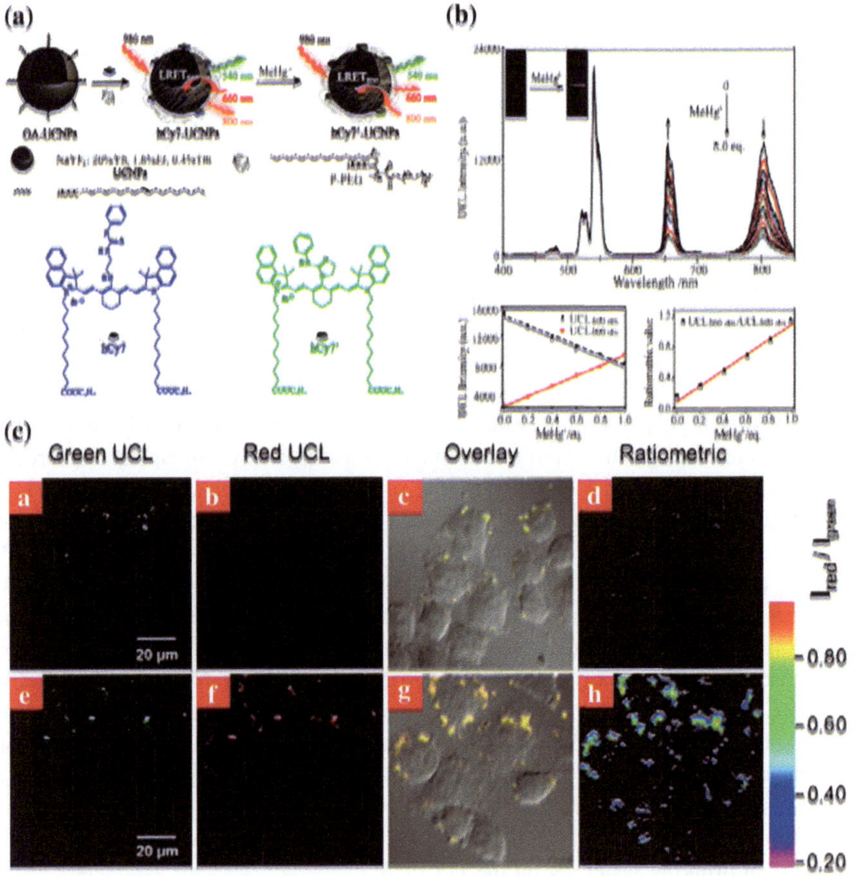

Fig. 8.4 **a** Schematic illustration of UCNPs-hCy7 and its sensing to $MeHg^+$ with a change in UCL emission. **b** Upconversion luminescence spectra of hCy7–UCNPs in the aqueous solution upon gradual addition of $MeHg^+$. **c** Ratiometric UCL images in living HeLa cells (*top*, *a–d*) and $MeHg^+$-pretreated HeLa cells (*bottom*, *e–h*) incubated with hCy7-UCNPs for 2 h at 37 °C. Reprinted with the permission from Ref. [76]. Copyright 2013 American Chemical Society

upconversion nanocrystals ($NaYF_4$:Yb/Er/Tm) and hCy7 were chosen as the energy donor and acceptor, respectively. This dye-assembled nanosensor for MeHg+ presented an impressive LOD of 0.18 ppb in aqueous solution (Fig. 8.4b), lower than the blood level (5.8 ppb) set by the US EPA. Based on the same $NaYF_4$:Yb/Er/Tm UCNPs, Li et al. also confirmed the monitoring of $MeHg^+$ by UC luminescence bioimaging in living cells (Fig. 8.4c) and small animals.

Li et al. reported the sensing of ions such as Hg^{2+} [72, 77–81], Pb^{2+} [81], Cr^{3+} [82, 83], Cu^{2+} [74, 75], Fe^{3+} [84], K^+ [85] by utilizing the UCNPs-based FRET or LRET nanosystems.

8.3.2 Biosensing of Gas Molecules

Determination of the concentration of gas molecules such as oxygen, carbon dioxide, ammonia, and many gas molecules in aqueous samples or biological fluids, has great significance in bio-analytical chemistry, clinical medical diagnosis, and bio-process monitoring [71, 86]. The use of UCNPs for gas molecules sensing has attracted extensive interest because of their excellent properties. The first oxygen sensor based on UCNPs was developed by Wolfbeis's group, in which $NaYF_4$:Yb/Tm UCNPs acted as nanolamps and iridium(III) complexes served as oxygen indicators were both incorporated in an ethyl cellulose thin film [71, 87]. These UCNPs, on excitation with a 980 nm diode laser, gave a dual emission with bands in the blue and red part of the spectrum. The iridium(III) complexes were chosen as the quenchable probe for oxygen because its absorbance has a maximum at 468 nm that strongly overlaps the two short-wavelength emissions (at 455 and 475 nm) of the UCNPs [71]. The luminescence of iridium(III) complex, produced by energy transfer from UCNPs, was sensitive to molecular oxygen concentration, which can quantify the content of oxygen molecules. Liu et al. developed a novel multifunctional nanocomposite sensor, which involved $NaYF_4$:Yb/Tm UCNPs, oxygen indicator of Ru (II) complex, and bio-compatible silica shell. It could precisely determine the concentration of O_2 from 0 to 100 % [71, 88].

However, until now, detection of oxygen is limited to oxygen molecules in gas environment. Quantification of molecular oxygen in physiological environment will be of much more importance and interest, which is essential for investigations of cellular respiration, ATP generation during oxidative phosphorylation, as well as for the investigation of hemoglobin [71]. Molecular carbon dioxide and ammonia are toxic to many aquatic organisms at low concentrations; optical detection of them is non-trivial [71]. It is known that they can interact with water molecules and thus vary the pH value of biological fluids. This provides an opportunity to quantify their concentration through optical evaluations of the pH variations. Inspired by the first optical pH sensor based on UCNPs [89], the pertinent colorimetrics or fluorometrics have been developed to meet this need. Wolfbeis and coworkers reported on a CO_2 optical sensor utilizing the $NaYF_4$:Yb/Er UCNPs in combination with a long-wavelength absorption pH probe bromothymol blue (BTB) [90]. The LOD is measured to be about 0.11 % for the concentration of CO_2.

According to a similar principle, they also designed an ammonia sensor utilizing the combination of $NaYF_4$:Yb/Er nanoparticles and a pH probe of the phenol red, which are both immobilized in a polystyrene matrix [91]. The ratiometric change of the intensity of the green to the red UC emission of the nanoparticles could determine the content of ammonia reaching a LOD as low as 400 μM [91].

8.3.3 Biosensing of Biomolecules

Some biomolecules, which can be proteins, fragments of proteins, DNA/RNA, or organic chemicals made by abnormal cells, are indicators of diseases at a specific time [7]. Early detection of disease markers enables effective treatment in order to reduce the suffering and cost to society associated with the diseases. The rapid development of nanotechnology in the past decade has led to several novel and sensitive UCNPs-based bio-detection, which will open up a new era of early disease diagnosis [1, 4, 7, 22]. The interaction of biomolecules pairs such as biotin–avidin and antibody–antigen with specific high affinity exists prevailingly in nature. If one of the biomolecular entities conjugates with luminescence nanoparticles, the resulting bio-functional luminescence nanoparticles can specifically bind to the other biomolecular entity. Based on this rudimentary principle, $NaYF_4$:Yb/Er UCNPs as bio-probes have been used for the detection of DNA [31, 72, 92, 93] and avidin [33, 41].

For example, Li et al. developed a simple and versatile strategy for converting hydrophobic β-$NaYF_4$:Yb^{3+}/Er^{3+} UCNPs into water-soluble and carboxylic acid-functionalized analogues by directly oxidizing oleic acid to azelaic acid with the Lemieux–von Rudloff reagent ($KMnO_4$/$NaIO_4$ mixture), which resulted in the generation of free carboxylic acid group on the surface [92]. The newly formed carboxylic acid groups on the surface of UCNPs facilitate further conjugation with streptavidin to form streptavidin-functionalized UCNPs by using typical EDC chemistry (Fig. 8.5a) which was used to detect DNA. A sandwich-type hybridization format was adopted in this DNA sensor, by using two shorter DNA to capture a longer target DNA. One of the two shorter oligonucleotides was labeled with biotin (capture DNA), and the other was labeled with TAMRA (reporter DNA) whose excitation spectrum overlay the green emission of the $NaYF_4$:Yb/Er nanoparticles (Fig. 8.5a). Biotin-labeled capture DNA bonded to streptavidin-functionalized β-$NaYF_4$:Yb/Er UCNPs through the specific interaction between streptavidin and biotin. With addition of target DNA, a broad emission band (580 nm) of the reporter DNA gradually appeared, and signal intensities of green emissions of UCNPs decreased significantly, while those of red emissions between 635 and 680 nm remain largely unchanged (Fig. 8.5b) due to FRET from UCNPs to reporter DNA. Thus, they quantified the target DNA by monitoring the intensity ratio upon continuous wave excitation at 980 nm. In addition, the luminescence intensity ratio (I_{580}/I_{540} or I_{540}/I_{654}) was found to vary linearly with the concentration of target

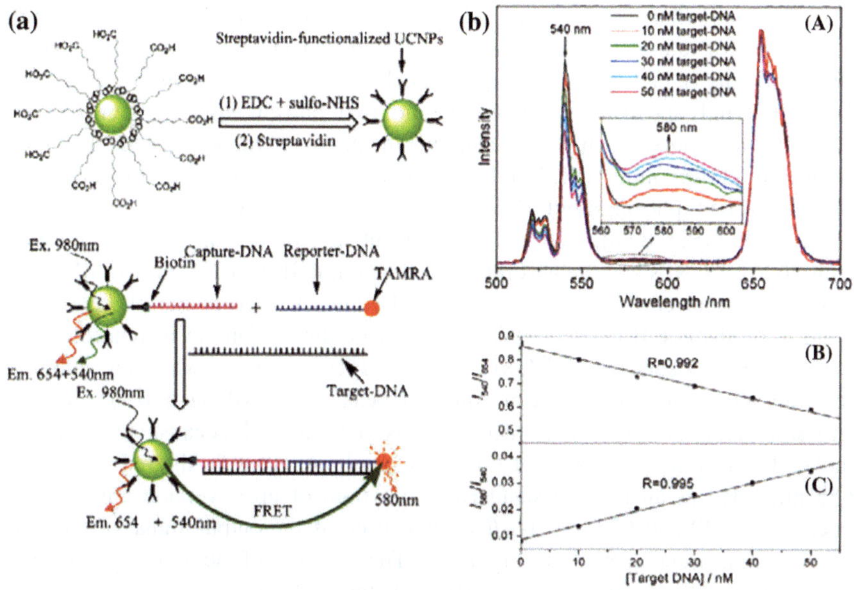

Fig. 8.5 **a** Schematic illustration of synthesis of streptavidin-functionalized UCNPs and DNA nanosensors based on UCNPs. **b** Luminescence spectra of a mixture of streptavidin-functionalized $NaYF_4$:Yb/Er nanoparticles, capture DNA, and reporter DNA in the presence of different concentrations of target DNA under continuous wave excitation at 980 nm (*A*), and the linear relationships between target DNA concentration and the intensity ratios I_{540}/I_{654} (*B*) or I_{580}/I_{540} (*C*). Reprinted with the permission from Ref. [92]. Copyright 2008 American Chemical Society

DNA in the measured range of 10–50 nM (Fig. 8.5b), suggesting high sensitivity [3, 92].

Liu and coworkers presented a new aptamer biosensor for thrombin which was based on FRET from UCNPs to CNPs [67]. In this work, the PAA functionalized UCNPs were covalently linked with a TBA, which bound to the surface of CNPs through π–π stacking interaction. On the basis of such interactions, the donor and the acceptor were taken into close proximity, which resulted in the occurrence of FRET and a maximum of the luminescence quenching rate of 89 % was observed in the absence of thrombin. With the addition of thrombin to the sensing system, the formation of quadruplex–thrombin complex weakened the π–π stacking interaction and led to the CNPs separation from UCNPs. Therefore, the FRET process was blocked, and the luminescence of UCNPs was restored. The sensor provided a linear range from 0.5 to 20 nM for thrombin with a detection limit of 0.18 nM in an aqueous buffer, and the same linear range was obtained in spiked human serum samples with a slightly higher detection limit (0.25 nM).

PLD is a critical component of intracellular signal transduction and has been implicated in many important biological processes. It has been observed that there are abnormalities in PLD expression in many human cancers, and PLD is thus recognized as a potential diagnostic biomarker as well as a target for drug discovery

[94]. Recently, Chu and coworkers designed a phospholipid-modified upconversion nanoprobe for ratiometric luminescence sensing and imaging of phospholipase D (PLD) activity in cell lysate and in living cells for the first time [94]. UCNPs with green and red upconversion emission and two phospholipid components with different functions were chosen to construct the nanosystem, one was the rhodamine B-labeled phospholipid (PE-Rhod), as an energy acceptor, and the other was the PEGylated phospholipid (DSPE-PEG) to afford water solubility and biocompatibility (Fig. 8.6a). Due to the significant spectral overlap between the green emission of UCNPs and the absorption of rhodamine B, FRET from UCNPs to dye molecules occurred. In the presence of PLD, the cleavage of the phosphodiester bond in rhodamine B-labeled phospholipid made the rhodamine B molecules keep away from the surface of UCNP, and the FRET process was thus inhibited, resulting in an increase in green emission peak of UCNPs at 540 nm. The red emission peak of UCNPs at 655 nm was not affected because rhodamine B molecules have no absorption at this wavelength. As a result, the red emission at 655 nm acted as an internal standard, and a ratio of upconversion luminescence intensity at 540 and 655 nm (I_{540}/I_{655}) was used as the output signal of the ratiometric luminescence detection (Fig. 8.6b). The intensity of the green upconversion emission at 540 nm increased linearly with the PLD concentration in the range 0–400 U L^{-1}, with a detection limit of 4 U L^{-1}. Furthermore, the PLD activity in cell lysates was also determined by the nanoprobe, confirming that PLD activity in a breast cancer cell was at least 7-fold higher than in a normal cell. Importantly, the nanoprobe was capable of monitoring PLD activity in living cells by UC luminescence bio-imaging (Fig. 8.6c) [94].

In another work we aforementioned, Li and coworkers designed a FRET-based biological detection for avidin using $NaYF_4$:Yb/Er UCNPs and gold nanoparticles (Au NPs) [33]. The intensity of upconversion luminescence varied linearly with the concentration of avidin in the solution from 0.5 to 370 nm. Liu and coworkers presented a biosensing platform using conA-modified UCNPs as energy donor and CS-functionalized GO as energy acceptor [54]. The competition between glucose and CS for binding to conA was the foundation of glucose quantification. The biosensor provided a linear relationship in the range from 0.56 to 2.0 mm.

A great deal of biomolecules was detected by using different kinds of UCNPs-based FRET or LRET biosensors, such as DNA [31, 60, 62, 92, 93, 95, 96], protein or enzyme [34, 59, 67, 94, 97–99], glutathione [61, 100], glucose [54, 101, 102], and avidin [41].

8.4 Detection Based on Magnetic Separation

Magnetic materials have attracted much attention due to their importance in the fields of chemistry, biology, medical sciences, and biotechnology. The UCNPs-based nanosensors, without using FRET principle, always perform with the assistance of magnetic nanoparticles. By combination of magnetic-field-assisted

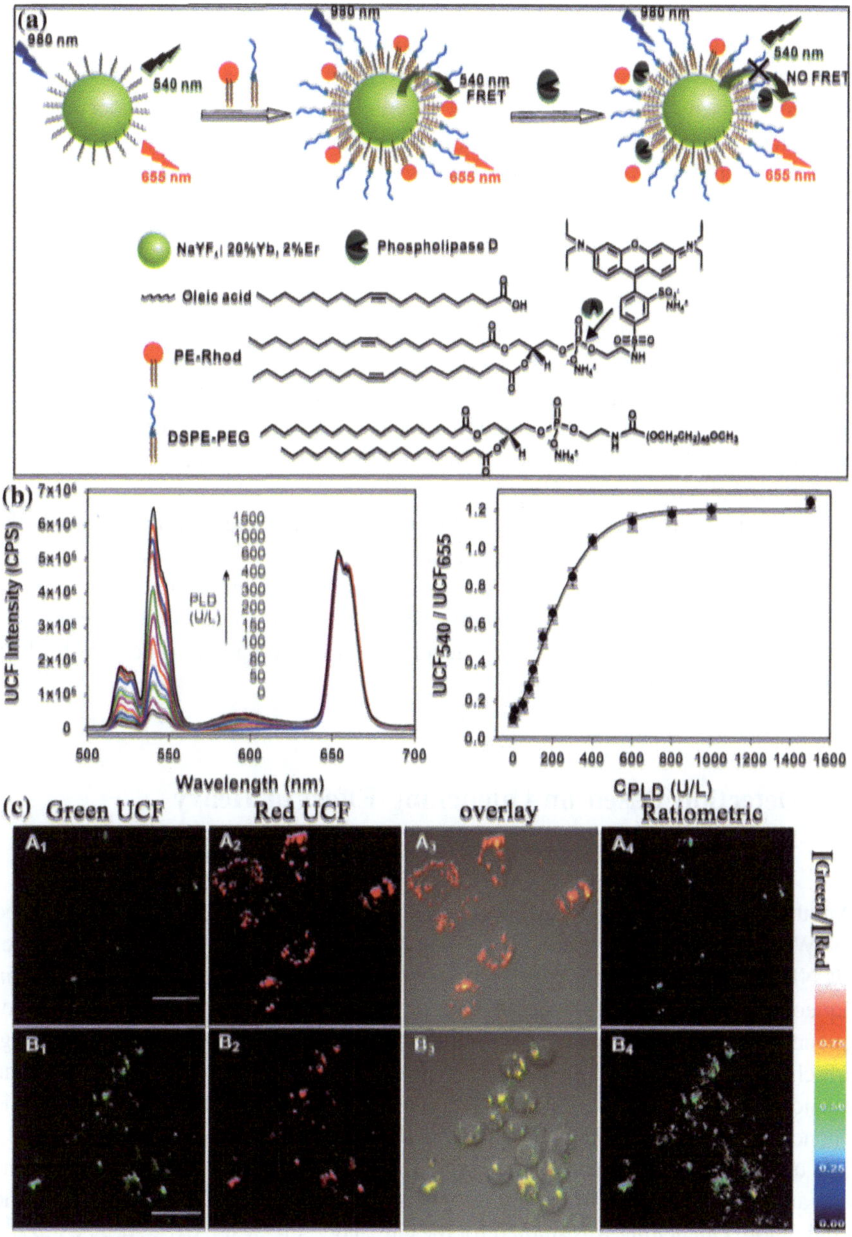

Fig. 8.6 **a** Schematic illustration of the design of Rhod-Lipo-UCNP nanoprobe and its sensing principle to phospholipase D with a change in upconversion emission. **b** Luminescence spectra of the Rhod-Lipo-UCNPs after incubation with PLD of varying concentrations and the ratio of the upconversion luminescence intensity at 540 and 655 nm (UCF_{540}/UCF_{655}) as a function of PLD concentration. Reprinted with the permission from Ref. [94]. Copyright 2014 American Chemical Society

biochemical separation and concentration technology, novel composite materials of Fe_3O_4 and $NaYF_4$:Yb/Er have been applied to the sensitive detection [2].

Wang et al. [103] applied $NaYF_4$:Yb/Er UCNPs for the sensitive detection of trace amounts of DNA with the aid of magnetic bio-separation and concentration technology. Firstly, the polymer-fabricated magnetite nanoparticles were linked with the propylthiol-terminated capture DNA, and the polyelectrolyte functionalized upconversion luminescent nanoparticles were covalently conjugated with probe DNA to monitor the presence of specific target DNA strands. As in a typical three-component sandwich-type sensor, the short capture DNA on the surface of magnetic NPs was first hybridized with long target DNA [103]. Then, the overhanging region of target DNA was matched to UCNPs-labeled probe DNA. As a result, the UCNPs were captured by the magnetic NPs through hybridization of the oligonucleotides and then purified by magnetic separation (Fig. 8.7a). The UC luminescence intensity of the captured UCNPs was found to present a linear relationship with the concentration of target DNA in the range of 7.8–78.0 nm (Fig. 8.7b).

From their results, it was reasonable to expect that the low background, the improved photostability and chemical stability of these UCNPs, could allow real-time observations of ligand–receptor interactions and of molecular interaction in living cells [103–106]. This magnetic separation and concentration technology could serve as a useful model for the development of bio-analytical devices for bio-detection in the near future [107].

8.5 Detection Based on Quenching Effect of Heavy Metal Ions

Another nanosensor based on UCNPs, without using FRET principle, was studied by Wolfbeis and coworkers [108]. They first reported that the luminescence of UCNPs rather surprisingly was quenched by heavy metal ions and halide ions in aqueous solution without relating to external quenchers. The mechanism of quenching at low metal ion concentrations most likely involved dynamic quenching, which resulted from collisions between heavy metal ions and the UCNPs. The quenching constants are in the order of 60–180 M^{-1}, indicative of dynamic quenching. Static quenching also occured, but only at higher concentrations [108]. The quenching of the luminescence of UCNPs (0.2 mg mL^{-1}) by heavy metal ions deviated from linearity at concentrations higher than 5–10 mM, depending on the kind of ion. Quenching was studied for the Cu^{II}, Hg^{II}, Pb^{II}, Cd^{II}, Co^{II}, Ag^{I}, Fe^{III}, Zn^{II}, bromide, and iodide ions and is found to be particularly strong for Hg^{II}. Stern–Volmer plots are virtually linear up to quencher concentrations of 10–25 mM [108].

The quenching of the luminescence of such nanoparticles by heavy metal and halide ions has several implications [108]. One aspect relates to the fact that the luminescence of UCNPs covered with a shell of another material (such as silica) is

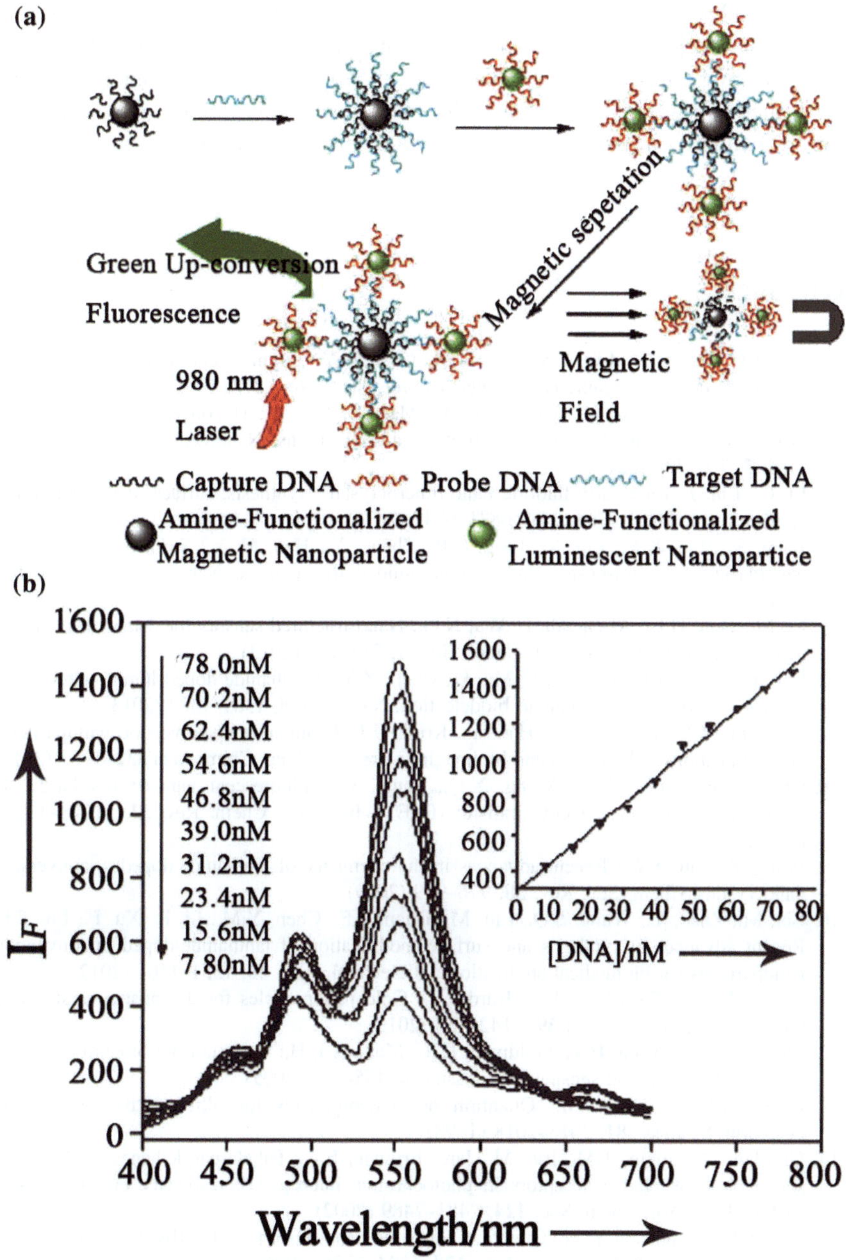

Fig. 8.7 **a** Scheme of the DNA-sensing method combining UC luminescence with magnetic separation. **b** Luminescence spectra of binary nanocomposite in the presence of different concentration of target DNA; and the linear relationship between luminescence intensity and target DNA content. Reprinted with the permission from Ref. [103]. Copyright 2006 the Royal Society of Chemistry

no longer quenched by heavy metal ions. This can form the basis for a simple test to evaluate the integrity or completeness of coatings on the surface of UCNPs [108]. The quenching effect by heavy metal ions can also serve to sense and even image heavy metal ions, such as inside cells (and in a nondestructive manner) or in microstructures or small volumes where other methods cannot be easily applied [108].

References

1. Tu, D.T., Zheng, W., Liu, Y.S., Zhu, H.M., Chen, X.Y.: Luminescent biodetection based on lanthanide-doped inorganic nanoprobes. Coord. Chem. Rev. **273**, 13–19 (2014)
2. Wang, M., Abbineni, G., Clevenger, A., Mao, C., Xu, S.: Upconversion nanoparticles: synthesis, surface modification and biological applications. Nanomed. Nanotechnol. Biol. Med. **7**, 710–729 (2011)
3. Li, C., Lin, J.: Rare earth fluoride nano-/microcrystals: synthesis, surface modification and application. J. Mater. Chem. **20**, 6831–6847 (2010)
4. Jiang, S., Win, K.Y., Liu, S., Teng, C.P., Zheng, Y., Han, M.-Y.: Surface-functionalized nanoparticles for biosensing and imaging-guided therapeutics. Nanoscale **5**, 3127–3148 (2013)
5. Li, M., Gou, H.L., Al-Ogaidi, I., Wu, N.Q.: Nanostructured sensors for detection of heavy metals: a review. ACS Sustain. Chem. Eng. **1**, 713–723 (2013)
6. Liu, Y.S., Tu, D.T., Zhu, H.M., Ma, E., Chen, X.Y.: Lanthanide-doped luminescent nano-bioprobes: from fundamentals to biodetection. Nanoscale **5**, 1369–1384 (2013)
7. DaCosta, M.V., Doughan, S., Han, Y., Krull, U.J.: Lanthanide upconversion nanoparticles and applications in bioassays and bioimaging: a review. Anal. Chim. Acta **832**, 1–33 (2014)
8. Chen, G.W., Song, F.L., Xiong, X.Q., Peng, X.J.: Fluorescent nanosensors based on fluorescence resonance energy transfer (FRET). Ind. Eng. Chem. Res. **52**, 11228–11245 (2013)
9. Wang, F., Liu, X.G.: Recent advances in the chemistry of lanthanide-doped upconversion nanocrystals. Chem. Soc. Rev. **38**, 976–989 (2009)
10. Lin, M., Zhao, Y., Wang, S.Q., Liu, M., Duan, Z.F., Chen, Y.M., Li, F., Xu, F., Lu, T.J.: Recent advances in synthesis and surface modification of lanthanide-doped upconversion nanoparticles for biomedical applications. Biotechnol. Adv. **30**, 1551–1561 (2012)
11. Lu, F., Doane, T.L., Zhu, J.-J., Burda, C.: Gold nanoparticles for diagnostic sensing and therapy. Inorg. Chim. Acta **393**, 142–153 (2012)
12. Medintz, I.L., Uyeda, H.T., Goldman, E.R., Mattoussi, H.: Quantum dot bioconjugates for imaging, labelling and sensing. Nat. Mater. **4**, 435–446 (2005)
13. Chan, W.C.W., Nie, S.M.: Quantum dot bioconjugates for ultrasensitive nonisotopic detection. Science **281**, 2016–2018 (1998)
14. Giordano, L., Jovin, T.M., Irie, M., Jares-Erijman, E.A.: Diheteroarylethenes as thermally stable photoswitchable acceptors in photochromic fluorescence resonance energy transfer (pcFRET). J. Am. Chem. Soc. **124**, 7481–7489 (2002)
15. Zhang, C.Y., Johnson, L.W.: Quantum-dot-based nanosensor for RRE IIB RNA-Rev peptide interaction assay. J. Am. Chem. Soc. **128**, 5324–5325 (2006)
16. Gnach, A., Bednarkiewicz, A.: Lanthanide-doped up-converting nanoparticles: merits and challenges. Nano Today **7**, 532–563 (2012)
17. Wang, F., Banerjee, D., Liu, Y.S., Chen, X.Y., Liu, X.G.: Upconversion nanoparticles in biological labeling, imaging, and therapy. Analyst **135**, 1839–1854 (2010)

18. Hardman, R.: A toxicologic review of quantum dots: toxicity depends on physicochemical and environmental factors. Environ. Health Perspect. 165–172 (2006)
19. Auffan, M., Rose, J., Bottero, J.Y., Lowry, G.V., Jolivet, J.P., Wiesner, M.R.: Towards a definition of inorganic nanoparticles from an environmental, health and safety perspective. Nat. Nanotechnol. **4**, 634–641 (2009)
20. Shen, J., Sun, L.-D., Yan, C.-H.: Luminescent rare earth nanomaterials for bioprobe applications. Dalton Trans. 5687–5697 (2008)
21. Wang, J.S., Wu, L., Ren, J.S., Qu, X.G.: Visual detection of telomerase activity with a tunable dynamic range by using a gold nanoparticle probe-based hybridization protection strategy. Nanoscale **6**, 1661–1666 (2014)
22. Zhang, C., Sun, L.D., Zhang, Y.W., Yan, C.H.: Rare earth upconversion nanophosphors: synthesis, functionalization and application as biolabels and energy transfer donors. J. Rare Earths **28**, 807–819 (2010)
23. Zhou, J., Liu, Z., Li, F.Y.: Upconversion nanophosphors for small-animal imaging. Chem. Soc. Rev. **41**, 1323–1349 (2012)
24. Liu, J.B., Yang, X.H., He, X.X., Wang, K.M., Wang, Q., Guo, Q.P., Shi, H., Huang, J., Huo, X.Q.: Fluorescent nanoparticles for chemical and biological sensing. Sci. China Chem **54**, 1157–1176 (2011)
25. Xu, C.T., Svensson, N., Axelsson, J., Svenmarker, P., Somesfalean, G., Chen, G., Liang, H., Liu, H., Zhang, Z., Andersson-Engels, S.: Autofluorescence insensitive imaging using upconverting nanocrystals in scattering media. Appl. Phys. Lett. **93**, 171103–171106 (2008)
26. Chen, J., Zhao, J.X.: Upconversion nanomaterials: synthesis, mechanism, and applications in sensing. Sensors **12**, 2414–2435 (2012)
27. Feng, W., Han, C.M., Li, F.Y.: Upconversion-nanophosphor-based functional nanocomposites. Adv. Mater. **25**, 5287–5303 (2013)
28. Sapsford, K.E., Berti, L., Medintz, I.L.: Materials for fluorescence resonance energy transfer analysis: Beyond traditional donor-acceptor combinations. Angew. Chem. Int. Ed. **45**, 4562–4588 (2006)
29. Ray, P.C., Fan, Z., Crouch, R.A., Sinha, S.S., Pramanik, A.: Nanoscopic optical rulers beyond the FRET distance limit: fundamentals and applications. Chem. Soc. Rev. **43**, 6370–6404 (2014)
30. Yuan, L., Lin, W.Y., Zheng, K.B., Zhu, S.S.: FRET-based small-molecule fluorescent probes: rational design and bioimaging applications. Acc. Chem. Res. **46**, 1462–1473 (2013)
31. Zhang, P., Rogelj, S., Nguyen, K., Wheeler, D.: Design of a highly sensitive and specific nucleotide sensor based on photon upconverting particles. J. Am. Chem. Soc. **128**, 12410–12411 (2006)
32. Liu, J.L., Liu, Y., Liu, Q., Li, C.Y., Sun, L.N., Li, F.Y.: Iridium(III) complex-coated nanosystem for ratiometric upconversion luminescence bioimaging of cyanide anions. J. Am. Chem. Soc. **133**, 15276–15279 (2011)
33. Wang, L.Y., Yan, R.X., Hao, Z.Y., Wang, L., Zeng, J.H., Bao, J., Wang, X., Peng, Q., Li, Y. D.: Fluorescence resonant energy transfer biosensor based on upconversion-luminescent nanoparticles. Angew. Chem. Int. Ed. **44**, 6054–6057 (2005)
34. Yuan, F., Chen, H., Xu, J., Zhang, Y., Wu, Y., Wang, L.: Aptamer-based luminescence energy transfer from near-infrared-to-near-infrared upconverting nanoparticles to gold nanorods and its application for the detection of thrombin. Chem. Eur. J. **20**, 1–8 (2014)
35. Yao, L.M., Zhou, J., Liu, J.L., Feng, W., Li, F.Y.: Iridium-complex-modified upconversion nanophosphors for effective LRET detection of cyanide anions in pure water. Adv. Funct. Mater. **22**, 2667–2672 (2012)
36. De, M., Ghosh, P.S., Rotello, V.M.: Applications of nanoparticles in biology. Adv. Mater. **20**, 4225–4241 (2008)
37. Mayilo, S., Kloster, M.A., Wunderlich, M., Lutich, A., Klar, T.A., Nichtl, A., Kürzinger, K., Stefani, F.D., Feldmann, J.: Long-range fluorescence quenching by gold nanoparticles in a sandwich immunoassay for cardiac troponin T. Nano Lett. **9**, 4558–4563 (2009)

38. Pons, T., Medintz, I.L., Sapsford, K.E., Higashiya, S., Grimes, A.F., English, D.S., Mattoussi, H.: On the quenching of semiconductor quantum dot photoluminescence by proximal gold nanoparticles. Nano Lett. **7**, 3157–3164 (2007)
39. Li, M., Cushing, S.K., Wang, Q., Shi, X., Hornak, L.A., Hong, Z., Wu, N.: Size-dependent energy transfer between CdSe/ZnS quantum dots and gold nanoparticles. J. Phys. Chem. Lett. **2**, 2125–2129 (2011)
40. Saha, K., Agasti, S.S., Kim, C., Li, X.N., Rotello, V.M.: Gold nanoparticles in chemical and biological sensing. Chem. Rev. **112**, 2739–2779 (2012)
41. Zhang, S., Wang, J., Xu, W., Chen, B.T., Yu, W., Xu, L., Song, H.W.: Fluorescence resonance energy transfer between $NaYF_4$:Yb, Tm upconversion nanoparticles and gold nanorods: near-infrared responsive biosensor for streptavidin. J. Lumin. **147**, 278–283 (2014)
42. Huang, Y.-F., Chang, H.-T., Tan, W.: Cancer cell targeting using multiple aptamers conjugated on nanorods. Anal. Chem. **80**, 567–572 (2008)
43. Chen, C.-C., Lin, Y.-P., Wang, C.-W., Tzeng, H.-C., Wu, C.-H., Chen, Y.-C., Chen, C.-P., Chen, L.-C., Wu, Y.-C.: DNA-gold nanorod conjugates for remote control of localized gene expression by near infrared irradiation. J. Am. Chem. Soc. **128**, 3709–3715 (2006)
44. Rao, C.N.R., Sood, A.K., Subrahmanyam, K.S., Govindaraj, A.: Graphene: the new two-dimensional nanomaterial. Angew. Chem. Int. Ed. **48**, 7752–7777 (2009)
45. Novoselov, K.S., Geim, A.K., Morozov, S., Jiang, D., Zhang, Y., Dubonos, S., Grigorieva, I., Firsov, A.: Electric field effect in atomically thin carbon films. Science **306**, 666–669 (2004)
46. Mohanty, N., Berry, V.: Graphene-based single-bacterium resolution biodevice and DNA transistor: interfacing graphene derivatives with nanoscale and microscale biocomponents. Nano Lett. **8**, 4469–4476 (2008)
47. Zhang, M., Yin, B.-C., Tan, W., Ye, B.-C.: A versatile graphene-based fluorescence "on/off" switch for multiplex detection of various targets. Biosens. Bioelectron. **26**, 3260–3265 (2011)
48. Chen, K., Lu, G., Chang, J., Mao, S., Yu, K., Cui, S., Chen, J.: Hg(II) ion detection using thermally reduced graphene oxide decorated with functionalized gold nanoparticles. Anal. Chem. **84**, 4057–4062 (2012)
49. Chang, H., Tang, L., Wang, Y., Jiang, J., Li, J.: Graphene fluorescence resonance energy transfer aptasensor for the thrombin detection. Anal. Chem. **82**, 2341–2346 (2010)
50. Li, J.L., Bao, H.C., Hou, X.L., Sun, L., Wang, X.G., Gu, M.: Graphene oxide nanoparticles as a nonbleaching optical probe for two-photon luminescence imaging and cell therapy. Angew. Chem. Int. Ed. **51**, 1830–1834 (2012)
51. Wen, Y.Q., Xing, F.F., He, S.J., Song, S.P., Wang, L.H., Long, Y.T., Li, D., Fan, C.H.: A graphene-based fluorescent nanoprobe for silver(I) ions detection by using graphene oxide and a silver-specific oligonucleotide. Chem. Commun. **46**, 2596–2598 (2010)
52. Swathi, R., Sebastian, K.: Long range resonance energy transfer from a dye molecule to graphene has (distance)—4 dependence. J. Chem. Phys. **130**, 086101 (2009)
53. Swathi, R., Sebastian, K.: Resonance energy transfer from a dye molecule to graphene. J. Chem. Phys. **129**, 054703 (2008)
54. Zhang, C.L., Yuan, Y.X., Zhang, S.M., Wang, Y.H., Liu, Z.H.: Biosensing platform based on fluorescence resonance energy transfer from upconverting nanocrystals to graphene oxide. Angew. Chem. Int. Ed. **50**, 6851–6854 (2011)
55. Li, M., Zhou, X., Guo, S., Wu, N.: Detection of lead (II) with a "turn-on" fluorescent biosensor based on energy transfer from CdSe/ZnS quantum dots to graphene oxide. Biosens. Bioelectron. **43**, 69–74 (2013)
56. Lu, C.H., Yang, H.H., Zhu, C.L., Chen, X., Chen, G.N.: A graphene platform for sensing biomolecules. Angew. Chem. **121**, 4879–4881 (2009)
57. He, S., Song, B., Li, D., Zhu, C., Qi, W., Wen, Y., Wang, L., Song, S., Fang, H., Fan, C.: A graphene nanoprobe for rapid, sensitive, and multicolor fluorescent DNA analysis. Adv. Funct. Mater. **20**, 453–459 (2010)

58. Zhao, X.-H., Kong, R.-M., Zhang, X.-B., Meng, H.-M., Liu, W.-N., Tan, W., Shen, G.-L., Yu, R.-Q.: Graphene–DNAzyme based biosensor for amplified fluorescence "turn-on" detection of Pb^{2+} with a high selectivity. Anal. Chem. **83**, 5062–5066 (2011)
59. Wang, Y.H., Shen, P., Li, C.Y., Wang, Y.Y., Liu, Z.H.: Upconversion fluorescence resonance energy transfer based biosensor for ultrasensitive detection of matrix metalloproteinase-2 in blood. Anal. Chem. **84**, 1466–1473 (2012)
60. Wang, Y.H., Wu, Z.J., Liu, Z.H.: Upconversion fluorescence resonance energy transfer biosensor with aromatic polymer nanospheres as the lable-free energy acceptor. Anal. Chem. **85**, 258–264 (2013)
61. Deng, R.R., Xie, X.J., Vendrell, M., Chang, Y.T., Liu, X.G.: Intracellular glutathione detection using MnO_2-nanosheet-modified upconversion nanoparticles. J. Am. Chem. Soc. **133**, 20168–20171 (2011)
62. Wu, S.J., Duan, N., Ma, X.Y., Xia, Y., Wang, H.G., Wang, Z.P., Zhang, Q.: Multiplexed fluorescence resonance energy transfer aptasensor between upconversion nanoparticles and graphene oxide for the simultaneous determination of mycotoxins. Anal. Chem. **84**, 6263–6270 (2012)
63. Liu, C.H., Wang, Z., Jia, H.X., Li, Z.P.: Efficient fluorescence resonance energy transfer between upconversion nanophosphors and graphene oxide: a highly sensitive biosensing platform. Chem. Commun. **47**, 4661–4663 (2011)
64. Li, H., Sun, D.E., Liu, Y.J., Liu, Z.H.: An ultrasensitive homogeneous aptasensor for kanamycin based on upconversion fluorescence resonance energy transfer. Biosens. Bioelectron. **55**, 149–156 (2014)
65. Wu, Y.M., Cen, Y., Huang, L.J., Yu, R.Q., Chu, X.: Upconversion fluorescence resonance energy transfer biosensor for sensitive detection of human immunodeficiency virus antibodies in human serum. Chem. Commun. **50**, 4759–4762 (2014)
66. Wei, W., He, T.C., Teng, X., Wu, S.X., Ma, L., Zhang, H., Ma, J., Yang, Y.H., Chen, H.Y., Han, Y., Sun, H.D., Huang, L.: Nanocomposites of graphene oxide and upconversion rare-earth nanocrystals with superior optical limiting performance. Small **8**, 2271–2276 (2012)
67. Wang, Y.H., Bao, L., Liu, Z.H., Pang, D.W.: Aptamer biosensor based on fluorescence resonance energy transfer from upconverting phosphors to carbon nanoparticles for thrombin detection in human plasma. Anal. Chem. **83**, 8130–8137 (2011)
68. Liu, Y.S., Tu, D.T., Zhu, H.M., Chen, X.Y.: Lanthanide-doped luminescent nanoprobes: controlled synthesis, optical spectroscopy, and bioapplications. Chem. Soc. Rev. **42**, 6924–6958 (2013)
69. Yuan, Y.X., Wu, S.F., Shu, F., Liu, Z.H.: An MnO_2 nanosheet as a label-free nanoplatform for homogeneous biosensing. Chem. Commun. **50**, 1095–1097 (2014)
70. Omomo, Y., Sasaki, T., Wang, L., Watanabe, M.: Redoxable nanosheet crystallites of MnO_2 derived via delamination of a layered manganese oxide. J. Am. Chem. Soc. **125**, 3568–3575 (2003)
71. Chen, G., Qiu, H., Prasad, P.N., Chen, X.: Upconversion nanoparticles: design, nanochemistry, and applications in theranostics. Chem. Rev. **114**, 5161–5214 (2014)
72. Kumar, M., Zhang, P.: Highly sensitive and selective label-free optical detection of mercuric ions using photon upconverting nanoparticles. Biosens. Bioelectron. **25**, 2431–2435 (2010)
73. Liu, Q., Peng, J.J., Sun, L.N., Li, F.Y.: High-efficiency upconversion luminescent sensing and bioimaging of Hg(II) by chromophoric ruthenium complex-assembled nanophosphors. ACS Nano **5**, 8040–8048 (2011)
74. Li, C.X., Liu, J.L., Alonso, S., Li, F.Y., Zhang, Y.: Upconversion nanoparticles for sensitive and in-depth detection of Cu^{2+} ions. Nanoscale **4**, 6065–6071 (2012)
75. Zhang, J., Li, B., Zhang, L.M., Jiang, H.: An optical sensor for Cu(II) detection with upconverting luminescent nanoparticles as an excitation source. Chem. Commun. **48**, 4860–4862 (2012)
76. Liu, Y., Chen, M., Cao, T.Y., Sun, Y., Li, C.Y., Liu, Q., Yang, T.S., Yao, L.M., Feng, W., Li, F.Y.: A cyanine-modified nanosystem for in vivo upconversion luminescence bioimaging of methylmercury. J. Am. Chem. Soc. **135**, 9869–9876 (2013)

77. Chen, H.Q., Yuan, F., Wang, S.Z., Xu, J., Zhang, Y.Y., Wang, L.: Near-infrared to near-infrared upconverting $NaYF_4$:Yb^{3+}, Tm^{3+} nanoparticles-aptamer-Au nanorods light resonance energy transfer system for the detection of mercuric(II) ions in solution. Analyst **138**, 2392–2397 (2013)
78. Li, H., Wang, L.Y.: $NaYF_4$:Yb^{3+}/Er^{3+} nanoparticle-based upconversion luminescence resonance energy transfer sensor for mercury(II) quantification. Analyst **138**, 1589–1595 (2013)
79. Wang, M.N., An, X.J., Gao, J.: An "off-on" Hg(II) sensor excited by near-infrared to visible upconversion nanorods. J. Lumin. **144**, 91–97 (2013)
80. Li, X.H., Wu, Y.Q., Liu, Y., Zou, X.M., Yao, L.M., Li, F.Y., Feng, W.: Cyclometallated ruthenium complex-modified upconversion nanophosphors for selective detection of Hg^{2+} ions in water. Nanoscale **6**, 1020–1028 (2014)
81. Wu, S., Duan, N., Shi, Z., Fang, C., Wang, Z.: Dual fluorescence resonance energy transfer assay between tunable upconversion nanoparticles and controlled gold nanoparticles for the simultaneous detection of Pb^{2+} and Hg^{2+}. Talanta **128**, 327–336 (2014)
82. Chen, H.Q., Ren, J.C.: Sensitive determination of chromium(VI) based on the inner filter effect of upconversion luminescent nanoparticles ($NaYF_4$:Yb^{3+}, Er^{3+}). Talanta **99**, 404–408 (2012)
83. Liu, B.X., Tan, H.L., Chen, Y.: Upconversion nanoparticle-based fluorescence resonance energy transfer assay for Cr(III) ions in urine. Anal. Chim. Acta **761**, 178–185 (2013)
84. Ding, Y.J., Zhu, H., Zhang, X.X., Zhu, J.J., Burda, C.: Rhodamine B derivative-functionalized upconversion nanoparticles for FRET-based Fe^{3+}-sensing. Chem. Commun. **49**, 7797–7799 (2013)
85. Xie, L.X., Qin, Y., Chen, H.Y.: Direct fluorescent measurement of blood potassium with polymeric optical sensors based on upconverting nanomaterials. Anal. Chem. **85**, 2617–2622 (2013)
86. Achatz, D.E., Ali, R., Wolfbeis, O.S.: Luminescent chemical sensing, biosensing, and screening using upconverting nanoparticles. Lumin. Appl. Sens. Sci. **300**, 29–50 (2011)
87. Achatz, D.E., Meier, R.J., Fischer, L.H., Wolfbeis, O.S.: Luminescent sensing of oxygen using a quenchable probe and upconverting nanoparticles. Angew. Chem. Int. Ed. **50**, 260–263 (2011)
88. Liu, L., Li, B., Ying, J., Wu, X., Zhao, H., Ren, X., Zhu, D., Su, Z.: Synthesis and characterization of a new trifunctional magnetic–photoluminescent–oxygen-sensing nanomaterial. Nanotechnology **19**, 495709 (2008)
89. Sun, L.-N., Peng, H., Stich, M.I., Achatz, D., Wolfbeis, O.S.: pH sensor based on upconverting luminescent lanthanide nanorods. Chem. Commun. 5000–5002 (2009)
90. Ali, R., Saleh, S.M., Meier, R.J., Azab, H.A., Abdelgawad, I.I., Wolfbeis, O.S.: Upconverting nanoparticle based optical sensor for carbon dioxide. Sens. Actuators B Chem. **150**, 126–131 (2010)
91. Mader, H.S., Wolfbeis, O.S.: Optical ammonia sensor based on upconverting luminescent nanoparticles. Anal. Chem. **82**, 5002–5004 (2010)
92. Chen, Z.G., Chen, H.L., Hu, H., Yu, M.X., Li, F.Y., Zhang, Q., Zhou, Z.G., Yi, T., Huang, C.H.: Versatile synthesis strategy for carboxylic acid-functionalized upconverting nanophosphors as biological labels. J. Am. Chem. Soc. **130**, 3023 (2008)
93. Kumar, M., Guo, Y., Zhang, P.: Highly sensitive and selective oligonucleotide sensor for sickle cell disease gene using photon upconverting nanoparticles. Biosens. Bioelectron. **24**, 1522–1526 (2009)
94. Cen, Y., Wu, Y., Kong, X., Wu, S., Yu, R., Chu, X.: Phospholipid-modified upconversion nanoprobe for ratiometric fluorescence detection and imaging of phospholipase D in cell lysate and in living cells. Anal. Chem. **86**, 7119–7127 (2014)
95. Kumar, M., Zhang, P.: Highly sensitive and selective label-free optical detection of DNA hybridization based on photon upconverting nanoparticles. Langmuir **25**, 6024–6027 (2009)
96. Liu, J.L., Cheng, J.T., Zhang, Y.: Upconversion nanoparticle based LRET system for sensitive detection of MRSA DNA sequence. Biosens. Bioelectron. **43**, 252–256 (2013)

97. Wang, M., Hou, W., Mi, C.C., Wang, W.X., Xu, Z.R., Teng, H.H., Mao, C.B., Xu, S.K.: Immunoassay of goat antihuman immunoglobulin G antibody based on luminescence resonance energy transfer between near-infrared responsive $NaYF_4$:Yb, Er upconversion fluorescent nanoparticles and gold nanoparticles. Anal. Chem. **81**, 8783–8789 (2009)
98. Zhang, J.P., Mi, C.C., Wu, H.Y., Huang, H.Q., Mao, C.B., Xu, S.K.: Synthesis of $NaYF_4$: Yb/Er/Gd up-conversion luminescent nanoparticles and luminescence resonance energy transfer-based protein detection. Anal. Biochem. **421**, 673–679 (2012)
99. Chen, H.Q., Yuan, F., Wang, S.Z., Xu, J., Zhang, Y.Y., Wang, L.: Aptamer-based sensing for thrombin in red region via fluorescence resonant energy transfer between $NaYF_4$:Yb, Er upconversion nanoparticles and gold nanorods. Biosens. Bioelectron. **48**, 19–25 (2013)
100. Zhai, Y.L., Zhu, C.Z., Ren, J.T., Wang, E.K., Dong, S.J.: Multifunctional polyoxometalates-modified upconversion nanoparticles: integration of electrochromic devices and antioxidants detection. Chem. Commun. **49**, 2400–2402 (2013)
101. Peng, J.H., Wang, Y.H., Wang, J.L., Zhou, X., Liu, Z.H.: A new biosensor for glucose determination in serum based on up-converting fluorescence resonance energy transfer. Biosens. Bioelectron. **28**, 414–420 (2011)
102. Zhang, Y., Tang, Y.R., Liu, X., Zhang, L.C., Lv, Y.: A highly sensitive upconverting phosphors-based off-on probe for the detection of glutathione. Sens. Actuators B Chem. **185**, 363–369 (2013)
103. Wang, L.Y., Li, Y.D.: Green upconversion nanocrystals for DNA detection. Chem. Commun. 2557–2559 (2006)
104. Wu, S.J., Duan, N., Shi, Z., Fang, C.C., Wang, Z.P.: Simultaneous aptasensor for multiplex pathogenic bacteria detection based on multicolor upconversion nanoparticles labels. Anal. Chem. **86**, 3100–3107 (2014)
105. Wu, S.J., Duan, N., Wang, Z.P., Wang, H.X.: Aptamer-functionalized magnetic nanoparticle-based bioassay for the detection of ochratoxin a using upconversion nanoparticles as labels. Analyst **136**, 2306–2314 (2011)
106. Yang, W.X., Wang, Y.C., Chang, L.J., Liu, C.H., Bai, J., Li, Z.P.: Highly sensitive detection of protein kinase activity using upconversion luminescent nanoparticles. RSC Adv. **4**, 14546–14549 (2014)
107. Wang, G.F., Peng, Q., Li, Y.D.: Lanthanide-doped nanocrystals: synthesis, optical-magnetic properties, and applications. Acc. Chem. Res. **44**, 322–332 (2011)
108. Saleh, S.M., Ali, R., Wolfbeis, O.S.: Quenching of the luminescence of upconverting luminescent nanoparticles by heavy metal ions. Chem. Eur. J. **17**, 14611–14617 (2011)
109. Chen, J.G., Chen, H.Q., Zhou, C.L., Xu, J., Yuan, F., Wang, L.: An efficient upconversion luminescence energy transfer system for determination of trace amounts of nitrite based on $NaYF_4:Yb^{3+}$, Er^{3+} as donor. Anal. Chim. Acta **713**, 111–114 (2012)
110. He, X.X., Li, Z.X., Jia, X.K., Wang, K.M., Yin, J.J.: A highly selective sandwich-type FRET assay for ATP detection based on silica coated photon upconverting nanoparticles and split aptamer. Talanta **111**, 105–110 (2013)
111. Arppe, R., Nareoja, T., Nylund, S., Mattsson, L., Koho, S., Rosenholm, J.M., Soukka, T., Schaferling, M.: Photon upconversion sensitized nanoprobes for sensing and imaging of pH. Nanoscale **6**, 6837–6843 (2014)
112. Harvey, P., Oakland, C., Driscoll, M.D., Hay, S., Natrajan, L.S.: Ratiometric detection of enzyme turnover and flavin reduction using rare-earth upconverting phosphors. Dalton Trans. **43**, 5265–5268 (2014)
113. Hazra, C., Adusumalli, V.N.K.B., Mahalingam, V.: 3,5-dinitrobenzoic acid-capped upconverting nanocrystals for the selective detection of melamine. ACS Appl. Mat. Interfaces **6**, 7833–7839 (2014)
114. Liu, C.H., Chang, L.J., Wang, H.H., Bai, J., Ren, W., Li, Z.P.: Upconversion nanophosphor: an efficient phosphopeptides-recognizing matrix and luminescence resonance energy transfer donor for robust detection of protein kinase activity. Anal. Chem. **86**, 6095–6102 (2014)

115. Liu, J., Liu, Y., Bu, W., Bu, J., Sun, Y., Du, J., Shi, J.: Ultrasensitive nanosensors based on upconversion nanoparticles for selective hypoxia imaging in vivo upon near-infrared excitation. J. Am. Chem. Soc. **136**, 9701–9709 (2014)
116. Ma, Y.X., Wang, L.Y.: Upconversion luminescence nanosensor for TNT selective and label-free quantification in the mixture of nitroaromatic explosives. Talanta **120**, 100–105 (2014)
117. Zhao, L., Peng, J., Chen, M., Liu, Y., Yao, L., Feng, W., Li, F.: Yolk-shell upconversion nanocomposites for LRET sensing of cysteine/homocysteine. ACS Appl. Mat. Interfaces **6**, 11190–11197 (2014)
118. Zhou, Y., Pei, W., Wang, C., Zhu, J., Wu, J., Yan, Q., Huang, L., Huang, W., Yao, C., Loo, J.S.C.: Rhodamine-modified upconversion nanophosphors for ratiometric detection of hypochlorous acid in aqueous solution and living cells. Small **10**, 3560–3567 (2014)
119. Zhang, S.Z., Sun, L.D., Tian, H., Liu, Y., Wang, J.F., Yan, C.H.: Reversible luminescence switching of $NaYF_4$:Yb,Er nanoparticles with controlled assembly of gold nanoparticles. Chem. Commun. 2547–2549 (2009)
120. Chen, H.Q., Yuan, F., Wang, L.: Simple and sensitive turn-on luminescent detection of biothiols based on energy transfer between green-emitting upconversion nanocrystals and gold nanoparticles. Anal. Methods **5**, 2873–2879 (2013)
121. Lin, F.B., Yin, B.D., Li, C.Z., Deng, J.H., Fan, X.Y., Yi, Y.H., Liu, C., Li, H.T., Zhang, Y.Y., Yao, S.Z.: Fluorescence resonance energy transfer aptasensor for platelet-derived growth factor detection based on upconversion nanoparticles in 30 % blood serum. Anal. Methods **5**, 699–704 (2013)
122. Tu, N.N., Wang, L.Y.: Surface plasmon resonance enhanced upconversion luminescence in aqueous media for TNT selective detection. Chem. Commun. **49**, 6319–6321 (2013)
123. Wang, Z.J., Wu, L.N., Shen, B.Z., Jiang, Z.H.: Highly sensitive and selective cartap nanosensor based on luminescence resonance energy transfer between $NaYF_4$:Yb,Ho nanocrystals and gold nanoparticles. Talanta **114**, 124–130 (2013)
124. Wu, S.J., Duan, N., Li, X.L., Tan, G.L., Ma, X.Y., Xia, Y., Wang, Z.P., Wang, H.X.: Homogenous detection of fumonisin B-1 with a molecular beacon based on fluorescence resonance energy transfer between $NaYF_4$:Yb,Ho upconversion nanoparticles and gold nanoparticles. Talanta **116**, 611–618 (2013)
125. Zeng, L.Y., Yuan, Y.X., Shen, P., Wong, K.Y., Liu, Z.H.: Graphitic carbon-nanoparticle-based single-label nanobeacons. Chem. Eur. J. **19**, 8063–8067 (2013)
126. Wang, Y.H., Gao, D.Y., Zhang, P.F., Gong, P., Chen, C., Gao, G.H., Cai, L.T.: A near infrared fluorescence resonance energy transfer based aptamer biosensor for insulin detection in human plasma. Chem. Commun. **50**, 811–813 (2014)

Chapter 9
Upconversion Nanoparticles for Light-Activated Therapy

Abstract Light-activated therapy (LAT) has attracted enormous attention over the past decades because light enables noninvasive activation or release of various therapeutic compounds, such as photosensitizers, chemotherapeutic drugs, proteins, and genes with high spatial and temporal resolution. However, most currently used photosensitive compounds in LAT are sensitive to ultraviolet (UV) or visible light which suffers from some limitations such as low tissue penetration and photo-damage to living organisms. Upconversion nanoparticles (UCNPs) that can convert near-infrared (NIR) light to visible/UV light have been proven to be effective for LAT in vivo in recent years, owing to the high tissue penetration and minimal photocytotoxicity of NIR light. Furthermore, hydrophilic UCNPs with targeting moieties can transport hydrophobic drugs in biological media and achieve active targeting, thus enhancing the therapeutic efficacy. In this chapter, we review the recent advances in the field of UCNP-based LAT with focus on photodynamic therapy (PDT) and NIR light-triggered release and uncaging.

Keywords Light-activated therapy · Upconversion nanoparticles · Near-infrared excitation · Photodynamic therapy · Photosensitizer · Light-triggered drug release · Uncaging

9.1 Introduction

The past decades have seen considerable interest and progress in light-activated therapy (LAT), including photodynamic therapy (PDT), chemotherapy, gene therapy, and photothermal therapy (PTT) as light is noninvasive and controllable both spatially and temporally compared to other stimuli such as temperature, pH, ultrasound, and applied magnetic or electric fields [1, 2]. Owing to these

---

Zhenzhen Guo and Fan Zhang contributed together to this chapter.

F. Zhang, *Photon Upconversion Nanomaterials*,
Nanostructure Science and Technology, DOI 10.1007/978-3-662-45597-5_9

advantages, this effective therapeutic approach is beneficial to control drug dosing in terms of quantity, location, and time, hence maximizing therapeutic effect while minimizing side effects [3]. However, the excitation of most commonly used photosensitizers for PDT and photoactive molecules for the release of chemotherapeutic drugs and genes relies on high-energy ultraviolet (UV) or visible light irradiation [4, 5]. The tissue penetration depth of UV–visible light is shallow because of the scattering and absorption by tissue chromophores [6]. What is worse, the heterocyclic bases of DNA are major UV-absorbing chromophores in the skin. The absorption leads to damage of the DNA, which may result in the generation of tumors [7]. These limitations would greatly hamper their applications in vivo. Compared to UV and visible light, near-infrared (NIR) light has a larger penetration depth, lower levels of the auto-fluorescence , and photodamage effect [8] and is thus better suited for biomedical applications. Therefore, another component that is capable of converting NIR excitation into UV or visible emission should be introduced into current LAT systems in order to expand their applications to living tissues.

Although the utilization of two-photon absorption (TPA) phenomena in PDT [9–14] and drug delivery [15–20] enables the use of NIR light in the tissue-transparent window (750–1,000 nm) [11], it remains limited in generality and practicality on account of the low conversion efficiency resulting from narrow absorption cross sections [14] and the use of powerful pulsed lasers focused on small areas [21].

In contrast to TPA in which a molecule absorbs two photons at the same time, the NIR-to-visible/UV upconversion is based on sequential absorption and energy transfer steps involving several metastable excited states and is thus more efficient. This increased efficiency permits the use of more economical, low-power, and continuous-wave (CW) diode lasers to generate the upconverted luminescence, avoiding two-photon photoluminescence background from biomolecules [22]. Moreover, upconversion nanoparticles (UCNPs) exhibit no blinking, good photostability, and low cytotoxicity and have a well-developed set of surface chemistry properties for conjugation with therapeutic agents and targeting agents [23]. All of these features make UCNPs good candidates for LAT in vivo.

9.2 UCNPs for Photodynamic Therapy

9.2.1 History and Mechanism of Photodynamic Therapy

The therapeutic effect of light itself has been known for thousands of years since people used light to treat various diseases, including psoriasis, rickets, and skin cancer in ancient Egypt, China, and India [24, 25]. In the early 1900s, some researchers observed that a combination of light and certain chemicals could induce cell death and two of them, Herman Von Tappeiner and A. Jodlbauer, described this phenomenon as "photodynamic action" [24]. The modern era of PDT started in the

1960s with the discovery by Lipson and Baldes [26–29] that "hematoporphyrin derivative" (HPD), which was initially prepared by Schwartz [30] through the acid treatment of hematoporphyrin (HP), had tumor-localizing properties and could be administered at much smaller doses than HP. A further study by Dougherty et al. [31] in 1975 showed that the combination of HPD and red light suppressed mammary tumor growth in mice. Human trials with HPD were performed by different research groups in the next 20 years [24]. But approval for the clinical use of PDT was not achieved until Photofrin, a more purified version of HPD prepared by chromatographic isolation, was first approved for the treatment of bladder cancer in Canada in 1993 [25]. Subsequent approvals for PDT with Photofrin were obtained in the Netherlands, France, Germany, Japan, and the USA for various types of cancers, including lung, esophageal, gastric, and cervical cancers [24]. Because of the success and approvals of Photofrin, there has been an increasing interest in the field of PDT over the past two decades. Presently, PDT is mainly used for the treatment of various oncological diseases such as head and neck, brain, lung, breast, bladder and skin cancer, and this research is focused on the optimization of parameters and the development of new photosensitizers, improved light sources, and better localization methods. Moreover, researchers are also exploiting its potential to kill bacteria and fungi for treating infections and cardiovascular, ophthalmic, and dermatological diseases [1].

PDT uses the combination of light and a photosensitizing drug called photosensitizer (PS) to damage the target tissue by generating reactive oxygen species (ROS) in the presence of an adequate concentration of molecular oxygen in the tissue [32]. The three essential components of PDT are oxygen, photosensitizers, and light. None of these elements is toxic individually in an ideal PDT system, but together they initiate the excitation of PS molecules, followed by the generation of cytotoxic ROS and finally the destruction of diseased tissues [33, 34] (Fig. 9.1).

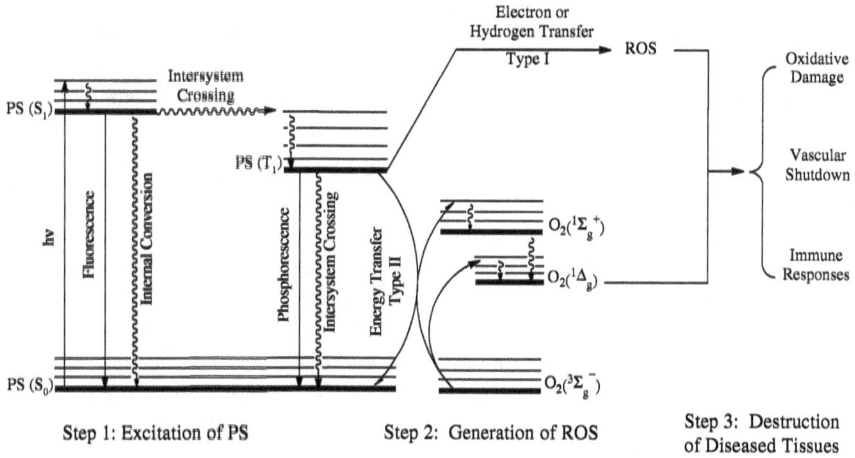

Fig. 9.1 Physical, chemical and biological processes involved in PDT

In the first step, upon light absorption, the photosensitizer is excited to a short-lived excited singlet state (S_1) from the ground state (S_0) and subsequently to a relatively long-lived excited triplet state (T_1) by intersystem crossing. Alternatively, the excited photosensitizer molecule can return to the S_0 state by a radiative process of emitting the absorbed energy as fluorescence or by a non-radiative process (internal conversion). From T_1 state, the photosensitizer can also relax back to S_0 state by emitting phosphorescence or by a non-radiative process (intersystem crossing). The resulting fluorescence and phosphorescence may be used to image photosensitizer localization in the tissue [25].

In the second step, the sufficiently long-lived T_1 state can react in two ways to generate ROS, defined as type I and type II mechanisms [35, 36]. It can undergo electron (or hydrogen) transfer reactions with a biological substrate, such as the cell membrane, molecule oxygen, or another sensitizer, to form radicals and radical ions. These reactive radical species can further interact with oxygen to produce ROS such as superoxide anion radical ($O_2^-\cdot$), hydrogen peroxide (H_2O_2), and hydroxyl radical ($OH\cdot$) [37–39] (type I, see below. $^3PS^*$ and ^1PS are photosensitizers at excited triplet state and ground state, respectively; SOD is superoxide dismutase).

$$
\begin{aligned}
&{}^3PS^* + {}^3PS^* && \longrightarrow && PS^-\cdot + PS^+\cdot \\
&{}^3PS^* + \text{Substrate} && \longrightarrow && PS^-\cdot + \text{Substrate}^+\cdot \\
&PS^-\cdot + O_2\left({}^3\textstyle\sum_g{}^-\right) && \longrightarrow && {}^1PS + O_2{}^-\cdot \\
&{}^3PS^-\cdot + O_2\left({}^3\textstyle\sum_g{}^-\right) && \longrightarrow && PS^+\cdot + O_2{}^-\cdot \\
&2O_2{}^-\cdot + 2H^+ && \xrightarrow{\text{SOD}} && H_2O_2 + O_2\left({}^3\textstyle\sum_g{}^-\right) \\
&O_2{}^-\cdot + H_2O_2 && \longrightarrow && OH\cdot + OH^- + O_2\left({}^3\textstyle\sum_g{}^-\right) \\
&O_2{}^-\cdot + Fe^{3+} && \longrightarrow && Fe^{2+} + O_2\left({}^3\textstyle\sum_g{}^-\right) \\
&Fe^{2+} + H_2O_2 && \longrightarrow && OH\cdot + OH^- + Fe^{3+}
\end{aligned}
$$

Type II mechanism involves the energy transfer from the excited triplet state of the sensitizer to the ground-state oxygen $O_2(^3\Sigma_g{}^-)$, to generate the first excited state of molecule oxygen, singlet oxygen $O_2(^1\Delta_g)$. Meanwhile, the second excited singlet state of molecule oxygen, $O_2(^1\Sigma_g{}^+)$, can also be formed in competition with $O_2(^1\Delta_g)$ and rapidly deactivated to the relatively long-lived $O_2(^1\Delta_g)$ species in solution [36, 40]. The first excited state $O_2(^1\Delta_g)$ has an energy of 94 kJ mol^{-1} above the ground state $O_2(^3\Sigma_g{}^-)$ [40], thus requiring that photosensitizers have a triplet-state energy higher than 94 kJ mol^{-1}. The singlet oxygen ($^1O_2$) has a lifetime of 0.01–0.04 μs in biological environment, corresponding to a short diffusion distance of 0.01–0.02 μm [41]. Therefore, the oxidative damage resulting from $^1O_2$ is limited to sites of photosensitizer localization. Notably, photosensitizers should act catalytically as they can return to the ground state ($^1PS \rightarrow {}^1PS^* \rightarrow {}^3PS^* + {}^3O_2 \rightarrow {}^1PS + {}^1O_2$) [32]. However, some studies show that $^1O_2$ can result in photobleaching of photosensitizers by an oxidative process [42].

Both type I and type II mechanisms occur simultaneously, and type II mechanism is generally considered to be dominant in most cases [35, 43]. However, type I mechanism becomes more important at low oxygen and high substrate concentrations [43].

The destruction of diseased tissues can be achieved through three mechanisms [44, 45]. Firstly, ROS can interact with biological substrates, such as cellular membranes, endoplasmic reticulum, and mitochondria, resulting in oxidative damage and ultimately cell death (apoptosis, autophagic cell death, and necrosis [46, 47]). Secondly, PDT can induce thrombosis and hemorrhage in tumor blood vessels, leading to vascular shutdown and subsequent tumor death via lack of oxygen and nutrients [48]. Finally, PDT-activated immune responses against diseased cells such as inflammation and release of cytokines contribute to tissue destruction [49]. These three mechanisms can influence each other, and the relative importance of each is yet to be defined [24].

9.2.2 Photosensitizers

As one of the three essential components in PDT, photosensitizers can convert oxygen in the tissue to cytotoxic ROS after absorbing light energy. An ideal photosensitizer should be a single substance with known and constant composition and a high degree of chemical purity and stability [43]. It should have no dark toxicity and relatively rapid clearance from healthy tissues, thereby minimizing side effects [34]. A high photochemical reactivity, with high triplet-state yield and long triplet-state lifetime, is also needed to generate ROS efficiently [35].

Photosensitizers currently in clinical trials or late preclinical development are summarized in Table 9.1. The first-generation photosensitizers developed in the 1970s and early 1980s are HPDs based on chemically modified natural HP such as Photofrin [43, 50]. Despite its successes in the treatment of early cancers of the stomach, lung, and head and neck, HPDs suffer from several limitations, including the weak absorbance at the activation wavelength of 630 nm and a long retention time up to 10 weeks in cutaneous tissue [35]. The latter would cause a prolonged skin photosensitivity. The second-generation photosensitizers developed since the late 1980s are mainly based on synthetic and semisynthetic porphyrins and porphyrin derivatives, including chlorins, texaphyrins, and phthalocyanines [43, 50]. They have shorter periods of photosensitivity, longer activation wavelengths, and higher yields of $^1O_2$ [51], and some of them such as Foscan and Visudyne have been already approved for PDT in the USA or Europe [34, 52]. Although most currently used photosensitizers are porphyrin-based compounds, several non-porphyrin dyes possess triplet states of appropriate energies for sensitization of oxygen and can also be used as photosensitizers [53]. These non-porphyrin-based photosensitizers include anthraquinones (hypericin), phenothiazines [methylene blue (MB)], xanthenes (rose bengal, 4,5-dibromorhodamine methyl ester), cyanines (merocyanine 540), and curcuminoids (curcumin) [53, 54]. However, none of them

Table 9.1 Clinically applied photosensitizers [24, 34, 35]

PS	Skeleton structure	Trade name	Activation wavelength (nm)	Cancer types
Porfimer sodium (HPD)	Porphyrin	Photofrin	630	Lung, esophagus, bile duct, bladder, brain, ovarian
BPD-MA (verteporfin)	Chlorin	Visudyne	690	Ophthalmic, pancreatic, skin
Temoporfin (mTHPC)	Chlorin	Foscan	652	Head and neck, lung, brain, skin, bile duct
5-ALA	Porphyrin precursor	Levulan	635	Skin, bladder, brain, esophagus
5-ALA methyl ester	Porphyrin precursor	Metvix	635	Basal cell skin
5-ALA benzyl ester	Porphyrin precursor	Benzvix	635	Gastrointestinal
5-ALA hexyl ester	Porphyrin precursor	Hexvix	375–400	Bladder
SnEt2	Chlorin	Purlytin	660	Skin, breast
HPPH	Chlorin	Photochlor	665	Head and neck, esophagus, lung
Lutetium texaphyrin (Lu-Tex)	Texaphyrin	Antrin (Lutrin, Optrin)	732	Breast
Silicon phthalocyanine	Phthalocyanine	Pc 4	675	Cutaneous T-cell lymphoma
Talaporfin sodium (MACE)	Chlorin	Laserphyrin	664	Liver, colon, brain
Ce6-PVP	Chlorin	Photolon (Fotolon)	660	Nasopharyngeal, sarcoma, brain
Padoporfin	Bacteriochlorin	Tookad	762	Prostate

HPD hematoporphyrin derivative; *BPD-MA* benzoporphyrin derivative monoacid ring A; *mTHPC* *meta*-tetrahydroxyphenylchlorin; *5-ALA* 5-aminolevulinic acid; *SnEt2* tin ethyl etiopurpurin; *HPPH* 2-(1-hexyloxyethyl)-2-devinyl pyropheophorbide-a; *MACE* mono-L-aspartyl chlorin-e6; *Ce6-PVP* chlorin e6-polyvinylpyrrolidone

have received Food and Drug Administration (FDA) approval for use in PDT [53]. The third-generation photosensitizers, which consist of photosensitizer moiety linked to biomolecules such as antibodies, polypeptide chains, and liposomes [43, 54] for selective accumulation within tumor tissue, are developed on the basis of the former two generations and still under investigation. Notably, dividing photosensitizers into three generations does not mean that new photosensitizers are always better than the old ones as most reports on newer photosensitizers are presented in abstract form on very limited numbers of patients with limited follow-up [50].

Furthermore, some organometallic compounds, such as ruthenium(II) and platinum(II) complexes with bipyridine (bpy), undergo efficient $^1O_2$ generation [55]. The semiconducting TiO_2, which can generate holes in the valence band and electrons in the conduction band followed by reactive radical species (OH·) upon light activation, can also induce oxidative damage to proteins and amino acids [56]. That is to say, they can also be used as photosensitizers. But these complexes have received little attention compared to porphyrin-based photosensitizers and non-porphyrin dyes.

9.2.3 UCNP-Based Photodynamic Therapy

As mentioned above, the photosensitizers used in traditional PDT are mostly activated by visible light which has a shallow penetration depth and are thus ineffective for deep-seated or large tumors. Stable NIR-absorbing photosensitizers are still rare [57] and notoriously difficult to synthesize [58]. Moreover, new photosensitizers would take a long time and many clinical trials to be approved. The utilization of TPA is also restricted on account of its low conversion efficiency and the use of pulsed lasers. By contrast, UCNPs would be a better choice as they can convert NIR excitation to visible emission efficiently using a CW diode laser. Additionally, UCNPs are well suited for use in PDT since (i) hydrophilic UCNPs prepared by one-pot synthetic method [59–61] or modification of hydrophobic nanocrystals [62] can transport photosensitizers which are mainly hydrophobic or have limited water solubility [63] and avoid the aggregation of photosensitizers in biological media; (ii) the passive targeting of tumors can be achieved by using nanoparticles instead of conventional small-molecule drugs because of the enhanced permeability and retention (EPR) effect [64], and the targeting ability can be further enhanced by means of active targeting due to tumor-specific molecules attached to UCNPs; and (iii) clinically approved photosensitizers can continue to be used in UCNP-based PDT systems.

In order to realize their full potential in NIR-induced PDT, high-quality UCNPs should be fabricated with high phase purity, uniform morphology, and good dispersibility. It is noteworthy that hexagonal-phase $NaYF_4$ is the most popular host material in UCNP-based PDT because $NaYF_4$ has been reported as the most efficient NIR-to-visible upconversion host so far and the upconversion efficiency of hexagonal-phase nanocrystals is much higher than that of cubic-phase nanocrystals [65, 66]. Optimal amounts of photosensitizers and/or targeting moieties should be attached to hydrophilic UCNPs by either non-covalent interactions or covalent conjugation. A large spectral overlap between the absorption of photosensitizers and the luminescence emission of UNCPs is also required to achieve efficient energy transfer. Considering that synthesis, modification, bio-conjugation, and emission tuning of UCNPs have been discussed in detail in other chapters of this book, these subjects will not be introduced separately in this chapter.

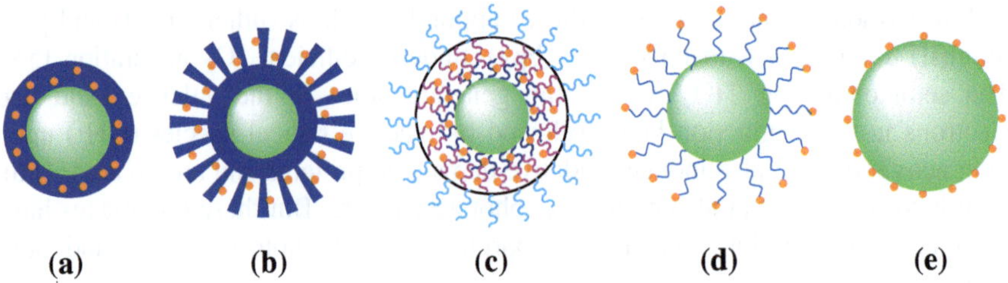

Fig. 9.2 Approaches for loading photosensitizers onto UCNPs

In general, current UCNP-based PDT designs can be divided into five types by PS-loading approaches: (i) photosensitizers are encapsulated into silica shell coated onto the surface of UCNPs (Fig. 9.2a); (ii) photosensitizers are deposited in the pores of mesoporous silica shell via physical adsorption or covalent conjugation (Fig. 9.2b); (iii) photosensitizers are loaded into "hydrophobic pocket" on the UCNP surface, which is formed by non-covalent interactions between the hydrophobic segments of polymers and UCNP surface ligands [e.g., oleic acid (OA)] (Fig. 9.2c); (iv) photosensitizers are covalently bonded to UCNPs capped with amino-/carboxyl-functionalized capping agents (Fig. 9.2d); (v) photosensitizers are directly attached to the positively charged surface of UCNPs (Fig. 9.2e) or loaded onto porous UCNPs (not shown in Fig. 9.2) via non-covalent interactions without an interlayer between UCNPs and the loaded photosensitizers. In the following paragraphs, we will review UCNP-based PDT systems (Table 9.2) to date according to this classification.

9.2.3.1 Silica Encapsulation

In 2007, Zhang et al. [67] first reported using UCNPs for PDT under NIR excitation. In this work, $NaYF_4$:Yb/Er UCNPs with emission at around 537 and 635 nm were first coated with a thin layer of silica via Stöber method, and merocyanine 540 (MC540, absorption peak at ~560 nm) was used as photosensitizer and doped into the silica layer with a loading capacity (defined as the weight ratio of encapsulated photosensitizers to PS-UCNP composites) of ~0.45 wt% during the coating process. Finally, a mouse monoclonal antibody, anti-MUC1/episialin, which is highly specific toward the MCF-7/AZ cells, was covalently attached to the UCNP@SiO_2 nanoparticles. When illuminating the mixture of as-prepared nanoparticles and a small amount of ADPA by a 974-nm laser (0.06 W), the fluorescence intensity of 9,10-anthracenedipropionic acid (ADPA) which could be bleached by $^1O_2$ decreased and followed an exponential decay over time, indicating that $^1O_2$ could be generated by these MC540-UCNP composites. However, the $^1O_2$ generation efficiency was low as the intensity decrease was only 3 % after 16-min irradiation. Despite the low efficiency, in vitro PDT test showed that antibody-functionalized MC540-UCNPs induced cell death after incubation with MCF-7/AZ breast cancer

Table 9.2 A summary of reported UCNP-based PDT systems up to now [1]

UCNPs	PS (Abs. max.)[a]	Stabilizers	Targeting moiety	Loading strategy	Loading capacity	References
$NaYF_4$:Yb/Er	MC540 (~560 nm)	Silica	Episialin antibody	(i)	0.45 wt%[b]	[67]
$NaYF_4$:Yb/Tm	$Ru(bpy)_3^{2+}$ (~450 nm)	Silica	None	(i)	N.M.[c]	[68]
$NaYF_4$:Yb/Er/Gd	MB (654 nm)	Silica	None	(i)	0.12 mol%[b]	[69]
$NaGdF_4$:Yb/Er@$NaGdF_4$	AlC_4Pc (685 nm)	Silica	None	(i)	~3.3 wt%	[70]
$NaYF_4$:Yb/Tm	DHMA (~440 nm)	Silica	FA	(i)	N.M.	[71]
$NaYF_4$:Yb/Er	Hypericin (550 nm)	Silica	FA	(i)	12.1 %[b]	[72]
$NaYF_4$:Yb/Er	ZnPc (~670 nm)	$mSiO_2$	None	(ii)	0.1 wt%	[73]
$NaYF_4$:Yb/Er	ZnPc (~670 nm)	$mSiO_2$	None	(ii)	0.1 wt%	[74]
$NaYF_4$:Yb/Er	MC540 (~560 nm) and ZnPc (~670 nm)	$mSiO_2$ and PEG	FA	(ii)	0.25 wt% (M-540) and 0.34 wt% (ZnPc)	[75]
$NaGdF_4$:Yb,Er@CaF_2	SPCD (672 nm)	$mSiO_2$	None	(ii)	N.M.	[76]
$NaYF_4$:Yb/Er	ZnPc (~670 nm)	PEI	FA	(iii)	0.0064 wt%	[77]
$NaYF_4$:Yb/Er	ZnPc (~670 nm)	PEI	DENV2 antibody	(iii)	N.M.	[78]
$NaYF_4$:Yb/Er	TPP (550 nm)	PEG-*b*-PCL	None	(iii)	33.3 wt%	[79]
$NaYF_4$:Yb/Er	TPP (550 nm)	PEG-*b*-PLA	None	(iii)	10 wt%	[80]
$NaYF_4$:Yb/Er	Ce6 (660 nm)	PMAO-PEG	None	(iii)	5–6 wt%	[81]
$NaYF_4$:Yb/Er/Mn	Ce6 (660 nm)	PAA, PAH, and PEG	None	(iii)	11.0 wt%[d]	[82]
$NaYF_4$:Yb/Tm@$NaYF_4$:Yb/Er	Ce6 (660 nm)	PEG–phospholipid	None	(iii)	23 wt%	[83]
$NaYF_4$:Yb/Er	ZnPc (~670 nm)	PAA	None	(iii)	12.3 wt%	[84]
$NaYF_4$:Yb/Er	ZnPc (~670 nm)	SOC	None	(iii)	10.8 wt%	[85]

(continued)

Table 9.2 (continued)

UCNPs	PS (Abs. max.)[a]	Stabilizers	Targeting moiety	Loading strategy	Loading capacity	References
$NaYF_4$:Yb/Er	ZnPc (~670 nm)	SOC	FA	(iii)	10 wt%[d]	[86]
$NaYF_4$:Yb/Er	ZnPc (~670 nm)	OQPGA	RGD/TAT peptides	(iii)	1.6 wt%	[87]
$NaYF_4$:Yb/Er	MC540 (~560 nm)	OQLCS	FA/TAT peptide	(iii)	9.8 wt%[d]	[88]
$NaGdF_4$:Yb/Er	ZnPc (~670 nm)	PMAO-PEG	None	(iii)	N.M.	[89]
$NaYF_4$:Yb/Er/Mn	Ce6 (660 nm), ZnPc (~670 nm), or MB (660 nm)	α-CD	None	(iii)	9.4 wt% (Ce6), 9.5 wt% (ZnPc), or 4.1 wt% (MB)	[90]
$NaGdF_4$:Yb/Er	RB (549 nm)	PAA and BSA	None	(iii)	~7.6 wt%	[91]
$NaGdF_4$:Yb/Er	Ce6 (660 nm)	PAA, PEG, and PEI	None	(iii)	11 wt%	[92]
Fe_3O_4@$NaYF_4$:Yb/Er	$AlPcS_4$ (676 nm)	PEG	None	(iii)	N.M.	[93]
$NaYbF_4$:Gd/Tm@$NaGdF_4$	HA (467 nm)	Tween 20	None	(iii)	5.5 wt%	[94]
$NaLuF_4$:Yb/Er/Gd	TMPyP4 (422 nm)	PEI	Sgc8 aptamer	(iii)	N.M.	[95]
$NaYF_4$:Yb/Er	IBDP (525 nm)	PEG	None	(iii)	28.1 wt%[b]	[96]
$NaYF_4$:Yb/Er	PP-a (668 nm)	PEI and OCMC	RGD peptide c(RGDyK)	(iii)	0.53 wt%	[97]
$NaYF_4$:Yb/Er	RB (549 nm)	PEG	FA	(iv)	100 RB per particle	[98]

(continued)

Table 9.2 (continued)

UCNPs	PS (Abs. max.)[a]	Stabilizers	Targeting moiety	Loading strategy	Loading capacity	References
$NaYF_4$:Yb/Er@$NaYF_4$:Yb/Tm	C_{60}MA[e]	PAAm and PEG	FA	(iv)	10.5 wt%	[99]
$NaYF_4$:Yb/Er	ZnPc-$(COOH)_4$ (680 nm)	PAAm and PEG	FA	(iv)	6 wt%	[100]
NaYF4:Yb/Er@NaGdF4	Ce6 (660 nm)	PEG–phospholipid	None	(iv)	$>10^3$ Ce6 per particle	[101]
$NaYF_4$:Yb/Er/Ce	Bodipy dyes[f]	PEI	None	(iv)	N.M.	[102]
$NaYF_4$:Yb/Tm	N-TiO_2[e]	PEI	cAngptl4 antibody	(iv)	N.M.	[103]
$NaYF_4$:Yb/Gd/Tm	HAC_{60}[e]	APBA	APBA and HA	(iv)	3.8 wt%	[104]
$LiYF_4$:Yb/Er	ZnPc-(COOH) (~676 nm)	PVP	None	(v)	1.1 wt%	[105]
ZnO:Yb/Gd/Er	MB (660 nm)	DEG	None	(v)	3.3 wt%	[106]
$NaYF_4$:Yb/Er	MB (660 nm)	GMP	None	(v)	0.014 wt%	[107]
NaYF4:Yb/Er/Sc	PpIX[g] (~400 nm)	None	None	None[h]	None	[108]

[a] Some photosensitizers have more than one characteristic absorption peak. The peak that overlaps the emission spectrum of UCNPs is listed in this table
[b] Calculated based on the experimental data as the author did not give the value of loading capacity
[c] N.M. = not mentioned
[d] Highest loading capacity with a decrease of $^1O_2$ generation efficiency or an increase of particle size. The loading amount employed in PDT test was lower than the value listed in this table
[e] Broad absorption spectrum
[f] Three types of carboxyl-functionalized Bodipy dyes with major absorption peaks at 638, 508, and 509 nm, respectively
[g] PpIX (protoporphyrin IX) was generated enzymatically in MKN45 cells from 5-aminolevulinic acid which was actually added
[h] Physically mixing

cells and exposure to 974-nm laser irradiation, while dyes without IR irradiation and IR irradiation alone both cannot cause cell death. Later, a similar work, [68] in which tris(bipyridine)ruthenium(II) ($Ru(bpy)_3^{2+}$, absorbance maximum at ~450 nm) and $NaYF_4$:Yb/Tm nanoparticle with blue emission were used as photosensitizer and NIR-to-blue upconverter, respectively, was reported by the same group. However, the author did not carry out in vivo PDT studies in either case.

Shi et al. [69] encapsulated MB into the biocompatible silica matrix of $NaYF_4$:Yb/Er/Gd@SiO_2 nanoparticles via a reverse microemulsion method. The extremely slight mismatch between MB monomer absorption (654 nm) and $NaYF_4$:Yb/Er/Gd red emission peak (651 nm) was beneficial for the full utilization of UCNP emission. Parameters such as suitable MB concentration (~0.12 mol%), reaction time (less than 24 h), silica shell thickness (<6.0 nm), and the addition sequence of precursors (adding MB before ammonia) were sufficiently controlled to inhibit the dimerization of MB molecules. No apparent MB leakage in water, ethanol, or nonpolar solvents such as cyclohexane was observed because of the strong electrostatic attraction between positively charged MB and the silica network and the small micropore channels in the silica matrix. Control experiments demonstrated that $^1O_2$ could only be generated when appropriately combining MB, UCNPs, and a 980-nm NIR laser. These Gd^{3+}-doped nanoparticles were also capable of magnetic resonance imaging (MRI). But NIR-induced PDT based on these MB-UCNP composites has been demonstrated neither in vitro nor in vivo. Moreover, the dimerization of MB molecules limited the loading capacity of the MB-UCNP composites.

In order to increase the loading amount of photosensitizers, Zheng et al. [70] incorporated tetrasubstituted carboxyl aluminum phthalocyanine (AlC_4Pc, absorption peak at 685 nm) into the silica shell via covalent bonding. AlC_4Pc was first treated with (3-aminopropyl) triethoxysilane (APTES) to form an APTES-AlC_4Pc conjugate and then cohydrolyzed with tetraethyl orthosilicate (TEOS) to form a silica shell on $NaGdF_4$:Yb/Er@$NaGdF_4$ UCNPs. Because of the improved loading capacity of ~3.3 wt%, nearly 40 % BNL 1 ME A. 7R.1 cells were killed after incubation with UCNP@SiO_2-AlC_4Pc and exposure to 980-nm irradiation (0.5 W cm^{-2}, 5 min). These nanoparticles also demonstrated good magnetic resonance (MR) contrast because of the presence of Gd^{3+}. However, the cubic-phase UCNP adopted in this work is known to be less efficient than the hexagonal-phase nanocrystal.

Ouyang et al. utilized (3-isocyanatopropyl) triethoxysilane (TESPIC) instead of APTES to encapsulate 1,8-dihydroxy-3-methylanthraquinone (DHMA, absorption peak at ~440 nm) [71] and hypericin (absorption peak at ~550 nm) [72] into the silica shells of $NaYF_4$:Yb/Tm and $NaYF_4$:Yb/Er UCNPs, respectively. Compared to the UCNP@SiO_2(PS) composites where photosensitizers were simply deposited into the silica shell in the absence of TESPIC, the UCNP@SiO_2@PS composites formed via covalent conjugation exhibited higher $^1O_2$ generation efficiency. Moreover, folic acid (FA) was covalently attached to UCNP@SiO_2@PS as a targeting agent. As a result, stronger upconversion luminescence (UCL) could be observed in the HeLa cells after incubation with UCNP@SiO_2@PS-FA than in both the 293T cells which express less folate receptors than the HeLa cells and the HeLa cells incubated with non-FA-modified UCNP@SiO_2@PS. Upon 980-nm irradiation, UCNP@SiO_2@PS-FA could induce the apoptosis of HeLa cells, as confirmed by both fluorescence imaging and flow cytometry analysis. Furthermore, it was found that the expression of caspase-3 decreased with increasing exposure time, while the expression of cleaved caspase-3 increased. As a family of cysteine proteases, the caspases exist in almost all cell types as inactive proenzymes. When the caspase-3 protein as an effector caspase of the family was cleaved, it was activated, indicating that the apoptotic process was triggered. Therefore, it was concluded that the PDT effect of UCNP@SiO_2@DHMA (or hypericin)-FA was associated with caspase-3 activation.

9.2.3.2 Mesoporous Silica-Based Loading

Although some studies [109–111] have demonstrated that a dense silica network has tiny pores for the diffusion of small molecules such as water and oxygen, mesoporous silica ($mSiO_2$) would be a better choice for the loading of photosensitizers and the release of ROS on account of its large surface area and mesoporous structure. In view of this, Zhang et al. [73] fabricated mesoporous silica-coated $NaYF_4$:Yb/Er UCNPs with a water-insoluble photosensitizer zinc(II) phthalocyanine (ZnPc, absorbance maximum at 670 nm) in the mesoporous silica shell in 2009. These nanoparticles had a Brunauer–Emmet–Teller (BET) surface area of 770 m^2 g^{-1}, and the amount of ZnPc adsorbed into the porous silica was about 0.1 wt%. ZnPc could be retained in porous silica in phosphate-buffered saline (PBS), deionized water, and cell culture medium, whereas it could be released in ethanol within 1 h. The viability of MB49-PSA cells incubated with the nanoparticles loaded with ZnPc and exposed to NIR laser for 5 min was significantly lower as compared to the nanoparticles without ZnPc. They further studied singlet oxygen-induced apoptosis of cancer cells using these UCNP@SiO_2@$mSiO_2$ nanoparticles [74]. In this study, MB49-PSA cells exposed to ZnPc-loaded nanoparticles and then irradiated by a 980-nm laser showed condensation of chromatin in the nucleus stained by 4′-6-diamidino-2-phenylindole (DAPI, blue fluorescence), as apparent from its irregularly shaped nuclear membrane. Distinctly segregated green fluorescence from cytochrome *c* and red fluorescence from the mitochondria were also clearly observed, suggesting the release of cytochrome *c* from mitochondria. Furthermore, DNA extracted from MB49-PSA cells after incubation with ZnPc-loaded nanoparticles and exposure to a 980-nm laser displayed more distinct DNA ladders as compared with non-ZnPc-loaded ones or those that merely received laser irradiation alone. Both changes in cell nuclear morphology and loss of DNA integrity are indicators of apoptosis. The released cytochrome *c* can activate caspase-3 to induce apoptosis. In earlier years, conventional PDT-induced cell death was thought to be attributed to necrosis, but it was later found to be caused mainly by apoptosis. The above results confirmed the involvement of apoptosis in UCNP-based PDT-induced cell death.

Two years later, Zhang's group [75] incorporated two photosensitizers, MC540 and ZnPc, into mesoporous silica-coated $NaYF_4$:Yb/Er UCNPs simultaneously (Fig. 9.3a). The loading amount was 0.25 and 0.34 wt% for MC540 and ZnPc, respectively. The two main emissions of $NaYF_4$:Yb/Er UCNPs around 540 and 660 nm matched well with the absorption of MC540 and ZnPc, respectively, thus enhancing the energy transfer efficiency. The PDT effect on tumor growth was first studied by preincubating melanoma cells with UCNPs before injecting them under the skin of the C57BL/6 mice. The growth of the tumor cells was significantly inhibited with injection of coloaded-UCNP-labeled cells and 980-nm irradiation at a low power density of 0.415 W cm^{-2}, and the population of apoptotic cells in histological sections increased compared to control groups. It is noteworthy that treatment with a laser alone also elicited some degree of apoptosis, which can be attributed to the mild thermal effect of the 980-nm laser. After the proof-of-principle study, they further confirmed the PDT efficacy in C57BL/6 mice bearing melanoma tumors. A significantly slower tumor growth rate could also be observed at 6 d and 8 d after treatment with coloaded UCNPs (intratumoral injection) and 980-nm irradiation. A bio-distribution study at 11 d after treatment on the mice that received only intratumoral injection of coloaded UCNPs revealed the presence of yttrium only in the tumor tissues, indicating that UCNPs accumulated in the tumor after injection and that minimal diffusion from the tumor region to other organs in the body had occurred. In addition, the therapeutic efficacy could be enhanced by modifying the surfaces of these nanoparticles with poly(ethylene glycol) (PEG) and FA (Fig. 9.3b, intravenous injection). However, the exposure time (2 h) to laser light in their PDT study was quite long because of the limited loading efficiency of photosensitizers, which might induce photodamage to living organisms.

In addition to physical adsorption, covalent bonding is also employed to load photosensitizers into mesoporous silica shells. Yan et al. [76] utilized silicon phthalocyanine dihydroxide (SPCD, absorption peak at 672 nm) which contains two silanol hydroxyls to fabricate UCNP@SiO_2@mSiO_2-PS composites via co-hydrolysis with TEOS. After loading, the BET surface area of the nanomaterial decreased from 270 to 104 m^2 g^{-1}. It was found that SPCD molecules covalently grafted onto the silica shell were not easily leached by ethanol, exhibiting higher stability than those attached to silica shell by physical adsorption [73]. The fluorescent intensity of 9,10-anthracenediylbis(methylene)dimalonic acid (ABMD), which could be quenched by $^1O_2$, dropped to 57 % after 20 min of 980-nm irradiation in the presence of SPCD-UCNP composites. In vitro PDT tests on HeLa cells showed that a critical value of irradiation intensity was required for an obvious PDT effect. The viability dropped to 84 and 80 % after incubation with 100-μM SPCD-UCNPs and 5-min irradiation of 1.0 and 1.7 W cm^{-2}, respectively. When the power density increased to 2.5 W cm^{-2}, the viability decreased obviously to 30 %. This was presumably due to the host material of cubic-phase $NaGdF_4$ in spite of an inert CaF_2 shell.

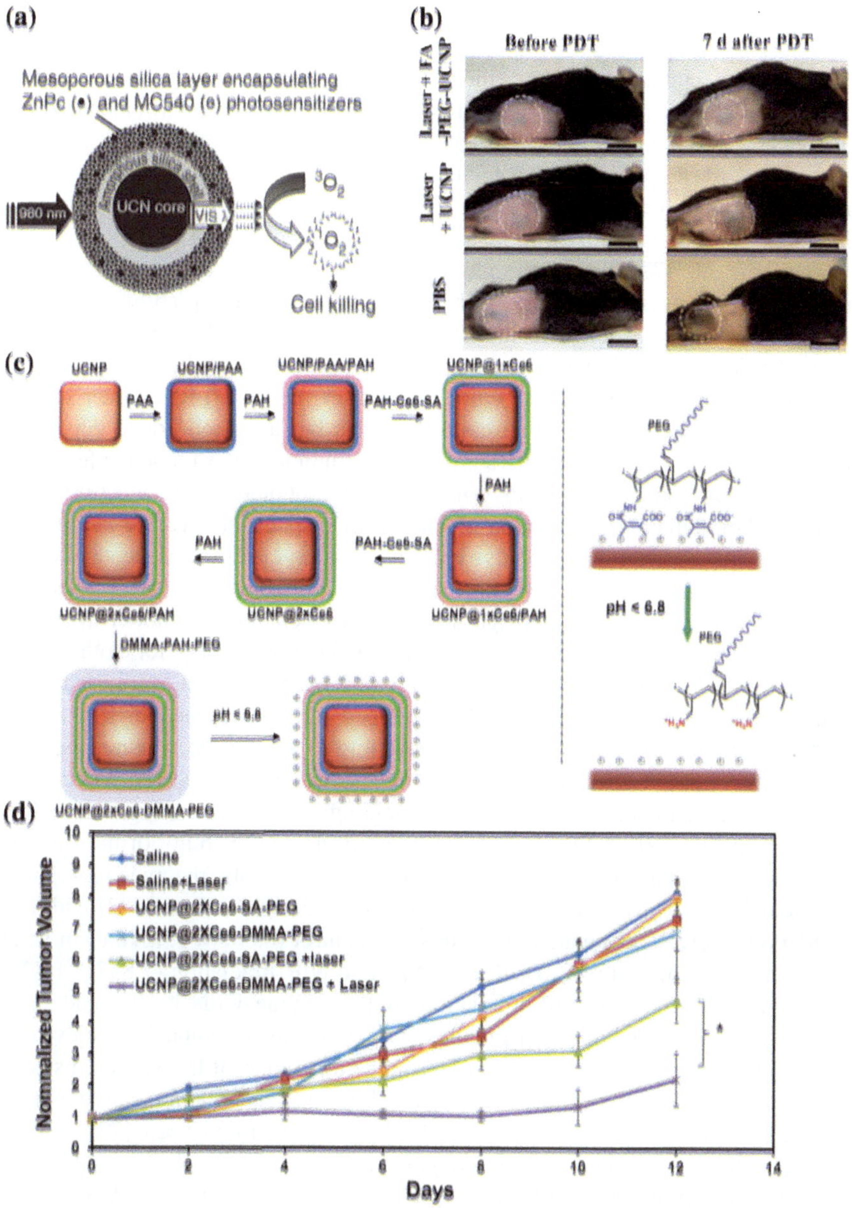

Fig. 9.3 **a** Schematic illustration of mesoporous silica-coated UCNPs coloaded with ZnPc and MC540 for PDT. **b** Representative photographs showing the change in tumor size (highlighted by *dashed white circles*) of different groups of mice before and 7 d after various treatments, as indicated ("FA-PEG-UCNP" is ZnPc/MC540-UCNP conjugated with FA and PEG). *Scale bars* 10 mm. **c** Schematic illustration of the fabrication process of pH-sensitive charge-reversible UCNPs with multi-layers of Ce6 loading (*left*) and the detachment of PAH-DMMA-PEG from the nanoparticle surface under pH 6.8 (*right*). **d** Tumor growth curves of different groups of mice after various treatments, as indicated. $^*P < 0.05$. **a, b** Reprinted with permission from Ref. [75]. Copyright 2012 Nature America, Inc; **c, d** Reprinted with permission from Ref. [82]. Copyright 2013 Wiley-VCH

9.2.3.3 Polymer-Based Loading

Although the introduction of a silica or mesoporous silica shell could protect photosensitizers from being degraded by the harsh outside environment [75], it hampers the encapsulated photosensitizers from interacting with relatively abundant oxygen dissolved in solution to some extent. More researchers utilized hydrophilic or amphiphilic polymers to convert hydrophobic UCNPs into hydrophilic ones and then attached photosensitizers to these polymer-modified UCNPs.

Zhang et al. [77] employed a one-pot hydrothermal method to synthesize hydrophilic, poly(ethylene imine) (PEI, 25 kD)-modified $NaYF_4$:Yb/Er UCNPs and then attached FA and ZnPc to these nanoparticles via covalent conjugation and hydrophobic interaction, respectively. Different amounts of ZnPc/FA-PEI-UCNPs were incubated with HT29 cells, which carry folate receptor (FR) on their surface. Cell viability was reduced for all concentrations, but appeared not to depend on nanoparticle concentration. Zhang's group [78] continued to examine the feasibility of using these ZnPc-PEI-UCNPs without FA for photodynamic inactivation of viruses. Photodynamic inactivation of viruses in suspension was first carried out, and the results showed that photodynamic inactivation effect was dependent on the concentration of nanoparticles, light fluence, and irradiation distance. Higher concentration (44 μg mL^{-1}), higher light fluence (14 kJ cm^{-2}), and shorter irradiation distance (5 cm) gave better photodynamic inactivation efficacy for both dengue virus serotype 2 (DENV2) and adenovirus type 5 (Ad5V), which were used as the virus models for enveloped and non-enveloped viruses, respectively. A murine model-based study on DENV2 was done to further investigate whether the inactivation effect could affect the pathogenicity of viruses in vivo. The inoculations were done intracranially into the Balb/c suckling mice. When a high ZnPc-PEI-UCNP concentration of 440 mg mL^{-1} was used, the survivability of the suckling mice was 100 % throughout the 15-day observation. Notably, the mice took longer to succumb to the virus pathogenesis (mean day of death 11.17 ± 1.722) with NIR irradiation solely, as compared to the positive control (6.667 ± 1.211), even better than that with 4 mg mL^{-1} ZnPc-PEI-UCNPs and NIR irradiation (7.833 ± 1.722). The results showed that NIR irradiation of the dengue virus itself could cause reduction in its pathogenicity. Moreover, target specificity could be achieved by bio-conjugation of DENV2 antibody. This is the first report of UCNP-based system for photodynamic inactivation of viruses. Unfortunately, the number of ZnPcs attached via this method was too low (0.0064 wt% in the former case [77]), and the photosensitizers were not stable on the nanoparticle surface as ZnPc was found to be detached from the nanoparticles [73]. Furthermore, the cationic PEI polymer is known to be cytotoxic [112, 113].

Compared to PEI, PEG is biocompatible and can prevent recognition and uptake of nanoparticles by reticuloendothelial system (RES) and mononuclear phagocyte system (MPS) [114–116], thus improving the blood circulation half-life of nanoparticles in vivo. Hence, it is not surprising that PEG has been the most commonly used polymer to modify the surface of UCNPs for NIR-induced PDT.

Austin et al. [79] first reported PEG-modified UCNPs for the generation of $^1O_2$ in 2009. They used PEG-*b*-PCL (polycaprolactone) block copolymer to encapsulate trioctylphosphine/OA-capped $NaYF_4$:Yb/Er UCNPs and tetraphenylporphyrin (TPP) simultaneously. Although a high loading amount of 33.3 wt% could be achieved, the products were actually composed of two kinds of particles: a larger composite with a hydrodynamic diameter of 280 nm which was presumably the coencapsulated product and smaller particles of roughly 60 nm which were presumably micelles formed by polymer surrounding a core of TPP. These composites were stable in aqueous media but unstable in buffer or serum. Later, they utilized the amorphous polylactide (PLA) block instead of the crystallizable PCL block to provide greater stability in serum [80]. Although the loading efficiency decreased to 10 wt%, these PEG-*b*-PLA-protected TPP-UCNPs remained stable in deionized water, PBS buffer, and the culture medium containing proteins for at least 3 months. Incubation of HeLa cells with TPP-UCNPs for 48 h and subsequent exposure to 978-nm laser (134 W cm^{-2}, 45 min) led to 75 % cell death, while the three control experiments (TPP-UCNPs alone, 134 W cm^{-2} NIR laser alone, without TPP-UCNPs and NIR irradiation) showed a maximum of 14 % cell death. Furthermore, the rate of cell death was dependent on NIR intensity. When the NIR intensity decreased to 39 W cm^{-2}, the rate of cell death with TPP-UCNPs was even lower than the maximum death rate of aforementioned three control groups. This phenomenon was consistent with the quadratic dependence of UCNP emission on NIR excitation intensity [117]. Another reason for high excitation power density was that the absorbance of TPP around 550 nm, which could be excited by 560 nm upconverted light, is weaker than that around 515 nm. In other words, the combination of NaYF4:Yb/Er UCNPs and TPP is not good enough to achieve high energy transfer efficiency.

Chlorin e6 (Ce6) is a widely used photosensitizer for experimental PDT with a relatively intense absorption peak of 660 nm which matches very well with the red emission of $NaYF_4$:Yb/Er UCNPs, thus potentially offering effective energy transfer. In 2011, Liu et al. [81] loaded Ce6 with a saturated loading capacity of 6–8 wt% onto $NaYF_4$:Yb/Er UCNPs functionalized with PEG-grafted poly(maleic anhydride-*alt*-1-octadecene) (PMAO-PEG) amphiphilic polymer (Fig. 9.3c) and demonstrated NIR-induced PDT by PS-UCNP composites in vivo for the first time. 4T1 murine breast tumor-bearing Balb/c mice were intratumorally injected with 40–50 ml of Ce6-UCNPs (20 mg mL^{-1} UCNPs, $\sim$1.5 mg mL^{-1} Ce6) or saline and then exposed to the 980-nm laser for 30 min (0.5 W cm^{-2}, 1-min interval after 1-min irradiation). While neither Ce6-UCNPs injection by itself nor NIR irradiation alone was able to delay the tumor growth, tumors on 7 out of 10 mice completely disappeared in 2 weeks after treatment with both Ce6-UCNPs and NIR light (Fig. 9.3d), leading to a prolonged survival time of over 60 days compared to the control mice, which had average life spans of only $\sim$23 days post-treatment. Meanwhile, the partially damaged tumors on the other 3 mice grew back and led to the animal death on day 32 (2 mice) and 42. Quantitative bio-distribution measurement of UCNPs by inductively coupled plasma atomic emission spectroscopy (ICP-AES) showed that Ce6-UCNPs were accumulated in RES organs including liver and spleen 15 days post-injection and was nearly completely excreted after 60 days. They further blocked tumors growing on mice by an 8-mm-thick pork tissue during PDT treatment, to mimic internal tumors located inside the body. The mice treated with both Ce6-UCNPs and 980 nm light showed significantly reduced tumor growth rate, while no appreciable delay in tumor progression was observed for the control group of Ce6-UCNPs and 660 nm light, owing to the difference of penetration depth between visible light and NIR light. Notably, in vivo UCL imaging showed that the red emission of UCNPs was largely quenched by the loaded Ce6 on days 0 and 1 after treatment and showed up 2 days later, indicating the release of Ce6 from UCNPs.

Considering that the tumor extracellular environment is more acidic (pH 6.5–6.8) than blood and normal tissues (pH *ca.* 7.4), Liu et al. [82] further used charge-reversible UCNPs loaded with Ce6 to realize pH-responsive PDT. In this study, Mn^{2+}-doped $NaYF_4$:Yb/Er UCNPs with an improved intensity of red emission were utilized to achieve more efficient energy transfer and MRI. Ce6 loading was realized via the layer-by-layer (LbL) process. In brief, negatively charged poly(acrylic acid) (PAA)-modified UCNPs were first mixed with positively charged poly(allylamine hydrochloride) (PAH) to form positively charged UCNP/PAA/PAH. Then, the counter-polyelectrolyte PAH-Ce6-SA, which was synthesized by the reaction between the amino groups of PAH and Ce6 as well as succinic anhydride (SA), was added to form a negatively charged UCNP/PAA/PAH/PAH-Ce6-SA (UCNP@1xCe6). Repeating the above process of coating, UCNP@2xCe6, and UCNP@3xCe6, with two and three layers of PAH/PAH-Ce6-SA, could be obtained. It is worth noting that this strategy combined the covalent conjugation of Ce6 with PAA and the physical adsorption of polymer on UCNPs and that hardly any Ce6 was found to be released from UCNPs within 2 days under physiological conditions. The loading efficiency of Ce6 onto UCNPs was determined to be 3.4, 7.7, and 11.0 wt%, for UCNP@1xCe6, UCNP@2xCe6, and UCNP@3xCe6, respectively. However, UCNP@2xCe6 showed the highest efficiency in terms of $^1O_2$ production under 980-nm laser irradiation at the same UCNP concentration, likely due to the aggregation and self-quenching of Ce6 molecules in the case of UCNP@3xCe6. Hence, UCNP@2xCe6 was chosen and further covered with PAH and a negatively charged and pH-sensitive polymer which was synthesized by cografting deprotonated PAH with PEG and 2,3-dimethylmaleic anhydride (DMMA), to fabricate UCNP@2xCe6-DMMA-PEG (Fig. 9.3e). Owing to the cleavage of the amide bond formed between the amino group and DMMA (Fig. 9.3f), obvious zeta potential increasing from around −18 to +10 mV was observed for UCNP@2xCe6-DMMA-PEG at pH 6.8. Because of the charge reversibility, UCNP@2xCe6-DMMA-PEG exhibited strong binding to negative membranes of cancer cells, leading to a higher cytotoxicity toward HeLa cells upon 980-nm irradiation (0.5 W cm^{-2}, 10 min) at pH 6.8 than that at pH 7.4, while UCNP@2xCe6-SA-PEG without charge reversibility resulted in insignificant cell death in vitro at either pH 6.8 or pH 7.4. Obviously, improved in vivo tumor retention and antitumor efficacy (Fig. 9.3g) could also be observed for UCNP@2xCe6-DMMA-PEG upon 980-nm irradiation (0.5 W cm^{-2}, 30 min, 1-min interval after 1-min irradiation), compared with UCNP@2xCe6-SA-PEG via intratumoral injection. However, in the case of systemic administration, the majority of injected UCNP@2xCe6-DMMA-PEG was still accumulated in RES organs, such as the liver and spleen, after intravenous injection, leading to an unsatisfactory PDT efficacy.

Zheng et al. [83] reported the synthesis of $NaYF_4$:Yb/Tm@$NaYF_4$:Yb/Er core–shell nanocrystals with emission peaks at 520, 539, 654, and 802 nm. The luminescence of these core–shell UCNPs was more intense than that of their codoping counterparts. The NIR-to-visible emission was used to activate Ce6 loaded into the hydrophobic layer of PEG–phospholipid-modified UCNPs, and the NIR-to-NIR emission could be utilized for image tracking.

In addition to Ce6, ZnPc with absorbance maximum of ~670 nm also matches well with the red emission of $NaYF_4$:Yb/Er UCNPs. Lin et al. [84] utilized the green emission of PAA-modified $NaYF_4$:Yb/Er UCNPs to initiate the polyaddition reaction of the C=C double bond in *N*-vinylpyrrolidone (NVP) and PEG diacrylate (PEGDA) to form cross-linked hydrogels in the presence of photoinitiator Eosin Y. Because of hydrogen bonding between UCNP capping agents of PAA and PEGDA hydrogels, the UCNPs tended to aggregate to nanoclusters as the NIR irradiation time increased from 0 to 0.5 h. The nanoclusters further aggregated to submicroflakes and finally to microspheres as the irradiation time increased to 4 h. ZnPc was successfully encapsulated into the hydrophobic cores of UCNP-PEGDA hybrid submicroflakes and microspheres through hydrophobic interaction with high loading efficiency of 8.7 and 12.3 wt%, respectively. 1,3-diphenylisobenzofuran (DPBF) bleaching experiment showed that these UCNP-PEGDA submicroflakes and microspheres both generated $^1O_2$ efficiently upon 980-nm excitation. But the PDT efficacy of these composites has not been proven, presumably owing to the large sizes of the hybrid submicroflakes and microspheres (0.5–1 and 3–10 μm, respectively).

Gu et al. [85] prepared hydrophilic UCNPs by coating the biodegradable and amphiphilic chitosan (*N*-succinyl-*N′*-octyl chitosan, SOC) on the surface of OA-capped $NaYF_4$:Yb/Er UCNPs and then loaded ZnPc into the SOC-coated UCNPs via hydrophobic interaction (Fig. 9.4a). The SOC-UCNPs possessed a high loading capacity (10.8 wt%) of ZnPc and high resistance to photobleaching, leading to good performance in $^1O_2$ generation. The absorption of DPBF in the ZnPc-SOC-UCNP solution showed a rapid decrease (~45 %) within 30 min of 980-nm irradiation and then declined slowly to ~65 % after 60-min irradiation. ZnPc-SOC-UCNP composites had a low cytotoxicity as the cell viability decreased slightly at a high concentration of 800 μg mL^{-1}. Moreover, SOC-UCNPs selectively accumulated in the tumor tissue as no fluorescence signal of NIR fluorescence dye ICG derivative-loaded SOC-UCNPs was detected in other organs primarily because of the EPR effect (ICG derivative was utilized because its NIR fluorescence enabled deeper tissue penetration than the luminescence of UCNPs). The combination of ZnPc-SOC-UCNPs (33 mg kg^{-1} UCNPs containing 3.86 mg kg^{-1} ZnPc) and 980 nm light (0.4 W cm^{-2}, twice for 15 min at 3 h and 24 h after intratumoral injection,

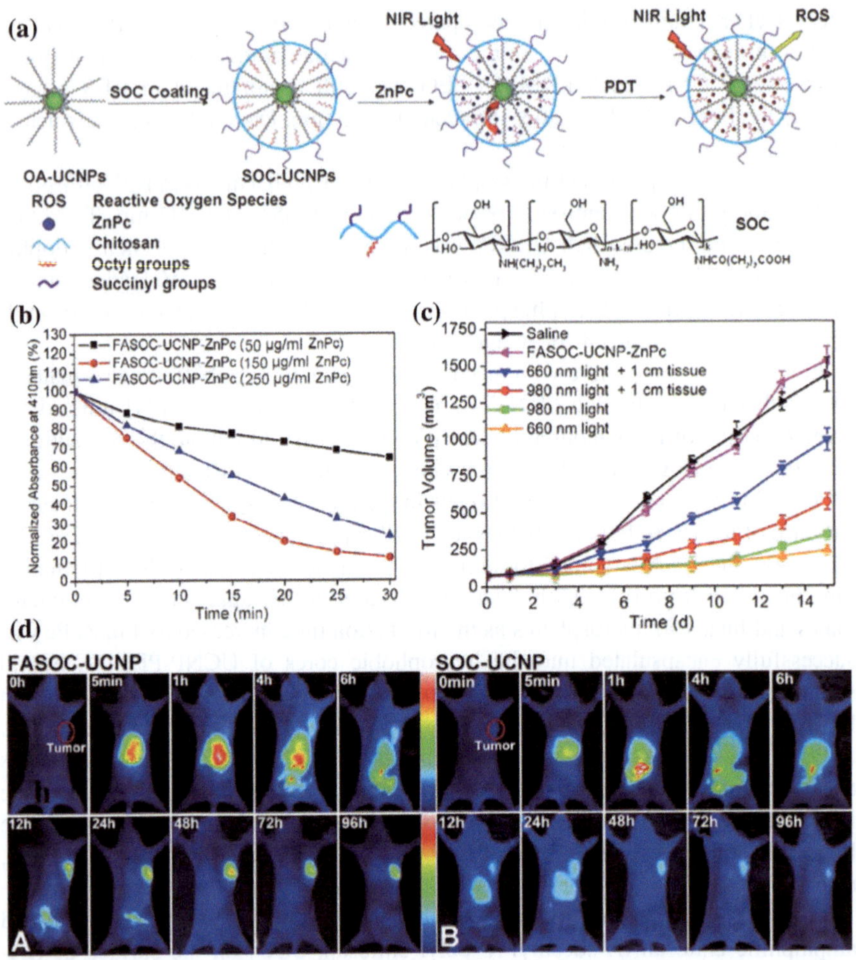

Fig. 9.4 **a** Schematic illustration of the synthesis of ZnPc-SOC-UCNPs. **b** $^1O_2$ production of ZnPc-FASOC-UCNPs loaded with different concentrations of ZnPc, determined by DPBF bleaching experiments. **c** Tumor growth curves of S180 tumor-bearing mice in different treatment groups (saline without irradiation, ZnPc-FASOC-UCNPs without NIR light irradiation, ZnPc-FASOC-UCNPs with 1-cm pork tissue and 660 nm or 980 nm light, ZnPc-FASOC-UCNPs with 660 nm or 980 nm light, respectively). **d** In vivo fluorescence image of NIR fluorescence dye ICG derivative-loaded *A* FASOC-UCNPs and *B* SOC-UCNPs in the nude mice bearing Bel-7402 tumors by intravenous injection. **a** Reprinted with permission from Ref. [85]. Copyright 2012 Royal Society of Chemistry; **b–d** Reprinted with permission from Ref. [86]. Copyright 2012 American Chemical Society

1-min interval after 2-min irradiation) resulted in a significant inhibitory effect on tumor growth, leading to a 100 % survival rate of S180 tumor-bearing mice.

On the basis of the aforementioned study, they introduced FA as the targeting agent via the condensation reaction with amino groups of SOC to construct a

FASOC-UCNP nanoplatform [86]. The ZnPc loading capacity increased with ZnPc concentration and attained saturation at 10 wt% when ZnPc concentration reached 346 μg mL^{-1}, accompanied by a decrease of encapsulation efficiency (defined as the weight ratio of encapsulated PSs to overall PSs added). However, the higher loading capacity of ZnPc did not guarantee better $^1O_2$ generation efficiency (Fig. 9.4b), which was also demonstrated in Liu's work mentioned above [82], and could be attributed to the self-quenching. The optimal concentration of ZnPc in the FASOC-UCNPs was determined to be 150 μg mL^{-1}. Although amphiphilic chitosan can dissolve in acidic solutions of pH below 6.5, less than 20 % of ZnPc was detached from FASOC-UCNPs even at pH 5.7 after 50-h incubation at 37 °C, indicating that ZnPc-FASOC-UCNPs had a good tolerance against slightly acidic tumor microenvironment. The results of acute toxicity study demonstrated that the acute toxicity of ZnPc-FASOC-UCNPs was dose dependent and the dose for PDT in vivo should be no more than 150 mg kg^{-1}. Then, the ZnPc-FASOC-UCNPs were incubated with three kinds of cancer cells expressing different levels of FR (Bel-7402 > MDA-MB-231 > A549). Red fluorescence of ZnPc and blue emission of Hoechst for cell nucleus staining were simultaneously observed in Bel-7402 and MDA-MB-231 cells with elevated FR expression (FR positive), while the ZnPc fluorescence signals in FR-negative A549 cells were much lower. Moreover, fluorescence signals of ZnPc dramatically reduced in FR-positive cancer cells when FRs were initially blocked by excess FA, indicating the importance of FA for active targeting. The in vivo bio-distribution of SOC-UCNPs with or without FA modification after intravenous injection further confirmed the high tumor-targeting ability of FASOC-UCNP in tumor-bearing mice (Fig. 9.4d). The ZnPc-FASOC-UCNPs irradiated by 660 nm light produced more ROS (~68 %) than 980 nm light (~53 %) for 5 min at the same optical power. However, once the ZnPc-FASOC-UCNP solution was covered by 1-cm pork tissues, ROS generation under 660-nm excitation significantly decreased by ~53 %, whereas it only decreased by ~17 % upon 980-nm light irradiation. The considerable advantage of NIR irradiation was also confirmed by in vivo PDT. 660 nm light possessed better therapeutic efficacy for subcutaneous tumors at the same concentration of ZnPc-FASOC-UCNPs (via intravenous injection), but 980-nm irradiation (0.2 W cm^{-2}, 30 min) demonstrated superior PDT efficacy toward simulated 1-cm-deep tumors (Fig. 9.4c), which was similar to an earlier report [81] mentioned above.

The enhanced targeting ability due to targeting moieties was also demonstrated in another work by Zhang et al. [87]. They encapsulated ZnPc and $NaYF_4$:Yb/Er UCNPs into OQPGA-PEG/RGD/TAT lipid micelles simultaneously. The amphiphilic octadecyl-quaternized small-lipid-grafted poly(γ-glutamic acid) (OQPGA) chain could form polymeric lipid micelles when mixed with ZnPc and UCNPs in chloroform. A thin lipid layer of the coencapsulated lipid micelles could be obtained after the evaporation of chloroform. In accordance with Gu's report [86], with the increase of ZnPc concentration, the loading efficiency increased and reached 1.6 wt%, but the encapsulation efficiency decreased and ZnPc was more likely to leak from the nanoparticles. Functional moieties in the lipid, including PEG, RGD peptide, and cell-penetrating TAT peptide, could improve the biocompatibility of the lipid micelles, the ability to target cancer cells, and the ability to penetrate the cell membrane. After incubation with B16F1 melanoma cells overexpressing $\alpha v\beta 3$ integrin, the red fluorescence of the OQPGA-PEG/RGD/TAT lipid micelle-encapsulated UCNPs and the green fluorescence of 5-(and 6-)carboxy-2′,7′-dichlorodihydro-fluorescein diacetate (carboxy-H_2DCFDA) marker for ROS were much stronger than those of the lipid micelle nanoparticles without the TAT and RGD peptides, suggesting that more nanoparticles were taken up by B16F1 melanoma cells with the help of TAT and RGD. Furthermore, after 30-min exposure to 980-nm laser, the cell viability was 15 % for the OQPGAPEG/RGD/TAT lipid micelle nanoparticles containing both ZnPcs and UCNPs, much lower than that of the nanoparticles without TAT and RGD.

Very recently, Chang et al. [88] encapsulated MC540 and $NaYF_4$:Yb/Er UCNPs into octadecyl-quaternized lysine-modified chitosan (OQLCS) liposome via this evaporation method. FA and TAT peptide were also employed to enhance the targeting ability and cell uptake efficiency. Dai et al. [89] adopted a modified strategy to achieve one-step coencapsulation of ZnPc and $NaGdF_4$:Yb/Er UCNPs in oil-in-water emulsions using an amphiphilic PMAO-PEG and the surfactant sodium dodecyl sulfate. Confocal imaging and cell viability (about 50 % decrease upon 980-nm irradiation of 0.3 W cm^{-2} for 20 min) performed on HeLa cells exhibited the cell internalization and the good PDT effect of the ZnPc-UCNP nanoclusters. However, the average size of nanoclusters increased to 340 nm after the self-assembly process. By contrast, PS-UCNP composites in the former two reports [87, 88] still dispersed as individuals with small particle sizes (<30 nm).

Zhao et al. [90] synthesized Mn^{2+}-doped $NaYF_4$:Yb/Er UCNPs featuring single-band emission and MRI capability and loaded three photosensitizers, Ce6, ZnPc and MB, onto α-cyclodextrin (α-CD)-modified $NaYF_4$:Yb/Er/Mn UCNPs with saturated loading amounts of 9.4, 9.5, and 4.1 wt%, respectively. Among the three PS-UCNP composites, Ce6-UCNPs exhibited best performance of $^1O_2$ generation, mainly owing to the large spectral overlap and high loading capacity. When A-549 cells were exposed to the 980-nm laser (1 W cm^{-2}, 5 min), the cell viability decreased significantly upon increasing the concentration of PS-UCNPs and Ce6-UCNPs showed best PDT efficacy as expected. They further loaded Ce6 and doxorubicin (DOX), a commonly used chemotherapy drug, onto α-CD-UCNPs simultaneously for synergistic therapy. Although the loading capacity of Ce6 in coloaded UCNPs decreased to 4.6 %, the cell killing ability of Ce6/DOX-UCNPs was greater, indicating the improved efficacy of the combinational therapeutic system.

Liu et al. [91] reported another combinational therapeutic system in which RB for PDT and an NIR-absorbing dye, IR825, for PTT were simultaneously loaded into the bovine serum albumin (BSA) layer of $NaGdF_4$:Yb/Er UCNPs with saturated loading capacities of ~7.6 and ~22 wt%, respectively. The BSA protein was covalently conjugated to PAA-coated UCNPs to avoid aggregation in PBS. The tumor growth of 4T1 tumor-bearing mice after intratumoral injection of coloaded UCNPs and exposure to 980-nm (0.4 W cm^{-2}, 30-min, 1-min interval for every 2–3 min) and 808-nm (0.5 W cm^{-2}, 5 min) lasers could be remarkably inhibited, whereas either PDT or PTT treatment alone could only slightly delay the tumor growth. Owing to the existence of Gd^{3+}, these nanoparticles could serve as a T_1 MR contrasting agent. Very recently, they demonstrated the synergistic anticancer effect of PDT and gene therapy [92]. In this study, $NaGdF_4$:Yb/Er UCNPs were functionalized via LbL coating of counter-charged polymers and then loaded with Ce6 via hydrophobic interaction. The loading capacity was determined to be 11 wt%. Pololike kinase 1 (Plk1) small interfering RNA (siRNA) was further loaded onto positively charged UCNP-PEG@2xPEI-Ce6 composites for gene therapy. It is known that Plk1 siRNA (siPlk1) can cause degradation of targeted mRNA, leading to the silence of Plk1 expression and subsequent cell apoptosis. As expected, obviously decreased Plk1 expression and a concentration-dependent reduction of cell viabilities were observed for HeLa cells treated with as-prepared UCNP-PEG@2xPEI-Ce6-siPlk1 composites. The cell viabilities also gradually decreased with the increase of UCNP-PEG@2xPEI-Ce6 concentrations. But the combination of PDT and gene therapy for HeLa cells treated with UCNP-PEG@2xPEI-Ce6-siPlk1 and then exposed to a 980-nm laser (0.8 W cm^{-2}, 20-min, 1-min interval after 1-min irradiation) showed much lower remaining cell viabilities compared to single PDT or gene therapy alone.

Wu et al. [93] utilized magnetic Fe_3O_4@$NaYF_4$:Yb/Er core–shell nanoparticles to achieve multi-functional T_2-weighted MR/UCL imaging and NIR-induced PDT. Although the Fe_3O_4 core was in cubic phase, most $NaYF_4$ shell remained to be in hexagonal phase. The tetrasulfonic phthalocyanine aluminum ($AlPcS_4$, absorption peak at 676 nm) was embedded in the PEG layer of Fe_3O_4@UCNPs. The cell death of MCF-7 cells incubated with Fe_3O_4@UCNP-$AlPcS_4$ composites (150 μg mL^{-1}) and exposed to 980-nm irradiation (0.02 W cm^{-2}, 5 min) could reach about 70 %.

Although most current PS-UCNP nanoplatforms are based on the green and red emissions of UCNPs, some blue light-excited photosensitizers have been exploited for UCNP-based PDT. For example, Zhao et al. [94] utilized hypocrellin A (HA, absorption peak at 467 nm) and Tween 20-modified $NaYbF_4$:Gd/Tm@$NaGdF_4$ UCNPs for PDT in vitro. The saturated loading amount of HA was determined to be 5.5 wt%. Owing to the presence of Yb^{3+} and Gd^{3+}, the as-synthesized HA-UCNPs also provided high-contrast computed tomography (CT) imaging and MRI modality.

Tan et al. [95] reported a specific aptamer-guided G-quadruplex (GQ) DNA nanoplatform for targeted PDT (Fig. 9.5a). In brief, a guanine-rich DNA segment was linked to a 26-mer DNA aptamer, sgc8, to form a bifunctional DNA sequence, termed a G4-aptamer. Thiol modification on the G4-aptamer enabled linking with PEI-modified $NaLuF_4$:Yb/Er/Gd UCNPs. Additionally, two G4-aptamer sequences folded into GQs with the aptamers unaffected in the presence of K^+. Then, upon the addition of 5,10,15,20-tetrakis-(1-methyl-4-pyridyl)-21H, 23H-porphine (TMPyP4, absorption peak at 422 nm), this cationic porphyrin-based photosensitizer could bind to GQs through π–π interaction with G-quartets and through electrostatic interaction with the anionic phosphate groups on GQs (Fig. 9.5c). Furthermore, sgc8, which binds to the CCRF-CEM cell membrane receptor protein tyrosine kinase 7 (PTK7) with high affinity and selectivity, could enhance the targeting

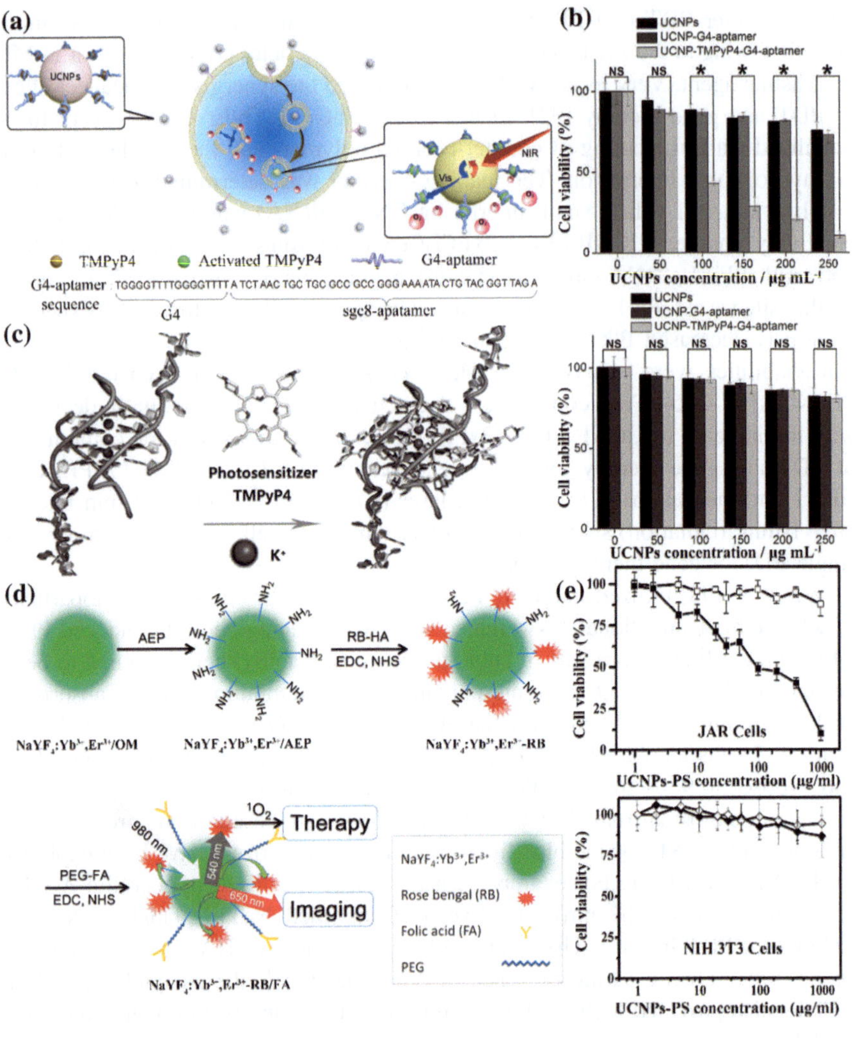

Fig. 9.5 **a** Schematic illustration of aptamer-guided GQ DNA-UCNP nanoplatform for targeted PDT. **b** Cell viability of CEM cells (*top*) and Ramos cells (*bottom*) treated with UCNPs, UCNP-G4-aptamer, and UCNP-TMPyP4-G4-aptamer and then exposed to NIR irradiation. *NS* not significant, $*P < 0.05$. **c** Interaction between the G4-aptamer and TMPyP4. **d** Schematic illustration of the covalent conjugation of $NaYF_4$:Yb/Er UCNPs, photosensitizer RB, and target molecule FA. **e** Cell viability of JAR cells (*top*) and NIH 3T3 cells (*bottom*) treated with RB/FA-AEP-UCNPs of different concentrations with (*solid*) or without (*open*) 980-nm exposure. **a–c** Reprinted with permission from Ref. [95]. Copyright 2013 Wiley-VCH; **d, e** Reprinted with permission from Ref. [98]. Copyright 2012 American Chemical Society

ability of this TMPyP4-G4-aptamer-UCNP composite. Significantly reduced cell viabilities were observed for CCRF-CEM cells incubated with the TMPyP4-G4-aptamer-UCNPs of more than 100 mg mL^{-1} after exposure to the 980-nm laser

(0.5 W cm^{-2}, 10-min, 1-min intervals). The cell viability of CCRF-CEM cells reduced to (10.4 ± 2.7) % at a concentration of 250 mg mL^{-1} (Fig. 9.5b). By contrast, a much less pronounced photocytotoxicity to Ramos cells (acute lymphoblastic leukemia B cells) without PTK7 receptor was observed for TMPyP4-G4-aptamer-UCNPs (Fig. 9.5b).

As mentioned in Sect. 9.2.1, the generation of $^1O_2$ relies on the energy transfer between photosensitizers at T_1 state, which is formed via intersystem crossing form S_1, and molecular oxygen at the ground state. The probability of the intersystem crossing can be enhanced in the presence of atoms with high atomic number such as iodine and bromine, which is known as the heavy atom effect [118]. Pérez-Prieto et al. [96] reported the preparation of PEG-modified $NaYF_4$:Yb/Er UCNPs loaded with a diiodo-substituted boron dipyrromethene (IBDP, absorption peak at 525 nm) photosensitizer. Control experiments showed that the addition of IBDP, bearing a carboxylic group to coordinate lanthanides, to oleate-capped UCNPs, led to the precipitation of the nanoparticles as a pink solid and these precipitates could not be dispersed in water, while the supernatant remained colorless. Hence, it was concluded that IBDP was anchored to the UCNP surface and, at the same time, embedded in the PEG layer. Values of 43 and 10 wt% with respect to the UCNP core in the IBDP/PEG-UCNP composites were estimated for IBDP and the PEG, respectively, indicating a high loading capacity due to the dual interaction. After incubation with the IBDP/PEG-UCNP composites at a low concentration of 25 μg mL^{-1} and exposure to 975-nm irradiation (0.239 W, 45 min), *ca.* 50 % human neuroblastoma SH-SY5Y cell death was observed. However, the incorporation of heavy atoms may also lead to an increased dark toxicity [102].

9.2.3.4 Covalent Conjugation

In general, silica or mesoporous silica-based loading strategy for photosensitizers is restricted by low drug loading capacity, while polymer-based strategy suffers from leakage of photosensitizers from the composites. To address this issue, covalent conjugation has been utilized to improve the stability of photosensitizers loaded onto UCNPs, as well as guarantee the loading efficiency.

In 2012, Xing et al. [97] attached pyropheophorbide-a (PP-a, absorption peak at 668 nm) and RGD peptide c(RGDyK) to PEI-capped $NaYF_4$:Yb/Er UCNPs via the condensation reaction. In this study, *O*-carboxymethyl chitosan (OCMC) with abundant amino and carboxyl groups was first introduced by reacting with amino groups of PEI. Then, it can further undergo condensation reactions with the amino groups of c(RGDyK) and the carboxyl groups of PP-a. Although the loading capacity of 0.53 wt% was low, fluorescence spectra showed that there was no free PP-a detected in the supernatants of PP-a/RGD-UCNPs with different pH after centrifugation, indicating a high stability of the nanoparticles. Because of the presence of targeting peptide, a strong red fluorescence could be observed in $\alpha v\beta 3$ integrin-positive cells (U87-MG) incubated with PP-a/RGD-UCNPs, while a weak fluorescence showed up in $\alpha v\beta 3$ integrin-negative cells (MCF-7) incubated with

PP-a/RGD-UCNPs, U87-MG cells incubated with PP-a-UCNPs, and U87-MG cells incubated with PP-a/RGD-UCNPs after treatment with excess free c(RGDyK). Without NIR irradiation, the cell viability of U87-MG cell was still beyond 85 % even at 400 μg mL^{-1} dose of PP-a/RGD-UCNPs. By contrast, the cell viability decreased by 30–50 % at a reduced concentration of 100 μg mL^{-1} upon 980-nm irradiation (0.5 W cm^{-2}, 5 min). Moreover, the viability of MCF-7 cells treated by PP-a/RGD-UCNPs and both cells incubated with PP-a-UCNPs all showed a slight decrease of about 8 % under the same conditions.

Kong et al. [98] used 2-aminoethyl dihydrogen phosphate (AEP) to replace the original oleylamine (OM) ligands of $NaYF_4$:Yb/Er UCNPs and then loaded a water-soluble photosensitizer, rose bengal (RB, absorbance maximum at 549 nm) via amidation (Fig. 9.5d). The optimal number of RB for individual UCNP was determined to be ~100 by DPBF bleaching experiment, since the dimerization of RB would become serious at a higher RB concentration. Absorption spectra of RB eluate after washing with DMSO showed that the amount of RB eluted from covalently bonded RB-UCNPs was negligible compared to that from RB-UCNP composites formed via electrostatic interaction, confirming the stability of covalent bonding. The energy transfer efficiency, calculated by the value of $E = (I_0 - I_1)/I_0$, where I_0 and I_1 are green emission intensities of UCNPs and RB-AEP-UCNP composites, respectively, was determined to be 83 %. They further monitored the characteristic phosphorescence of $^1O_2$ at 1,270 nm to confirm the highly efficient $^1O_2$ generation. This was the first time that $^1O_2$ from upconversion nanoplatforms was directly probed. In order to enhance the targeting ability and biocompatibility in cellular-level studies, FA was covalently functionalized to RB-AEP-UCNPs via a dual-functional PEG (NH_2-PEG-COOH). These composites were mainly located in cytoplasm and perinuclear regions after incubation with JAR choriocarcinoma cells, illustrating the endocytosis of the UCNP nanoplatform mediated by FRs. By contrast, only a few UCNPs were observed on the NIH 3T3 cells, which are poor in expressing FR. When JAR cancer cells were exposed to 980 nm light (1.5 W cm^{-2}, 10 min), the cell viability decreased significantly with an increase of RB/FA-AEP-UCNP concentration (Fig. 9.5e). In marked contrast to JAR cells, 980-nm irradiation had no obvious effect on cell viability of NIH 3T3 fibroblast cells (Fig. 9.5e). It is worth noting that the unemployed red emission of $NaYF_4$:Yb/Er UCNPs could be simultaneously used for bio-imaging since the red band is close to NIR region and affords relatively high tissue penetration.

They further applied the covalent approach to other photosensitizers including monomalonic fullerene (C_{60}MA) [99] and ZnPc [100]. In the former case, a strategy of separately doping core and shell with Er^{3+} and Tm^{3+} instead of homogeneous codoping was employed to achieve strong multi-color upconversion. Hydrophilic amino-functionalized UCNPs were prepared via ligand exchange between poly (allylamine) (PAAm) and OA ligand on the surface of UCNPs. PEG–succinimidyl carbonate (PEG-SC) was used to stabilize the nanocomposites in biological media and enhance the biocompatibility, while FA was used to enhance the targeting ability. The broad absorption spectrum of C_{60}MA, which was covalently attached to UCNPs with a saturated loading capacity of 10.5 wt%, overlapped well with the multi-color UCL bands (450, 475, 540, 650 nm) of $NaYF_4$:Yb/Er@$NaYF_4$:Yb/Tm UCNPs. After incubation with C_{60}MA-UCNPs and exposure to 980-nm laser irradiation (1.37 W cm^{-2}, 10 min), the viability of HeLa cells declined rapidly with an increase in the concentration of C_{60}MA-UCNPs. Moreover, C_{60}MA-UCNPs ($NaYF_4$:Yb/Er@$NaYF_4$:Yb/Tm) exhibited better efficacy than C_{60}MA-UCNPs ($NaYF_4$:Yb/Er), C_{60}MA-UCNPs ($NaYF_4$:Yb/Tm), and RB-UCNPs ($NaYF_4$:Yb/Er@$NaYF_4$:Yb/Tm), owing to the intense UCL of core–shell structures and large spectral overlap.

In the latter work, carboxyl-modified ZnPc, 2, 9, 16, 23-tetracarboxylic zinc phthalocyanine ($ZnPc(COOH)_4$) was used instead of ZnPc, so as to conjugate with the amino group of PAAm. $ZnPc(COOH)_4$ exhibited a redshift in the peak absorbance from ~670 to ~680 nm compared to ZnPc. Maybe because of this, the doping concentration of Yb^{3+} in this study was increased from 20 to 25 % to strengthen the intensity of red emission. The energy transfer efficiency was determined to be 80.9 %. The optimal loading amount of $ZnPc(COOH)_4$ was 6 wt%, and higher loading amount led to the decrease of $^1O_2$ generation efficiency, which was analogous to some other reports based on non-covalently attachment approaches [82, 86]. Before the PDT test in vivo, the overheating effect of 980-nm irradiation and cytotoxicity of $ZnPc(COOH)_4$/FA-PAAm-UCNP was checked. It was found that the overheating effect was negligible at low power density of less than 0.39 W cm^{-2} and that the cell toxicity became obvious at a concentration higher than 200 μg mL^{-1}. Hence, Hepa1-6 tumor-bearing C57/6J mice were intratumorally injected with UCNPs-ZnPc/FA (10 mg mL^{-1}, 100 μL) and then exposed to 980-nm laser (0.39 W cm^{-2}, 15-min, 3-min interval after 3-min irradiation). The tumor inhibitory ratio was calculated to be approximately 80.1 % within 2 weeks in comparison with the control group treated by saline alone. Furthermore, markedly increased apoptotic and necrotic tumor cells were demonstrated in PDT treatment group by the histological analysis, while no pathological changes in the heart, lung, kidney, liver, or spleen were observed.

Hyeon et al. [101] combined two approaches of physical adsorption and covalent amide bonding to maximize the loading amount of photosensitizers. In this study, Ce6 was loaded onto the surface of $NaYF_4$:Yb/Er@$NaGdF_4$ UCNPs functionalized with amino group-modified PEG–phospholipid. The $NaGdF_4$ shell could enhance the upconversion efficiency, as well as T_1 MRI contrast. The amount of Ce6 calculated from the UV–visible absorption spectra was found to be $>10^3$ Ce6 per particle. Ce6-UCNPs exhibited better colloidal stability in cell culture media and longer blood circulation half-life (21.6 min) than UCNPs without Ce6 (4.2 min). Bio-distribution of UCNPs and Ce6-UCNPs indicated that the amount of Ce6-UCNPs in the liver and spleen where these nanoparticles accumulated at first was much lower than that of UCNPs 7 days after intravenous injection. The facile excretion of Ce6-UCNPs would decrease the possibility of long-term toxicity. U87MG tumor-bearing nude mice were injected with Ce6-UCNPs through the tail vein. Under 980-nm irradiation (~0.6 W cm^{-2}, 5-min, 10-s interval after 5-s irradiation), tumor growth was inhibited, while non-treated tumors grew significantly. Histological examination also confirmed the tumor necrosis induced by Ce6-UCNPs and laser irradiation.

Some uncommonly used photosensitizers have been attached to UCNPs via covalent conjugation. For instance, Akkaya et al. [102] attached carboxyl-functionalized boron dipyrromethene (Bodipy) dyes to PEI-modified $NaYF_4$:Yb/Er/Ce UCNPs in the presence of DCC/DMAP coupling reagents. Tan et al. [103] prepared thioglycolic acid-modified N-TiO_2 for the conjugation with PEI-modified $NaYF_4$: Yb/Tm, and the PDT effect was mainly attributed to OH· instead of $^1O_2$.

Very recently, Yan et al. [104] reported a dual-targeting UCNP nanoplatform for imaging-guided PDT. In this report, the inorganic photosensitizer fullerene was functionalized with hyaluronic acid (HA) and then attached to aminophenylboronic acid (APBA)-modified UCNPs with a saturated loading capacity of 3.8 wt% via the diol–borate condensation. Furthermore, these two compounds, HA and APBA, had specificity toward polysialic acid (PSA) and cluster determinant 44 (CD44) receptors overexpressed in cancer cells, respectively, and thus could also be used as targeting ligands. $NaYF_4$:Yb/Gd/Tm UCNP with multi-color UCL at 475, 650, 700, and 800 nm was adopted in order to fully utilize the broad absorption of hyaluronated fullerene (HAC_{60}) from 200 to 700 nm. The unemployed 800-nm emission of UCNPs, as well as the red emission at 640 nm from HAC_{60}, could be used for two-color luminescence imaging. The HAC_{60}-UCNP dual-targeting nanoplatform showed significantly improved targeting capacity as compared to the single-targeting probes of APBA-modified UCNPs and HAC_{60}. The mean fluorescence intensity of HAC_{60}-UCNP-incubated PC12 cells overexpressing PSA and CD44 receptors was about two times that of APBA-modified UCNPs or HAC_{60}-incubated PC12 cells. Moreover, the amount of HAC_{60}-UCNPs taken up by PC12 cells was reduced by 34.8 % after APBA blocking and 31.6 % after HA blocking via UCL determination. The above results demonstrated the synergistic targeting effects. It was further found that HAC_{60}-UCNPs internalized into the cytoplasm of PC12 cells through dual-receptor-mediated endocytosis up to 24 h. The treatment of HAC_{60}-UCNPs (50 or 100 μg mL^{-1}) under 980-nm irradiation (0.5 W cm^{-2}, 15 min) led to a concentration-dependent decrease of the cell viability by 32–55 %, whereas APBA-modified UCNPs (50 or 100 μg mL^{-1}) resulted in little decrease and HAC_{60} (2 or 4 μg mL^{-1}) led to a slight decrease.

9.2.3.5 Direct Conjugation via Non-covalent Interactions

All of aforementioned loading approaches introduced an interlayer between UCNPs and the loaded photosensitizers, which would increase the distance between them and thus reduce the distance-dependent energy transfer efficiency. In view of this, Chen and Wang [105] attached β-carboxyphthalocyanine zinc (ZnPc(COOH), absorption peak at ~676 nm) directly to ligand-free $LiYF_4$:Yb/Er UCNPs in a very recent report. Owing to the removal of surface oleate ligands by acid treatment, positively charged Ln^{3+} ions were exposed on the surface of UCNPs, which was confirmed by a positive zeta potential of +49.5 mV. Thus, ZnPc(COOH) could be conjugated to the surface of UCNPs via electrostatic interaction with a loading amount of 1.1 wt%. Considering that π–π stacking interaction of the phthalocyanine aromatic ring of ZnPc(COOH) [119] may result in aggregation of the UCNPs in aqueous solution, they further decorated ZnPc(COOH)-UCNPs with poly(vinyl pyrrolidone) (PVP). These composites were stable in aqueous solution with varying pH (3–11) and PBS. Although the loading capacity was relatively low compared with some reports based on polymer or covalent bonding (Table 9.2), an extremely high energy transfer efficiency of 96.3 % could be achieved in this study. As a result, the absorbance of DPBF incubated with PVP-ZnPc(COOH)-UCNPs decreased exponentially with the time upon 980-nm irradiation ($\sim$1 W cm^{-2}), with approximately 91 % decrease in 26 min. Furthermore, the growth rate of tumor in H22 tumor-bearing mice intratumorally injected with PVP-ZnPc(COOH)-UCNPs (0.5 mg mL^{-1}, 30 μL) and exposed to 980-nm irradiation (0.5 W cm^{-2}, 20-min, 5-min interval after 1-min irradiation) was significantly attenuated as evidenced by the tumor volume change from $\sim$240 to $\sim$501 mm^3 over a period of 2 weeks, while the tumor volume in the control group treated with PVP-ZnPc(COOH)-UCNPs alone increased remarkably from $\sim$248 to $\sim$1,282 mm^3. Histological analysis of tumor tissues at 14 days post-treatment reconfirmed the therapeutic efficacy of PVP-ZnPc(COOH)-UCNPs as obvious apoptotic and necrotic tumor cells could be observed in tissues from the experiment group.

Chen and Wei [106] fabricated rare earth (Er^{3+}, Yb^{3+}, Gd^{3+})-doped ZnO particles by a homogenous precipitation method. The as-prepared ZnO:Yb/Gd/Er particles were further sintered at high temperatures (>700 °C) for the enhancement of UCL. After calcination, hydrophilic groups (e.g., hydroxyl) attached to the particles as the products were prepared in the solution of diethylene glycol (DEG) could be removed partially. When HepG2 cells were incubated with ZnO:Yb/Gd/Er nanoparticles, transferrin, which came from the fetal bovine serum (FBS) in the culture medium, could be absorbed on the surface of these charged particles as confirmed by FT-IR spectra. On account of the transferrin–receptor interaction, it may promote the endocytosis process of HepG2 cells for ZnO:Yb/Gd/Er nanoparticles. Furthermore, owing to the electrostatic interaction between negative charge and positive charge on the surface of MB and UCNPs, MB could be attached directly to the RE-doped ZnO nanoparticles, as evidenced by the conversion of zeta potential from +22 to −15 mV. The maximum MB-loading capacity was $\sim$3.3 wt%. When incubated with MB-loading ZnO:Yb/Gd/Er nanoparticles, the discrepancy in cell viabilities of HepG2 cells with and without 980-nm irradiation (0.5 W, 3 min) was about 50 %, larger than that of non-MB-loading ones ($\sim$15 %). However, the particle size sintered at 800 °C was $\sim$327 nm according to the statistical analysis of their SEM images, as it is comprised of numerous fine nanocrystals with average size of 20 nm. The large size is not suitable for PDT in vivo. Besides, MB is formally positively charged because of the quaternary ammonium structure and thus not favorable for the direct attachment to lanthanide-doped UCNPs, as compared with photosensitizers bearing carboxyl groups.

Qu et al. [107] loaded MB onto porous UCNPs to avoid the introduction of an interlayer between UCNPs and photosensitizers. $NaYF_4$:Yb/Er UCNPs possessing a porous structure were fabricated by a hydrothermal reaction in the presence of guanosine monophosphate (GMP) as a bio-template. The as-prepared UCNPs had a BET surface area of 15.76 m^2 g^{-1} and an average pore size of 6.1 nm. GMP acted not only as a template, but also as a stabilizer to make UCNPs stable in various physiological solutions. Photosensitizer loading was accomplished by soaking UCNPs in a solution of MB to allow the molecules to diffuse into the pores of the UCNPs, and the loading amount was calculated to be 0.014 wt%. It was found that the viability of A549 cells that incubated with MB-UCNPs remained about above 90 % without NIR irradiation, whereas upon 980-nm irradiation (0.75 W cm^{-2}), the cell viability decreased gradually with the increase of the irradiation time or the concentration of MB-UCNPs.

Although the approach of direct attachment via non-covalent interactions is simple and low cost, the stability and loading capacity of as-prepared PS-UCNP composites remain to be further confirmed by more reports. Moreover, in order to achieve a long blood circulation half-life and active targeting, these UCNPs still need to be functionalized with PEG and targeting moieties.

9.3 UCNPs for Chemotherapy and Gene Therapy

A Holy Grail of medical therapy is to deliver therapeutic compounds into individual diseased tissues for maximum treatment efficacy and minimum side effects [120]. However, the molecules administered systemically are typically distributed to the whole body through blood circulation and suffer from hydrolysis and enzymatic degradation, thus posing great challenges for both traditional chemotherapy and emerging gene therapy utilizing microRNA (miRNA) or small interfering RNA (siRNA) [121]. Like the PS-UCNP composites used for PDT, the UCNP-based nanocarrier can improve the solubility, stability, half-life, and targeting ability of various pharmaceutical payloads, such as chemotherapeutic drugs, genes, and other related biomolecules, thus enhancing their therapeutic efficacy [62, 122]. The payloads can also be loaded into the UCNP-based carriers via several methods: (i) deposited in the pores of mesoporous silica shell or porous UCNP@silica fiber composites; (ii) loaded into the hydrophobic pockets; (iii) loaded into a hollow sphere coated with a mesoporous UCNP shell; and (iv) attached to the surface of UCNPs via covalent bonding or the coordination of carboxyl [23, 123].

However, it is not necessary to release photosensitizers from PS-UCNP composites since the actually effective $^1O_2$ can diffuse out of the composites. Furthermore, the release of photosensitizers is unfavorable for the energy transfer from UCNPs to photosensitizers. By contrast, the release of drugs or genes is necessary in the case of drug and gene delivery. In view of this, UCNP-based carriers can be categorized into two types. In the first case, UCNPs are utilized only as luminescent indicators to track nanocarriers or to monitor the release of payloads in the presence of the energy transfer process between drugs and UCNPs [124]. The release of drugs is triggered by internal or external stimuli, such as pH [125–134], temperature

[135, 136], and redox species [137], or the concentration gradient of drug molecules [138–144]. For example, Lin et al. [130] attached DOX, a widely used anticancer drug, to hydrazine-functionalized $NaYF_4$:Yb/Tm UCNPs via covalent conjugation. The release rate of DOX from UCNPs at pH 5.0 was significantly faster than that at pH 7.4. The pH-dependent release was attributed to the cleavage of the hydrazone bond, which is quite stable in normal physiological environment (pH 7.4) but cleaves quickly in mildly acidic environment (pH 4.5–5.0). Owing to the energy transfer from UCNPs to DOX, the blue emission intensity of UCNPs decreased after DOX loading and gradually recovered along with the release of DOX and thus could be utilized to monitor the release process.

The second type involves photoswitchable or photocaged molecules that respond to UV or visible light, and UCNPs perform as NIR-to-UV/visible upconverters in the system. That is to say, the liberation of cargo from nanocarriers or the uncaging of caged drugs is triggered by the upconversion luminescence of UCNPs. Therefore, it can be achieved with noninvasive property, high spatiotemporal precision and high penetration depth. Generally speaking, the published reports on this type (Table 9.3) encompass three directions [123]: (i) NIR light-triggered photoisomerization of photoswitchable molecules between two structurally and electronically different isomers; (ii) NIR light-triggered photolysis of photocaged compounds; and (iii) NIR light-triggered photoreduction of Pt(IV) prodrugs.

9.3.1 NIR Light-Triggered Photoisomerization

9.3.1.1 Azobenzene

Azobenzenes (azo), which have "–N=N–" structure with phenyl rings on either side, have been widely used for light-triggered release systems [164, 165]. The *trans* isomer of the azobenzene molecules will transform into the *cis* isomer under UV light irradiation and further transform back to the *trans* isomer under irradiation of visible light.

Shi et al. [145] reported a general strategy for NIR light-triggered anticancer drug release based on the *cis-trans* photoisomerization of azobenzenes (Fig. 9.6a). After installing azobenzenes into the mesopores of mesoporous silica shell coated on the surface of ligand-free $NaYF_4$:Yb/Tm@$NaYF_4$ UCNPs, DOX was loaded into the mesopores. The reversible transformation of azobenzenes by simultaneous UV and visible light emitted by the UCNPs created a continuous rotation–inversion movement. The back and forth wagging motion of the azobenzene molecules acted as a molecular impeller that propelled the release of DOX, leading to a controllable release of DOX. The DOX release amount reached 40 wt% in 16 h under intermittent 980-nm exposures (2.4 W cm^{-2}) and reached a maximum of 80 wt% in 16 h of 980-nm laser irradiation of 8.9 W cm^{-2} (Fig. 9.6b). Less than 5 wt% of DOX was leached in 17 h without NIR irradiation (Fig. 9.6b). They further carried out controlled drug release studies by incubating the composites with HeLa cells.

Table 9.3 A summary of NIR light-triggered UCNP-based carriers

UCNPs	Stabilizers	Photosensitive moieties	Payloads	References
$NaYF_4$:Yb/Tm@$NaYF_4$	$mSiO_2$	Azobenzene	DOX	[145]
$NaYF_4$:Yb/Tm@$NaGdF_4$	$mSiO_2$	Azobenzene	DOX	[146]
$NaYF_4$:Yb/Er	Herring sperm DNA	Spiropyran	β-Galactosidase	[147]
$NaYF_4$:Yb/Tm@$NaYF_4$	*o*-NB-modified block copolymer	*o*-Nitrobenzyl (*o*-NB)	Nile red	[148]
$NaYF_4$:Yb/Tm@$NaYF_4$	*o*-NB-modified hydrogel	*o*-NB	FITC-BSA	[149]
$NaYF_4$:Yb/Tm@$NaYF_4$	*o*-NB-modified polymer	*o*-NB	Coumarin 153	[150]
$NaYF_4$:Yb/Tm@$NaYF_4$	NPE-modified OEG	1-(2-Nitrophenyl)- ethyl (NPE)	DOX	[151]
$NaYF_4$:Yb/Tm@$NaYF_4$	None	3′,5′-Di(carboxy-methoxy)benzoin	Benzoin-caged acetic acid	[152]
$NaYF_4$:Yb/Tm@$NaYF_4$	Silica and PEG	NPE	NPE-caged D-luciferin	[153]
$NaYF_4$:Yb/Tm@$NaLuF_4$	$mSiO_2$	Amino-coumarin (AC)	AC-caged chlorambucil	[154]
$NaYF_4$:Yb/Tm	SiO_2 and PEG	*o*-NB	DOX	[155]
$NaYbF_4$:Tm@ CaF_2	PEI	*o*-NB	*o*-NB-caged fluorescein	[156]
$NaYF_4$:Yb/Tm	$mSiO_2$	4,5-Dimethoxy-2-nitroacetophenone (DMNPE)	DMNPE-caged plasmid DNA/siRNA	[157]
$NaYF_4$:Yb/Tm	SiO_2 and *o*-NB-modified PEG	*o*-NB	siRNA	[158]
$NaYF_4$:Yb/Tm@$NaYF_4$:Yb/Er	$mSiO_2$	Photomorpholino and TPPS2a	Photomorpholino	[159]
$NaYF_4$:Yb/Er@$NaYF_4$	$mSiO_2$ and PAH	$Fe_4S_3(NO)_7^-$ (NO precursor)	$Na[Fe_4S_3(NO)_7]$	[160]
$NaYF_4$:Gd/Yb/(Er or Tm) @$NaYF_4$	None	$Fe_4S_3(NO)_7^-$	$[NH_4][Fe_4S_3(NO)_7]$	[161]
$NaYF_4$:Yb/Tm@$NaGdF_4$:Yb	PEI and PEG	Pt(IV) prodrugs	Pt(IV) prodrugs	[162]
$NaYF_4$:Yb/Tm	SiO_2 and OEG	Pt(IV) prodrugs	Pt(IV) prodrugs	[163]

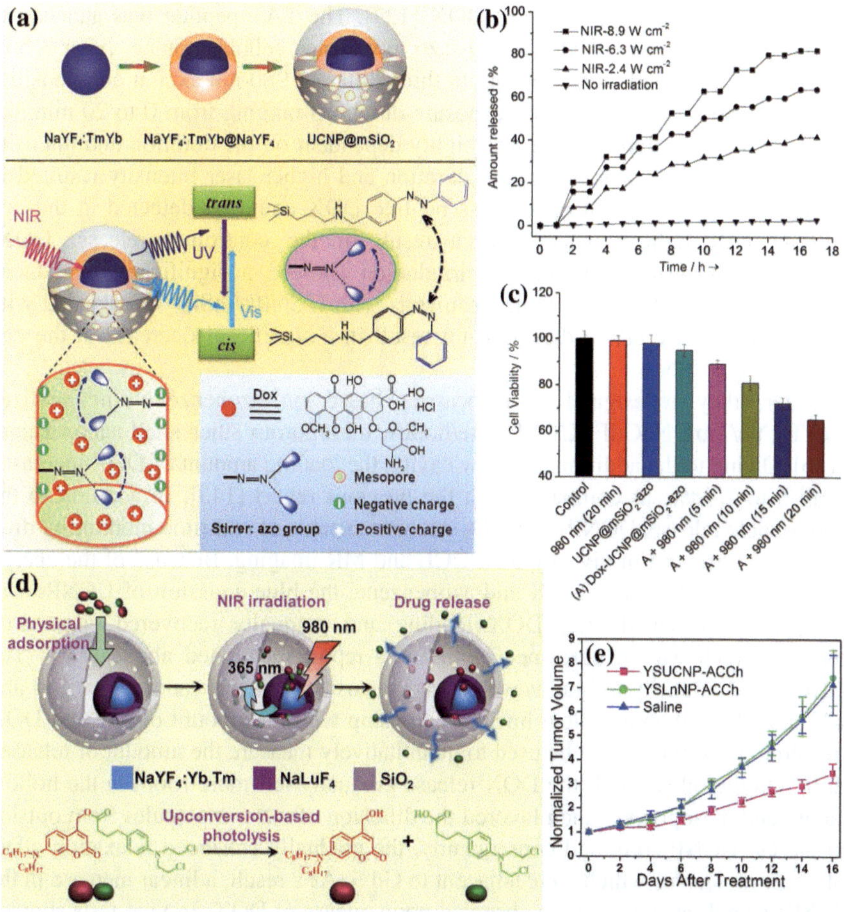

Fig. 9.6 **a** Schematic illustration of the synthesis of mesoporous silica-coated UCNPs and NIR light-triggered dox release due to the *trans-cis* photoisomerization of azo molecules grafted in the mesopore network of the mesoporous silica layer. **b** DOX release from UCNP@mSiO_2-azo in PBS under NIR light irradiation and dark conditions, alternatively. **c** Inhibition of HeLa cell growth in the presence of 980-nm irradiation (20 min), UCNP@mSiO_2-azo, and Dox-UCNP@mSiO_2-azo without or with 980-nm irradiation for varied durations. **d** Schematic illustration of NIR light-triggered photolysis and release of amino-coumarin-caged chlorambucil (ACCh) from yolk–shell $NaYF_4$:Yb/Tm@$NaLuF_4$ UCNPs (YSUCNP). **e** The normalized S180 tumor volume in mice intratumorally injected with YSUCNP-ACCh, YSLnNP-ACCh (YSLnNP: yolk–shell $NaYF_4$:Yb@$NaLuF_4$ UCNPs without Tm^{3+} activator) and saline. In all mice groups, the tumors were exposed to 980-nm laser. **a–c** Reprinted with permission from Ref. [145]. Copyright 2013 Wiley-VCH; **d**, **e** Reprinted with permission from Ref. [154]. Copyright 2013 Wiley-VCH

Considering that the released DOX exhibited much stronger fluorescence compared with DOX loaded onto UCNPs, the fluorescence of which was quenched owing to the aggregation of DOX molecules, the fluorescence of DOX could be used to determine the amount of released DOX [127]. The TAT peptide was attached to amino-functionalized UCNP@mSiO_2-azo to enhance cellular uptake. After 3 h of incubation in the dark and exposure to three different 980-nm excitation intensities (2.4, 6.3, and 8.9 W cm^{-2}) with exposure durations ranging from 0 to 20 min, the intensity of DOX fluorescence was highly dependent on the duration and intensity of NIR exposure. Longer exposure duration and higher laser intensity resulted in stronger DOX fluorescence, whereas no free DOX could be detected in the cell without NIR light exposure. As a result of the controlled release, DOX-UCNP@mSiO_2-azo upon 980-nm irradiation showed a significantly enhanced cytotoxicity to HeLa cells, while treatments with UCNP@mSiO_2-azo loaded with and without DOX in the dark did not result in a significant decrease in the cell viability (Fig. 9.6c).

Later, they redesigned a nanocarrier based on azobenzene-functionalized $NaYF_4$:Yb/Tm@$NaGdF_4$ UCNP core/hollow mesoporous silica shell nanoparticles [146]. Owing to the additional hollow cavity, the loading amount of DOX increased from 5 to 7 wt%, as compared with the previous report [145]. In addition to the controllable release, UCNP@hmSiO_2-azo was capable of real-time monitoring drug release in vitro and in vivo by both UCL and MR imaging. Because of the energy transfer from UCNPs to DOX and azobenzene, the blue emission of UCNPs was completely quenched after DOX loading and gradually recovered upon drug release, which was in agreement with Lin's report mentioned above [130]. The UCL signal intensity at blue wavelengths (intensity arithmetic addition at 450 and 477 nm) showed an excellent linear relationship with the amount of released DOX in solution and thus could be used to quantitatively measure the amount of released drug. On the other hand, the DOX release gradually left more room in the hollow cavity and mesopores, which favored the diffusion of water molecules from outside to the Gd-UCNP core, and consequently, the gradually enhanced relaxation effect of water molecules which were adjacent to Gd^{3+}. As a result, a linear increase in the T_1 MR signal intensity can be observed upon release of DOX. In vivo tests showed that no apparent UCL signals were found in the vascular system of zebra fish embryos injected with UCNP@hmSiO_2-azo composites into their heart ventricles, whereas significantly brightened UCL signals showed up in the zebra fish heart regions after exposure to a 980-nm laser (1.5 W cm^{-2}, 60 min). Correspondingly, zebra fish exposed to NIR irradiation demonstrated an approximately 51 % enhancement in the TS_1 MR signal intensity over those of the control group. It is worth noting that although DOX absorbed large amount of upconverted visible light, the unaffected UV light and the remaining visible light were still effective to trigger the photoisomerization of azobenzene.

9.3.1.2 Spiropyran

Another strategy for obtaining nanoparticles with photoswitchable drug release is to take advantage of the spiropyran–merocyanine isomerization [165]. Colorless spiropyrans can undergo ring-opening reactions upon UV irradiation to yield their isomeric merocyanine forms with strong absorption in the visible region. In turn, the "open" merocyanine forms can transform back to the "closed" spiro forms either thermally (relatively slow rates, $t_{1/2} \approx 10^3$ s) or by exposure to visible light [166].

Qu et al. [147] reported NIR-triggered release of proteins from spiropyran-modified hollow UCNPs. In this study, carboxyl-functionalized spiropyrans were attached to amino-modified $NaYF_4$:Yb/Er UCNPs via covalent conjugation. The isomerization from the spiro forms to the merocyanine forms was almost complete after irradiation for 15 min upon UV light. β-galactosidase, a membrane-impermeable negatively charged protein, was encapsulated into the hollow structure and stabilized by the electrostatic interaction with the positively charged merocyanine forms. Upon 980-nm irradiation (0.5 W, 20 min), UCNPs emitted photons in the visible region, resulting in the photoisomerization of merocyanine forms to neutral spiro forms. Owing to the absence of the electrostatic interaction, β-galactosidase was released with a release amount of 70 wt% in 720 min. The in vitro studies further showed that the UCNP platforms were capable of the transportation of proteins into HeLa cells and that the released proteins also retained its biological activity. Thus, it would be a promising method for protein therapy [167].

9.3.2 NIR Light-Triggered Photolysis

9.3.2.1 Caging Group

"Caging" refers to the strategy of masking the activity of a compound with photolabile protecting groups [168]. Upon light irradiation (mainly UV light), these caging groups can be removed from caged compounds to make them functional, termed as "uncaging" [169]. After incorporating caging groups into UCNP-based carriers, photolysis reactions can lead to the liberation of cargo from nanocarriers by either the disruption of caging group-modified polymers utilized to encapsulate unmodified drugs or the bond cleavage of caging group caged drugs.

Zhao et al. incorporated a photolabile moiety, *o*-nitrobenzyl group, into block copolymer (BCP) [148] and hydrogel [149] structure, respectively, and then loaded UCNPs and payload molecules into as-prepared photosensitive BCP micelle and hydrogel to achieve NIR light-triggered release. In a prior work, $NaYF_4$:Yb/Tm@$NaYF_4$ core–shell UCNPs were loaded into micelles made from a photosensitive diblock copolymer composed of hydrophilic poly(ethylene oxide) (PEO) and hydrophobic polymethacrylate (PMA) bearing *o*-nitrobenzyl groups [148]. In order to absorb the upconverted blue light around 350 nm, the nitrobenzyl moiety was further functionalized with two methoxy groups to shift the absorption maximum from ~300 to 350 nm. Upon 980-nm irradiation, 4,5-dimethoxy-2-nitrobenzyl groups were cleaved, thus converting the hydrophobic PMA block into hydrophilic poly(methacrylic acid), which shifted the hydrophilic–hydrophobic balance toward the destabilization of micelles. When loading Nile red (NR) as a model payload into the micelle containing UCNPs, it was found that exposure to the

980-nm laser (5 W) resulted in a continuous decrease in fluorescence intensity of NR within 4 h, indicating micellar dissociation and subsequent release of NR molecules, the fluorescence of which would be quenched in water. In comparison, no change in the fluorescence intensity was observed for micelles containing only NR. Furthermore, the release of NR was faster with a higher NIR laser power or with more UCNPs in the micelle. In a follow-up study, the hydrogel utilized to encapsulate UCNPs and fluorescein isothiocyanate bovine serum albumin (FITC-BSA) had a cross-linked hybrid polyacrylamide–PEG structure held together by *o*-nitrobenzyl groups [149]. Upon NIR irradiation, the resulting UV emission cleaved the *o*-nitrobenzyl groups, resulting in the breakdown of the cross-linked structure of the gel (gel-sol transition), followed by the release of FITC-BSA. The release after 50 min of 980-nm irradiation was ~67 % at a laser power of 5 W and ~38 % at 3.1 W. The latter experiment was repeated twice, and the release of proteins was found to be ~40 and ~33 %, respectively, suggesting quite good reproducibility. Each time the light was turned off, there was no change in the emission intensity of FITC-BSA released into the solution, indicating the on-demand release of proteins.

Later, Almutairi et al. [150] and Xing et al. [151] applied this strategy to induce the disruption of nanocarriers and subsequent release of coumarin 153 and DOX, respectively. In Almutairi's report, the NIR-induced disruption was achieved by a cascade of cyclization and rearrangement reactions of the polymer chains consisting of *o*-nitrobenzyl-functionalized cresol monomer. As for Xing's work, the release of DOX was attributed to the cleavage of cross-linked 1-(2-nitrophenyl)ethyl photo-caged oligo(ethylene glycol) (OEG) linker done by upconverted UV light from $NaYF_4$:Yb/Tm@$NaYF_4$ UCNPs. Furthermore, FA was attached onto the surface of cross-linked nanocarriers to enhance the selective photocontrolled drug release in the tumor cells expressing FRs.

As mentioned above, drugs can also be caged by the caging groups and then loaded onto UCNPs. In other words, drugs are liberated from caged forms upon NIR irradiation.

In 2010, Branda et al. [152] reported remote control release of caged compounds using NIR light and $NaYF_4$:Yb/Tm@$NaYF_4$ UCNPs. Carboxylic acid as a model compound was covalently linked to 3′,5′-di(carboxymethoxy)benzoin. The as-prepared 3′,5′-di(carboxymethoxy)benzoin acetate bearing two carboxylic acid groups was attached to the surface of UCNPs via coordination. Exposure of the benzoin cage-decorated UCNPs in CH_3CN to a 980-nm laser (4.37 W, 556 W cm^{-2}) resulted in the photolysis of 3′,5′-di(carboxymethoxy)benzoin acetate, of which the spectrum partially overlapped (λ = 282 nm) with the emission spectrum of $NaYF_4$:Yb/Tm@$NaYF_4$ UCNPs (λ = 290 nm), leading to the generation of 5,7-di(carboxymethoxy)-2-phenylbenzo[*b*]furan and the release of acetic acid as evidenced by the UV–visible absorption spectrum. However, the emission of UCNPs at 290 nm was much weaker than that at 333–355 nm. In addition, the furan species as a product of photolysis reaction had an absorption maximum at 283–292 nm. Therefore, it is no wonder high laser intensity was required.

Xing et al. [153] utilized caged D-luciferin-UCNP composites for photouncaging and bioluminescence imaging (BLI). D-luciferin can recognize firefly luciferase (fLuc) reporter genes and produce bioluminescence in the presence of O_2, Mg^{2+}, and adenosine triphosphate (ATP) and thus can be used for BLI. In this study, D-luciferin, which was photocaged with a 1-(2-nitrophenyl)ethyl group, was modified with a maleimide-functionalized PEG linker and then attached to thiol-modified $NaYF_4$:Yb/Tm@$NaYF_4$ UCNPs via covalent conjugation. Upon 980-nm irradiation, the fluorescence of the solution containing the D-luciferin-UCNP composites increased significantly, while both the caged D-luciferin and the D-luciferin-UCNP composites did not exhibit D-luciferin fluorescence without NIR light. The photouncaging of D-luciferin-UCNP composites was further demonstrated in vivo via BLI in the presence of fLuc enzyme. The composites and the enzymes were subcutaneously injected into living mice at a tissue depth of 2 mm, and there was no detectable bioluminescence signal without NIR irradiation. After exposure to a 980-nm laser (0.8 W, 255 W cm^{-2}) for 1 h, strong bioluminescence shown up. By contrast, no notable bioluminescence was detected after UV irradiation (10 min), which was presumably attributed to the shallow penetration depth. However, in spite of the high power density, the signal detected after NIR irradiation was still weaker than that detected in a mouse directly injected with D-luciferin.

Li et al. [154] prepared a yolk–shell-structured nanocarrier with core–shell $NaYF_4$:Yb/Tm@$NaLuF_4$ nanocrystals as inner yolks and porous silica as the hollow shells and loaded amino-coumarin-caged chlorambucil (ACCh) inside the hollow inner cavity by physical adsorption (Fig. 9.6d). In order to avoid undesired premature release in vivo, amino-coumarin, the phototrigger moiety of ACCh, was further hydrophobized with two octanyl chains. Additionally, the mesoporous silica shell prevented contact between the adsorbed prodrug molecules and endogenous enzymes, avoiding the release triggered by enzymolysis. The loading ratio of ACCh was measured to be 710 μmol g^{-1} (49 wt%, 5.3×10^6 molecules per particle). The high loading capacity gave rise to a sustainable release profile. In addition to the long-term release pattern, the nanocarriers exhibited on–off pattern of chlorambucil release. The UCNP-based carriers were intratumorally injected into S180 tumor-bearing mice on day 1 and day 9, followed by exposure of 980-nm irradiation (0.05 W cm^{-2}) for 20 min each day for 16 d. These mice showed a much slower tumor growth rate (Fig. 9.6e) and a prolonged survival time as compared to the mice treated by saline or $NaYF_4$:Yb@$NaLuF_4$ UCNP-based carriers without the Tm^{3+} activator.

Yeh et al. [155] utilized caging group to address the issue of the heterogeneity in receptor expression among different cancer cells. In general, active targeting can be achieved by using specific markers, such as monoclonal antibodies, peptides, and aptamers. However, the expression of receptors responding to these makers varies for different cancer cells and even happens in pluripotent or young healthy cells [170]. For instance, a moderate-to-low frequency of FR expression is found in lung (66 %, non-small cell), endometrium (64 %), thyroid (48 %), breast (29 %), and lung (25 %, small cell) carcinoma, and the expression also occurs in normal kidney cells [171]. In this work, *o*-nitrobenzyl-caged FA was attached to the surface of silica-coated $NaYF_4$:Yb/Tm UCNPs via covalent conjugation to achieve NIR light-triggered active targeting. Enzyme was utilized as the stimulus instead of light to trigger the release of payload molecules. Specifically, DOX was thiolated on the surface of UCNPs, forming a disulfide bond that was cleaved by lysosomal enzymes within the cells.

Recently, Han et al. [156] reported the synthesis of α-$NaYbF_4$:Tm@CaF_2 UCNPs with an enhanced luminescence. These UCNPs were further employed for NIR light-triggered uncaging of 5-carboxymethoxy-2-nitrobenzyl-caged fluorescein that was attached to PEI-modified UCNPs via covalent conjugation. HeLa cells were incubated with these UCNPs loaded with caged fluorescein and then exposed to a 975-nm CW laser diode (0.011 W). The fluorescence signal from fluorescein was recorded in real time. It showed that photoactivation was able to be monitored in a real-time manner and was completed after 90 s of NIR irradiation. It is noteworthy that the cubic-phase α-$NaYbF_4$:Tm@CaF_2 UCNPs had even stronger UV emissions than the hexagonal-phase β-$NaYF_4$:Yb/Tm@$NaYF_4$ UCNPs which were most commonly used in NIR light-triggered photolysis of caged compounds as mentioned above. The UV emission quantum yield was measured to be 0.1 %, enabling the use of low laser intensity. In addition, compared to the typically used $NaYF_4$ shell, the biocompatible CaF_2 coating was found to be more effective in resisting quenching of luminescence in aqueous medium.

Gene therapy is a promising and rapidly developing therapeutic approach for the treatment of various diseases, ranging from birth defects to neurological disorders and from cancer to infectious diseases [172, 173].Whereas most gene therapy approaches have focused on a gain of function to alter the phenotype of a cell, there is increasing interest in developing therapies based on inhibition of function using nucleic acid molecules [174]. For instance, siRNAs are utilized to trigger the degradation of target mRNA, leading to the inhibition of gene expression [175]. Although viral vectors are very effective to deliver the nucleic acid molecules into living cells and tissues, they are limited by unpredictable cytotoxicity and immune responses [23]. To this end, some non-viral vectors, such as cationic lipids and polymers, have been developed for gene delivery [176, 177]. Moreover, caging these functional nucleic acids before they reach the target area is undoubtedly beneficial to minimize side effects to normal organs. Recently, some researchers have demonstrated NIR light-triggered uncaging of caged nucleic acids by virtue of UCNP-based carriers.

In an earlier work by Zhang et al. [157], 4,5-dimethoxy-2-nitroacetophenone (DMNPE)-caged DNA/siRNA (Fig. 9.7a) was loaded into the mesopores of $NaYF_4$:Yb/Tm UCNP@SiO_2@mSiO_2, which protected them from harmful enzymatic environment. After exposure to a 980-nm laser, DMNPE-caged siRNA loaded in the mesopores showed a similar shift in the trend of the absorbance spectra as compared with that directly exposed to UV light, thereby confirming the ability of UCNPs to uncage nucleic acids. In order to monitor NIR light-triggered uncaging, caged plasmid DNA-encoding enhanced green fluorescent protein (EGFP) was loaded into the mesoporous silica shell and then incubated with B16-F0 cells. Gradually enhanced green fluorescence showed up with the increase of NIR light dosage. When the dosage reached 2,000 J cm^{-2}, the fluorescence intensity was even equal to that in the group of UCNPs loaded with uncaged

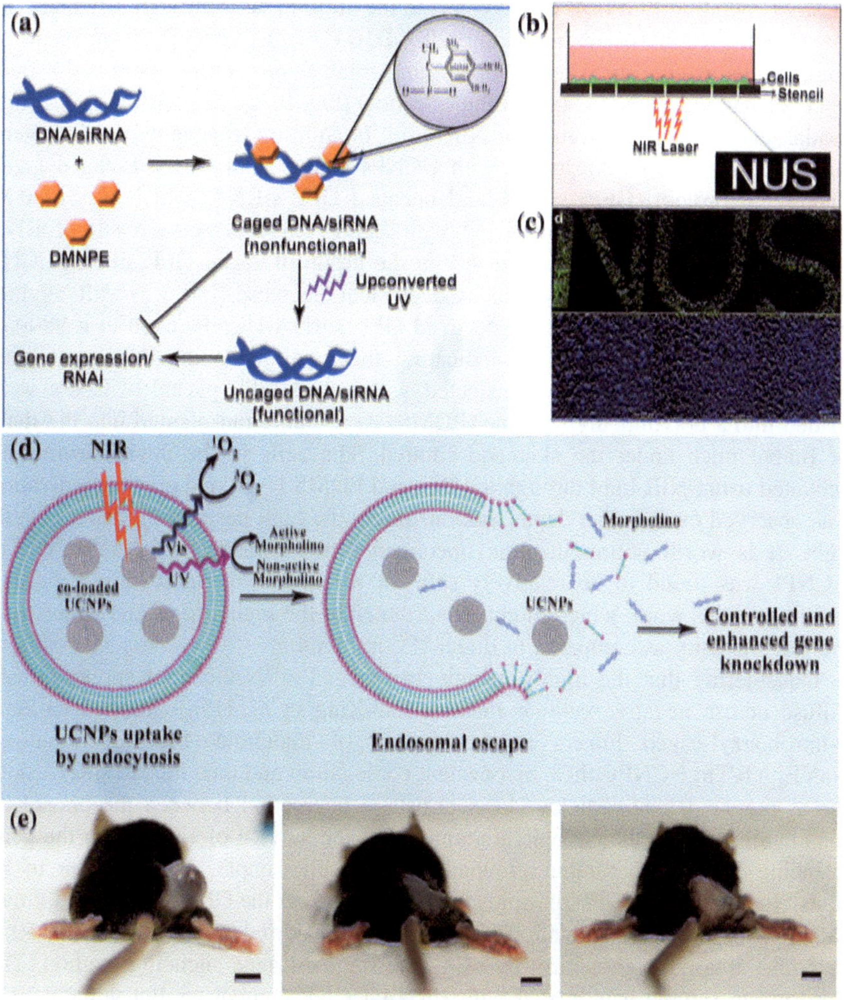

Fig. 9.7 **a** Schematic illustration for caging plasmid DNA and siRNA with DMNPE and then uncaging caged nucleic acids by upconverted UV light from UCNPs. **b** Schematic illustration of setup showing the stencil and the position of the laser for patterning of cells transfected with caged EGFP plasmid. Inset shows the pattern (NUS) on the stencil. **c** Composite image showing the GFP fluorescence (*top*) from three different wells (one letter in each well of a 96-well plate) and live-cell DAPI staining (*bottom*) of the same to show the cell confluence. *Scale bar* 200 μm. **d** Schematic illustration of NIR light-triggered endosomal escape and gene knockdown by using photomorpholino- and TPPS2a-coloaded UCNPs. **e** Representative photographs of mice after various treatments (*left* saline; *middle* photomorpholino-loaded UCNPs and NIR irradiation; *right* photomorpholino- and TPPS2a-coloaded UCNPs and NIR irradiation.). *Scale bar* 1 cm. **a–c** Reprinted with permission from Ref. [157]. Copyright 2012 National Academy of Sciences; **d, e** Reprinted with permission from Ref. [159]. Copyright 2014 American Chemical Society

plasmid encoding EGFP. When exposed to the 980-nm laser through a stencil with the pattern "NUS" for 12 min (Fig. 9.7b), B16-F0 cells, which distributed in the whole wells as confirmed by DAPI staining (blue fluorescence), exhibited a fluorescence pattern of "NUS" (green fluorescence) (Fig. 9.7c), demonstrating the high spatial and temporal resolution of NIR light. In addition to gene expression, gene silencing was also demonstrated with UCNPs loaded with DMNPE-caged green fluorescent protein (GFP) siRNA and uncaged GFP siRNA, which was used to transfect B16F0 cells pretransfected with EGFP plasmid. As a result of RNA interference, the fluorescence intensity in the group of caged GFP siRNA-UCNP composites showed a laser intensity-dependent decrease, but was still slightly higher than that in the group of uncaged GFP siRNA-UCNPs even at a 980-nm light dosage of 2,000 J cm^{-2}, indicating the incomplete knockdown of gene expression. After that, cells transfected with caged GFP plasmid-UCNPs were loaded into a polydimethylsiloxane (PDMS) device and transplanted into the flank of Balb/c mice under the skin and sutured. The cells in the device were then activated using NIR light through the skin and PDMS layer, and efficient activation was observed even at that depth, demonstrating the high penetration depth of NIR light. It is worth noting that the upconverted UV emission of $NaYF_4$:Yb/Tm UCNPs was found to be relatively safe for the cells without lowering the cell viability in this work, which might be attributed to the highly localized feature and the low intensity as compared to direct UV irradiation.

Considering that the highly negatively charged siRNA cannot spontaneously diffuse across negative cellular membranes, Xing et al. [158] attached cationic *o*-nitrobenzyl-caged linkers to the surface of maleimide–PEG-functionalized $NaYF_4$:Yb/Tm UCNP@SiO_2 by covalent conjugation and then loaded siRNA onto the as-prepared nanoparticles via electrostatic interaction between the phosphate ions in siRNA and the cationic ammonium on the surface of UCNPs. In the gene silencing tests, four parallel groups of HeLa cell samples were chosen to be transferred with the EGFP plasmids in order to express the GFP. The second group sample was also treated with EGFP siRNA transferred by lipofectamine 2000 (Lipo2k) as a positive control by using the standard gene silencing method. The third group sample was treated with siRNA-UCNP composites but without light irradiation as a negative control. The last cell samples (fourth group) were treated with siRNA-UCNPs and further exposed to a 980-nm laser for 2 h. As expected, the strong green fluorescence was observed in the first group and no fluorescence signal was observed in the second group. Meanwhile, very strong green fluorescence intensity was detected from the third group, whereas cells of the last group displayed very weak fluorescence signals, indicating the uncaging of *o*-nitrobenzyl - caged linkers and subsequent release of siRNA upon NIR irradiation. Moreover, the cell viability observed in the samples with NIR light irradiation was quite similar with those treated with siRNA/Lipo2k, demonstrating the safety of the NIR light-triggered UCNP-based approach.

In addition to siRNA, the photomorpholino duplex, which consists of a sense morpholino oligonucleotide (MO) with a photolabile moiety and an antisense MO that is unmodified, can also be utilized to achieve phototriggered gene knockdown

[159, 178]. The sense photomorpholino sequesters the antisense morpholino and prevents it from participating in RNA interference (RNAi). Upon UV light irradiation, the sense photomorpholino is cleaved, leading to the release of the antisense morpholino and subsequent targeted RNAi-mediated knockdown. Very recently, Zhang et al. [159] combined this approach with UCNPs to achieve NIR light-triggered gene knockdown (Fig. 9.7d). Moreover, photochemical internalization (PCI) was utilized to enhance endosomal escape of photomorpholino in this study. PCI involves the use of photosensitizers which localize primarily in the endosomes and lysosomes of cells. Light activation of photosensitizers results in the localized generation of ROS, thus destroying the membranes of endosomes and lysosomes and liberating the nanoparticles taken up by cells via endocytosis, whereas the contents of the organelles remain intact [179]. It is noteworthy that PCI is different from PDT, the purpose of which is to cause cell death. They simultaneously loaded a photosensitizer, *meso*-tetraphenylporphyrin disulfonate (TPPS2a, absorption peak at 413 nm), and a photomorpholino, anti-STAT3, into the pores of mesoporous silica-coated $NaYF_4$:Yb/Tm@$NaYF_4$:Yb/Er UCNPs. The core–shell UCNPs had multiple UV and visible emission peaks, two of which coincided with the photomorpholino and TPPS2a absorption maxima, respectively. The loading of TPPS2a and photomorpholino for coloaded UCNPs was found to be 1.61-μg TPPS2a/mg UCNP and 49.2-μg photomorpholino/mg UCNP, respectively. In vitro tests showed that there was an increase in cellular concentration of the UCNPs loaded with TPPS2a when compared to those without TPPS2a. Meanwhile, the extracellular concentration of UCNPs (in the media) in the control group treated with UCNPs alone was higher than that in the group incubated with UCNPs loaded with TPPS2a. After exposure to a 980-nm laser (2.8 W cm^{-2}, 8 min), the STAT3 knockdown was significantly higher in the sample with TPPS2a and photomorpholino as compared to sample with only photomorpholino (p = 0.0063), while control groups (cells treated with NIR irradiation alone, photomorpholino-loaded UCNPs alone, and TPPS2a and photomorpholino-coloaded UCNPs alone) did not cause significant changes. UCNPs loaded with STAT3 photomorpholino and TPPS2a or not were further administered to melanoma-bearing mice on day 1 and day 4, and the tumors were irradiated with a NIR laser (0.3 W cm^{-2}, 40 min, on day 1 and day 4). Both groups showed a significant reduction in tumor volume as compared to the control group where mice were only injected with saline (Fig. 9.7e). Furthermore, there was a marked decline in tumor volume in mice treated with UCNPs coloaded with photomorpholino and TPPS2a as compared to those treated with photomorpholino-loaded UCNPs alone (p = 0.00383).

9.3.2.2 Nitric Oxide Precursor

Nitric oxide (NO), the first identified gasotransmitter, regulates a variety of physiological and pathological processes, such as vascular and neurological processes (blood flow regulation, smooth muscle relaxation, platelet reactivity, neurotransmission) at low concentration, and cytotoxic processes (cytostasis and cytolysis) at relatively high concentration [180–182]. Hence, the controlled therapeutic delivery of NO is highly desired. Zhang and Ford et al. [160, 161] demonstrated the use of UCNPs to facilitate NO uncaging from a well-characterized NO precursor, iron/sulfur/nitrosyl cluster Roussin's black salt anion $Fe_4S_3(NO)_7^-$ (RBS) (see below).

$$Fe_4S_3(NO)_7^- \xrightarrow[O_2]{\text{NIR light, UCNPs}} 3.9Fe^{2+} + 5.9NO + 3S^{2-} + \text{other products}$$

In a prior work, $NaYF_4$:Yb/Er@$NaYF_4$ UCNPs was first coated with an amino-modified silica shell and then treated with acidic solution to obtain cationic $-NH^{3+}$ groups which enabled the attraction of RBS anion [160]. The quantitative data showed that the amount of NO generated from the solution containing as-prepared RBS-UCNP composites upon 980-nm irradiation (~1 W) was a linear function of the irradiation interval from 2 to 160 s. By contrast, NO release was not detected for the solution containing only RBS upon 980-nm irradiation. After the preliminary experiments that confirmed the ability of NIR light to trigger photolysis of NO precursor in the presence of UCNPs, they loaded RBS (Na^+ salt) into the pores of mesoporous silica-coated UCNP@SiO_2 (i.e., UCNP@SiO_2@mSiO_2). The as-synthesized nanoparticles were further coated with PAH to inhibit diffusion of the NO precursor back into the solution. When these composites were subjected to a 980-nm laser (4.0 W), the NO release responded linearly to the irradiation time but displayed a nonlinear response to the laser power intensity, presumably attributed to the nonlinear nature of upconversion process.

Later, they demonstrated that UCNP-based carriers were capable of generating physiologically relevant levels of NO upon NIR irradiation through skin, muscle, and fat tissue [161]. In this study, UCNPs and RBS $(NH)_4{}^+$ salt) were incorporated into a PDMS matrix to form a polymer disk (PD) (Fig. 9.8a). The resulting PDs were 5 mm in diameter and 2 mm thick but displayed some inhomogeneity in UCNP distribution. In order to average sample heterogeneity and minimize localized heating effects, a novel optical train that scanned the focused excitation beam rapidly and repeatedly over the PD surface was utilized. Specifically, it was achieved by reflecting the NIR beam off two mirrors, each mounted on a speaker oscillating under the control of a tunable function generator (Fig. 9.8b). The tissue sample was placed before the PD in the optical train. It is noteworthy that without the oscillating device, the focused beam (400 W cm^{-2}) damaged the tissue samples. Three types of porcine tissue, skin, muscle, and fat and two types of UCNPs, $NaYF_4$:Gd/Yb/Er@$NaYF_4$, and $NaYF_4$:Gd/Yb/Tm@$NaYF_4$ were used for these experiments in the presence of oscillating focused beam. It was found that the transmittance of muscle (~2.7 mm thick), fat (~2 mm), and skin (~2.6 mm) samples was 30, 16, and 13 %, respectively. Therefore, the signals were significantly decreased as compared to the "unfiltered" experiments without tissue samples. However, NO was still generated in very significant quantities, comparable to the physiologically relevant levels.

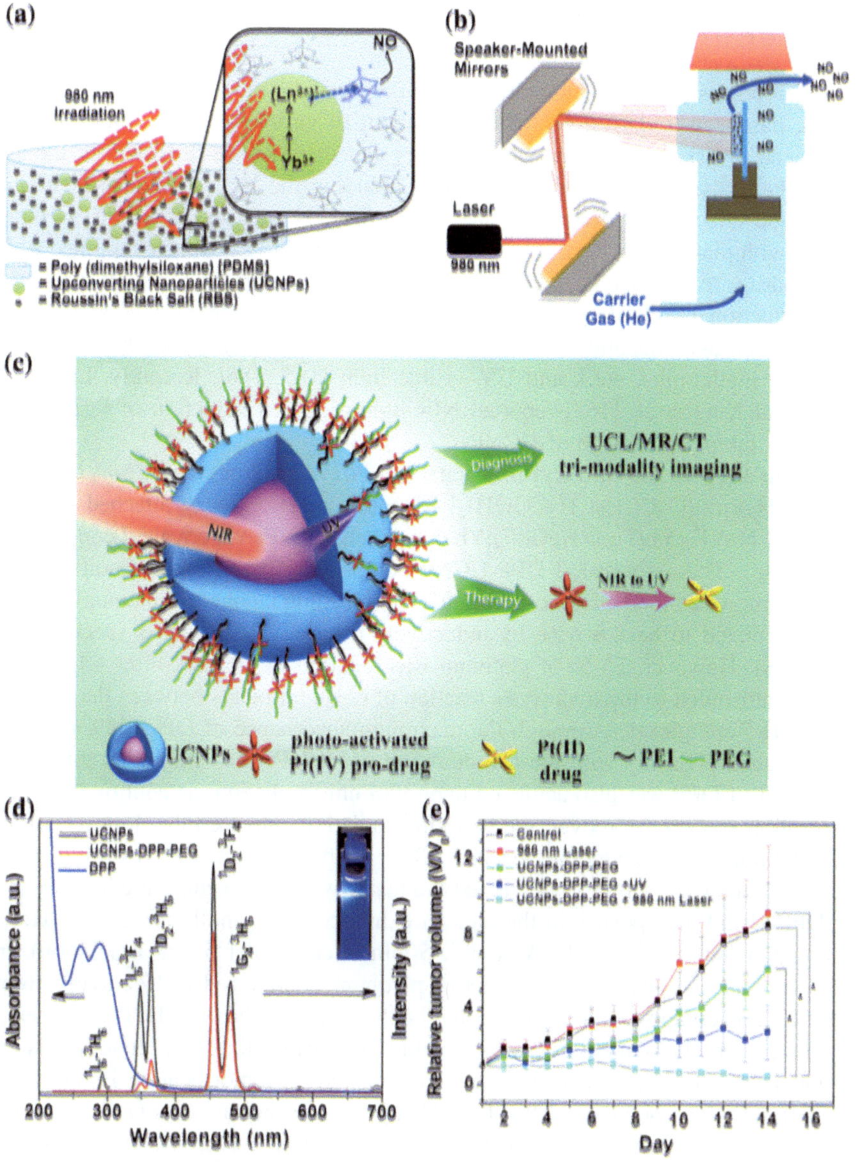

Fig. 9.8 **a** Schematic illustration of PDMS PDs embedded with UCMPs and RBS as NIR light transducer and NO precursor, respectively. **b** Schematic illustration of the optical train for photolysis of the PDs. The beam is moved in the "X" and "Y" directions using the two speaker-mounted mirrors. Also depicted is the flow cell in which the sample is mounted where He carries any NO released to the NO analyzer for detection. **c** Schematic illustration of DPP-UCMP composites for therapy and multimodal imaging. **d** Absorption spectrum of the DPP (*blue line*), emission spectra of pure UCNPs (*black line*), and DPP-UCNPs (*red line*) under 980-nm laser. The luminescence photograph of UCNPs under 980-nm laser in the dark is given in the inset. **e** In vivo tumor volume changes of Balb/c mice on different groups after various treatments, as indicated. *$P < 0.05$. **a**, **b** Reprinted with permission from Ref. [161]. Copyright 2013 American Chemical Society; **c–e** Reprinted with permission from Ref. [162]. Copyright 2013 American Chemical Society

9.3.3 NIR Light-Triggered Photoreduction

Cis-dichlorodiammineplatinum(II) (cisplatin), one of the most successful anticancer drugs, has been widely used for the treatment of various cancers, such as lung, colorectal, ovarian, breast, and genitourinary cancers [183, 184]. However, the use of cisplatin also suffers from some issues including the lack of tumor specificity, toxic side effects, and drug resistance [123]. To this end, less toxic Pt(IV) complexes with platinum in the more inert +IV oxidation state have been exploited as Pt (II) prodrugs to improve the therapeutic effect and minimize toxic side effects [183]. The reduction of Pt(IV) prodrugs can be achieved by using of several types of stimuli such as extracellular and intracellular reducing agents (ascorbic acid, glutathione, methionine, etc.), and UV–visible light [185, 186]. Recently, Lin et al. [162] and Xing et al. [163] reported NIR light-triggered reduction of Pt(IV) prodrugs by taking advantage of UCNPs.

In Lin's report, the carboxyl-functionalized Pt(IV) prodrug, *trans,trans,trans*-[Pt$(N_3)_2(NH_3)(py)(O_2CCH_2CH_2COOH)_2$] (denoted as DPP), was attached to PEI-modified $NaYF_4$:Yb/Tm@$NaGdF_4$:Yb core–shell UCNPs via covalent conjugation (Fig. 9.8c). The as-prepared DPP-UCNP composites were further modified with PEG to improve blood circulation half-life. The loading amount of platinum content was calculated to be 2.6 wt% by inductively coupled plasma mass spectrometry (ICP-MS). However, 25 % of platinum was released in the dark after 4 h, which may be attributed to the hydrolysis reaction of covalent bonds between the UCNPs and Pt(IV) complexes. Although the maximum absorbance of DPP (289 nm) was not matched precisely with the UV emission of UCNPs (365 nm) (Fig. 9.8d), the reduction of DPP was also achieved upon 980-nm irradiation, as confirmed by the absorption spectra. Moreover, the 980-nm irradiation enhanced the drug release as the platinum release amount reached 48.4 % after 30 min of NIR irradiation. Although the cell viabilities decreased gradually with increasing the concentration of DPP-UCNP composites in the dark, it exhibited better inhibition against HeLa cells upon 980-nm laser (4.5 W cm^{-2}, 5-min interval after 5-min irradiation). In vitro tests also showed that the mass of platinum internalized in HeLa cells from DPP-UCNP composites was much higher than that of the free DPP after being incubated for the same time, which could be attributed to the endocytosis of nanoparticles. Further experiments were conducted on tumor-bearing Balb/c mice by intratumoral injection of the DPP-UCNP composites to evaluate the tumor inhibition efficacy. The group treated with DPP-UCNPs and 980-nm irradiation (2.5 W cm^{-2}, 5-min interval after 5-min irradiation) demonstrated better tumor growth inhibition efficacy as compared with the control groups treated with DPP-UCNPs and direct UV light or not (Fig. 9.8e), confirming the large tissue penetration depth of NIR light. It was also found that the nanoparticles accumulated in the lungs with 1 h post-injection and were transferred to the spleen within 24 h post-injection. Finally, they were nearly completely excreted from the mice after 7 days. In addition, the UCNPs were used as contrast agents for UCL/MR/CT trimodality imaging.

As for Xing's study, *trans,trans,trans*-[Pt(N_3)$_2$(OH)(py)$_2$($O_2CCH_2CH_2CO_2H$)] (donated as MCP) with one carboxyl group was used as Pt(IV) prodrug and covalently attached to $NaYF_4$:Yb/Tm UCNP@SiO_2, which was first functionalized with a bridge peptide through an OEG linker. Furthermore, a short peptide probe with a flanking activatable fluorescence resonance energy transfer (FRET) pair consisting of a far-red fluorescence donor, Cy5, and a NIR acceptor, Qsy21 (Cy5-acp-CGDEVDAK-Qsy21), for specific recognition of caspase-3, was also attached to UCNP@SiO2 to conduct real-time apoptosis imaging. The optimal amounts of Pt content on the nanoparticle surface were determined to be 3.5 wt% by ICP-MS. Prior to NIR light irradiation, platinum release was not observed from both the free Pt(IV) complex and the MCP-UCNP composites in the dark. Upon 980-nm laser excitation, the time-dependent release of the Pt component was detected from the solution of MCP-UCNP composites, whereas no notable Pt release was detected in the case of the free Pt(IV) prodrug. After that, a well-characterized tumor cell line pair, cisplatin-susceptible human ovarian carcinoma A2780 cell, and its derivative cisplatin-resistant subline A2780cis cell were chosen to incubate with the MCP-UCNP composites. After exposure to a 980-nm laser (1.5 W cm^{-2}), both A2780 and A2780cis cells displayed increased cytotoxicity with increasing the concentration of MCP-UCNPs or the duration of NIR light irradiation. In addition, the combination of MCP-UCNPs and NIR light exhibited higher antitumor efficacy for cisplatin-resistant A2780cis cells as compared with cisplatin. They further carried out a cytotoxic mechanism study by taking advantage of the FRET pair attached to the surface of UCNPs. NIR light irradiation of A2780 and A2780cis cells incubated with MCP-UCNPs resulted in a remarkable Cy5 fluorescence in both cells. By contrast, obvious fluorescence was not observed in the control experiments for cells that were either treated with additional caspase-3 inhibitors or not exposed to laser irradiation. Therefore, the Cy5 fluorescence was attributed to the activated caspase-3, which cleaved the sensing peptide and thus resulted in the release of Qsy21 acceptor and subsequent disruption of the FRET process. Considering that caspase-3 is an apoptosis-related cysteine peptidase, it was concluded that NIR light-triggered photoreduction of Pt(IV) prodrugs initiated apoptosis.

9.4 UCNPs for Photothermal Therapy

PTT employs photoabsorbers which can generate heat from the absorption of light to achieve thermal ablation of diseased cells. UCNPs generally have limited capability to convert light directly into heat because of the low extinction coefficient of lanthanide ions [62] and are thus coupled to photoabsorbers possessing high photothermal conversion efficiency, including Au nanomaterials [187–189], Ag nanomaterials [190], CuS nanoparticles [191], and organic dyes [91, 192] in PTT treatment of diseases. In this context, UCNPs are mainly responsible for multimodal imaging as nanoprobes. For example, Liu et al. [187, 188] employed a LbL self-assembly approach to synthesize multi-functional nanoparticles consisting a

$NaYF_4$:Yb/Er UCNP particle as the core, a layer of ultrasmall iron oxide nanoparticles (IONPs) as the intermediate shell and a thin layer of gold as the outer shell. The intermediate shell of IONPs not only afforded the UCNP-IONP-Au composites' magnetic properties but also significantly reduced the luminescence quenching effects of the gold shell to UCNPs. The gold shell possessing surface plasmon absorption with large cross sections endowed the UCNP-IONP-Au composites with a wide absorption band from the visible to the NIR region. Hence, these composites were capable of UCL/MR bimodal imaging and killing cancer cells upon 808-nm irradiation.

In another work by Shi et al. [191], CuS nanoparticles were employed as PTT agents because of their low cost, low cytotoxicity, and high photothermal conversion efficiency, and the negatively charged citrate-stabilized CuS nanoparticles were attached to the positively charged surface of amino-modified $NaYbF_4$:Gd/Er UCNPs via an electrostatic adsorption method. The UCNP inner core offered the UCNP-CuS composites with three imaging modalities, including UCL, MR, and CT. Furthermore, the presence of high Z elements (Yb, Gd, and Er) caused a large local radiation dose enhancement around the UCNP-CuS composites, leading to a synergistic therapeutic effect of radiotherapy (RT) and PTT.

UCNPs can act not only as imaging agents and radiosensitizers, but also as temperature sensors in PTT. Song et al. [190] fabricated $NaYF_4$:Yb/Er UCNP@Ag core–shell nanocomposites and tuned the surface plasmon absorption of Ag shell to NIR region. The UCL intensity ratio of $^2H_{11/2}$–$^4I_{15/2}$ to $^4S_{3/2}$–$^4I_{15/2}$ was sensitive to temperature and thus was utilized to achieve thermal detection in real time, while the Ag shell was employed for NIR-induced PTT.

9.5 Conclusion and Perspectives

In this chapter, we summarized advances in UCNP-based NIR-induced LAT including PDT, chemotherapy, and gene therapy. The precise spatial and temporal control of these therapeutic approaches can be achieved by taking advantage of NIR light that possessing high tissue penetration and minimal photocytotoxicity, while the activation of photosensitive compounds (photosensitizers, photoswitchable or photocaged molecules, and photoactivated prodrugs) involved in the therapy is realized by virtue of NIR-to-UV/visible emissions of UCNPs. Moreover, incorporating photosensitizers and other therapeutic payloads into UCNP-based carriers can enhance their solubility, stability, blood circulation half-life, and targeting ability. Hence, the UCNP-based systems possess improved therapeutic efficacy while reducing side effects.

Nevertheless, several issues still remain to be resolved in order to fully realize the potential of the UCNP-based therapeutic approach.

1. As the fundamental of this approach, UCNPs still suffer from unsatisfactory quantum yield. As demonstrated in previous paragraphs, most in vivo tests (especially in the cases of chemotherapy and gene therapy) still need high laser power intensity and/or a long irradiation time, which may cause an overheating effect to damage living organisms. Hence, UCNPs with high quantum yield are strongly required for minimizing the overheating effect. Alternatively, Nd^{3+} ion, with a large absorption cross section at 808 nm, can serve as a sensitizer for the upconversion process and thus significantly reduce overheating as the absorption of water at 808 nm is much lower relative to that at 980 nm [123].
2. In addition to upconversion efficiency, some other parameters including the spectral overlap between photosensitizers and UCNPs, the loading amount, and the extinction coefficient of photosensitizers also greatly affect the generation of $^1O_2$. However, the higher loading capacity of photosensitizers does not guarantee a better PDT efficiency owing to the aggregation and self-quenching of photosensitizer molecules. Thus, utilizing photosensitizers with large spectral overlap and extinction coefficient is preferred. Most currently used photosensitizers in the clinic are porphyrins and porphyrin derivatives which have a strong absorption band in the 400–450 nm region (Soret band) and four weak absorption bands in the 500–700 nm region (Q-bands) [193]. However, NIR-to-blue upconversion luminescence has been seldom utilized to activate porphyrin-based photosensitizers (Table 9.2).
3. Although some researchers have demonstrated the activate targeting of UCNP-based carriers by using several types of targeting agents including FA, peptide, antibody, and aptamer, most in vivo tests are still conducted by intratumoral injection, while intravenous injection is found to result in an unsatisfactory PDT efficacy [82]. Actually, the intravenous injection is more practical in the clinic. Hence, an improved approach that combines several targeting agents and even magnetic targeting is highly desired.

References

1. Idris, N.M., Jayakumar, M.K.G., Bansal, A., Zhang, Y.: Upconversion nanoparticles as versatile light nanotransducers for photoactivation applications. Chem. Soc. Rev. (2014). doi:10.1039/c4cs00158c
2. Mayer, G., Heckel, A.: Biologically active molecules with a "light switch". Angew. Chem. Int. Ed. **45**, 4900–4921 (2006)
3. Langer, R.: Drugs on target. Science **293**, 58–59 (2001)
4. Chatterjee, D.K., Fong, L.S., Zhang, Y.: Nanoparticles in photodynamic therapy: an emerging paradigm. Adv. Drug Deliver. Rev. **60**, 1627–1637 (2008)
5. Shao, Q., Xing, B.: Photoactive molecules for applications in molecular imaging and cell biology. Chem. Soc. Rev. **39**, 2835–2846 (2010)
6. Anderson, R.R., Parrish, J.A.: The optics of human skin. J. Invest. Dermatol. **77**, 13–19 (1981)

7. Clydesdale, G.J., Dandie, G.W., Muller, H.K.: Ultraviolet light induced injury: immunological and inflammatory effects. Immunol. Cell Biol. **79**, 547–568 (2001)
8. Wang, F., Liu, X.: Recent advances in the chemistry of lanthanide-doped upconversion nanocrystals. Chem. Soc. Rev. **38**, 976–989 (2009)
9. Dichtel, W.R., Serin, J.M., Edder, C., Fréchet, J.M.J., Matuszewski, M., Tan, L., Ohulchanskyy, T.Y., Prasad, P.N.: Singlet oxygen generation via two-photon excited FRET. J. Am. Chem. Soc. **126**, 5380–5381 (2004)
10. Karotki, A., Khurana, M., Lepock, J.R., Wilson, B.C.: Simultaneous two-photon excitation of photofrin in relation to photodynamic therapy. Photochem. Photobiol. **82**, 443–452 (2006)
11. Kim, S., Ohulchanskyy, T.Y., Pudavar, H.E., Pandey, R.K., Prasad, P.N.: Organically modified silica nanoparticles co-encapsulating photosensitizing drug and aggregation-enhanced two-photon absorbing fluorescent dye aggregates for two-photon photodynamic therapy. J. Am. Chem. Soc. **129**, 2669–2675 (2007)
12. Shen, X., He, F., Wu, J., Xu, G.Q., Yao, S.Q., Xu, Q.: Enhanced two-photon singlet oxygen generation by photosensitizer-doped conjugated polymer nanoparticles. Langmuir **27**, 1739–1744 (2011)
13. Goyan, R.L., Cramb, D.T.: Near-infrared two-photon excitation of protoporphyrin IX: photodynamics and photoproduct generation. Photochem. Photobiol. **72**, 821–827 (2000)
14. Gary-Bobo, M., Mir, Y., Rouxel, C., Brevet, D., Basile, I., Maynadier, M., Vaillant, O., Mongin, O., Blanchard-Desce, M., Morère, A., Garcia, M., Durand, J., Raehm, L.: Mannose-functionalized mesoporous silica nanoparticles for efficient two-photon photodynamic therapy of solid tumors. Angew. Chem. Int. Ed. **50**, 11425–11429 (2011)
15. Härtner, S., Kim, H., Hampp, N.: Phototriggered release of photolabile drugs via two-photon absorption-induced cleavage of polymer-bound dicoumarin. J. Polym. Sci. Part A Polym. Chem. **45**, 2443–2452 (2007)
16. Zheng, Q., Bonoiu, A., Ohulchanskyy, T.Y., He, G.S., Prasad, P.N.: Water-soluble two-photon absorbing nitrosyl complex for light-activated therapy through nitric oxide release. Mol. Pharm. **5**, 389–398 (2008)
17. Croissant, J., Maynadier, M., Gallud, A., Peindy N'Dongo, H., Nyalosaso, J.L., Derrien, G., Charnay, C., Durand, J., Raehm, L., Serein-Spirau, F., Cheminet, N., Jarrosson, T., Mongin, O., Blanchard-Desce, M., Gary-Bobo, M., Garcia, M., Lu, J., Tamanoi, F., Tarn, D., Guardado-Alvarez, T.M., Zink, J.I.: Two-photon-triggered drug delivery in cancer cells using nanoimpellers. Angew. Chem. Int. Ed. **52**, 13813–13817 (2013)
18. Ji, W., Li, N., Chen, D., Qi, X., Sha, W., Jiao, Y., Xu, Q., Lu, J.: Coumarin-containing photo-responsive nanocomposites for NIR light-triggered controlled drug release via a two-photon process. J. Mater. Chem. B **1**, 5942–5949 (2013)
19. Croissant, J., Chaix, A., Mongin, O., Wang, M., Clément, S., Raehm, L., Durand, J., Hugues, V., Blanchard-Desce, M., Maynadier, M., Gallud, A., Gary-Bobo, M., Garcia, M., Lu, J., Tamanoi, F., Ferris, D.P., Tarn, D., Zink, J.I.: Two-photon-triggered drug delivery via fluorescent nanovalves. Small **10**, 1752–1755 (2014)
20. Banerjee, S.S., Chen, D.H.: A multifunctional magnetic nanocarrier bearing fluorescent dye for targeted drug delivery by enhanced two-photon triggered release. Nanotechnology **20**, 185103–185111 (2009)
21. Wang, C., Cheng, L., Liu, Z.: Upconversion nanoparticles for photodynamic therapy and other cancer therapeutics. Theranostics **3**, 317–330 (2013)
22. Wu, S., Han, G., Milliron, D.J., Aloni, S., Altoe, V., Talapin, D.V., Cohen, B.E., Schuck, P. J.: Non-blinking and photostable upconverted luminescence from single lanthanide-doped nanocrystals. Proc. Natl. Acad. Sci. USA **106**, 10917–10921 (2009)
23. Shen, J., Zhao, L., Han, G.: Lanthanide-doped upconverting luminescent nanoparticle platforms for optical imaging-guided drug delivery and therapy. Adv. Drug Deliver. Rev. **65**, 744–755 (2013)
24. Dolmans, D.E.J.G., Fukumura, D., Jain, R.K.: Photodynamic therapy for cancer. Nat. Rev. Cancer **3**, 380–387 (2003)

25. Celli, J.P., Spring, B.Q., Rizvi, I., Evans, C.L., Samkoe, K.S., Verma, S., Pogue, B.W., Hasan, T.: Imaging and photodynamic therapy: mechanisms, monitoring, and optimization. Chem. Rev. **110**, 2795–2838 (2010)
26. Lipson, R.L., Baldes, E.J.: The photodynamic properties of a particular hematoporphyrin derivative. Arch. Dermatol. **82**, 508–516 (1960)
27. Lipson, R.L., Baldes, E.J., Olsen, A.M.: The use of a derivative of hematoporphyrin in tumor detection. J. Natl Cancer Inst. **26**, 1–11 (1961)
28. Lipson, R.L., Baldes, E.J., Olsen, A.M.: Hematoporphyrin derivative: a new aid for endoscopic detection of malignant disease. J. Thorac. Cardiovasc. Surg. **42**, 623–629 (1961)
29. Lipson, R.L., Baldes, E.J., Olsen, A.M.: Further evaluation of the use of hematoporphyrin derivative as a new aid for the endoscopic detection of malignant disease. Dis. Chest. **46**, 676–679 (1964)
30. Schwartz, S., Absolon, K., Vermund, H.: Some relationships of porphyrins, X-rays and tumors. Univ. Minn. Med. Bull. **27**, 7–8 (1955)
31. Dougherty, T.J., Grindey, G.B., Fiel, R., Weishaupt, K.R., Boyle, D.G.: Photoradiation therapy. II. cure of animal tumors with hematoporphyrin and light. J. Natl Cancer Inst. **55**, 115–121 (1975)
32. Brown, S.B., Brown, E.A., Walker, I.: The present and future role of photodynamic therapy in cancer treatment. Lancet Oncol. **5**, 497–508 (2004)
33. Shibu, E.S., Hamada, M., Murase, N., Biju, V.: Nanomaterials formulations for photothermal and photodynamic therapy of cancer. J. Photochem. Photobiol. C **15**, 53–72 (2013)
34. Agostinis, P., Berg, K., Cengel, K.A., Foster, T.H., Girotti, A.W., Gollnick, S.O., Hahn, S. M., Hamblin, M.R., Juzeniene, A., Kessel, D., Korbelik, M., Moan, J., Mroz, P., Nowis, D., Piette, J., Wilson, B.C., Golab, J.: Photodynamic therapy of cancer: an update. CA Cancer J. Clin. **61**, 250–281 (2011)
35. Sharman, W.M., Allen, C.M., van Lier, J.E.: Photodynamic therapeutics: basic principles and clinical applications. Drug Discov. Today **4**, 507–517 (1999)
36. Ochsner, M.: Photophysical and photobiological processes in the photodynamic therapy of tumours. J. Photochem. Photobiol. B Biol. **39**, 1–18 (1997)
37. Plaetzer, K., Krammer, B., Berlanda, J., Berr, F., Kiesslich, T.: Photophysics and photochemistry of photodynamic therapy: fundamental aspects. Laser. Med. Sci. **24**, 259–268 (2009)
38. Foote, C.S.: Definition of type I and type II photosensitized oxidation. Photochem. Photobiol. **54**, 659 (1991)
39. Castano, A.P., Demidova, T.N., Hamblin, M.R.: Mechanisms in photodynamic therapy: part one-photosensitizers, photochemistry and cellular localization. Photodiag. Photodyn. Ther. **1**, 279–293 (2004)
40. Schweitzer, C., Mehrdad, Z., Noll, A., Grabner, E., Schmidt, R.: Mechanism of photosensitized generation of singlet oxygen during oxygen quenching of triplet states and the general dependence of the rate constants and efficiencies of O_2. J. Phys. Chem. A **107**, 2192–2198 (2003)
41. Moan, J., Berg, K.: The photodegradation of porphyrins in cells can be used to estimate the lifetime of singlet oxygen. Photochem. Photobiol. **53**, 549–553 (1991)
42. Bonnett, R., Martínez, G.: Photobleaching of sensitisers used in photodynamic therapy. Tetrahedron **57**, 9513–9547 (2001)
43. Szaciłowski, K., Macyk, W., Drzewiecka-Matuszek, A., Brindell, M., Stochel, G.: Bioinorganic photochemistry: frontiers and Mechanisms. Chem. Rev. **105**, 2647–2694 (2005)
44. Henderson, B.W., Dougherty, T.J.: How does photodynamic therapy work? Photochem. Photobiol. **55**, 145–157 (1992)
45. Robertson, C.A., Evans, D.H., Abrahamse, H.: Photodynamic therapy (PDT): a short review on cellular mechanisms and cancer research applications for PDT. J. Photochem. Photobiol. B Biol. **96**, 1–8 (2009)

46. Buytaert, E., Dewaele, M., Agostinis, P.: Molecular effectors of multiple cell death pathways initiated by photodynamic therapy. Biochim. Biophys. Acta **1776**, 86–107 (2007)
47. Castano, A.P., Demidova, T.N., Hamblin, M.R.: Mechanisms in photodynamic therapy: part two-cellular signaling, cell metabolism and modes of cell death. Photodiag. Photodyn. Ther. **2**, 1–23 (2005)
48. Castano, A.P., Demidova, T.N., Hamblin, M.R.: Mechanisms in photodynamic therapy: Part three-photosensitizer pharmacokinetics, biodistribution, tumor localization and modes of tumor destruction. Photodiag. Photodyn. Ther. **2**, 91–106 (2005)
49. Dougherty, T.J., Gomer, C.J., Henderson, B.W., Jori, G., Kessel, D., Korbelik, M., Moan, J., Peng, Q.: Photodynamic therapy. J. Natl Cancer Inst. **90**, 889–905 (1998)
50. Allison, R.R., Downie, G.H., Cuenca, R., Hu, X., Childs, C.J., Sibata, C.H.: Photosensitizers in clinical PDT. Photodiag. Photodyn. Ther. **1**, 27–42 (2004)
51. Hopper, C.: Photodynamic therapy: a clinical reality in the treatment of cancer. Lancet Oncol. **1**, 212–219 (2000)
52. Ethirajan, M., Chen, Y., Joshi, P., Pandey, R.K.: The role of porphyrin chemistry in tumor imaging and photodynamic therapy. Chem. Soc. Rev. **40**, 340–362 (2010)
53. Ormond, A., Freeman, H.: Dye sensitizers for photodynamic therapy. Materials **6**, 817–840 (2013)
54. O Connor, A.E., Gallagher, W.M., Byrne, A.T.: Porphyrin and nonporphyrin photosensitizers in oncology: preclinical and clinical advances in photodynamic therapy. Photochem. Photobiol. **85**, 1053–1074 (2009)
55. DeRosa, M.C., Crutchley, R.J.: Photosensitized singlet oxygen and its applications. Coordin. Chem. Rev. **233–234**, 351–371 (2002)
56. Muszkat, L., Feigelson, L., Bir, L., Muszkat, K.A.: Titanium dioxide photocatalyzed oxidation of proteins in biocontaminated waters. J. Photochem. Photobiol. B Biol. **60**, 32–36 (2001)
57. Lovell, J.F., Liu, T.W.B., Chen, J., Zheng, G.: Activatable photosensitizers for imaging and therapy. Chem. Rev. **110**, 2839–2857 (2010)
58. Kachynski, A.V., Pliss, A., Kuzmin, A.N., Ohulchanskyy, T.Y., Baev, A., Qu, J., Prasad, P. N.: Photodynamic therapy by in situ nonlinear photon conversion. Nat. Photon. **8**, 455–461 (2014)
59. Li, Z., Zhang, Y.: Monodisperse silica-coated polyvinylpyrrolidone/$NaYF_4$ nanocrystals with multicolor upconversion fluorescence emission. Angew. Chem. Int. Ed. **45**, 7732–7735 (2006)
60. Zhou, J., Yao, L., Li, C., Li, F.: A versatile fabrication of upconversion nanophosphors with functional-surface tunable ligands. J. Mater. Chem. **20**, 8078–8085 (2010)
61. Wang, Z., Liu, C., Chang, L., Li, Z.: One-pot synthesis of water-soluble and carboxyl-functionalized β-$NaYF_4$:Yb, Er(Tm) upconversion nanocrystals and their application for bioimaging. J. Mater. Chem. **22**, 12186–12192 (2012)
62. Chen, G., Qiu, H., Prasad, P.N., Chen, X.: Upconversion nanoparticles: design, nanochemistry, and applications in theranostics. Chem. Rev. **114**, 5161–5214 (2014)
63. Rai, P., Mallidi, S., Zheng, X., Rahmanzadeh, R., Mir, Y., Elrington, S., Khurshid, A., Hasan, T.: Development and applications of photo-triggered theranostic agents. Adv. Drug Deliver. Rev. **62**, 1094–1124 (2010)
64. Barreto, J.A., OMalley, W., Kubeil, M., Graham, B., Stephan, H., Spiccia, L.: Nanomaterials: applications in cancer imaging and therapy. Adv. Mater. **23**, 18–40 (2011)
65. Krämer, K.W., Biner, D., Frei, G., Güdel, H.U., Hehlen, M.P., Lüthi, S.R.: Hexagonal sodium yttrium fluoride based green and blue emitting upconversion phosphors. Chem. Mater. **16**, 1244–1251 (2004)
66. Li, Z., Zhang, Y.: An efficient and user-friendly method for the synthesis of hexagonal-phase $NaYF_4$:Yb, Er/Tm nanocrystals with controllable shape and upconversion fluorescence. Nanotechnology **19**, 345606–345610 (2008)
67. Zhang, P., Steelant, W., Kumar, M., Scholfield, M.: Versatile photosensitizers for photodynamic therapy at infrared excitation. J. Am. Chem. Soc. **129**, 4526–4527 (2007)

68. Guo, Y., Kumar, M., Zhang, P.: Nanoparticle-based photosensitizers under CW infrared excitation. Chem. Mater. **19**, 6071–6072 (2007)
69. Chen, F., Zhang, S., Bu, W., Chen, Y., Xiao, Q., Liu, J., Xing, H., Zhou, L., Peng, W., Shi, J.: A uniform sub-50 nm-sized magnetic/upconversion fluorescent bimodal imaging agent capable of generating singlet oxygen by using a 980 nm laser. Chem. Eur. J. **18**, 7082–7090 (2012)
70. Zhao, Z., Han, Y., Lin, C., Hu, D., Wang, F., Chen, X., Chen, Z., Zheng, N.: Multifunctional core-shell upconverting nanoparticles for imaging and photodynamic therapy of liver cancer cells. Chem. Asian J. **7**, 830–837 (2012)
71. Wang, F., Yang, X., Ma, L., Huang, B., Na, N.E.Y., He, D., Ouyang, J.: Multifunctional upconverting nanocomposites with multimodal imaging and photosensitization at near-infrared excitation. J. Mater. Chem. **22**, 24597–24604 (2012)
72. Yang, X., Xiao, Q., Niu, C., Jin, N., Ouyang, J., Xiao, X., He, D.: Multifunctional core-shell upconversion nanoparticles for targeted tumor cells induced by near-infrared light. J. Mater. Chem. B **1**, 2757–2763 (2013)
73. Qian, H.S., Guo, H.C., Ho, P.C., Mahendran, R., Zhang, Y.: Mesoporous-silica-coated upconversion fluorescent nanoparticles for photodynamic therapy. Small **5**, 2285–2290 (2009)
74. Guo, H., Qian, H., Idris, N.M., Zhang, Y.: Singlet oxygen-induced apoptosis of cancer cells using upconversion fluorescent nanoparticles as a carrier of photosensitizer. Nanomed. Nanotechnol. Biol. Med. **6**, 486–495 (2010)
75. Idris, N.M., Gnanasammandhan, M.K., Zhang, J., Ho, P.C., Mahendran, R., Zhang, Y.: In vivo photodynamic therapy using upconversion nanoparticles as remote-controlled nanotransducers. Nat. Med. **18**, 1580–1585 (2012)
76. Qiao, X., Zhou, J., Xiao, J., Wang, Y., Sun, L., Yan, C.: Triple-functional core-shell structured upconversion luminescent nanoparticles covalently grafted with photosensitizer for luminescent, magnetic resonance imaging and photodynamic therapy in vitro. Nanoscale **4**, 4611–4623 (2012)
77. Chatterjee, D.K., Yong, Z.: Upconverting nanoparticles as nanotransducers for photodynamic therapy in cancer cells. Nanomedicine **3**, 73–82 (2008)
78. Lim, M.E., Lee, Y., Zhang, Y., Chu, J.J.H.: Photodynamic inactivation of viruses using upconversion nanoparticles. Biomaterials **33**, 1912–1920 (2012)
79. Ungun, B., Prud'Homme, R.K., Budijon, S.J., Shan, J., Lim, S.F., Ju, Y., Austin, R.: Nanofabricated upconversion nanoparticles for photodynamic therapy. Opt. Expr. **17**, 80–86 (2009)
80. Shan, J., Budijono, S.J., Hu, G., Yao, N., Kang, Y., Ju, Y., Prud'Homme, R.K.: Pegylated composite nanoparticles containing upconverting phosphors and meso-tetraphenyl porphine (TPP) for Photodynamic therapy. Adv. Funct. Mater. **21**, 2488–2495 (2011)
81. Wang, C., Tao, H., Cheng, L., Liu, Z.: Near-infrared light induced in vivo photodynamic therapy of cancer based on upconversion nanoparticles. Biomaterials **32**, 6145–6154 (2011)
82. Wang, C., Cheng, L., Liu, Y., Wang, X., Ma, X., Deng, Z., Li, Y., Liu, Z.: Imaging-guided pH-sensitive photodynamic therapy using charge reversible upconversion nanoparticles under near-infrared light. Adv. Funct. Mater. **23**, 3077–3086 (2013)
83. Chen, X., Zhao, Z., Jiang, M., Que, D., Shi, S., Zheng, N.: Preparation and photodynamic therapy application of $NaYF_4$:Yb, Tm-$NaYF_4$:Yb, Er multifunctional upconverting nanoparticles. New J. Chem. **37**, 1782–1788 (2013)
84. Xiao, Q., Ji, Y., Xiao, Z., Zhang, Y., Lin, H., Wang, Q.: Novel multifunctional $NaYF_4:Er^{3+}$, Yb^{3+}/PEGDA hybrid microspheres: NIR-light-activated photopolymerization and drug delivery. Chem. Commun. **49**, 1527–1529 (2013)
85. Cui, S., Chen, H., Zhu, H., Tian, J., Chi, X., Qian, Z., Achilefu, S., Gu, Y.: Amphiphilic chitosan modified upconversion nanoparticles for in vivo photodynamic therapy induced by near-infrared light. J. Mater. Chem. **22**, 4861–4873 (2012)
86. Cui, S., Yin, D., Chen, Y., Di, Y., Chen, H., Ma, Y., Achilefu, S., Gu, Y.: In vivo targeted deep-tissue photodynamic therapy based on near-infrared light triggered upconversion nanoconstruct. ACS Nano **7**, 676–688 (2013)

87. Wang, H.J., Shrestha, R., Zhang, Y.: Encapsulation of photosensitizers and upconversion nanocrystals in lipid micelles for photodynamic therapy. Part. Part. Syst. Char. **31**, 228–235 (2014)
88. Wang, H., Liu, Z., Wang, S., Dong, C., Gong, X., Zhao, P., Chang, J.: MC540 and upconverting nanocrystal coloaded polymeric liposome for near-infrared light-triggered photodynamic therapy and cell fluorescent imaging. ACS Appl. Mater. Interfaces **6**, 3219–3225 (2014)
89. Wang, X., Zhang, Q., Zhao, J., Dai, J.: One-step self-assembly of ZnPc/$NaGdF_4$:Yb, Er nanoclusters for simultaneous fluorescence imaging and photodynamic effects on cancer cells. J. Mater. Chem. B **1**, 4637–4643 (2013)
90. Tian, G., Ren, W., Yan, L., Jian, S., Gu, Z., Zhou, L., Jin, S., Yin, W., Li, S., Zhao, Y.: Red-emitting upconverting nanoparticles for photodynamic therapy in cancer cells under near-infrared excitation. Small **9**, 1929–1938 (2013)
91. Chen, Q., Wang, C., Cheng, L., He, W., Cheng, Z., Liu, Z.: Protein modified upconversion nanoparticles for imaging-guided combined photothermal and photodynamic therapy. Biomaterials **35**, 2915–2923 (2014)
92. Wang, X., Liu, K., Yang, G., Cheng, L., He, L., Liu, Y., Li, Y., Guo, L., Liu, Z.: Near-infrared light triggered photodynamic therapy in combination with gene therapy using upconversion nanoparticles for effective cancer cell killing. Nanoscale **6**, 9198–9205 (2014)
93. Zeng, L., Xiang, L., Ren, W., Zheng, J., Li, T., Chen, B., Zhang, J., Mao, C., Li, A., Wu, A.: Multifunctional photosensitizer-conjugated core-shell Fe_3O_4@$NaYF_4$:Yb/Er nanocomplexes and their applications in T_2-weighted magnetic resonance/upconversion luminescence imaging and photodynamic therapy of cancer cells. RSC Adv. **3**, 13915–13925 (2013)
94. Jin, S., Zhou, L., Gu, Z., Tian, G., Yan, L., Ren, W., Yin, W., Liu, X., Zhang, X., Hu, Z., Zhao, Y.: A new near infrared photosensitizing nanoplatform containing blue-emitting up-conversion nanoparticles and hypocrellin A for photodynamic therapy of cancer cells. Nanoscale **5**, 11910–11918 (2013)
95. Yuan, Q., Wu, Y., Wang, J., Lu, D., Zhao, Z., Liu, T., Zhang, X., Tan, W.: Targeted bioimaging and photodynamic therapy nanoplatform using an aptamer-guided G-quadruplex DNA carrier and near-infrared light. Angew. Chem. Int. Ed. **52**, 13965–13969 (2013)
96. González-Béjar, M., Liras, M., Francés-Soriano, L., Voliani, V., Herranz-Pérez, V., Duran-Moreno, M., Garcia-Verdugo, J.M., Alarcon, E.I., Scaiano, J.C., Pérez-Prieto, J.: NIR excitation of upconversion nanohybrids containing a surface grafted Bodipy induces oxygen-mediated cancer cell death. J. Mater. Chem. B **2**, 4554–4563 (2014)
97. Zhou, A., Wei, Y., Wu, B., Chen, Q., Xing, D.: Pyropheophorbide A and c(RGDyK) comodified chitosan-wrapped upconversion nanoparticle for targeted near-infrared photodynamic therapy. Mol. Pharm. **9**, 1580–1589 (2012)
98. Liu, K., Liu, X., Zeng, Q., Zhang, Y., Tu, L., Liu, T., Kong, X., Wang, Y., Cao, F., Lambrechts, S.A.G., Aalders, M.C.G., Zhang, H.: Covalently assembled NIR nanoplatform for simultaneous fluorescence imaging and photodynamic therapy of cancer cells. ACS Nano **6**, 4054–4062 (2012)
99. Liu, X., Zheng, M., Kong, X., Zhang, Y., Zeng, Q., Sun, Z., Buma, W.J., Zhang, H.: Separately doped upconversion-C_{60} nanoplatform for NIR imaging-guided photodynamic therapy of cancer cells. Chem. Commun. **49**, 3224–3226 (2013)
100. Xia, L., Kong, X., Liu, X., Tu, L., Zhang, Y., Chang, Y., Liu, K., Shen, D., Zhao, H., Zhang, H.: An upconversion nanoparticle—zinc phthalocyanine based nanophotosensitizer for photodynamic therapy. Biomaterials **35**, 4146–4156 (2014)
101. Park, Y.I., Kim, H.M., Kim, J.H., Moon, K.C., Yoo, B., Lee, K.T., Lee, N., Choi, Y., Park, W., Ling, D., Na, K., Moon, W.K., Choi, S.H., Park, H.S., Yoon, S., Suh, Y.D., Lee, S.H., Hyeon, T.: Theranostic probe based on lanthanide-doped nanoparticles for simultaneous in vivo dual-modal imaging and photodynamic therapy. Adv. Mater. **24**, 5755–5761 (2012)
102. Topel, S.D., Cin, G.T., Akkaya, E.U.: Near IR excitation of heavy atom free Bodipy photosensitizers through the intermediacy of upconverting nanoparticles. Chem. Commun. **50**, 8896–8899 (2014)

103. Xu, Q.C., Zhang, Y., Tan, M.J., Liu, Y., Yuan, S., Choong, C., Tan, N.S., Tan, T.T.Y.: Anti-cAngptl4 Ab-Conjugated N-TiO_2/$NaYF_4$:Yb, Tm nanocomposite for near infrared-triggered drug release and enhanced targeted cancer cell ablation. Adv. Healthcare Mater. **1**, 470–474 (2012)
104. Wang, X., Yang, C., Chen, J., Yan, X.: A dual-targeting upconversion nanoplatform for two-color fluorescence imaging-guided photodynamic therapy. Anal. Chem. **86**, 3263–3267 (2014)
105. Wang, M., Chen, Z., Zheng, W., Zhu, H., Lu, S., Ma, E., Tu, D., Zhou, S., Huang, M., Chen, X.: Lanthanide-doped upconversion nanoparticles electrostatically coupled with photosensitizers for near-infrared-triggered photodynamic therapy. Nanoscale **6**, 8274–8282 (2014)
106. Wei, X., Wang, W., Chen, K.: ZnO:Er, Yb, Gd particles designed for magnetic-fluorescent imaging and near-infrared light triggered photodynamic therapy. J. Phys. Chem. C **117**, 23716–23729 (2013)
107. Zhou, L., Li, Z., Liu, Z., Yin, M., Ren, J., Qu, X.: One-step nucleotide-programmed growth of porous upconversion nanoparticles: application to cell labeling and drug delivery. Nanoscale **6**, 1445–1452 (2014)
108. Shimoyama, A., Watase, H., Liu, Y., Ogura, S., Hagiya, Y., Takahashi, K., Inoue, K., Tanaka, T., Murayama, Y., Otsuji, E., Ohkubo, A., Yuasa, H.: Access to a novel near-infrared photodynamic therapy through the combined use of 5-aminolevulinic acid and lanthanide nanoparticles. Photodiag. Photodyn. Ther. **10**, 607–614 (2013)
109. Guo, H., Li, Z., Qian, H., Hu, Y., Muhammad, I.N.: Seed-mediated synthesis of $NaYF_4$:Yb, Er/$NaGdF_4$ nanocrystals with improved upconversion fluorescence and MR relaxivity. Nanotechnology **21**, 125602–125607 (2010)
110. Wong, Y.J., Zhu, L., Teo, W.S., Tan, Y.W., Yang, Y., Wang, C., Chen, H.: Revisiting the stöber method: inhomogeneity in silica shells. J. Am. Chem. Soc. **133**, 11422–11425 (2011)
111. Chen, F., Bu, W., Zhang, S., Liu, X., Liu, J., Xing, H., Xiao, Q., Zhou, L., Peng, W., Wang, L., Shi, J.: Positive and negative lattice shielding effects co-existing in Gd (III) ion doped bifunctional upconversion nanoprobes. Adv. Funct. Mater. **21**, 4285–4294 (2011)
112. Moghimi, S., Symonds, P., Murray, J., Hunter, A., Debska, G., Szewczyk, A.: A two-stage poly(ethylenimine)-mediated cytotoxicity: implications for gene transfer/therapy. Mol. Ther. **11**, 990–995 (2005)
113. Brunot, C., Ponsonnet, L., Lagneau, C., Farge, P., Picart, C., Grosgogeat, B.: Cytotoxicity of polyethyleneimine (PEI), precursor base layer of polyelectrolyte multilayer films. Biomaterials **28**, 632–640 (2007)
114. Gref, R., Domb, A., Quellec, P., Blunk, T., Müller, R.H., Verbavatz, J.M., Langer, R.: The controlled intravenous delivery of drugs using PEG-coated sterically stabilized nanospheres. Adv. Drug Deliver. Rev. **16**, 215–233 (1995)
115. Chen, F., Zhang, L., Chen, Q., Zhang, Y., Zhang, Z.: Synthesis of a novel magnetic drug delivery system composed of doxorubicin-conjugated Fe_3O_4 nanoparticle cores and a PEG-functionalized porous silica shell. Chem. Commun. **46**, 8633–8635 (2010)
116. Perry, J.L., Reuter, K.G., Kai, M.P., Herlihy, K.P., Jones, S.W., Luft, J.C., Napier, M., Bear, J.E., DeSimone, J.M.: PEGylated PRINT nanoparticles: the impact of PEG density on protein binding, macrophage association, biodistribution, and pharmacokinetics. Nano Lett. **12**, 5304–5310 (2012)
117. Hehlen, M.P., Güdel, H.U., Lüthi, S.R., Gamelin, D.R., Pollnau, M.: Power dependence of upconversion luminescence in lanthanide and transition-metal-ion systems. Phys. Rev. B **61**, 3337–3346 (2000)
118. Adarsh, N., Avirah, R.R., Ramaiah, D.: Tuning photosensitized singlet oxygen generation efficiency of novel Aza-BODIPY dyes. Org. Lett. **12**, 5720–5723 (2010)
119. Chen, J., Chen, N., Huang, J., Wang, J., Huang, M.: Derivatizable phthalocyanine with single carboxyl group: Synthesis and purification. Inorg. Chem. Commun. **9**, 313–315 (2006)

120. Liu, Z., Chen, K., Davis, C., Sherlock, S., Cao, Q., Chen, X., Hongjie, D.: Drug delivery with carbon nanotubes for in vivo cancer treatment. Cancer Res. **68**, 6652–6660 (2008)
121. Bao, G., Mitragotri, S., Tong, S.: Multifunctional nanoparticles for drug delivery and molecular imaging. Annu. Rev. Biomed. Eng. **15**, 253–282 (2013)
122. Shi, J., Votruba, A.R., Farokhzad, O.C., Langer, R.: Nanotechnology in drug delivery and tissue engineering: from discovery to applications. Nano Lett. **10**, 3223–3230 (2010)
123. Yang, D., Ma, P., Hou, Z., Cheng, Z., Li, C., Lin, J.: Current advances in lanthanide ion (Ln^{3+})-based upconversion nanomaterials for drug delivery. Chem. Soc. Rev. (2014). doi:10.1039/c4cs00155a
124. Feng, W., Han, C., Li, F.: Upconversion-nanophosphor-based functional nanocomposites. Adv. Mater. **25**, 5287–5303 (2013)
125. Xu, H., Cheng, L., Wang, C., Ma, X., Li, Y., Liu, Z.: Polymer encapsulated upconversion nanoparticle/iron oxide nanocomposites for multimodal imaging and magnetic targeted drug delivery. Biomaterials **32**, 9364–9373 (2011)
126. Li, C., Yang, D., Ma, P., Chen, Y., Wu, Y., Hou, Z., Dai, Y., Zhao, J., Sui, C., Lin, J.: Multifunctional upconversion mesoporous silica nanostructures for dual modal imaging and in vivo drug delivery. Small **9**, 4150–4159 (2013)
127. Wang, C., Cheng, L., Liu, Z.: Drug delivery with upconversion nanoparticles for multi-functional targeted cancer cell imaging and therapy. Biomaterials **32**, 1110–1120 (2011)
128. Xu, Z., Ma, P.A., Li, C., Hou, Z., Zhai, X., Huang, S., Lin, J.: Monodisperse core-shell structured up-conversion $Yb(OH)CO_3$@$YbPO_4:Er^{3+}$ hollow spheres as drug carriers. Biomaterials **32**, 4161–4173 (2011)
129. Chen, Y., Ai, K., Liu, Y., Lu, L.: Tailor-made charge-conversional nanocomposite for pH-responsive drug delivery and cell imaging. ACS Appl. Mater. Interfaces **6**, 655–663 (2014)
130. Dai, Y., Yang, D., Ma, P., Kang, X., Zhang, X., Li, C., Hou, Z., Cheng, Z., Lin, J.: Doxorubicin conjugated $NaYF_4:Yb^{3+}/Tm^{3+}$ nanoparticles for therapy and sensing of drug delivery by luminescence resonance energy transfer. Biomaterials **33**, 8704–8713 (2012)
131. Dai, Y., Ma, P.A., Cheng, Z., Kang, X., Zhang, X., Hou, Z., Li, C., Yang, D., Zhai, X., Lin, J.: Up-conversion cell imaging and pH-induced thermally controlled drug release from $NaYF_4:Yb^{3+}/Er^{3+}$@hydrogel core-shell hybrid microspheres. ACS Nano **6**, 3327–3338 (2012)
132. Yang, D., Dai, Y., Liu, J., Zhou, Y., Chen, Y., Li, C., Ma, P., Lin, J.: Ultra-small $BaGdF_5$-based upconversion nanoparticles as drug carriers and multimodal imaging probes. Biomaterials **35**, 2011–2023 (2014)
133. Li, L., Liu, C., Zhang, L., Wang, T., Yu, H., Wang, C., Su, Z.: Multifunctional magnetic-fluorescent eccentric-(concentric-Fe_3O_4@SiO_2)@polyacrylic acid core-shell nanocomposites for cell imaging and pH-responsive drug delivery. Nanoscale **5**, 2249–2253 (2013)
134. Zhang, F., Braun, G.B., Pallaoro, A., Zhang, Y., Shi, Y., Cui, D., Moskovits, M., Zhao, D., Stucky, G.D.: Mesoporous multifunctional upconversion luminescent and magnetic "nanorattle" materials for targeted chemotherapy. Nano Lett. **12**, 61–67 (2012)
135. Zhang, X., Yang, P., Dai, Y., Ma, P., Li, X., Cheng, Z., Hou, Z., Kang, X., Li, C., Lin, J.: Multifunctional up-converting nanocomposites with smart polymer brushes gated mesopores for cell imaging and thermo/pH dual-responsive drug controlled release. Adv. Funct. Mater. **23**, 4067–4078 (2013)
136. Cheng, Z., Chai, R., Ma, P., Dai, Y., Kang, X., Lian, H., Hou, Z., Li, C., Lin, J.: Multiwalled carbon nanotubes and $NaYF_4:Yb^{3+}/Er^{3+}$ nanoparticle-doped bilayer hydrogel for concurrent NIR-triggered drug release and up-conversion luminescence tagging. Langmuir **29**, 9573–9580 (2013)
137. Zhao, P., Zhu, Y., Yang, X., Shen, J., Jiang, X., Zong, J., Li, C.: Multifunctional MnO_2 nanosheet-modified Fe_3O_4@$SiO_2/NaYF_4$:Yb, Er nanocomposites as novel drug carriers. Dalton Trans. **43**, 451–457 (2013)

138. Hou, Z., Li, C., Ma, P., Li, G., Cheng, Z., Peng, C., Yang, D., Yang, P., Lin, J.: Electrospinning preparation and drug-delivery properties of an up-conversion luminescent porous $NaYF_4:Yb^{3+}$, Er^{3+}@Silica fiber nanocomposite. Adv. Funct. Mater. **21**, 2356–2365 (2011)
139. Zhao, Q., Lei, Z., Huang, S., Han, X., Shao, B., Lü, W., Jia, Y., Lv, W., Jiao, M., Wang, Z., You, H.: Facile fabrication of single-phase multifunctional $BaGdF_5$ nanospheres as drug carriers. ACS Appl. Mater. Interfaces **6**, 12761–12770 (2014)
140. Tian, G., Gu, Z., Liu, X., Zhou, L., Yin, W., Yan, L., Jin, S., Ren, W., Xing, G., Li, S., Zhao, Y.: Facile fabrication of rare-earth-doped Gd_2O_3 hollow spheres with upconversion luminescence, magnetic resonance, and drug delivery properties. J. Phys. Chem. C **115**, 23790–23796 (2011)
141. Yang, D., Dai, Y., Ma, P., Kang, X., Cheng, Z., Li, C., Lin, J.: One-step synthesis of small-sized and water-soluble $NaREF_4$ upconversion nanoparticles for in vitro cell imaging and drug delivery. Chem. Eur. J. **19**, 2685–2694 (2013)
142. Gai, S., Yang, P., Li, C., Wang, W., Dai, Y., Niu, N., Lin, J.: Synthesis of magnetic, up-conversion luminescent, and mesoporous core-shell-structured nanocomposites as drug carriers. Adv. Funct. Mater. **20**, 1166–1172 (2010)
143. Xu, Z., Li, C., Ma, P., Hou, Z., Yang, D., Kang, X., Lin, J.: Facile synthesis of an up-conversion luminescent and mesoporous Gd_2O_3: Er^{3+}@$nSiO_2$@$mSiO_2$ nanocomposite as a drug carrier. Nanoscale **3**, 661–667 (2011)
144. Kang, X., Cheng, Z., Li, C., Yang, D., Shang, M., Ma, P.A., Li, G., Liu, N., Lin, J.: Core-shell structured up-conversion luminescent and mesoporous $NaYF_4:Yb^{3+}/Er^{3+}$@$nSiO_2$@$mSiO_2$ nanospheres as carriers for drug delivery. J. Phys. Chem. C **115**, 15801–15811 (2011)
145. Liu, J., Bu, W., Pan, L., Shi, J.: NIR-triggered anticancer drug delivery by upconverting nanoparticles with integrated azobenzene-modified mesoporous silica. Angew. Chem. Int. Ed. **52**, 4375–4379 (2013)
146. Liu, J., Bu, J., Bu, W., Zhang, S., Pan, L., Fan, W., Chen, F., Zhou, L., Peng, W., Zhao, K., Du, J., Shi, J.: Real-time in vivo quantitative monitoring of drug release by dual-mode magnetic resonance and upconverted luminescence imaging. Angew. Chem. Int. Ed. **53**, 4551–4555 (2014)
147. Zhou, L., Chen, Z., Dong, K., Yin, M., Ren, J., Qu, X.: DNA-mediated construction of hollow upconversion nanoparticles for protein harvesting and near-infrared light triggered release. Adv. Mater. **26**, 2424–2430 (2014)
148. Yan, B., Boyer, J., Branda, N.R., Zhao, Y.: Near-infrared light-triggered dissociation of block copolymer micelles using upconverting nanoparticles. J. Am. Chem. Soc. **133**, 19714–19717 (2011)
149. Yan, B., Boyer, J., Habault, D., Branda, N.R., Zhao, Y.: Near infrared light triggered release of biomacromolecules from hydrogels loaded with upconversion nanoparticles. J. Am. Chem. Soc. **134**, 16558–16561 (2012)
150. Viger, M.L., Grossman, M., Fomina, N., Almutairi, A.: Low power upconverted near-IR light for efficient polymeric nanoparticle degradation and cargo release. Adv. Mater. **25**, 3733–3738 (2013)
151. Yang, Y., Velmurugan, B., Liu, X., Xing, B.: NIR photoresponsive crosslinked upconverting nanocarriers toward selective intracellular drug release. Small **9**, 2937–2944 (2013)
152. Carling, C., Nourmohammadian, F., Boyer, J., Branda, N.R.: Remote-control photorelease of caged compounds using near-infrared light and upconverting nanoparticles. Angew. Chem. Int. Ed. **49**, 3782–3785 (2010)
153. Yang, Y., Shao, Q., Deng, R., Wang, C., Teng, X., Cheng, K., Cheng, Z., Huang, L., Liu, Z., Liu, X., Xing, B.: In vitro and in vivo uncaging and bioluminescence imaging by using photocaged upconversion nanoparticles. Angew. Chem. Int. Ed. **51**, 3125–3129 (2012)
154. Zhao, L., Peng, J., Huang, Q., Li, C., Chen, M., Sun, Y., Lin, Q., Zhu, L., Li, F.: Near-infrared photoregulated drug release in living tumor tissue via yolk-shell upconversion nanocages. Adv. Funct. Mater. **24**, 363–371 (2014)

155. Chien, Y., Chou, Y., Wang, S., Hung, S., Liau, M., Chao, Y., Su, C., Yeh, C.: Near-infrared light photocontrolled targeting, bioimaging, and chemotherapy with caged upconversion nanoparticles. ACS Nano **7**, 8516–8528 (2013)
156. Shen, J., Chen, G., Ohulchanskyy, T.Y., Kesseli, S.J., Buchholz, S., Li, Z., Prasad, P.N., Han, G.: Tunable near infrared to ultraviolet upconversion luminescence enhancement in (α-$NaYF_4$:Yb, Tm)/CaF_2 core/shell nanoparticles for in situ real-time recorded biocompatible photoactivation. Small **9**, 3213–3217 (2013)
157. Jayakumar, M.K.G., Idris, N.M., Zhang, Y.: Remote activation of biomolecules in deep tissues using near-infrared-to-UV upconversion nanotransducers. Proc. Natl. Acad. Sci. USA **109**, 8483–8488 (2012)
158. Yang, Y., Liu, F., Liu, X., Xing, B.: NIR light controlled photorelease of siRNA and its targeted intracellular delivery based on upconversion nanoparticles. Nanoscale **5**, 231–238 (2012)
159. Jayakumar, M.K.G., Bansal, A., Huang, K., Yao, R., Li, B.N., Zhang, Y.: Near-infrared-light-based nano-platform boosts endosomal escape and controls gene knockdown. ACS Nano **8**, 4848–4858 (2014)
160. Garcia, J.V., Yang, J., Shen, D., Yao, C., Li, X., Wang, R., Stucky, G.D., Zhao, D., Ford, P. C., Zhang, F.: NIR-triggered release of caged nitric oxide using upconverting nanostructured materials. Small **8**, 3800–3805 (2012)
161. Burks, P.T., Garcia, J.V., GonzalezIrias, R., Tillman, J.T., Niu, M., Mikhailovsky, A.A., Zhang, J., Zhang, F., Ford, P.C.: Nitric oxide releasing materials triggered by near-infrared excitation through tissue filters. J. Am. Chem. Soc. **135**, 18145–18152 (2013)
162. Dai, Y., Xiao, H., Liu, J., Yuan, Q., Ma, P.A., Yang, D., Li, C., Cheng, Z., Hou, Z., Yang, P., Lin, J.: In vivo multimodality imaging and cancer therapy by near-infrared light-triggered trans-platinum pro-drug-conjugated upconverison nanoparticles. J. Am. Chem. Soc. **135**, 18920–18929 (2013)
163. Min, Y., Li, J., Liu, F., Yeow, E.K.L., Xing, B.: Near-infrared light-mediated photoactivation of a platinum antitumor prodrug and simultaneous cellular apoptosis imaging by upconversion-luminescent nanoparticles. Angew. Chem. Int. Ed. **53**, 1012–1016 (2014)
164. Fomina, N., Sankaranarayanan, J., Almutairi, A.: Photochemical mechanisms of light-triggered release from nanocarriers. Adv. Drug Deliver. Rev. **64**, 1005–1020 (2012)
165. Mura, S., Nicolas, J., Couvreur, P.: Stimuli-responsive nanocarriers for drug delivery. Nat. Mater. **12**, 991–1003 (2013)
166. Zhu, M., Zhu, L., Han, J.J., Wu, W., Hurst, J.K., Li, A.D.Q.: Spiropyran-based photochromic polymer nanoparticles with optically switchable luminescence. J. Am. Chem. Soc. **128**, 4303–4309 (2006)
167. Schwarze, S.R.: In vivo protein transduction: delivery of a biologically active protein into the mouse. Science **285**, 1569–1572 (1999)
168. Kröck, L., Heckel, A.: Photoinduced transcription by using temporarily mismatched caged oligonucleotides. Angew. Chem. Int. Ed. **44**, 471–473 (2005)
169. Fedoryak, O.D., Dore, T.M.: Brominated hydroxyquinoline as a photolabile protecting group with sensitivity to multiphoton excitation. Org. Lett. **4**, 3419–3422 (2002)
170. Shanmugam, V., Selvakumar, S., Yeh, C.: Near-infrared light-responsive nanomaterials in cancer therapeutics. Chem. Soc. Rev. **43**, 6254–6287 (2014)
171. Fan, N., Cheng, F., Ho, J.A., Yeh, C.: Photocontrolled targeted drug delivery: photocaged biologically active folic acid as a light-responsive tumor-targeting molecule. Angew. Chem. Int. Ed. **51**, 8806–8810 (2012)
172. Verma, I.M., Somia, N.: Gene therapy-promises, problems and prospects. Nature **389**, 239–242 (1997)
173. Somia, N., Verma, I.M.: Gene therapy: trials and tribulations. Nat. Rev. Genet. **1**, 91–99 (2000)
174. Caplen, N.J.: Gene therapy progress and prospects. Downregulating gene expression: the impact of RNA interference. Gene Ther. **11**, 1241–1248 (2004)

175. Urban-Klein, B., Werth, S., Abuharbeid, S., Czubayko, F., Aigner, A.: RNAi-mediated gene-targeting through systemic application of polyethylenimine (PEI)-complexed siRNA in vivo. Gene Ther. **12**, 461–466 (2004)
176. Schmidt-Wolf, G.D., Schmidt-Wolf, I.G.H.: Non-viral and hybrid vectors in human gene therapy: an update. Trends Mol. Med. **9**, 67–72 (2003)
177. Morille, M., Passirani, C., Vonarbourg, A., Clavreul, A., Benoit, J.: Progress in developing cationic vectors for non-viral systemic gene therapy against cancer. Biomaterials **29**, 3477–3496 (2008)
178. Tallafuss, A., Gibson, D., Morcos, P., Li, Y., Seredick, S., Eisen, J., Washbourne, P.: Turning gene function ON and OFF using sense and antisense photo-morpholinos in zebrafish. Development **139**, 1691–1699 (2012)
179. Berg, K., Selbo, P.K., Prasmickaite, L., Tjelle, T.E., Sandvig, K., Moan, J., Gaudernack, G., Fodstad, O., Kjolsrud, S., Anholt, H., Rodal, G.H., Rodal, S.K., Hogset, A.: Photochemical internalization: a novel technology for delivery of macromolecules into cytosol. Cancer Res. **59**, 1180–1183 (1999)
180. Fukumura, D., Kashiwagi, S., Jain, R.K.: The role of nitric oxide in tumour progression. Nat. Rev. Cancer **6**, 521–534 (2006)
181. Mocellin, S., Bronte, V., Nitti, D.: Nitric oxide, a double edged sword in cancer biology: Searching for therapeutic opportunities. Med. Res. Rev. **27**, 317–352 (2007)
182. Wang, R.: Gasotransmitters: growing pains and joys. Trends Biochem. Sci. **39**, 227–232 (2014)
183. Graf, N., Lippard, S.J.: Redox activation of metal-based prodrugs as a strategy for drug delivery. Adv. Drug Deliver. Rev. **64**, 993–1004 (2012)
184. McWhinney, S.R., Goldberg, R.M., McLeod, H.L.: Platinum neurotoxicity pharmacogenetics. Mol. Cancer Ther. **8**, 10–16 (2009)
185. Hall, M.D., Hambley, T.W.: Platinum(IV) antitumour compounds: their bioinorganic chemistry. Coordin. Chem. Rev. **232**, 49–67 (2002)
186. Farrer, N.J., Woods, J.A., Salassa, L., Zhao, Y., Robinson, K.S., Clarkson, G., Mackay, F.S., Sadler, P.J.: A potent trans-diimine platinum anticancer complex photoactivated by visible light. Angew. Chem. Int. Ed. **49**, 8905–8908 (2010)
187. Cheng, L., Yang, K., Li, Y., Chen, J., Wang, C., Shao, M., Lee, S., Liu, Z.: Facile preparation of multifunctional upconversion nanoprobes for multimodal imaging and dual-targeted photothermal therapy. Angew. Chem. Int. Ed. **50**, 7385–7390 (2011)
188. Cheng, L., Yang, K., Li, Y., Zeng, X., Shao, M., Lee, S., Liu, Z.: Multifunctional nanoparticles for upconversion luminescence/MR multimodal imaging and magnetically targeted photothermal therapy. Biomaterials **33**, 2215–2222 (2012)
189. Shen, J., Li, K., Cheng, L., Liu, Z., Lee, S., Liu, J.: Specific detection and simultaneously localized photothermal treatment of cancer cells using layer-by-layer assembled multifunctional nanoparticles. ACS Appl. Mater. Interfaces **6**, 6443–6452 (2014)
190. Dong, B., Xu, S., Sun, J., Bi, S., Li, D., Bai, X., Wang, Y., Wang, L., Song, H.: Multifunctional $NaYF_4$: Yb^{3+}, Er^{3+}@Ag core/shell nanocomposites: integration of upconversion imaging and photothermal therapy. J. Mater. Chem. **21**, 6193–6200 (2011)
191. Xiao, Q., Zheng, X., Bu, W., Ge, W., Zhang, S., Chen, F., Xing, H., Ren, Q., Fan, W., Zhao, K., Hua, Y., Shi, J.: A core/satellite multifunctional nanotheranostic for in vivo imaging and tumor eradication by radiation/photothermal synergistic therapy. J. Am. Chem. Soc. **135**, 13041–13048 (2013)
192. Shan, G., Weissleder, R., Hilderbrand, S.A.: Upconverting organic dye doped core-shell nano-composites for dual-modality NIR imaging and photo-thermal therapy. Theranostics **3**, 267–274 (2013)
193. Cherian, S., Wamser, C.C.: Adsorption and photoactivity of tetra(4-carboxyphenyl)porphyrin (TCPP) on nanoparticulate TiO_2. J. Phys. Chem. B **104**, 3624–3629 (2000)

Chapter 10
Upconversion Nanoparticles for Thermal Sensing

Abstract Thermometer has been developed for several centuries from contact to non-contact. With the development of nanotechnology, it becomes possible to use nanothermometry to detect the temperature within a single cell. Upconversion nanoparticle (UCNPs) is one of the promising candidates for developing non-contact nanothermometry. There are several parameters that define the luminescence properties which are dependent on temperature: intensity, band shape, spectral position, bandwidth, polarization, and lifetime. Among these parameters, band shape is the most widely used one. So we introduce the FIR theory which is used to figure out the relationship between the band shape and temperature. We also compare UCNPs with quantum dots, gold nanoparticles (GNPs), and green fluorescent protein (GFP) to show each other's merits and demerits. We can foresee that UCNPs could be very useful in the near future even though there are still many challenges need to conquer.

Keywords Non-contact thermometer · FIR theory · Band shape · Spectral position · Upconversion nanoparticle · Nanocomposites · Quantum dots · Gold nanoparticles · Green fluorescent protein · Upconversion nanothermometry

10.1 Introduction

A thermometer is a device which measures temperature or temperature gradient. It has two elements (1) temperature sensor in which some physical changes occur with the change of temperature and (2) means to convert these changes into a numerical value. The first thermometer was developed by several scientists in sixteenth and seventeenth centuries, notably Galileo. At that time, these devices were developed based on the principle that certain substances such as air and water

---

Chengli Wang and Fan Zhang contributed together to this chapter.

F. Zhang, *Photon Upconversion Nanomaterials*,
Nanostructure Science and Technology, DOI 10.1007/978-3-662-45597-5_10

expand when heated. From then on, various kinds of thermometers have been invented (see Table 10.1).

Among the methods for temperature determination, one may distinguish between contact and non-contact methods. The contact methods (thermistors and thermoelements) require electrical wiring which might produce strong electromagnetic noise and cannot be applied in corrosive environments. To conquer these limitations, non-contact methods have been developed, such as infrared light (IR), thermoreflectance, optical interferometry, Raman spectroscopy, and luminescence (see Table 10.2). The most widespread method is based on the use of IR. An infrared thermometer is a laser thermometer which infers temperature from a portion of the thermal radiation sometimes called blackbody radiation emitted by the object being measured. By knowing the amount of infrared energy emitted by the object and its emissivity, the object's temperature can often be determined. Infrared thermometers can measure temperature from a distance and allow fast imaging, but requires the kind of material investigated to be known and its spatial resolution is in the micrometer range and limited by the size of the sensor pixels [1]. Luminescence thermometry in the visible and near-infrared (NIR) range can overcome many of these limitations. There are several sensing schemes, most based on the use of molecular probes, sometimes in combination with optical fibers [2, 3] and sometimes with fluorescence microcopy, which can provide high spatial [4]. Table 10.2 shows some general applications of high-resolution thermal techniques.

Thermometers are used in a wide variety of applications. They can be easily seen in our daily life. They can also be seen in steel industry, manufacturing, climatology, and so on. What's more, temperature is a fundamental parameter that plays important role not only in industrial processes, but also in scientific research, especially in biology. For example, metabolism and enzyme reaction in a single live cell are marked by temperature change [6]. An accurate and non-invasive determination of temperature is of particular importance for the investigation of the dynamics of cellular heat [7] production and propagation in the different intracellular compartments [8]. It is also well known that the cellular pathogenesis of diseases like cancer is characterized by heat production [7]. Consequently, thermal mapping could provide not only the understanding of a variety of cellular events, but also the possibility of detecting disease at the cellular level, thus boosting the development of diagnostic and therapeutic techniques [8]. However, it need the probe to be small enough to get in cell (normal cell size is less than 100 μm) and good bio-compatibility.

The fast development of nanotechnology has made it possible to fabricate a variety of photostable nanometric fluorescent compounds, whose typical sizes range from a few nanometers to approximately 100 nm. The small size of these nanoprobes and the possibility of surface design facilitate their internalization in living cells and therefore the possibility of luminescence cell mapping. Consequently, those nanoprobes whose fluorescence is sensitive to tiny temperature changes are potentially useful as nanothermometers, providing the possibility of intracellular thermal mapping [8].

Table 10.1 Some common kinds of contact thermometer and their principles and applications

Name	Principle	Characteristic
Alcohol thermometer	Expanded when heated	Working range: −114 to 78 °C
Mercury-in-glass thermometer	Expanded when heated	Working range: −39 to 357 °C
Thermocouple	Thermoelectric (any conductor which is subjected to a thermal gradient will generate a voltage) $V_b - V_c = \int_{Tc}^{Th} (S_A(T) - S_B(T))\mathrm{d}T$ $S_A \cdot S_B$ = Seebeck coefficients of materials A and B as a function of temperature $T_h \cdot T_c$ = temperatures of the two junctions V_b and V_c = voltages measured at the cold ends of materials A and B	Application: steel industry, gas appliance safety, manufacturing, power production, process plants, etc.
Thermistor	Relationship between resistance and temperature is linear $\Delta R = k\Delta T$	Application: control systems such as air conditioners, freezers, refrigerators etc.; inexpensive
Liquid crystal thermometer	Temperature change can affect the color of liquid crystal	Measure body temperature; safe but not accurate
Silicon band gap temperature sensor	Forward voltage of a silicon diode, which may be the base–emitter junction of a bipolar junction transistor (BJT), is temperature-dependent $V_{\mathrm{BE}} = V_{G0}\left(1 - \frac{T}{T_0}\right) + V_{\mathrm{BE}_0}\left(\frac{T}{T_0}\right) + \left(\frac{nKT}{q}\right) \mathrm{In} \left(\frac{T_0}{T}\right) + \left(\frac{KT}{q}\right) \mathrm{In} \left(\frac{I_C}{I_{C0}}\right)$ T = temperature in Kelvin T_0 = reference temperature V_{G0} = band gap voltage at absolute zero V_{BE0} = band gap voltage at temperature T_0 and current I_{C0} K = Boltzmann's constant q = charge on an electron n = a device-dependent constant	Can be included in a silicon integrated circuit at very low cost
Quartz thermometer	It measures temperature by measuring the frequency of a quartz crystal oscillator. The oscillator contains a specially cut crystal that results in a linear temperature coefficient of frequency, so the measurement of the temperature is essentially reduced to measurement of the oscillator frequency	High precision; high accuracy; high resolution; working range: 0–100 °C

Table 10.2 Varieties kinds of non-contact thermometers and their advantages and disadvantages [5]

Method	Principle	Typical resolution (K)	Advantages	Disadvantages
Infrared thermograph	Planck blackbody emission	10^{-1}	1. Well implemented commercial technique 2. Provides temperature image profile of the surface	1. Detector saturation at high temperatures 2. Difficulties on the precise estimation of the emissivity of the surface materials 3. Spatial resolution for the temperature detection, which is rayleigh limited (not all "hot bodies" are perfect blackbodies, in the physical meaning of the term)
Thermoreflectance	Temperature dependence of the reflection	10^{-2}	1. High thermal and temporal resolution 2. Quantitative and qualitative measurement	1. Requires the calibration of the reflectivity index 2. Spatial resolution limited by the diffraction limit 3. Thermo reflectance coefficient is not available for every material and depends on excitation wavelength and thickness of the optical layer
Raman	Inelastic scattering of monochromatic light	10^{-1}	1. No sample preparation needed 2. Works in solids and liquids 3. Small volumes (<1 mm diameter) can be probed	1. Highly time-consuming technique implying image point analysis as slow as 0.5 point s^{-1} 2. Low signal and cross talk with fluorescent molecules
Fluorescence thermograph	Temperature dependence of quantum efficiency/lifetime/intensity	10^{-2}	1. Spatial resolution (at one dimension) of 25 nm 2. Diverse experimental techniques to measure temperature 3. High temperature sensitivity 4. Ratiometric algorithms are independent of illumination source	1. Voltage reflections and capacitive coupling limit the timescale for transient thermometry of the device 2. Photobleaching limits the long term intensity and lifetime determination 3. High-expensive excitation sources and detectors to measure temperature using lifetime-based algorithms

(continued)

Table 10.2 (continued)

Method	Principle	Typical resolution (K)	Advantages	Disadvantages
Near-field scanning optical microscopy	Use near field to improve optical resolution	10^{-1}	1. Spatial resolution below the Rayleigh limit (100 nm)	1. Depends on the surface characteristics 2. Only access to surface temperature 3. Slow temperature acquisition 4. Vacuum and/or cryogenic temperatures required
Scanning thermal microscopy	AFM with thermocouple or Pt thermistor tip	10^{-1}	1. Uses AFM tips to simultaneously measure temperature and determine the surface roughness 2. Submicrometric spatial resolution	1. Slow acquisition times 2. Limited to solid samples 3. Requires fundamental knowledge of tip-sample heat transfer
Transmission electron microscopy	Thermal expansion	10^{-1}	1. High spatial resolution 2. Compatible with different CNTs	1. Vacuum required 2. Difficult to transpose for practical applications 3. Temperature determination by image analysis
Optical interferometry	Thermal expansion or refractive index change	10^{-5}	1. Can be integrated in remote detection systems 2. All optical temperature determination	1. Cross talk with other stimulus as strain/stress and bending 2. Low spatial resolution in the transverse direction

There are several kinds of nanothermometry with high promising for application including rare earth-doped nanoparticles, quantum dots, and gold nanoparticles (GNPs). Each of them has its merits. Here, we mainly introduce rare earth-doped nanoparticles and simply discuss its advantage and disadvantage compared to other nanothermometry.

10.2 Rare Earth-Doped Upconversion Nanothermometry

Rare earth-doped upconversion nanoparticles (UCNPs) have attracted growing research interest recently thanks to their favorable properties, including high photostability, weak auto-fluorescence from the background environment, low toxicity, and high detection sensitivity. The luminescence of rare earth-doped nanoparticles is independent from the ions environment because of the particular electronic configurations of trivalent rare earth (RE) ions. However, it can be modified by temperature, thus can be used for monitoring temperature. The luminescence occurs when an orbital electron relaxes from excited state to ground state by emitting a photon of light. The properties of the emitted photons depend on the properties of the electronic states involved in photon emission. And the electronic states, in turn can be affected by temperature. That is why we can achieve thermal sensing from the spatial and spectral analysis of the light by exploiting the relationship between temperature and luminescence properties. There are several parameters that define the luminescence properties: intensity, band shape, spectral position, bandwidth, polarization, and lifetime. See in Fig. 10.1a.

However, UCNPs as thermosensor always used intensity and band shape two parameters to define the luminescence properties. So herein, we will focus on the rare earth-doped upconversion nanothermometry based on intensity, band shape, and lifetime separately. After that we simply introduce some rare earth-doped downconversion nanothermometry using spectral position and bandwidth to better understand how luminescence could be used to measure temperature.

10.2.1 Intensity Upconversion Nanothermometry

When temperature changes, there is an overall change in the number of emitted photons per second such that the emission spectrum becomes less (or more) intense. Temperature induced changes in the luminescence intensity are caused by complex factors, and there are a large number of mechanisms to explain including quenching mechanisms, the effect of non-radiative process, population redistribution due to Boltzmann statistics, and appearance of phonon assisted Auger conversion processes.

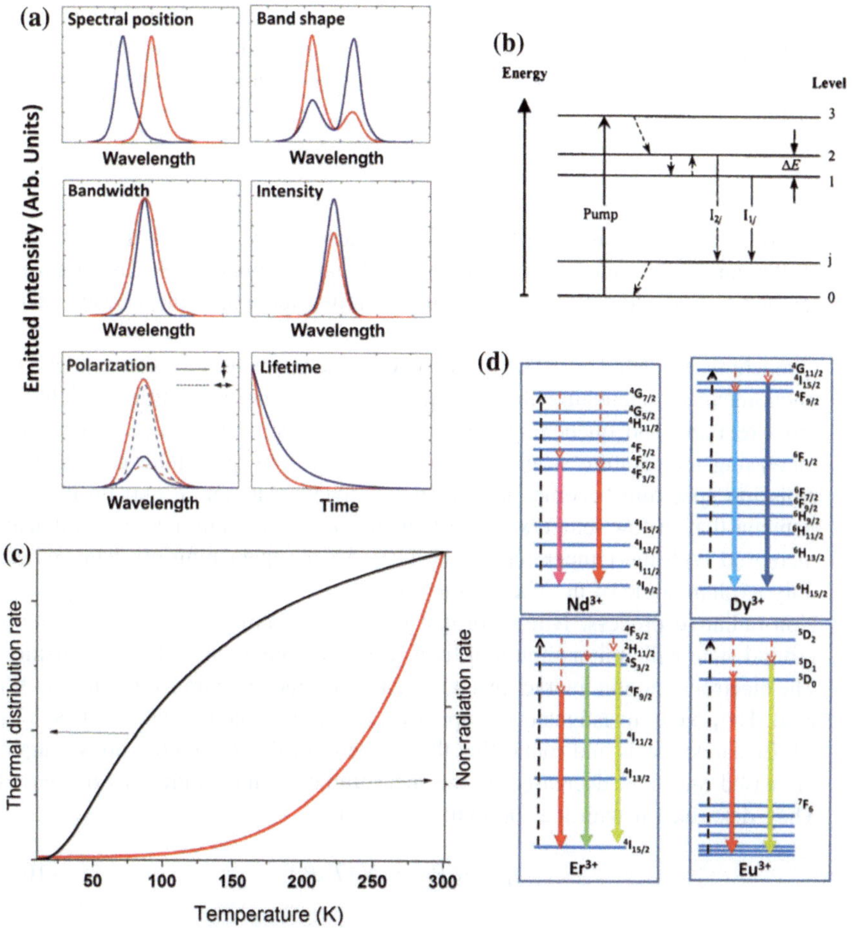

Fig. 10.1 **a** Schematic representation of the possible effects caused by a temperature increment on the luminescence: spectral position, band shape, bandwidth, intensity, polarization, and lifetime. **b** Simplified energy-level diagram showing the energy levels and transitions of interest in the fluorescence intensity ratio (FIR) temperature sensing technique. The *dashed lines* represent non-radiative decay, while the *solid arrows*, labeled I_{2j} and I_{1j}, represent the luminescence transitions used to measure the fluorescence intensity ratio. **c** Thermal distribution rate and non-radiation rate versus temperature using the simulated parameters. **d** Energy-level diagrams of the trivalent RE ions used for band-shaped luminescence nanothermometry. The transitions used in band-shaped luminescence nanothermometry are also indicated. Copyright 2012 Royal Society of Chemistry. **a** is reprinted with the permission from Ref. [9]. Copyright 2012 Royal Society of Chemistry. **b** is reprinted with the permission from Ref. [3]. Copyright 2003 American Institute of Physics. **c** is reprinted with the permission from Ref. [10]. Copyright 2003 American Institute of Physics. **d** is reprinted with the permission from Ref. [9]. In Figure **d**, neodymium is reprinted with the permission from Ref. [11]. Copyright 2010 Elsevier B.V. Dysprosium is reprinted with the permission from Ref. [12]. Copyright 2006 Elsevier B.V. Erbium is reprinted with the permission from Ref. [11]. Copyright 2010 Elsevier B.V. Europium is reprinted with the permission from Ref. [13]. Copyright 1994 American Institute of Physics

1. Population redistribution due to Boltzmann statistics: In statistical mechanics and mathematics, a Boltzmann distribution is a probability distribution, probability measure, or frequency distribution over various possible states of a system.

$$N_i = e^{-\varepsilon_i/kT} \tag{10.1}$$

where N_i is the population of state i, ε_i is state energy, and k is the product of Boltzmann's constant, and T is the thermodynamic temperature. The change of temperature would activate the population redistribution of the various energy states of RE.
2. Quenching mechanisms: There are several mechanisms that can explain the phenomena of quenching in RE doped luminescence nanoparticles including cross-relaxing and quenching centers (always refer to the lattice defect). The increasing temperature would activate the process of quenching such that the emission spectrum becomes less intense. However, it cannot explain the phenomena that in low temperature (less then 200 K), the luminescence intensity increased with the temperature rising [10]. So the quenching mechanisms can only explain some of the phenomena, not all.
3. Non-radiative process: Not all orbital electrons relaxing from excited state to ground state will emit a photon of light, which is called non-radiative transition. The electron–phonon interactions may cause non-radiative transition. Keyue Wu et al. [10] have investigated the upconversion luminescence (UCL) of $NaYF_4$: Yb,Er nanocrystals and found that the intensity of the green emissions initially increased and then decreased as the measurement temperature was decreased. They discussed it with the following derivation:

$$I_i = N_i R_i \tau = N_i(T) R_i \tau \tag{10.2}$$

where I_i is the UCL intensity, R_i the radiative rate, and τ_i the lifetime of the excitation states. The lifetime depends on the radiative rate R and non-radiative transition rate W_{NR} and can be written as follows:

$$\frac{1}{\tau} = R + W_{NR} \tag{10.3}$$

The non-radiative transition rate of multi-phonon relaxation in the RE ions can be expressed as

$$W_{NR}(T) = W_{NR}(0)\,[1 - \exp(-\hbar\omega/k_B T)]^{-p} \tag{10.4}$$

where W_{NR} (0) is the non-radiative rate at zero temperature and $\hbar\omega$ is the phonon energy. p the number of phonons involved. Combining Eqs. (10.2)–(10.4), the emission intensity is given by,

$$I_i = N_{i,\infty} \exp[-p_i \Delta E/k_B T] \, [\exp -k_B T/Eq]^{pi} \times \{1 - \tau W_{NR}(0)[1 - \exp(-\hbar\omega/k_B T)]^{-p}\} \tag{10.5}$$

Thermal distribution rate and non-radiation rate could be stimulated as discussed above, both thermal distribution and relaxation processes play an important role in the UCL. The non-radiation has to be taken into account for nanocrystals in order to explain the temperature-dependent upconversion intensity.

Indeed, the intensity of the luminescence lines of RE ions does not have a linear dependence on temperature. The thermal behavior of different lines could differ by a great amount to that observed for other lines. However, it could be regarded as linear within narrow range of temperature.

Zhang et al. [14] have demonstrated intense yellow UC luminescence (585 nm) which is identified as the $^4T_{1g}(G) \rightarrow {}^6A_{1g}(S)$ transition of octahedrally coordinated Mn^{2+} ions in $KZnF_3$:Yb^{3+},Mn^{2+} nanocrystals upon excitation with a 976 nm laser diode at RT. They proposed an UCL mechanism based on exchange coupled Yb^{3+}–Mn^{2+} dimer and the upconversion process consists of a sequence of ground-state absorption (GSA)/excited-state absorption (ESA) steps. Then, they investigated the temperature-dependent upconversion luminescence of $KZnF_3$:Yb^{3+},Mn^{2+} during heating and cooling cycles and found that the thermal quenching is reversible. The author thought that the good temperature-tunable UC emission properties of material showed the obtained nanocrystals might have potential applications in the fields of sensing, fluorescence imaging, and three-dimensional displays and so on. What's more the Mn^{2+} ion has been demonstrated as a useful T1 magnetic resonance imaging (MRI) contrast agent making it a new dual-mode imaging material.

10.2.2 Band-Shaped Upconversion Nanothermometry-FIR Theory

Here, the "band shape" means the relative intensity between the different spectral lines, namely fluorescence intensity ratio (FIR) method. In some works, they called it "self-referencing." It has been shown that the relative population of such "thermally coupled" levels follows a Boltzmann-type population distribution. Now, we assume that the two spectral lines we observe are from the state level 1 and 2. (See Fig. 10.1b) Then, the N_1 and N_2 will be given by

$$N_1 = \exp(-E_1/kT) \tag{10.6}$$

$$N_2 = \exp(-E_2/kT) \tag{10.7}$$

Obviously, not all transitions have the same temperature dependence. The intensity of the luminescence line corresponding to the deexcitations from state 1 and 2 down to the terminal state j is given by

$$I_{1j} = \varphi_1 N_1 \tag{10.8}$$

$$I_{2j} = \varphi_2 N_2 \tag{10.9}$$

$$\varphi_1 = g_1 \delta_1 \omega_1 \tag{10.10}$$

$$\varphi_2 = g_2 \delta_2 \omega_2 \tag{10.11}$$

where g is the degeneracy, δ is emission cross-section, and ω is the angular frequency.

Thus, the ratio between both intensities is given by

$$\begin{aligned} \mathrm{FIR} &= \frac{I_{2j}}{I_{1j}} = \frac{\varphi_2 N_2}{\varphi_1 N_1} = \frac{g_2 \delta_2 \omega_2}{g_1 \delta_1 \omega_1} \exp(-\Delta E/kT) = B\ \exp(-\Delta E/kT) \\ \Delta E &= E_2 - E_1 \\ B &= \frac{g_2 \delta_2 \omega_2}{g_1 \delta_1 \omega_1} \end{aligned} \tag{10.12}$$

By referring temperature to intensity ratios of selected dual emission bands rather than an individual emission peaks, it has many advantages. Firstly, the FIR theory is relatively well understood, and thus, the behavior is easier to predict. Secondly, this helps to reduce the dependence of the measurement technique on the excitation power and the sensor's concentration, so it is called "self-referencing." With these advantages, FIR method has become widely used in luminescence nanothermometry.

Two important parameters that measure the property of nanothermometry are sensitivity (S) and resolution (r).

Sensitivity Clearly, remarkable changes in FIR for small temperature changes will provide the largest thermal sensitivities.

$$S = \frac{\mathrm{dFIR}}{\mathrm{d}T} = \mathrm{FIR}\left(\frac{\Delta E}{kT^2}\right) \tag{10.13}$$

And relative sensitivity (S') is given by,

$$S' = \frac{S}{\mathrm{FIR}} = \frac{\Delta E}{kT^2} \tag{10.14}$$

From this equation, we can see that using pairs of energy levels with larger energy differences increases the FIR sensitivity. However, this is true to some extent, and care needs to be taken that the levels are not too far apart so that thermalization is no longer observed [3]. What's more, we should ensure that the luminescence intensity not too low to measure, which may be affected by the overlapping of luminescence peaks and stray light. Otherwise, the date may not be accurate.

The sensitivity to temperature also depends on the annealing temperature as it affects the energy gap between the two excited levels (in the case of Er in ZnO host) [15]. The UCNPs are more sensitive to temperature when annealed at higher temperatures. On the other hand, temperature has a larger effect on nanosized crystals than on the respective bulk material in general, probably because surface defects become increasingly important.

Resolution Resolution is the smallest change the sensor can detect.

$$r = \frac{\sigma_T}{S} \tag{10.15}$$

where σ_T is the standard deviation at a special T and S the slope of the fit.

Wolfbeis et al. [16] have investigated hexagonal $NaYF_4$ nanocrystals doped with Yb^{3+} as the sensitizer and $Er^{3+}/Ho^{3+}/Tm^{3+}$ as the activator with different synthesis methods displaying at least two emission lines that respond differently to temperature changes. The result turned out that core–shell UCNPs of $NaYF_4$:Yb^{3+},Er^{3+}@$NaYF_4$ were the best choice for temperature measurements in the physiologically important range between 20 and 45 °C because of their brightness and the excellent temperature resolution of 0.45, which was better than other lanthanide ions (e.g., holmium and thulium).

We can classify the rare earth-doped luminescence nanothermometry into two groups: Single-center and multi-center, which is from the fundamental mechanism, depend on whether the luminescence lines under analysis are generated by a single type of RE or by a combination of different RE ions.

10.2.2.1 Single-Centered Upconversion Nanothermometry

Many kinds of RE ions such as praseodymium (Pr), neodymium (Nd), samarium (Sm), europium (Eu), dysprosium (Dy), erbium (Er), and ytterbium (Y) have been used as single-center nanothermometry [3]. It is well known that the temperature is determined by the energy level as we have discussed, we show some energy-level diagrams of the RE ions which are favorite as thermosensor (Fig. 10.1d [9]).

Table 10.1 has showed various kinds of upconversion nanothermotry with different host and dopant ion. We can see that Er might be the most used one because Er^{3+} ions typically show a very intense green emission that consists of two

Table 10.3 Mo codoped YbAG showed higher quantum yield [18]

Host	Dopant	Quantum yield (%)		
		Blue	Green	Red
YbAG	Er		0.000033	0.00037
YbAG	Er-Mo		0.2	0.01
YbAG	Tm	0.000017	0.0001	
YbAG	Tm-Mo	0.18		

luminescence bands centered at 520 and 540 nm, whose relative intensity is strongly temperature dependent.

To ensure good property of temperature detecting, high quantum yield is needed, especially these two green luminescence bands. Dong et al. [17, 18] greatly improved the green upconversion emission intensity with $Yb_3Al_5O_{12}$ (YbAG) and $Yb_2Ti_2O_7$ by codoping Mo (Table 10.3). Figure 10.2a shows the energy-level diagram of the UC emission for the Er:$Yb_2Ti_2O_7$ and Er–Mo:$Yb_2Ti_2O_7$ by a 976 nm LD excitation. For the Er:$Yb_2Ti_2O_7$,Yb was excited to the $^2F_{5/2}$ level by the GSA. The green UC emission was produced by transitions of $^2H_{11/2}$, $^4S_{3/2} \rightarrow {}^4I_{15/2}$ of Er ions. The two emitting levels of $^2H_{11/2}$ and $^4S_{3/2}$ were mainly populated by the sequential energy transfers ET (I) and ET (III) and the non-radiative relaxations from $^4F_{7/2}$ to $^2H_{11/2}$ and $^4S_{3/2}$ levels. The red UC emission by the transition of $^4F_{9/2} \rightarrow {}^4I_{15/2}$ of Er ions was ascribed mainly to the cross-relaxation (CR) and ET (II) processes. However, the intermediate states of Er ions at the low energy level would be strongly affected by the strong lattice phonon couplings, thus the energy transferred from Yb at the earliest stage of ET (I) for Er could be relaxed to a lower energy level or be completely quenched; as a result, the green UC emission of Er ions became very dim. On the other hand, with Mo codoping, the sensitization through the Yb^{3+}–MoO_4^{2-} dimer complex formed, which entailed both GSA and the ESA. This two-photon absorption was followed by a high energy transfer (HET) to the Er^{3+}ions. The HET from $|^2F_{7/2},{}^3T_2\rangle$ state of the Yb^{3+}–MoO_4^{2-} dimer to the $^4F_{7/2}$ level of Er^{3+} occurred, resulting in strong green UC emissions. The HET process significantly avoided the lattice phonon quenching of oxide matrix, thus increasing the green UC emissions dramatically. Because of the red UC generating from non-radiative relaxation from $^4S_{3/2}$ to $^4F_{9/2}$ levels and subsequent transition of $^4F_{9/2} \rightarrow {}^4I_{15/2}$, the large energy gap between $^4S_{3/2}$ and $^4F_{9/2}$ levels made the red emission weak [18].

From the Table 10.4, we can also see that among these host materials, $NaLaF_4$ might be an optimal choice to be applied in vivo because their bio-compatibility. What's more, Er^{3+} and Yb^{3+} codoped $NaYF_4$ ($NaYF_4$:Er^{3+},Yb^{3+}) are widely used because of their outstanding luminescence brightness, their low toxicity, and their good chemical and physical stability. Here, the main use of Yb^{3+} ions is to provide efficient excitation of Er^{3+} ions via a two-photon absorption process.

Vetrone et al. [19] have used $NaYF_4$:Er^{3+},Yb^{3+} as intracellular nanothermotry. The UCNPs with size of 18 nm are efficiently internalized by HeLa cervical cancer cells. When excited with a 920 nm laser (with an excitation intensity below

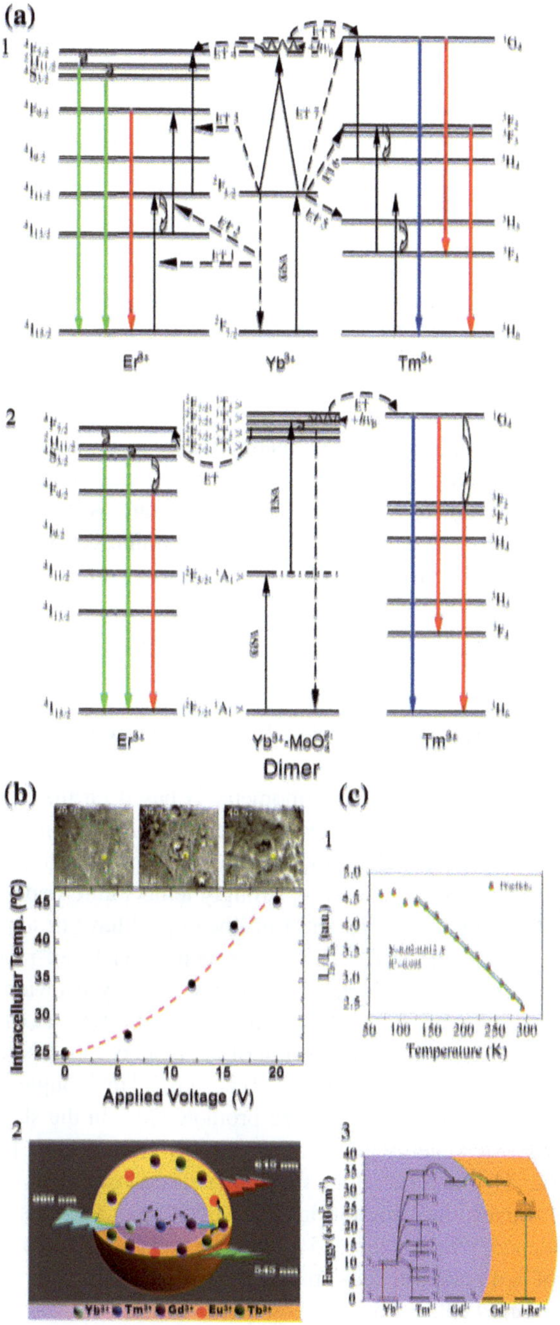
(a)
1
Er3+
Yb3+
Tm3+
GSA
2
Er3+
Yb3+-MoO4 2-
Dimer
GSA
ESA
(b)
Intracellular Temp. (ºC)
Applied Voltage (V)
2
(c)
1
Temperature (K)
3
Energy

◀**Fig. 10.2** **a** Single-centered upconversion nanothermometry YbAG's sensitization diagrams. (*1*) Er(Tm)-Yb codoped systems. (2) Er(Tm)-Yb-Mo codoped systems. **b** Single-centered upconversion nanothermometry $NaYF_4$:Er^{3+},Yb^{3+}. *top* optical transmission images of an individual HeLa cell at three inner temperatures. Cell death is observed at 45 °C. *bottom* temperature of the HeLa cell determined by the Er^{3+} ion luminescence in the $NaYF_4$:Er^{3+},Yb^{3+} NPs as a function of the applied voltage. **c** Multi-centered upconversion nanothermometry $NaGd_4$:Yb^{3+}/Tm^{3+}@Tb^{3+}/Eu^{3+}. (*1*) Temperature dependence of intensity ratio of Tb^{3+} (~545 nm) to Eu^{3+} (~615 nm) and the linear fitting curve for $NaGdF_4$:Tm^{3+}/Yb^{3+}@$NaGdF_4$:Tb^{3+}/Eu^{3+} core–shell nanoparticles. (*2*) Schematic design of lanthanide-doped $NaGdF_4$@$NaGdF_4$ core–shell nanoparticle. (*3*) The energy levels of the Ln^{3+} ions and energy transfer mechanisms in the core–shell nanoparticles. **a** is reprinted with the permission from Ref. [18]. Copyright 2012 WILEY-VCH Verlag GmbH & Co. KGaA, Weinheim. **b** is reprinted with the permission from Ref. [19]. Copyright 2010 American Chemical Society. **c** is reprinted with the permission from Ref. [20]. Copyright 2014 The Royal Society of Chemistry

0.5 kW cm^{-2} to avoid pump-induced heating), they displayed fairly strong green luminescence which was collected using a confocal microscope connected to a high-resolution spectrometer. The internal temperature was measured by determining the temperature-dependent ratio of the luminescence intensities of the UCNPs at 525 and 545 nm. Intracellular changes in temperature resulting from external heating were easily detectable, and cells were simultaneously imaged. Figure 10.2b shows a typical result [19].

10.2.2.2 Multi-centered Upconversion Nanothermometry

Multi-centered upconversion nanothermometry is based on the incorporation of a luminescent compound containing two RE ions, whose luminescence intensities follow very different thermal behaviors, in such a way that the luminescence intensity ratio between their emissions would be strongly temperature dependent. The most common used RE elements are Tb and Eu. Zheng [20] have designed a core–shell UCNPs ($NaGd_4$:Yb^{3+}/Tm^{3+}@Tb^{3+}/Eu^{3+}) for non-contact self-referencing multi-centered temperature sensors, which exhibit high sensitivity at a wide range from 125 to 300 K based on two emissions of Tb^{3+} at 545 nm and Eu^{3+} at 615 nm under the NIR laser excitation. The schematic design of lanthanide-doped $NaGdF_4$@$NaGdF_4$ core–shell nanoparticle can be seen in Fig. 10.2c. Yb^{3+}/Tm^{3+}couple was doped in the core which harvest NIR photons and then promote Gd^{3+} in the shell to the excited state through a two-step energy transfer (ET) via Yb^{3+}-Tm^{3+}-Gd^{3+} [31, 32]. The UC intensities of both Tb^{3+} and Eu^{3+} ions show obvious decrease with the rise in sample temperature. However, the emission intensity of 545 nm (Tb^{3+}) shows linear temperature dependence, whereas the 615 nm (Eu^{3+}) emission is much less affected by the changing of temperature, thus make it possible to use the intensity ratio to measure temperature [20].

Compared to single-centered system, these kinds of temperature sensors might have these advantages as below. (1) The large difference in energy between the emission of Tb^{3+} and Eu^{3+} makes the spectra easier to be detected and calculated

Table 10.4 Various kinds of UCNPs with different host materials and dopant ions

Host material	Dopant ion	Particle size (nm)	Excitation density	λ_{em} (nm)	S(max) (K^{-1})	R(average) (K)	Temperature range	References
ZrO_2	Er/Yb/Li		50 mW	517/562	0.0134		323–673	[19]
				646/678	0.0104			
YbAG	Er(Tm)/Mo	40–80	2 mW	522/546	0.0048	0.3	295–973	[18]
$Yb_2Ti_2O_7$	Er/Mo		5 mW	530/550	0.0074	0.1	290–610	[17]
$YNbO_4$	Yb/Er		2 W	520/557	0.0073		300–573	[20]
$NaYF_4$	Yb/Er	6,000	29 mW	525/545	0.012	0.4	160–300	[21]
$KZnF_3$	Yb/Mn	200		585			300–573	[14]
Gd_2O_3	Er/Yb	17–50		523/548	0.0039		300–900	[22]
CaF_2	Tm/Yb	2		538/522				[23]
				790/800				
$Ca_{12}Al_{14}O_{33}$	Yb/Ho	100	2 W	467/492			300–600	[24]
$BaTiO_3$	Er	26	2 kWc m^{-2}	526/547	0.0052		310–500	[25]
Al_2O_3	Er	–	0.8 W	523/545	0.0052	0.3	295–773	[26]
Al_2O_3	Er/Yb	–	0.35 W	523/545	0.0051	0.3	295–973	[27]
Al_2O_3	Er/Yb/Mo		2 mW	522/546	0.0051	0.3	294–973	[28]
ZnO	Er	80	27.5–51 kWc m^{-2}	536/553	0.0062		273–573	[15]
$NaYF_4$	Yb/Er	95–100		541/523		3.4	293–333	[16]
$NaYF_4$	Yb/Er	26–30		656/541		2.1	293–333	[16]

(continued)

Table 10.4 (continued)

Host material	Dopant ion	Particle size (nm)	Excitation density	λ_{em} (nm)	S(max) (K^{-1})	R(average) (K)	Temperature range	References
$NaYF_4$	Yb/Er @Y	31–34		541/523		0.45 (20–45 °C)	293–333	[16]
						0.69 (50–60 °C)		
$NaYF_4$	Yb/Ho	95–100		656/541		7.3	293–333	[16]
$NaYF_4$	Yb/Ho	26–30		656/541		11.4	293–333	[16]
$NaYF_4$	Yb/Tm	26–30		695/475		>20	293–333	[16]
$NaGdF_4$	Yb/Tm	25	1 W	545/615	0.012		125–300	[29]
	Tb/Eu							

based on relative intensity measurement. (2) Their small size and adjustable surface chemistry endow them high performance in nanoscale environment with good spatial and temperature resolution.

10.2.3 Lifetime Upconversion Nanothermometry

Luminescence lifetime τ_f is defined as the time that the emitted luminescence intensity decays down to $1/e$ of its initial value after a pulsed excitation. This is an indication of the total decay probability of the emitted intensity. Decay probabilities from electronic levels depend on a great variety of factors, and many of them are related to temperature (such as phonon assisted energy transfer processes and multi-phonon decays). This temperature dependence makes it possible to extract temperature readings from the determination of the luminescence lifetime [9]. Lifetime luminescence nanothermometry has many advantages such as temperature reading is not affected by luminescence intensity. In addition, unlike the different methods reviewed up, which need to employ large acquisition time to ensure an adequate signal-to-noise ratio in the luminescence spectrum aiming an accurate temperature reading, it just need to acquire the signal a time interval on the order of the luminescence lifetime (typically in the range of nanoseconds to milliseconds). And long illumination time may lead to the possibility of laser-induced local heating of the system, which will affect the accuracy. Furthermore, for the acquisition of luminescence decay curves, the use of high-resolution spectrometers is not necessary since spectral analysis is no longer required [9]. Temperature-induced luminescence lifetime variation in different rare earth-based systems. Figure 10.3a shows the experimental data of the change of lifetime induced by the change of temperature [9, 33–36]. Recently, Savchuk et al. [37] have studied the thermal sensing ability of fluoresncent lifetime of $NaYF_4$:Yb,Er and NaY_2F_5O:Yb,Er nanoparticles and found that the normalized lifetime values are linear to temperature and NaY_2F_5O:Yb,Er has a better sensitivity compared to $NaYF_4$:Yb,Er.

10.2.4 Bandwidth Nanothermometry and Spectral Position Nanothermometry

Rare earth-doped nanoparticles can be used as upconversion or downconversion luminescence materials. Because of their high quantum efficiency, downconversion luminescence (DCL) materials are also widely applied as nanothermometry. So far, few articles about upconversion nanothermometry using bandwidth or spectral position to determine temperature have been reported. The reason may be that these two parameters are not dependent on temperature or difficult to detect when using upconversion nanothermometry to measure temperature. However, there are many

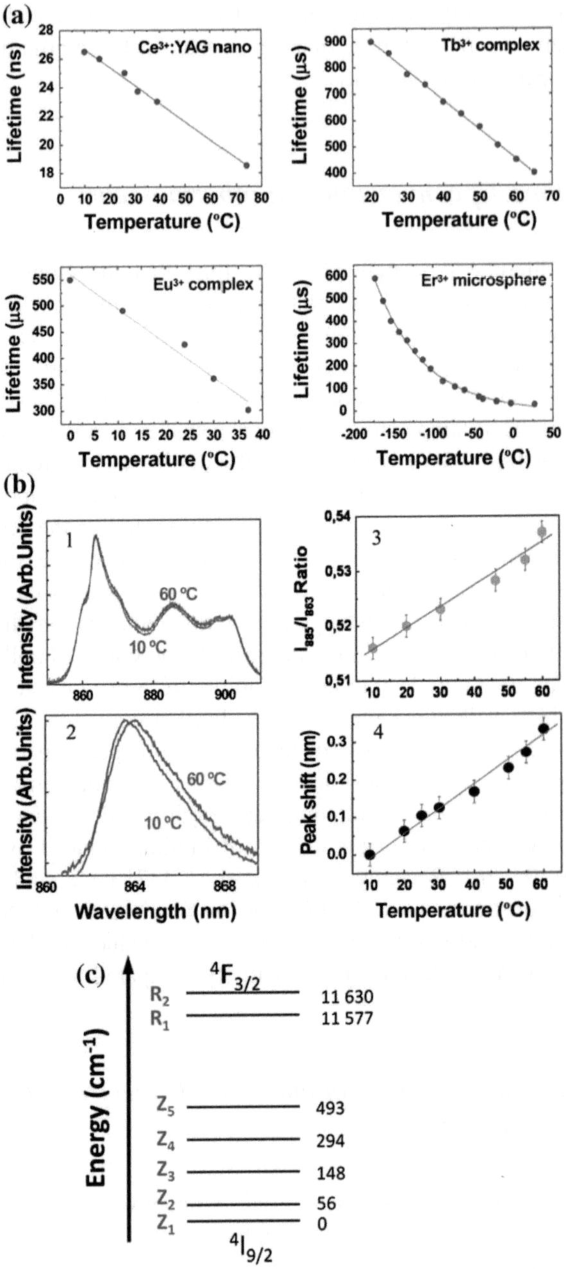
(a)
Ce³⁺:YAG nano
Lifetime (ns)
Temperature (°C)
Tb³⁺ complex
Lifetime (μs)
Temperature (°C)
Eu³⁺ complex
Lifetime (μs)
Temperature (°C)
Er³⁺ microsphere
Lifetime (μs)
Temperature (°C)
(b)
1
Intensity (Arb.Units)
60 °C
10 °C
2
Intensity (Arb.Units)
60 °C
10 °C
Wavelength (nm)
3
I885/I863 Ratio
4
Peak shift (nm)
Temperature (°C)
(c)
Energy (cm-1)
4F3/2
R2 11 630
R1 11 577
Z5 493
Z4 294
Z3 148
Z2 56
Z1 0
4I9/2

◀**Fig. 10.3 a** Lifetime upconversion nanothermometry. Temperature-induced luminescence lifetime variation in different RE (Ce^{3+}, Tb^{3+}, Eu^{3+}, Er^{3+})-based systems. **b** Spectral position nanothermometry Nd^{3+}:LaF_3 nanoparticles. (*1*) Emission spectra of Nd^{3+}:LaF_3 nanoparticles as obtained at two different temperatures (10 and 60 °C and normalized at the maximum). (*2*) Details of the emission spectra of Nd^{3+}:LaF_3 nanoparticles around 864 nm as obtained at 10 and 60 °C. The temperature-induced spectral shift is evidenced. (*3*) Temperature variation of the ratio between the emitted intensities at 863 and 885 nm. (*4*) Temperature-induced spectral shift of the main emission line around 864 nm as a function of temperature. *Dots* are experimental data, and the *solid line* is the best linear fit. **c** Spectral position nanothermometry Nd^{3+}:LaF_3 nanoparticles. The energy-level diagram of the emission line of the Nd^{3+}:LaF_3 nanoparticles. **a** is reprinted with the permission from Ref. [9]. Copyright 2012 Royal Society of Chemistry. In **a**, Tb^{3+} and Eu^{3+} are reprinted with the permission from Ref. [34]. Copyright 2010 Royal Society of Chemistry. Er^{3+} is reprinted with the permission from Ref. [35]. Copyright 2009 Elsevier B.V. **b** is reprinted with the permission from Ref. [36]. Copyright 2013 American Chemical Society. **c** is reprinted with the permission from Ref. [36]. Copyright 2013 American Chemical Society

articles have been reported indicate that DCL material could be a promising nanothermometry and some articles have pointed out that these downconversion nanothermometry determined temperature by detecting bandwidth or spectral position.

10.2.4.1 Bandwidth Nanothermometry

The width of the various emission lines that make up any luminescence spectrum is determined by the properties of the material (such as the degree of disorder) and temperature. The change in the bandwidth of the luminescence spectra is exploited in bandwidth luminescence nanothermometry to achieve a thermal reading. In fact, the bandwidth we measure is the half-peak width. In general, higher the temperature, wider the emission line because of the thermal vibration of the luminescent center and its neighboring atoms/molecules [9]. The magnitude of the temperature-induced luminescence line broadening is small, as a result it can be only observed in systems showing inherent narrow emission lines. For example, Nd:YAG [36] shows two hypersensitive luminescence lines at around 940 nm that correspond to the low energy emissions within the $^4F_{3/2} \rightarrow {}^4I_{9/2}$ transition, from which the bandwidth of these two transitions is strongly temperature dependent.

10.2.4.2 Spectral Position Nanothermometry

It is based on the analysis of the spectral positions of the emission lines, which are unequivocally determined by the energy separation between the two electronic levels involved in the emission. In turn, this depends on a large variety of temperature-dependent parameters of the emitting material including refractive index and inter-atomic distances (density). Thus, in any emitting material, the spectral positions of the luminescence lines are expected to be temperature dependent, and this is exploited by spectral luminescence nanothermometry to translate spectral shifts into temperature [9]. Its advantage is that temperature reading is not affected by luminescence intensity fluctuations caused by variations in the local concentration of emitting centers. QDs have been successfully used for high-resolution spectral position nanothermometry. However, due to QDs' shortness like poor bio-compatibility, other kinds of nanomaterials with high sensitivity to the change of temperature are needed. Surprisingly, in 2003, DCNPs which had a good function in temperature sensor also been reported [36]. They showed the multi-functional character of neodymium-doped LaF_3 core/shell nanoparticles. It has been found that the infrared emission bands of Nd^{3+}:LaF_3 nanoparticles show a remarkable thermal sensitivity, so that they can be advantageously used as luminescent nanothermometers for subtissue thermal sensing. Under infrared (808 nm) optical excitation, these nanoparticles show very intense Nd^{3+} emission bands at around 885 and 1,060 nm. The author used FIR (885 nm/863 nm) and spectral shift of around 864 nm. This temperature-induced spectral shift has been found a linear effect at an almost constant rate of 7×10^3 nm/°C, as can be observed Fig. 10.3c. To explain the mechanism causing this red shift, they have identified the Stark energy sublevels of both $^4F_{3/2}$ and $^4I_{9/2}$ manifold states that give place structure of the $^4F_{3/2} \rightarrow {}^4I_{9/2}$ emission spectrum [Fig. 10.3(4)]. This temperature-induced red shift was unequivocally associated with thermally induced strains in the Nd^{3+} environment that occur as a result of electron–phonon interaction. These strains are temperature dependent through the correlated changes in the amplitude of the lattice vibrations that produce a substantial effect in the line position [36].

10.3 UCNP-Based Nanocomposite Materials for Nanothermotry

Simple UCNPs may not meet the need for high-sensitivity and high-resolution nanothermometry, and people tried to design more complicated materials for better property.

To date, the temperature-dependent emission of UCNPs is always based on the common mechanism that temperature influences the excited-state population of RE ions. However, such Boltzmann distribution governed mechanism sets an inherent limit on the temperature sensitivity of pristine UCNPs. So, new design is needed.

A hybrid complexes of UCNPs ($NaYF_4$:Yb^{3+},Er^{3+}) and rhodamine 6G (R6G) dye molecules were proposed by Sun et al. [38]. They improved the sensitivity of the complexes due to resonance energy transfer. Förster resonance energy transfer (FRET) is a non-radiative process characterized by energy transfer between excited donors and acceptors through long-range dipole–dipole interaction. It is widely used as a spectroscopic ruler in biological detection and in photodynamic therapy. Here, the UCNPs were the energy donor and R6G the acceptor. Actually, for donor and acceptor with spectral overlap, the energy transfer could be realized by either non-radiative energy transfer (FRET) or radiative energy transfer (radiative recombination in the donor first and then reabsorption by the acceptor). The author combined the UCNPs and R6G to create a hemispherical microstructure based on the hydrophobic effect. Upconversion emission from UCNPs under NIR laser excitation, and the absorption and emission from R6G are presented in Fig. 10.4a. They found that the hemispherical microstructure with dye-doped UCNPs exhibited PL emission similar to that in solution. The FRET efficiency was temperature dependant. So the author set spectrometer to monitor the emission intensity from R6G at 575 nm, while the environmental temperature was changing and found the photoluminescence (PL) intensity decreased when temperature raise. From the linear fitting, the average temperature sensitivity was estimated to be 1 % per degree over the whole physiological temperature range (25–45 °C) and 2.7 % per degree in the higher temperature region (47–57 °C). Compared with RE ions, dye molecules possessed a stronger temperature-dependent absorption and emission. So these self-assembled hemispherical microstructures containing the hybrid complexes exhibited better sensitivity than pristine UCNPs [38].

Xiao et al. [39] also designed a hybrid complex based on FRET system. They utilized red/NIR dual emitting $NaLuF_4{:}Mn^{2+},Ln^{3+}$ (Ln^{3+} = Yb^{3+}, Er^{3+}, Tm^{3+}) UCNPs as energy donor and Au nanoparticles as acceptor. Noble metals such as Au and Ag NPs have been regarded as exceptionally sensitive antennas in modulating the optical properties of UCNPs due to strong surface plasmon resonance (SPR) absorption. Instead of simply combining them together to create a microstructure, they made use of the thermal responsive polymer poly (N-isopropylacrylamide) (PNIPAM) as a spacer between the donor and acceptor to modulate their distance. PNIPAM underwent a reversible phase transition from a hydrophilic, water-swollen state to a hydrophobic collapse state at its lower critical solution temperature (LCST, 32 °C). This process led to the chain collapse/extension of PNIPAM to regulate the distance (see in Fig. 10.4b1). Tuning the distance had been widely regarded as a general and facile way to affect the energy transfer efficiency in the FRET system [41]. In addition, SPR absorption of Au nanoparticles also was temperature dependant [42]. So this system thermally tuned the UCL intensity of UCNPs via changing both the donor–acceptor distance and the SPR absorption of Au NPs as the variation of temperature. The spectrometer was set to monitor the emission intensity from transitions of $^4F_{9/2} \rightarrow {}^4I_{15/2}$ (Er^{3+}) (660 nm) and $^3H_4 \rightarrow {}^3H_6$ (Tm^{3+}) (800 nm) in the temperature range of 25–55 °C. The NIR emission (800 nm) decreased monotonously with the increase in temperature, which was similar to that of PNIPAM-conjugated UCNPs. However, the red emission intensity was distinguished into two regimes, a fast decrease within the temperature range of 25–41 °C and a gentle slope at temperature above 41 °C (see in Fig. 10.4b2). The appearance of the inflection for the red UCL at 25 and 41 °C may reflect the occurrence of chain collapse/extension of PNIPAM at this temperature, which thus significantly regulated the FRET efficiency between UCNPs and Au

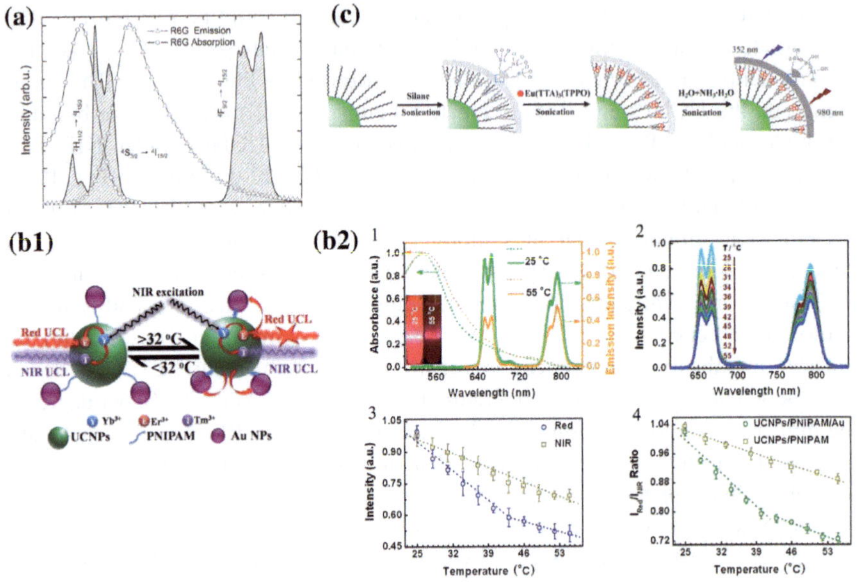

Fig. 10.4 **a** UCNP-based nanocomposite materials for nanothermotry (UCNPs + R6G) upconversion emission from UCNPs under NIR laser excitation and the absorption and emission from R6G. **b1** UCNP-based nanocomposite materials for nanothermotry (UCNPs + PNIPAM + Au particles). Schematic illustration of smart-polymer-switchable FRET in upconverting nanocomposites. **b2** UCNP-based nanocomposite materials for nanothermotry (UCNPs + PNIPAM + Au particles). Optical response of the upconverting nanocomposites to temperature variation: (*1*) absorption spectra, UCL spectra, and corresponding luminescent photos of the colloidal solutions at 25 and 55 °C, respectively; (*2*) detailed UCL spectra at various temperatures; (*3*) integrated intensity plots with the respective linear fits for the red and NIR emissions, respectively; (*4*) ratiometric calibration plots with the respective linear fits for the upconverting nanocomposites as well as PNIPAM-conjugated UCNPs. **c** Schematic illustration for the synthesis of silane-modified $NaYF_4$:Yb^{3+},Er^{3+} loaded with $Eu(TTA)_3(TPPO)_2$. **a** is reprinted with the permission from Ref. [38]. Copyright 2013 Wiley-VCH Verlag GmbH & Co. KGaA, Weinheim. **b1**, **b2** is reprinted with the permission from Ref. [39]. Copyright 2014 The Royal Society of Chemistry. **c** is reprinted with the permission from Ref. [40]. Copyright 2013 The Royal Society of Chemistry

NPs. By the ratiometric sensing, using the intensity ration of red to NIR emission, the upconverting nanocomposites yield a relatively high resolution of 0.9 °C [39].

On the other hand, UCNPs can also be modified to create complexes to realize their multi-functions.

Recently, Hongwei Song et al. [40] have proposed an amphiphilic silane-modified $NaYF_4$:Yb,Er loaded with $Eu(TTA)_3(TPPO)_2$ nanoparticles. Here, $NaYF_4$:Yb, Er displayed UCL while $Eu(TTA)_3(TPPO)_2$ DCL. Due to the UC and DC properties, they achieved the dual-mode physiological range temperature sensing and dual-mode cell imaging. Firstly, the oleophilic $NaYF_4$:Yb,Er was modified by amphiphilic silanes to be converted into hydrophilic ones (SUCNPs). Then, Eu

$(TTA)_3$ $(TPPO)_2$ was loaded into SUCNPs to create Eu-SUCNPs. This process could be seen in Fig. 10.4c. The author investigated both the UC and DC temperature sensing property. As is known that intensity-based measurements may suffer from variation of the sensor concentration and drifts of the optoelectronic system, the author used FIR of $^2H_{11/2} \rightarrow {}^4I_{15/2}$ to $^4S_{3/2} \rightarrow {}^4I_{15/2}$ of Er^{3+} and lifetime of $^5D_0$ state of Eu^{3+} to serve as the detection signals for temperature, which was more reliable. The temperature sensitivities of 0.89 % per °C for UC detection and 3.48 % per degree for DC detection were obtained in the physiological range from 26 to 46 °C. Because of the different detect methods of UC (FIR) and DC (lifetime), the Eu-SUCNPs display good physiological temperature sensing with high resolution from UCL and DCL which can make up for the drawbacks of each other. In addition, Eu-SUCNPs can be applied for dual-mode cell imaging as UC and DC luminescent biological labels [40].

In the same year, Carlos et al. [43] combined UCNPs and Au nanoparticles (Au NPs) together to form a multi-function nanoplatform integrating heaters and thermometers. Photothermal therapy was one promising method for cancer treatment. It transformed electromagnetic radiation to partial heat to kill target cells. In noble metals such as Au and Ag, the coherent collective oscillation of electrons in the conduction band induced large surface electric fields which greatly enhance the radiative properties when they interact with resonant electromagnetic radiation. This made the absorption cross-section of these nanoparticles orders of magnitude stronger than that of the most strongly absorbing molecules, which implied that they could be used in photothermal therapy [44]. However, it might cause overheating to injure healthy tissues when no thermometer to guide the therapy. So the author designed this complex aiming to solve the problem. Firstly, they synthesized $(Gd_{0.95}Yb_{0.03}Er_{0.02})_2O_3$ NRs via a simple wet-chemical route. Au NPs were adsorbed on the NRs by the in situ reduction of $HAuCl_4 \cdot 3H_2O$ with $NaBH_4$ in an aqueous dispersion of the NRs. The $(Gd_{0.95}Yb_{0.03}Er_{0.02})_2O_3$ NRs acted as thermometer and Au NPs as the heaters. The local temperature was ascertained from the Boltzmann's distribution of the Er^{3+} upconversion $^2H_{11/2} \rightarrow {}^4I_{15/2}$ and $^4S_{3/2} \rightarrow {}^4I_{15/2}$ intensity ratio, in the range 300–1050 K with resolution of 0.3–2.0 K. While in the range of high temperature (1,200–2,000 K), the temperature was calculated using Plank's law with resolution of 4–13 K. Changing the concentration of Au NPs could modulate the surface temperature. A maximum relative sensitivity up to 1.5% K (resolution 1 K) in the temperature range of physiological interest (301–350 K) was demonstrated under low-power (32–86 W cm^{-2}) NIR light (980 nm) excitation in the therapeutic window [43]. What's more, both the Au NPs and UCNPs could be applied in optical bio-imaging. So the author named it "All-In-One Optical Heater-Thermometer Nanoplatform."

10.4 Upconversion Nanoparticles Versus Quantum Dots Versus Gold Nanoparticles Versus Green Fluorescent Protein as Nanothermometry

Not only UCNPs, with the development of nanotechnology, many other nanomaterials are also good candidates for nanothermometry. Here, we will introduce some nanomaterials such as quantum dots, GNPs, and GFP which have been studied a lot and might be promising nanothermometry.

10.4.1 Quantum Dots (QDs)

Semiconductor QDs have several properties such as size-tunable absorption and emission frequencies, high quantum yield, and photostability, which made QDs an active field of research and development for imaging, labeling, and sensing. The fluorescence of QDs with certain materials and sizes is strongly dependent on temperature, which has been known for a long time, making QDs promising nanoprobes for thermal detection [45, 46]. With the increase in temperature, the fluorescence intensity will decrease (quenching) [45], accompanied by a spectral shift. And this shift can be assumed to be linear in the bio-physical range [47]. The spectral shift (0.105 nm K^{-1}) is very accurate. Vlaskin et al. firstly reported a core–shell dual emission material to avoid problems associated with the instability and uncertainty of excitation sources or photobleaching and/or photoblinking. Such dual processes from two excited states in the same QD enabled self-calibration ratiometric methods which were much more stable and accurate. Maestro et al. [48] reported on the first intracellular temperature measurement using temperature-induced spectral shift (0.1 nm/°C). One year after, Yang [4] used QDs to map intracellular heat generation in NIH/3T3 cells following Ca^{2+} stress and cold shock (see in Fig. 10.5a). The observation of inhomogeneous intracellular temperature progression raised new possibilities that subcellular temperature gradient might be signal for regulating in cell. In 2013, CdSe QDs were used to monitor intracellular temperature during plasmonic-mediated hyperthermia of HeLa cancer cells [49]. It showed that the photothermally induced damage was extended to several cells due to heat diffusion which means that single CdSe QDs were demonstrated to be capable of providing real temperature control during photothermal therapy carried out on individual cells. In addition, Maestro et al. [50] used QDs as thermometry combined with nanogold to investigate the absorption/heating efficiency of GNPs with different morphologies (nanorods, nanocages, nanoshells, and nanostars).

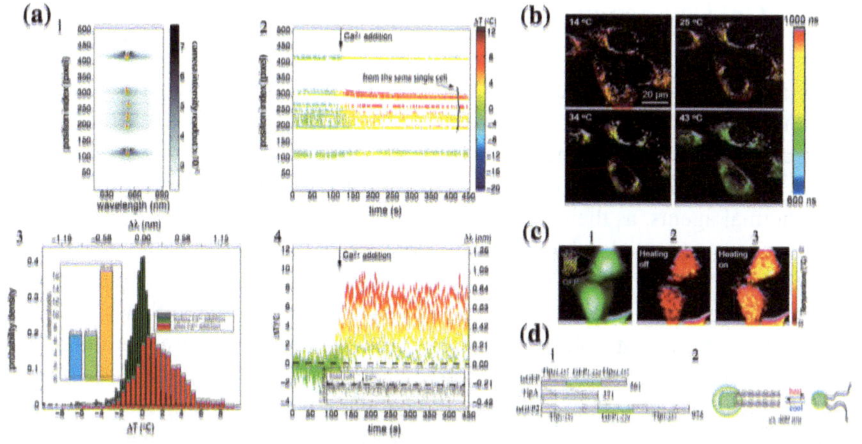

Fig. 10.5 **a** QD-based nanothermometry. Chemical perturbation: Ca^{2+} shock experiments. (*1*) Representative snapshot of the raw position-spectrum maps for the sample. The ordinate is the *quantum dot* position index in units of camera pixels, and the abscissa is the spectral index in units of pixels (0.144 nm/pixel). The positions of *quantum dots* were automatically detected and marked by *red dots* enclosed by *yellow circles*. The position and wavelength of each *quantum dot* spot were then determined by fitting the spectrum to a two-dimensional Gaussian profile. (*2*) Location-dependent intracellular temperature progression as a function of time of the same cell shown in (*1*). The time point at which ionomycin calcium complex was added, ~120 s, is marked by an *arrow*. The uncertainties of position localization as estimated from the Gaussian fitting are represented as *shades* of the temperature color; the more intense and narrower the vertical distribution, the more accurately the location is. Note the non-linear spacing in the color code for temperature shift. The same color code and scaling will be used for the remainder of this report for consistency. (*3*) Statistics of temperature shift for the 31 cells examined, showing significant broadening of the temperature shift distribution. Here, temperature shifts acquired between 30 and 100 s were categorized as "before Ca^{2+} addition," whereas those acquired between 350 and 450 s were categorized as "after Ca^{2+} addition." *Inset* Cell-by-cell variations in temperature response to Ca^{2+} shock, showing that the majority of the cells exhibited increased temperature. (*4*) Local temperature shift as a function of time for the cell shown in (*1*) and (*2*), showing heterogeneous temperature response following Ca^{2+} shock. The uncertainties of the temperature measurement as estimated from the Gaussian fitting are represented as *shades* of the temperature color; the more intense and the narrower the vertical distribution, the more accurately the measurement is. *Inset* Time-dependent local temperature shift of fixed cells, showing that no detectable temperature response upon addition of Ca^{2+}. **b** Gold nanocluster-based nanothermometry. Typical FLIM (luminescence lifetime imaging microscopy) images of HeLa cells with internalized AuNCs at four different temperatures. **c** GFP-based nanothermometry. Temperature measurements in HeLa cells while delivering local heat via photothermal approach. (*1*) Luminescence intensity of GFP transfected HeLa cells. (*2*) Temperature map while not heating. (*3*) Temperature map while heating the NR with a focused infrared laser with a power of $P = 50$ mW, located 50 μm to the right of center of the image. **d** Genetically encoded GFP-based nanothermometry. Design of tsGFP thermosensors. (*1*) Schematic diagram of the tsGFP thermosensors. Coiled-coil regions of TlpA are shown in *gray*. GFP (*green*) is inserted in-frame in between tandemly repeated coiled-coil regions of TlpA. (*2*) The tandem formation of the coiled-coil structure and the resulting luminescence change of tsGFP are associated with thermal changes. Ex, excitation. **a** is reprinted with the permission from Ref. [4]. Copyright 2011 The Royal Society of Chemistry. **b** is reprinted with the permission from Ref. [51]. Copyright 2013 Wiley-VCH Verlag GmbH & Co. KGaA, Weinheim. **c** is reprinted with the permission from Ref. [52]. Copyright 2012 American Chemical Society. **d** is reprinted with the permission from Ref. [53]. Copyright 2013 Nature America, Inc

10.4.2 Gold Nanoparticles (GNPs)

GNPs are widely used for cellular imaging. Their attractive properties including ultrasmall size, low toxicity, colloidal stability, facile synthesis, lack of intermittency, and low photobleaching make them attractive fluorescent probes for in vitro and in vivo imaging [54]. GNPs are also regarded as one of the most promising photothermal agents, as their light-to-heat conversion efficiencies are close to unity. Thus, these nanoparticles have already been used for both in vitro and in vivo photothermal therapy [55]. On the other hand, GNPs can also be applied as biosensor to detect ion [56], protein [57], and even cell [58]. Recently, Shang et al. [51] have demonstrated that gold nanoclusters (GNCs) can also be used as fluorescent nanothermometers. Shang et al. reported that the lifetime of the near-infrared luminescence of GNCs dramatically decreases by 67 % upon raising the temperature from 10 to 45 °C. The fluorescence lifetime of Au NCs in HeLa cells was not sharply defined but distributed. Unlike QDs, which displayed remarkable temperature-dependent spectral shifts, the emission spectra of Au NCs did not shift within the physiological thermal range. The origin of the thermally induced luminescence and lifetime reduction of GNCs was far from being fully understood. However, it suggested that it could be related to temperature-induced changes in the interactions between GNCs and the complexes at their surface because that they found that the magnitude of the temperature-induced lifetime reduction was dependent on the GNCs environment—that was, inside living cells, the lifetime versus temperature behavior was different to that obtained in saline buffers. The formation of a biomolecular corona around the internalized nanoparticles might modify their photophysical properties [59]. Recent studies revealed that protein adsorption can lead to significant changes of the photophysical properties of fluorescent nanoparticles [60–62]. Shang et al. also demonstrated the ability of GNCs for thermal sensing in HeLa cancer cells after proper calibration of intracellular luminescence lifetime measurements (see in Fig. 10.5b). GNCs have made it possible to determine intracellular temperature during homogenous and inhomogeneous external heating with an absolute thermal sensitivity estimated to be in the 0.3–0.5 °C range. This rather good thermal sensitivity is accompanied by a long luminescence lifetime (which implies a reduction in the auto-fluorescence background), complete reversibility, long-term stability, and minimum cell toxicity. On the contrary, an obvious disadvantage of these nanothermometers is the lifetime dependence on the environment, which requires an “in cell” calibration for each experiment [8].

10.4.3 Green Fluorescent Protein (GFP)

GFP has been extensively researched and found useful for multi-functional marker for gene expression, protein localization, pH level, and protein–protein interactions [63] because they can span the entire visible spectrum and have high quantum efficiency [64]. Donner et al. [52] demonstrated that GFP can also be used as a biocompatible fluorescent thermometer. They found that the polarization state of the GFP luminescence was strongly temperature dependent in the physiological range (20–60 °C). The method was tested on HeLa and U-87 MG cancer cell in which GFP was over expressed, and they monitored the heat delivery by photothermal heating of gold nanorods surrounding the cells (see in Fig. 10.5c). A spatial resolution of 300 nm and a temperature accuracy of about 0.4 °C were achieved.

Recently, genetically encoded GFPs with higher thermal sensitivity have been demonstrated and have been used to monitor intracellular temperature changes in specific organelles [53]. This GFP-based thermosensor (tsGFP) contained two parts: GFP and a tandem formation of coiled-coil structures by the Salmonella thermosensing protein TlpA. TlpA is an auto-regulatory repressor protein that senses temperature changes and controls transcription repression in a temperature-dependent manner in Salmonella (see in Fig. 10.5d1). Here, these proteins displayed reversible structural transitions when the temperature was varied around 37 °C, from a parallel coiled-coil dimer to the unfolded monomers (see in Fig. 10.5d2). These changes were transmitted to the GFP protein, as a result, giving rise to temperature-dependent changes in the two main bands (400 and 480 nm). The author also investigated fusion proteins which comprised GFP inserted between different kinds of TlpA (coiled-coil region $TlpA_{94-257}$ or full-length $TlpA_{1-371}$). They claimed that the temperature-sensing range of tsGFP can be controlled by selecting the suitable TlpA. What's more, by using this tsGFP, they visualized different thermogenesis processes in HeLa cells, brown adipocytes, and skeletal muscle myotubes.

10.4.4 Comparison Between UCNPs, GNPs, QDs, and GFP-Based Nanothermometers

Compared to GNPs and GFP, QDs and UCNP-based nanothermometers have been studied more extensively. As a result, the properties of the signal response to the change of temperature are better known to us. Lately, it is grabbing some serious attention that nanothermotry could be put into use within a single cell. All these four nanothermometers applied in different cancer cells have been reported (see in Table 10.5).

We made a simple summary of the advantages and disadvantages of the four nanothermometers in Table 10.6. As we want to put nanothermometers into use in cell, the cytotoxicity of the thermosensors should be low. From this point of view, UCNPs and GFP are much better than the others. From our point of view, nanocomposites which can make up each other's shortcomings may be good choices in the future development.

Table 10.5 Different nanothermometers applied in different cancer cells

Nanosensor	R (°C)	S (%/°C)	Cell line	λ_{ex} (nm)	λ_{em} (nm)	References
CdSe QD	1		HeLa	800	650	[65]
CdSe/ZnS QD		0.7	PC-3	<525	620	[46]
Polymer plus CdSe-CdS	0.2	2.4	HeLa, NiH3T3	400	650	[66]
Gold nanocluster	0.1–0.3	0.5	HeLa	580	710	[56]
GFP	0.4		HeLa/U-87MG	473	530	[52]
Genetically encoded GFP	<0.2		HeLa	400	510	[53]
$NaYF_4$:Yb, Er NP	1		HeLa	920	530	[19]

R resolution; *S* sensitivity; λ_{ex} wavelength of excitation light; λ_{em} wavelength of emission light

Table 10.6 Main detecting parameters, advantages, and disadvantages of UCNPs, QDs, GNPs, and GFP-based nanothermometers

Nanothermometers	Main detecting parameters	Main advantages	Main disadvantages
UCNPs	Band shape	1. Zero background 2. High resolution	Poor quantum yield
QDs	Band shape, spectral position	High quantum yield	Poor bio-compatibility
GNPs	Intensity	High sensitivity	Easy to be affected by the environment
GFP	Band shape	1. Good bio-compatibility 2. High quantum yield	design is too complex, always need to combine with dye or thermosensing proteins

10.5 Summary and Outlook

In this chapter, we have described recent developments of thermometer. It has been developed for several centuries from contact to non-contact. With the development of nanotechnology, it becomes possible to use nanothermometry to detect the temperature within a single cell. UCNPs are one of the promising candidates for developing non-contact nanothermometry. There are several parameters that define the luminescence properties which are dependent on temperature: intensity, band shape, spectral position, bandwidth, polarization, and lifetime. Among these parameters, band shape is most widely used. So we introduced the FIR theory which is used to figure out the relationship between the band shape and temperature. At last, we compared UCNPs with quantum dots, GNPs, and GFP to show each other's merits and demerits. We can foresee that UCNPs could be very useful in the near future even though there are still many challenges to conquer.

The most important advantage of UCNPs to be applied as nanothermotry might be their stability and zero background arising from NIR photoexcitation. However, major challenges remain including low quantum yields, the limitation of resolution, and short depth of luminescent penetration. Up to now, UCNPs using spectral position and bandwidth to measure temperature are still absent and wait for us to explore. In addition, upconverting nanocomposites might be a promising nanosensor to increase the resolution. Even though there are still many problems to solve, we can look forward to making full use of UCNPs in temperature measurements in various fields, including micro- and nanofluidic systems, micro/nano/femto-volume (bio)chemistry, thermo-induced drug release, microdroplets used for chemical synthesis, and wherever exothermal chemical or enzymatic reactions occur at a micro- or nanoscale [1].

References

1. Fischer, L.H., Harms, G.S., Wolfbeis, O.S.: Upconverting nanoparticles for nanoscale thermometry. Angew. Chem. Int. Ed. Engl. **50**, 4546–4551 (2011)
2. dos Santos, P.V., de Araujo, M.T., Gouveia-Neto, A.S., Medeiros Neto, J.A., Sombra, A.S.B.: Optical temperature sensing using upconversion fluorescence emission in Er^{3+}/Yb^{3+}-codoped chalcogenide glass. Appl. Phys. Lett. **73**, 578–581 (1998)
3. Wade, S.A., Collins, S.F., Baxter, G.W.: Fluorescence intensity ratio technique for optical fiber point temperature sensing. J. Appl. Phys. **94**, 4743–4757 (2003)
4. Yang, J.-M.: Quantum dot nano thermometers reveal heterogeneous local thermogenesis in living cells. ACS Nano **5**, 5067–5071 (2011)
5. Brites, C.D., Lima, P.P., Silva, N.J., Millan, A., Amaral, V.S., Palacio, F., Carlos, L.D.: Thermometry at the nanoscale. Nanoscale **4**, 4799–4829 (2012)
6. Hildebrandt, B., Wust, P., Ahlers, O., Dieing, A., Sreenivasa, G., Kerner, T., Riess, H.: The cellular and molecular basis of hyperthermia. Critical Rev. Oncol./Hematol. **43**, 33–56 (2002)
7. Kuruganti, P.T., Qi, H.: Asymmetry analysis in breast cancer detection using thermal infrared images. Proceedings of the Second Joint EMBSiBMES Conference **2**, 1155–1156 (2002)
8. Daniel Jaque, B., del Rosal, E.M., Rodríguez, L.M., Maestro, P.H., Solé, José García: Fluorescent nanothermometers for intracellular thermal sensing. Nanomedicine **9**, 1743–1758 (2014)
9. Jaque, D., Vetrone, F.: Luminescence nanothermometry. Nanoscale **4**, 4301–4326 (2012)
10. Wu, K., Cui, J., Kong, X., Wang, Y.: Temperature dependent upconversion luminescence of Yb/Er codoped $NaYF_4$ nanocrystals. J. Appl. Phys. **110**, 053510 (2011)
11. Haro-González, P., Martín, I.R., Martín, L.L., León-Luis, S.F., Pérez-Rodríguez, C., Lavín, V.: Characterization of Er^{3+} and Nd^{3+} doped Strontium Barium Niobate glass ceramic as temperature sensors. Opt. Mater. **33**, 742–745 (2011)
12. Heyes, A.L., Seefeldt, S., Feist, J.P.: Two-colour phosphor thermometry for surface temperature measurement. Opt. Mater. **38**, 257–265 (2006)
13. Bizzak, D.J., Chyu, M.K.: Rare-earth phosphor laser-induced fluorescence thermal imaging system. Rev. Sci. Instrum. **65**, 102–108 (1994)
14. Song, E.-H., Ding, S., Wu, M., Ye, S., Xiao, F., Dong, G.-P., Zhang, Q.-Y.: Temperature-tunable upconversion luminescence of perovskite nanocrystals $KZnF_3:Yb^{3+}, Mn^{2+}$. J. Phys. Chem. C **1**, 4209–4215 (2013)

15. Wang, X., Kong, X., Yu, Y., Sun, Y., Zhang, H.: Effect of annealing on upconversion luminescence of ZnO:Er_{3+} nanocrystals and high thermal sensitivity. J. Phys. Chem. C **111**, 15119 (2007)
16. Sedlmeier, A., Achatz, D.E., Fischer, L.H., Gorris, H.H., Wolfbeis, O.S.: Photon upconverting nanoparticles for luminescent sensing of temperature. Nanoscale **4**, 7090–7096 (2012)
17. Cao, B.S., He, Y.Y., Feng, Z.Q., Li, Y.S., Dong, B.: Optical temperature sensing behavior of enhanced green upconversion emissions from Er–Mo:$Yb_2Ti_2O_7$ nanophosphor. Sens. Actuat. B-Chem. **159**, 8–11 (2011)
18. Dong, B., Cao, B., He, Y., Liu, Z., Li, Z., Feng, Z.: Temperature sensing and in vivo imaging by molybdenum sensitized visible upconversion luminescence of rare-earth oxides. Adv. Mater. **24**, 1987–1993 (2012)
19. Vetrone, F., Sole, J.G., Naccache, R., Sanz-Rodríguez, F., Capobianco, J.A.: Temperature Sensing Using Fluorescent Nanothermometers. Acs Nano **4**, 3254-3258 (2010)
20. Zheng, S.: Lanthanide-doped NaGdF4 core-shell nanoparticles for non-contact self-referencing temperature sensors. Nanoscale **1**, 5 (2014)
21. Liu, L., Wang, Y., Zhang, X., Yang, K., Bai, Y., Huang, C., Song, Y.: Optical thermometry through green and red upconversion emissions in $Er^{3+}/Yb^{3+}/Li^{+}$:ZrO_2 nanocrystals. Opt. Commun. **284**, 1876–1879 (2011)
22. Singh, A.K., Singh, S.K., Gupta, B.K., Prakash, R., Rai, S.B.: Probing a highly efficient dual mode: down-upconversion luminescence and temperature sensing performance of rare-earth oxide phosphors. Dalton Trans. **42**, 1065–1072 (2013)
23. Zhou, S., Deng, K., Wei, X., Jiang, G., Duan, C., Chen, Y., Yin, M.: Upconversion luminescence of $NaYF_4$:Yb^{3+}, Er^{3+} for temperature sensing. Opt. Commun. **291**, 138–142 (2013)
24. Singh, S.K., Kumar, K., Rai, S.B.: Er^{3+}/Yb^{3+} codoped Gd_2O_3 nano-phosphor for optical thermometry. Sens. Actuat. A Phys. **149**, 16–20 (2009)
25. Dong, N.N., Pedroni, M., Piccinelli, F., Conti, G., Sbarbati, A., Ramírez-Hernández, J.E., Maestro, L.M., Iglesias-de la Cruz, M.C., Sanz-Rodriguez, F., Juarranz, A., Chen F., Vetrone, F., Capobianco, J.A., Sole, J.G., Bettinelli, M., Jaque, D., Speghini, A.: NIR-to-NIR Two-Photon Excited CaF_2: Tm^{3+},Yb^{3+} Nanoparticles: multifunctional nanoprobes for highly penetrating fluorescence bio-imaging. Acs Nano **5**, 8665–8671 (2011)
26. Verma, R.K., Rai, S.B.: Laser induced optical heating from Yb^{3+}/Ho^{3+}:$Ca_{12}Al_{14}O_{33}$ and its applicability as a thermal probe. J. Quant. Spectrosc. Radiat. Transfer **113**, 1594–1600 (2012)
27. Alencar, M.A.R.C., Maciel, G.S., de Araújo, C.B., Patra, A.: Er^{3+}-doped $BaTiO_3$ nanocrystals for thermometry: influence of nanoenvironment on the sensitivity of a fluorescence based temperature sensor. Appl. Phys. Lett. **84**, 4753–4756 (2004)
28. Dong, B., Yang, T., Lei, M.K.: Optical high temperature sensor based on green up-conversion emissions in Er^{3+} doped Al_2O_3. Sens. Actuat. B-Chem. **123**, 667–670 (2007)
29. Dong, B., Liu, D.P., Wang, X.J., Yang, T., Miao, S.M., Li, C.R.: Optical thermometry through infrared excited green upconversion emissions in Er^{3+}–Yb $^{3+}$ codoped Al_2O_3. Appl. Phys. Lett. **90**, 1–4 (2007)
30. Dong, B., Cao, B.S., Feng, Z.Q., Wang, X.J., He, Y.Y.: Optical temperature sensing through extraordinary enhancement of green up-conversion emissions for Er–Yb–Mo:Al_2O_3. Sens. Actuat. B-Chem. **165**, 34–37 (2012)
31. Su, Q., Han, S., Xie, X., Zhu, H., Chen, H., Chen, C.-K., Liu, R.-S., Chen, X., Wang, F., Liu, X.: The effect of surface coating on energy migration-mediated upconversion. J. Am. Chem. Soc. **134**, 20849–20857 (2012)
32. Wang, F., Deng, R., Wang, J., Wang, Q., Han, Y., Zhu, H., Chen, X., Liu, X.: Tuning upconversion through energy migration incore–shell nanoparticles. Nat. Mater. **10**, 1–6 (2011)
33. Allison, S.W., Gillies, G.T., Rondinone, A.J., Cates, M.R.: Nanoscale thermometry via the fluorescence of YAG:Ce phosphor particles: measurements from 7 to 77 °C. Nanotechnology **14**, 859–863 (2003)

34. Yu, J., Sun, L., Peng, H., Stich, M.I.J.: Luminescent terbium and europium probes for lifetime based sensing of temperature between 0 and 70 °C. J. Mater. Chem. **20**, 6975–6981 (2010)
35. Cai, Z.P., Xiao, L., Xu, H.Y., Mortier, M.: Point temperature sensor based on green decay in an Er:ZBLALiP microsphere. J. Lumin. **129**, 1994–1996 (2009)
36. Rocha, U., da Silva, C.J., Silva, W.F., Guedes, I., Benayas, A., Maestro, M.L.M., Bovero, E., van Veggel, F.C.J.M., Sole, J.A.G.S., Jaque, D.: Subtissue thermal sensing based on neodymium-doped LaF_3 nanoparticles. Acs Nano **7**, 1188–1192 (2013)
37. Savchuk, O.A., Carvajal, J.J., Pujol, M.C., Massons, J., Haro-González, P., Jaque, D., Díaz, F., Aguilò, M.: New strategies for luminescence thermometry in the biological range using upconverting nanoparticles. Biophotonics: Photonic Solutions for Better Health Care IV **9129**, doi:10.1117/12.2052074 (2014)
38. Chen, R., Ta, V.D., Xiao, F., Zhang, Q., Sun, H.: Multicolor hybrid upconversion nanoparticles and their improved performance as luminescence temperature sensors due to energy transfer. Small **9**, 1052–1057 (2013)
39. Xiao, Q., Li, Y., Li, F., Zhang, M., Zhang, Z., Lin, H.: Rational design of a thermalresponsive-polymer-switchable FRET system for enhancing the temperature sensitivity of upconversion nanophosphors. nanoscale (2014). doi:10.1039/C4NR02497D
40. Chen, B., Dong, B., Wang, J., Zhang, S., Xu, L., Yu, W., Song, H.: Amphiphilic silane modified $NaYF_4$:Yb, Er loaded with $Eu(TTA)_3$ $(TPPO)_2$ nanoparticles and their multi-functions: dual mode temperature sensing and cell imaging. Nanoscale **5**, 8541–8549 (2013)
41. Li, C., Hu, J., Liu, S.: Engineering FRET processes within synthetic polymers, polymeric assemblies and nanoparticles via modulating spatial distribution of fluorescent donors and acceptors. Soft Matter **8**, 7096–7102 (2012)
42. Liu, Y., Han, X., He, L., Yin, Y.: Thermoresponsive assembly of charged gold nanoparticles and their reversible tuning of plasmon coupling. Angew. Chem. Int. Ed. Engl. **51**, 6373–6377 (2012)
43. Debasu, M.L., Ananias, D., Pastoriza-Santos, I., Liz-Marzan, L.M., Rocha, J., Carlos, L.D.: All-in-one optical heater-thermometer nanoplatform operative from 300 to 2000 k based on Er (3+) emission and blackbody radiation. Adv. Mater. **25**, 4868–4874 (2013)
44. Huang, X., El-Sayed, I.H., Qian, W., El-Sayed, M.A.: Cancer cell imaging and photothermal therapy in the near-infrared region by using gold nanorods. J. Am. Chem. Soc. **128**, 2115-2120 (2006)
45. de Mello Donegá, C., Bode, M., Meijerink, A.: Size- and temperature-dependence of exciton lifetimes in CdSe quantum dots. Phys. Rev. B **74**, 1–9 (2006)
46. Han, B., Hanson, W.L., Bensalah, K., Tuncel, A., Stern, J.M., Cadeddu, J.A.: Development of quantum dot-mediated fluorescence thermometry for thermal therapies. Ann. Biomed. Eng. **37**, 1230–1239 (2009)
47. Li, S., Zhang, K., Yang, J.M., Lin, L., Yang, H.: Single quantum dots as local temperature markers. Nano Lett. **7**, 3102–3105 (2007)
48. Maestro, L.M., Rodriguez, E.M., Rodriguez, F.S., la Cruz, M.C., Juarranz, A., Naccache, R., Vetrone, F., Jaque, D., Capobianco, J.A., Sole, J.G.: CdSe quantum dots for two-photon fluorescence thermal imaging. Nano Lett. **10**, 5109–5115 (2010)
49. Maestro, L.M., Haro-González, P., Iglesias-de la Cruz, M.C., Sanz-Rodríguez, F., Juarranz, A., Solé, J.G., Jaque, D.: Fluorescent nanothermometers provide controlled plasmonic-mediated intracellular hyperthermia. Nanomedicine **8**, 379–388 (2013)
50. Maestro, L.M., Haro-Gonzalez, P., Sanchez-Iglesias, A., Liz-Marzan, L.M., Garcia Sole, J., Jaque, D.: Quantum dot thermometry evaluation of geometry dependent heating efficiency in gold nanoparticles. Langmuir **30**, 1650–1658 (2014)
51. Shang, L., Stockmar, F., Azadfar, N., Nienhaus, G.U.: Intracellular thermometry by using fluorescent gold nanoclusters. Angew. Chem. Int. Ed. Engl. **52**, 11154–11157 (2013)
52. Donner, J.S., Thompson, S.A., Kreuzer, M.P., Baffou, G., Quidant, R.: Mapping intracellular temperature using green fluorescent protein. Nano Lett. **12**, 2107–2111 (2012)

53. Kiyonaka, S., Kajimoto, T., Sakaguchi, R., Shinmi, D., Omatsu-Kanbe, M., Matsuura, H., Imamura, H., Yoshizaki, T., Hamachi, I., Morii, T., Mori, Y.: Genetically encoded fluorescent thermosensors visualize subcellular thermoregulation in living cells. Nat. Methods **10**, 1232–1238 (2013)
54. Shang, L., Dong, S., Nienhaus, G.U.: Ultra-small fluorescent metal nanoclusters: synthesis and biological applications. Nano Today **6**, 401–418 (2011)
55. Goodrich, G.P., Bao, L., Gill-Sharp, K., Sang, K.L., Wang, J., Payne, J.D.: Photothermal therapy in a murine colon cancer model using near-infrared absorbing gold nanorods. J. Biomed. Opt. **15**, 1–8 (2010)
56. Liu, Y., Ai, K., Cheng, X., Huo, L., Lu, L.: Gold-nanocluster-based fluorescent sensors for highly sensitive and selective detection of cyanide in water. Adv. Funct. Mater. **20**, 951–956 (2010)
57. Wang, Y., Wang, Y., Zhou, F., Kim, P., Xia, Y.: Protein-protected Au clusters as a new class of nanoscale biosensor for label-free fluorescence detection of proteases. Small **8**, 3769–3773 (2012)
58. Huang, C.C., Chen, C.T., Shiang, Y.C., Lin, Z.H., Chang, H.T.: Synthesis of fluorescent carbohydrate-protected au nanodots for detection of concanavalin A and escherichia coli. Anal. Chem. **81**, 875–882 (2009)
59. Shang, L., Brandholt, S., Stockmar, F., Trouillet, V., Bruns, M., Nienhaus, G.U.: Effect of protein adsorption on the fluorescence of ultrasmall gold nanoclusters. Small **8**, 661–665 (2012)
60. Amelia, M., Flamini, R., Latterini, L.: Recovery of CdS nanocrystal defects through conjugation with proteins. Langmuir **26**, 10129–10134 (2010)
61. Perevedentseva, E., Melnik, N., Tsai, C.Y., Lin, Y.C., Kazaryan, M., Cheng, C.L.: Effect of surface adsorbed proteins on the photoluminescence of nanodiamond. J. Appl. Phys. **109**, 034704 (2011)
62. Shang, L., Dörlich, R.M., Trouillet, V., Bruns, M., Nienhaus, U.G.: Ultrasmall fluorescent silver nanoclusters: protein adsorption and its effects on cellular responses. Nano Res. **5**, 531–542 (2012)
63. Chudakov, D.M., Lukyanov, S., Lukyanov, K.A.: Fluorescent proteins as a toolkit for in vivo imaging. Trends Biotechnol. **23**, 605–614 (2005)
64. Day, R.N., Davidson, M.W.: The fluorescent protein palette: tools for cellular imaging. Chem. Soc. Rev. **38**, 2887–2921 (2009)
65. Maestro, L.M., Rodriguez, E.M., Rodriguez, F.S., la Cruz, M.C., Juarranz, A., Naccache, R., Vetrone, F., Jaque, D., Capobianco, J.A., Sole, J.G.: CdSe quantum dots for two-photon fluorescence thermal imaging. Nano Lett. **10**, 5109–5115 (2010)
66. Albers, A.E., Chan, E.M., McBride, P.M., AjoFranklin, C.M., Cohen, B.E., Helms, B.A.: Dualemitting quantum dot/quantum rodbased nanothermometers with enhanced response and sensitivity in live cells. J. Am. Chem. Soc. **134**, 9565–9568 (2012)

Chapter 11
Upconversion Nanoparticles for Other Applications

Abstract Besides various bio-applications such as biomedicine, bioassay, photon dynamic treatment, and drug delivery, UCNPs have also been used in other applications. They can be explored as a NIR absorber in solar cells because they can absorb NIR light which has been wasted in traditional solar cells. They can be used as sensitizer in photocatalysis due to the energy transfer from UCNPs to the quantum dots or organic dyes. They are considered as next generation anti-counterfeiting materials due to their unique optical properties. In this chapter, the recent progress on other applications (besides bio-application) based on UCNPs is reviewed.

Keywords Upconversion nanoparticles · Solar cell · Photocatalysis · Anti-counterfeiting materials · NIR absorber · Light harvesting · Downconversion luminescence

11.1 Introduction

As was discussed in pervious chapters, UCNPs have attracted a great deal of attention from researchers because of their potential for optical manipulation in many fields (such as temperature sensors, bio-imaging agents, bio-analysis agents, and optical trigger for nanocapsules in medical therapy) [1–3]. In addition, UCNPs are widely used in light harvesting (for instance in photovoltaic cells and photosynthesis/photocatalytic reactions) and security applications due to their unique optical properties.

This chapter will focus on the applications of UCNPs other than their bio-applications and give a brief review on their mechanisms, manipulative methods, and future development.

Rui Wang and Fan Zhang contributed together to this chapter.

F. Zhang, *Photon Upconversion Nanomaterials*,
Nanostructure Science and Technology, DOI 10.1007/978-3-662-45597-5_11

11.2 UCNPs for Photovoltaic Cells

Photovoltaic cell is an electrical device that converts the energy of light directly into electricity by the photovoltaic effect. And when the irradiation source is sunlight, it is defined as solar cell. Solar cells are attracting a great deal of attention from researchers because the sunlight they are using is one of the cheapest and cleanest energy sources and it is easier to be obtained than other clean energy such as wind energy, terrestrial heat, tidal energy, and nuclear energy [4]. However, it is still a challenge to make use of sunlight, generally speaking, to convert it to electricity through photovoltaic effect.

The photovoltaic effect was first experimentally demonstrated by French physicist Edmond Becquerel. In 1839, he built the world's first photovoltaic cell in his father's laboratory. Willoughby Smith first described the "Effect of Light on Selenium during the passage of an Electric Current" in 1873. In 1883, Charles Fritts built the first solid-state photovoltaic cell by coating the semiconductor selenium with a thin layer of gold to form the junctions. The device was only around 1 % efficient. In 1888, Russian physicist, Aleksandr Stoletov, built the first cell based on the outer photoelectric effect discovered by Heinrich Hertz in 1887.

Albert Einstein explained the underlying mechanism of light instigated carrier excitation—the photoelectric effect—in 1905, for which he received the Nobel Prize in Physics in 1921. Russell Ohl patented the modern junction semiconductor solar cell in 1946 and discovered while working on the series of advances that would lead to the transistor.

The first practical photovoltaic cell was publicly demonstrated on April 25, 1954, at Bell Laboratories. The inventors were Daryl Chapin, Calvin Souther Fuller, and Gerald Pearson.

Solar cells gained prominence when they were proposed as an addition to the 1958 Vanguard I satellite. By adding cells to the outside of the body, the mission time could be extended with no major changes to the spacecraft or its power systems. In 1959, the USA launched Explorer 6, featuring large wing-shaped solar arrays, which became a common feature in satellites. These arrays consisted of 9,600 Hoffman solar cells [5–8].

Improvements were gradually made over the next two decades. The only significant use was in space applications where they offered the best power-to-weight ratio. However, this success was also the reason that costs remained high, because space users were willing to pay for the best possible cells, leaving no reason to invest in lower cost, less-efficient solutions. The price was determined largely by the semiconductor industry; their move to integrated circuits in the 1960s led to the availability of mass production at lower relative prices. As their price fell, the price of the resulting cells did as well. These effects lowered 1,971 cell costs to some $100 per watt [9].

11.2.1 Solar Cell Operation Principle and Their Obstacles

One important way to convert solar radiation into electricity occurs by the photovoltaic effect which was first observed by Becquerel. It is quite generally defined as the emergence of an electric voltage between two electrodes attached to a solid or liquid system upon shining light onto this system. Practically, all photovoltaic devices incorporate a pn-junction in a semiconductor across which the photovoltage is developed. These devices are also known as solar cells. A cross section through a typical solar cell is shown in Fig. 11.1a. The semiconductor material has to be able to absorb a large part of the solar spectrum. Depending on the absorption properties of the material, the light is absorbed in a region more or less close to the surface. When light quanta are absorbed, electron hole pairs are generated and if their recombination is prevented they can reach the junction where they are separated by an electric field. Even for weakly absorbing semiconductors such as silicon, most carriers are generated near the surface. This leads to the typical solar cell structure of Fig. 11.1a: The pn-junction which separates the emitter and base layer is very close to the surface in order to have a high collection probability for free carriers. The thin emitter layer above the junction has a relatively high resistance which requires a well-designed contact grid also shown in figure [10].

For practical use, solar cells are packaged into modules containing either a number of crystalline Si cells connected in series or a layer of thin-film material which is also internally series connected. The module serves two purposes: It protects the solar cells from the environment, and it delivers a higher voltage than a single cell, which develops only a voltage of less than 1 V.

Light-absorbing materials are the key to a solar cell. They must have characteristics that match the spectrum of available light in order to absorb it and then convert it. Solar cell can be divided into several subgroups based on which light-absorbing materials are used: silicon solar cell, dye-sensitized solar cell (DSSC) [5], organic/polymer solar cell [6], and quantum dots solar cell (QDSCs) [7]. By far, the silicon solar cell (which can be further divided into monocrystalline silicon, polycrystalline silicon, and amorphous silicon) is the most mature one among the above-mentioned solar cells. The reported conversion efficiency of $\sim$20 % makes it still dominant in the large-scale application and industrial production even through the high price. DSSCs are made of low-cost materials and do not need elaborate manufacturing equipment, so they can be made in a very standard equipment. In bulk, it should be significantly less expensive than older solid-state cell designs. DSSCs can be engineered into flexible sheets, and although its conversion efficiency is less than the best thin-film cells, its price/performance ratio may be high enough to allow them to compete with fossil fuel electrical generation. Organic solar cells and polymer solar cells are built from thin films (typically 100 nm) of organic semiconductors including polymers, such as polyphenylene vinylene and small molecular compounds like copper phthalocyanine (a blue or green organic pigment) and carbon fullerenes and fullerene derivatives. They can be processed from liquid solution, offering the possibility of a simple roll-to-roll printing process,

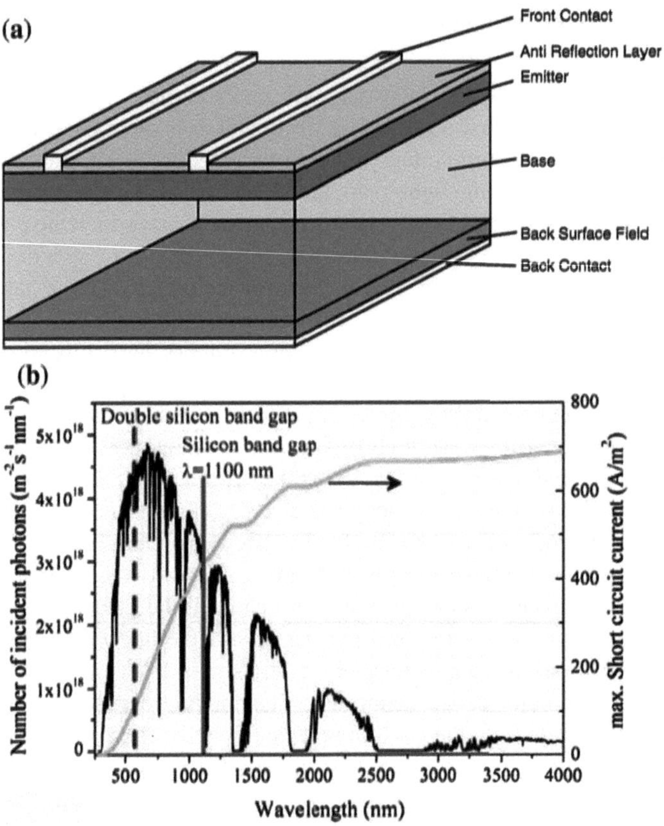

Fig. 11.1 **a** Structure of a typical solar cell. **b** The number of incident photons from the sun (AM1.5D) and the corresponding maximum short-circuit current (*green*) are displayed as a function of wavelength. The *vertical solid line* (*red*) and *dashed line* show the band gap and double band gap of silicon, respectively. The theoretically accessible parts of the spectrum for UC (>1,100 nm) and DC (<505 nm) are shown for silicon solar cells as well. **a** Reprinted with the permission from Ref. [10]. Copyright 2003 Elsevier. **b** Reprinted with the permission from Ref. [11]. Copyright 2011 Wiley-VCH Verlag GmbH & Co. KGaA

potentially leading to inexpensive, large-scale production. In addition, these cells can be beneficial for some applications where mechanical flexibility and disposability are important. QDSCs are based on the Gratzel cell or DSSC architecture, but employ low band gap semiconductor nanoparticles, fabricated with crystallite sizes small enough to form quantum dots (such as CdS, CdSe, Sb_2S_3, and PbS), instead of organic or organometallic dyes as light absorbers. QD's size quantization allows for the band gap to be tuned by simply changing particle size. They also have high extinction coefficients and have shown the possibility of multiple exciton generation. These four kinds of solar cells all have a small difference in mechanism and operation principles, resulting in differences in efficiency and feasible, which is not our topic in this book; additional details can be found in recent book and review papers [8].

Typically, the real efficiency of a solar cell must be lower than the predicted efficiency. To understand this phenomenon, two primary loss mechanisms in solar cells arise from the intrinsic properties of the semiconductor material, which account for the fundamental energy conversion efficiency limitation of solar cells: (a) Sub-band gap energy photon loss, where the photons' energy is not enough to excite the semiconductor and generate electron–hole pairs. For example, in the case of crystalline silicon solar cells, about 20 % of the sun's energy (AM1.5 solar spectrum) is lost owing to these losses. (b) Thermalization of charge carriers is caused by the absorption of high-energy photons with energies larger than the band gap of the solar cell. A high-energy photon can only generate one electron–hole pair, and its excess energy is transformed into heat. Besides these two primary losses linked to photon harvesting, there are many other loss mechanisms, which are, however, more related to the electronic properties of solar cells, including (c) contact voltage loss, (d) recombination loss as a result of poor interface or material quality, (e) junction loss, and (f) reflection loss from interfaces at the surface of a module. These fundamental losses directly lead to an efficiency limit of ca. 30 % for all single-junction cells under non-concentrated AM1.5 illumination [11, 12].

11.2.2 Recent Solutions to the Above Obstacles

To reduce thermalization losses, the concepts of rare earth-based downconversion (REDC) [13, 14] and downshift [15] (sometimes also called photoluminescence (PL) conversion) have been proposed for solar cell performance optimization (Fig. 11.2a) [11]. As was mentioned in Chap. 1, REDC is also a very unique mechanism found in rare earth ions like upconversion (UC). REDC has a higher potential for performance enhancement than PL conversion. For the DC process, it is demonstrated that RE-ion-doped phosphors are able to take up one high-energy incident photon and efficiently convert it into two or more low-energy photons (e.g., UV to visible) [16]. These DC layers are usually placed in front of a solar cell. The disadvantage of that geometry is the reflection loss at the air–downconverter interface. Although it was predicted that DC in conjunction with a silicon solar cell can achieve a theoretical conversion efficiency of up to 38.6 % [13], various loss mechanisms, such as the aforementioned reflection losses, have prevented new record efficiencies based on DC materials so far.

UC materials are, by definition, not transparent to photons non-resonant with the band gap. Therefore, the UC material layer can be integrated at the backside of a solar cell (Fig. 11.2a). In this position, the reflection losses limiting the performance for DC materials are no longer relevant. Figure 11.1b shows that about 52 % of the total solar energy flux is in the infrared range ($\lambda > 700$ nm), which suggests that a great performance gain can be achieved with efficient UC. RE-ion-doped phosphors offer a wide range of emission, from the near-infrared (NIR) through the visible to

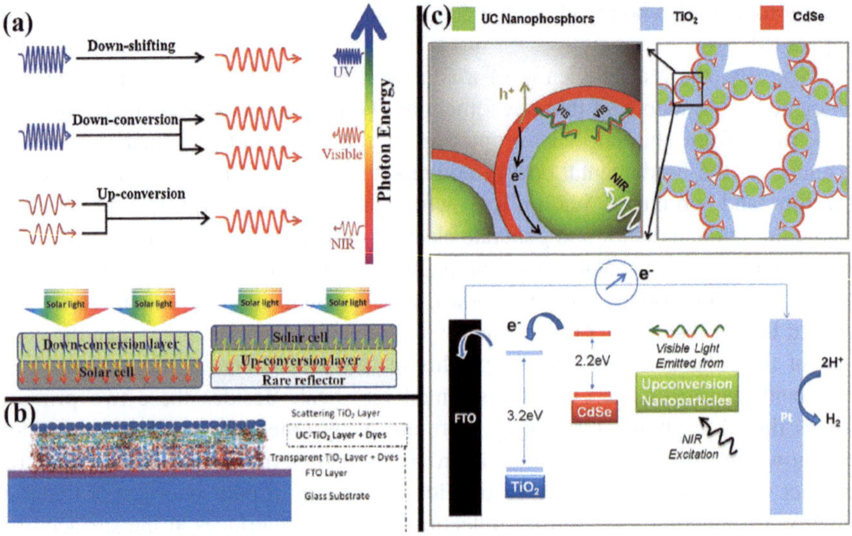

Fig. 11.2 **a** Mechanisms of photon luminescent conversion processes (*top*): downshifting and down- and upconversion. Only two-photon processes are shown for illustrative purposes. *Bottom* simplified schemes for combining solar cells with either a down- or an upconversion layer. **b** The structure of upconversion TiO_2-based work electrode. **c** Schematic of energy flow in the as-prepared hetero-nanostructures from UCNPs to VIS-absorbing CdSe nanoparticles to TiO_2 inverse opals upon NIR excitation (*top*). Energy diagram illustrating the charge injection from the excited CdSe into TiO_2 and the transportation of photoinjected electrons to the electrode surface for hydrogen generation (*bottom*). **a** Reprinted with the permission from Ref. [11]. Copyright 2011 Wiley-VCH Verlag GmbH & Co. KGaA. **b** Reprinted with the permission from Ref. [23]. Copyright 2010 Wiley-VCH Verlag GmbH & Co. KGaA. **c** Reprinted with the permission from Ref. [24]. Copyright 2013 Wiley-VCH Verlag GmbH & Co. KGaA

the ultraviolet (UV), and have increasingly moved into the focus of solar cell research in recent years [17, 18]. Besides UC, other device strategies have been developed to overcome the sub-band gap losses in solar cells: The first strategy is the spectral beam splitting technology, which was introduced in the early 1960s [19]. Here, the solar spectrum is split into two parts; the high-energy part is converted by a photovoltaic cell, while the low-energy part is converted into heat or again electricity via thermophotovoltaic conversion. The second strategy is the tandem or multi-band gap cells. This concept was originally proposed and investigated by Wolf in 1960 [20] and further investigated by Barnham and Duggan [21] and Keevers and Green [22]. In such systems, several cells with different band gaps are monolithically connected in series. The top cell absorbs the UV and visible photons, and the remaining IR photons are absorbed in the subsequent cells with successively lower band gap. In contrast to these two strategies, UC is the only strategy that does not require specific engineering of solar cells. The benefit of this is that UC materials can be added to any solar cell as introduced before: crystalline, thin-film, and organic/hybrid solar cells. An UC material layer is preferentially placed on the backside (or both sides) of a solar cell, where it takes up the sub-band gap part of the solar spectrum and reemits visible light, which now is resonant to the band gap of the solar cell. A reflective layer is frequently added to an UC layer to improve photon harvesting [17, 18].

Compared with UC bulk materials, UCNPs benefit from several advantages, such as size- and shape-dependent optical properties, and are playing outstanding roles in building new optoelectronic devices in the near future—especially for UCNPs, which have been coupled to a wide range of solar cells including crystalline silicon, amorphous silicon thin film, organic, and DSSCs. UCNPs can be easily introduced into some nanocomposite materials; these novel structure materials (fabrication method seen in chapter) formed by combining of two or more materials could present complementary properties that have shown important solar cell applications. The interaction between different compositions in the nanosize surface is beneficial to the energy transfer in the nanostructure, resulting in higher solar cell efficiency [12].

11.2.3 Recent Progress in Upconversion-Assisted Solar Cell

As was addressed before, for conventional solar cell, one of the major energy loss mechanisms is the sub-band gap energy photon loss. The photon UC of sunlight by lanthanide-doped materials is an emerging concept to improve solar energy conversion efficiency. The use of UC materials provides a solution to the sub-band gap energy photon loss by converting two sub-band gap photons into one above-band gap photon. In 1996, Zahraman et al. [25] developed a device based upon the concept of two-photon UC to use a part of the IR photons otherwise lost by transparency in a GaAs cell. An ultrathin GaAs cell was fabricated using the technique of epitaxial lift-off (ELO). This thin cell was placed on the top of a 100-um-thick vitroceramic doped with Yb^{3+} and Er^{3+}. Then, this cell was lighted by a Ti–sapphire IR laser at 1.391 V, a photon's energy below the band gap of GaAs, with an input power able to reach $\sim$1 W. The efficiency of 2.5 % was obtained even though the excitation wavelength (891 nm) was not resonant with the absorption peak of Yb^{3+} (980 nm). The authors have explained this mechanism: The two-photon UC process involved here was based on sequential absorption and energy transfer of two IR photons from Yb^{3+} to Er^{3+} (traditional energy transfer UC mechanism, see Chap. 1) which then emit one photon in the green. This green light then produced a photoresponse in the GaAs cell. It should be pointed out that the excitation density used here was extremely higher than the standard terrestrial (AM1.5) solar irradiation. Richards and Shalav [26] showed that UC under excitation density of 2.4 W/cm^2 (which corresponds to a solar concentration of 1,000×) reached 3.4 % external quantum efficiency (EQE) at 1,523 nm in a crystalline Si solar cell with Er^{3+} doped $NaYF_4$ as upconverter, which was similar to that reported by Shalav et al. [27]. Shan and Demopoulos [23] reported for the first time the application of an Yb^{3+}/Er^{3+} codoped LaF_3–TiO_2 layer to fabricate a "triple-layer" working electrode on fluorine-doped tin oxide (FTO) glass for a DSSC (Fig. 11.2b). Based on illumination with a 980-nm wavelength fiber laser with 2.5 W power supply, open-circuit voltage VOC = 0.40 V and short-circuit current ISC = 0. 036 mA were obtained. This unique FTO/transparent TiO_2/UC–TiO_2/scattering TiO_2 structure offered both opportunities and challenges. For example, because the NIR light was converted in situ to the dye-absorbable visible light, the loss of the converted light can be minimized. On the other hand, the red light (~655 nm) generated by the UC–TiO_2 UC nanocomposite is not efficiently absorbed by the dye N719. Another challenge is the avoidance of charge recombination that will allow full utilization of the upconverted light. Further optimization of the UC–TiO_2 photoanode design can be achieved by tailoring morphological features such as the structure and composition of the nanocomposites. Wang et al. [11] also demonstrated the feasibility of UCNPs in P3HT:PCBM organic solar cells. A photocurrent density of ~16.5 μA/cm^2 was obtained under a laser illumination intensity of ~25 mW/cm^2. More recently, Su et al. [24] have demonstrated a new method to utilize UCNPs in a hetero-nanostructured photoanode for enhancement of NIR light harvesting. In their work, rational design and fabrication of interconnected porous network of three-dimensional hetero-nanostructured photoanodes for enhanced NIR light harvesting were reported. As shown in Fig. 11.2c, the photoelectrode constitutes a titanium oxide (TiO_2) inverse opal, which optimally conducts electrons with minimal carrier losses due to its continuous electron conducting pathways. The three-dimensional hierarchical porous network of the inverse opal not only provides a large interfacial surface area for efficient loading of UCNPs and quantum dot sensitizers, but also allows an excellent interaction between the electrolyte and quantum dots. In addition, the inverse opal is known to improve the light–matter interaction because of the strong light scattering and increased path length travelled by the light in the electrode. Using the as-fabricated photoanode, they successfully demonstrated hydrogen generation via a photoelectro-chemical process driven by NIR irradiation. Their results show that UC was responsible for the photoresponse upon NIR exposure. This is the first verification of photoelectron chemical electrodes with UCNPs embedded in porous photonic crystals. The interconnected pores of photonic crystals render the nanoparticles spatially positioned in close proximity to CdSe QDs, leading to an enhanced energy transfer and light harvesting. Wang's group [28] also synthesized monodispersed and uniform active-core/active-shell/inert-shell $NaGdF_4:Er^{3+}@NaGdF_4:Ho^{3+}@NaGdF_4$ architecture UC core/shell nanocrystals. In these NCs, the intermediate shells not only play a passivation role to enhance the UC luminescence of Er^{3+} in the cores, but also act as the hosts to realize the UC luminescence of Ho^{3+}. Importantly, the doping of Er^{3+} and Ho^{3+} into the core and shell, respectively, can effectively suppress the adverse energy transfers between Er^{3+} and Ho^{3+}, resulting in the lessening of the UC emission quenching for both Er^{3+} and Ho^{3+} dopants. The novel core/shell nanocrystals can be excited by ultrabroadband NIR photons transparent to crystalline silicon and emit visible–NIR photons above crystalline silicon band gap due to two basic UC route: Er^{3+}, $^2H_{11/2}$, $^4S_{3/2} \rightarrow {}^4I_{15/2}$ (539 nm), $^4F_{9/2} \rightarrow {}^4I_{15/2}$ (655 nm), $^4I_{9/2} \rightarrow {}^4I_{15/2}$ (809 nm), and $^4I_{11/2} \rightarrow {}^4I_{15/2}$ (980 nm) under 1,480–1,600-nm excitation ($^4I_{15/2} \rightarrow {}^4I_{13/2}$); Ho^{3+}, $^5F_5 \rightarrow {}^5I_8$ (645 nm) and $^5I_5 \rightarrow {}^5I_8$ (900 nm) transitions under 1,140–1,250-nm excitation ($^5I_8 \rightarrow {}^5I_6$). All of the emission can be absorbed by the crystalline silicon, indicating this novel nanomaterial may find potential application in crystalline silicon solar cell to enhance the energy efficiency. More recently, Li et al. [29] have reported a simple approach for the preparation of UC $NaYF_4$:Yb/Er/Gd (18/2/30 mol%) nanorods coated with Au nanoparticles or Au shell nanostructures. This nanocomposite was placed on the front side of the flexible standard p–i–n hydrogenated amorphous silicon (a-Si:H) solar cell on a steel substrate, where they took up the sub-band gap part of the solar spectrum and in situ reemitted visible light, which was resonant with the band gap of the solar cell. Compared with the reference cell without UC nanomaterial, the solar cells based on $NaYF_4$:Yb/Er/Gd nanorods with different Au nanostructures showed significant improvement of short-circuit current (16- to 72-fold improvement). The device based on $NaYF_4$:Yb/Er/Gd nanorods with Au nanoparticles showed the best NIR photovoltaic performance, displaying a short-circuit current of 1.16 mA and an EQE of 0. 14 % under the irradiation of 980-nm laser with a excitation power of 1,100 mW. Moreover, there was no need to reform or destroy the intrinsic structure of the solar cells or change the solar cell fabrication processes in this approach. Table 11.1 illuminates a few selected reports on UC in various photovoltaic systems.

11.2.4 Conclusion and Future Opportunity for Upconversion-Assisted Solar Cell

In conclusion, the UC process is one of the most elegant concepts to generically increase the performance of all classes of solar cells. Significant improvement, including both theoretical analysis and experimental achievements, has been reported in the field of solar UC in recent years. However, until now, no significant performance enhancement was achieved for natural solar irradiation (AM1.5). Here, we present several reasons for this phenomenon:

1. The efficiency of the existing UC materials is still low. As was discussed in Chap. 3, the reported quantum yield of nanomaterials is always below 1 %. This number is less than 3.5 % even in the bulk UC materials. By comparison, the quantum yield of quantum dots and organic dyes is much higher (10–90 %).
2. The absorption cross section of lanthanide ions is very low, such as Nd^{3+}, 1.2×10^{-19} cm^2 at 790–810 nm; and Yb^{3+}, 1.2×10^{-20} cm^2 at 900–1,000 nm, the absorption cross section of other lanthanide ions (such as Er^{3+}, Tm^{3+}, and Ho^{3+}) is much lower. By comparison, the absorption cross section of quantum dots and organic dyes are always several orders higher. Therefore, the irradiation source in many reported works of the UC-assisted solar cell is NIR laser, rather than natural solar irradiation.

Table 11.1 Lanthanide-based upconverter/solar cell performance reported in the literatures

Upconverter	Solar cell type	Light source (nm)	Power density	Remarks	References
Vitroceramic: Yb^{3+}, Er^{3+}	GaAs	891	~25.6 W/cm^2	Efficiency: 2.5 %	[25]
$NaYF_4$:Er^{3+}	Bifacial Si	1,523	5.1 mW	EQE: 2.5 ± 0.2 %	[27]
$NaYF_4$:Er^{3+}	Bifacial Si	1,523	~2.4 W/cm^2	EQE: 3.4 %	[26]
$NaYF_4$: Er^{3+}	Bifacial Si	1,550	~200 mW/cm^2	EQE: ~1 %	[30]
$NaYF_4$:Yb/Er/Gd–Au	Amorphous Si	980	1,100 mW	EQE: ~0.14 %	[29]
β-$NaYF_4$:Yb^{3+},Er^{3+}	Amorphous Si	980	28 mW/cm^2	EQE: 0.03 %	[31]
β-$NaYF_4$:Yb^{3+},Er^{3+}	Amorphous Si	980	10 mW	10 μA/cm^2 (Enhancement)	[32]
LaF_3:Yb^{3+},Er^{3+}	Dye-sensitized solar cell	980	2.5 W	V_{oc}: 0.4 V I_{sc}: 0.036 mA	[23]
α-$NaYF_4$:Yb^{3+},Er^{3+}	Dye-sensitized	Simulated solar light with a UV cutoff filter	100 mW/cm^2	Conversion efficiency: 2.84 %	[33]
$NaYF_4$:Yb^{3+},Er^{3+} @ $NaYF_4$	QDs-sensitized	980	500 mW	Photocurrent: 0.02 mA (Enhancement)	[24]
YF_3:Yb,Er	Organic	980	25 mW/cm^2	Photocurrent: 16.5 μA/cm^2	[11]

Note EQE is abbreviation of external quantum efficiency

3. Besides the low quantum yield and absorption cross section, the absorption spectrum region is also very limited. The main absorption wavelength region of different rare earth ions are listed as following: Yb^{3+}: 900–1,000 nm ($^2F_{7/2} \rightarrow {}^2F_{5/2}$); Er^{3+}: 1,480–1,600 nm ($^4I_{15/2} \rightarrow {}^4I_{13/2}$); Ho^{3+}: 1,140–1,250-nm excitation ($^5I_8 \rightarrow {}^5I_6$), etc. Moreover, the absorption cross section is also very different as discussed before. Obviously, it is not enough to absorb all of the NIR light using only one or two kinds of lanthanide ions, because one kind of lanthanide ions can only absorb a small part of the NIR light. Although there are efforts have been devoted on the design and synthesis of multi-doped nanocomposite to realize broad absorption in NIR region, the quench effect (also called cross-relaxation) between lanthanide ions (such as Er^{3+} with Ho^{3+}, Er^{3+} with Nd^{3+}, and Tm^{3+} with Tm^{3+}) cannot be ignored. Therefore, complex nanostructure is required to contain multi-doped lanthanide-based nanocomposite.
4. The host matrix of most efficient UCNPs ($NaYF_4$, $NaGdF_4$) is insulating, which is not preferred in the solar cell, because the higher electrical resistance must result in lower efficiency.

Aiming at above drawbacks, here are some propositions as an outlook to the UC-assisted solar cell:

1. The quantum yield of UCNPs is very low; in other words, they also hold great potential to make an improvement. More reasonable and advanced nanostructures should be designed to reduce the crystal defects and modify the crystal field of the host. Furthermore, in order to obtain the highly efficient UCNPs, increasing the doping concentration without dropping down the quantum yield is also a thinkable idea, because the more light will be absorbed along with the increasing of doping concentration.
2. To deal with the poor electrical conductivity of the traditional UCNPs, more conductive host matrix with high-efficient UC property should be established. For instance, several semiconductor used in quantum dots would be a good choice. In the recent research, semiconductor nanocrystals such as ZnS have also been proposed as UC host materials. However, it is still debatable whether lanthanide ions in these nanocrystals are mainly located on the outermost layer of the nanocrystals due to largely mismatched ionic size between the host and dopant ions or incorporated homogeneously in the host lattice. Therefore, further investigation should be undertaken to improve the electrical conductivity of the UCNPs for better solar cell performance.
3. The absorption wavelength of UCNPs is determined by the doped lanthanide ions. Obviously, it is impossible to change the absorption wavelength of a specific lanthanide ion. However, if the lanthanide ions are not responsible for the absorption property, but merely upconvert the absorbed light, things will be changed. MS:Eu,Sm (M = Ca, Sr) is a kind of electron trapping material which has been studied since 1980s. The host matrix MS can absorb UV and blue light and cause a downconversion process which is also advantageous to the efficient solar cell part of this absorption energy being stored in the electron trapping energy levels made from doped lanthanide ions. The electrons trapped in these

states cannot be thermally stimulated to their excited states at room temperature because these electron trapping energy levels are so deep (1 eV); however, when stimulated by some infrared light, it can go back and then recombine with holes to emit visible light, which results in the NIR UC luminescence. It should be pointed out that the NIR excitation light here can locate in a broad spectral wavelength (800–1,600 nm) with low power density (mW) which is coincident with the natural sunlight.

Overall, significant progress, including theoretical analysis and experimental achievement, has occurred in the field of UC for photovoltaic applications. However, the performance of the currently investigated UC materials is not efficient enough to improve the practical efficiency of solar cells under normal solar irradiation. More efforts are required to design and develop higher efficiency UC materials. On the other hand, more sophisticated engineering concepts with respect to the architecture of the solar cells need to be developed to enable the exploitation of UC. Based on the recent improvements in UC efficiency of lanthanide-doped phosphors, a promising future for this concept can be foreseen.

11.3 UCNPs for Photocatalysis

Photocatalysis is an environmentally friendly technique to eliminate toxic organic substances in air and water. The process is initiated by the irradiation of semiconductor materials with a suitable light source. The subsequently formed free radicals with strong oxidizing ability could mineralize organic pollutants to CO_2 and H_2O. Photocatalysis has attracted much attention in recent years because it allows the utilization of clean, safe, and renewable solar energy to solve environmental problems. Fujishima and Honda [34] discovered the photocatalytic splitting of water on titanium dioxide (TiO_2) photoanodes under UV light in 1972. Since then, TiO_2 has been widely studied as an efficient photocatalyst for its appropriate electronic band structure, biological and chemical inertness, strong oxidizing power, and long-term stability.

11.3.1 Photocatalysis Mechanism

TiO_2 is a semiconductor material with a band gap of ~3.2 eV, corresponding to a wavelength of ~390 nm. TiO_2 has three major stable polymorphs, namely anatase, rutile, and brookite. Among those, anatase has generally shown the highest photocatalytic activity [35], with some crystallographic planes of anatase being particularly reactive [36]. The presence of rutile in an anatase sample, however, has been suggested to increase catalytic activity [37–39]. Degussa P25 TiO_2 is a typical commercially available material composed of 75 % anatase and 25 % rutile. It is

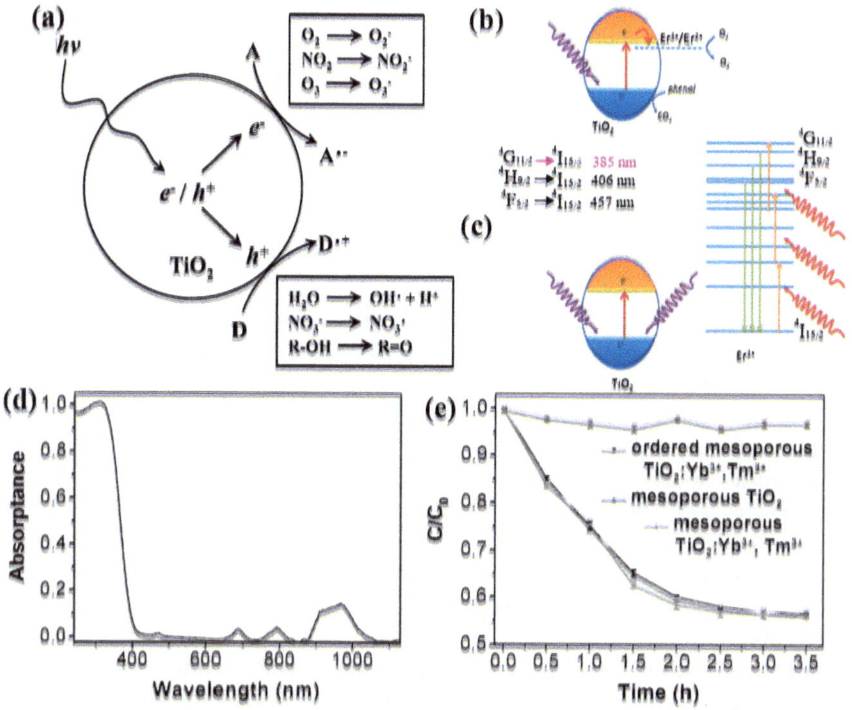

Fig. 11.3 **a** Schematic of TiO_2 photoexcitation to initiate electron–hole pairs (e^-/h^+) and redox chemistry at the TiO_2 surface. **b, c** Mechanism of the photocatalytic induction process in the Er^{3+}–TiO_2 heterostructures: **b** UV-assisted mechanism and **c** NIR upconversion-assisted mechanism. **d, e** The UV–Vis–NIR absorption **d** spectrum of the ordered mesoporous TiO_2:Yb^{3+},Tm^{3+}. **e** Photocatalytic degradation of RhB monitored as the normalized concentration change versus irradiation time. **a** Reprinted with the permission from Ref. [41]. Copyright 2007 American Chemical Society. **b, c** Reprinted with the permission from Ref. [57]. Copyright 2012 Royal Society of Chemistry. **d, e** Reprinted with the permission from Ref. [58]. Copyright 2011 Royal Society of Chemistry

generally accepted to exhibit better photocatalytic efficiency compared to other forms, and it has been widely used as a reference material in laboratory studies [40, 41].

When excited with light of wavelength equal to or greater than the band gap energy, there is the generation of electron–hole pairs in the conduction and valence bands (VB), respectively. The photogenerated electron–hole pairs can then follow several pathways: (i) recombination of electrons and holes on the surface or in the bulk of the semiconductor; (ii) reduction of electron acceptors by photogenerated electrons; and (iii) oxidation of electron donors by photogenerated holes. Electrons and holes transported to the particle surface can therefore initiate redox chemistry [42]. Some examples of oxidation and reduction processes with electron acceptors (A) and donors (D) are shown in Fig. 11.3a. Photogenerated electrons and holes, for example, can react with H_2O and O_2 molecules, leading to the formation of reactive oxygen species (ROS) [41].

The reaction mechanism can be depicted as follows (Eqs. 11.1–11.3):

$$TiO_2 + hv(\lambda < 390\ nm) \rightarrow e^- + h^+ \tag{11.1}$$

$$h^+ + H_2O \rightarrow OH^\bullet + H^+ \tag{11.2}$$

$$e^- + O_2 \rightarrow O_2^- \tag{11.3}$$

Photocatalytic reactions can be initiated by the resulting ROS [43]. Additionally, recent studies have shown that the resultant ROS can diffuse far from the surface and induce remote gas phase oxidation in the vicinity of the TiO_2 surface. Remote photoprocesses were first observed by Tatsuma et al. In their studies, aromatic and aliphatic substances were reported to be photo-oxidized into CO_2 by ROS that was generated at the TiO_2 surface and then desorbed into the gas phase. The impact distance has been shown to be quite high up to 2.2 mm. Direct detection of the presence of ROS confirmed the diffusion of ROS away from the TiO_2 surface. Furthermore, the diffusion of ROS was thought to occur in all media, including gas, liquid, and solid phases, and to be an important part of these photocatalytic processes [41].

Species that form upon irradiation can reduce and/or oxidize environmental pollutants adsorbed on the TiO_2 surface, including volatile organic compounds (VOCs), leading to the breakdown of these pollutants into low molecular weight products. The photolysis of TiO_2 is a main feature developed as a usage of TiO_2-based materials by adding TiO_2 or applying a TiO_2 coating to conventional building materials for environmental remediation. TiO_2 surface chemistry and photochemistry and its applications in environmental remediation have been extensively reviewed very nicely in several publications [41, 44, 45].

Besides pollutant degradation, hydrogen production is also a very important application of photocatalytic technology. The mechanism of photocatalytic hydrogen production is very similar with above-mentioned photocatalytic pollutant degradation mechanism. Moreover, several other semiconductors have also been utilized as photocatalysts such as CdS, $SrTiO_3$, and $NaTaO_3$. We will not introduce them in detail because they share the similar mechanism with TiO_2 [44].

11.3.2 Obstacles and Recent Solutions for Photocatalysis

Semiconductor-based photocatalysis has attracted great interest since the discovery of water splitting on a TiO_2 electrode. However, only the light whose energy matches or exceeds the band gap energy of the semiconductor can excite the photocatalyst to generate electron–hole pairs. The low usage of sunlight has restrained the photocatalytic efficiency of photocatalysts from real application [12]. For example,

TiO_2 possesses a wide band gap of 3.2 eV, allowing it only to be activated by UV light. In order to increase the utilization rate of solar energy, numerous methods have been adopted to modify the photocatalysts, such as doping of metal or nonmetal elements, surface modification, and coupling with other semiconductors [46, 47]. Li et al. [48] found that 1.5 % of Nd^{3+}-doped TiO_2 nanoparticles reduced the band gap by as much as 0.55 eV and that the band gap narrowing was primarily attributed to the substituted Nd^{3+} ions, which introduced electron states into the band gap of TiO_2 to form the new lowest unoccupied molecular orbital (LUMO). In Asahi's study, the substitutional doping of N was the most effective in the band gap narrowing [49], and Nakano et al. [50] found that, in N-doped TiO_2, deep levels located at approximately 1.18 and 2.48 eV below the conduction band (CB) were attributed to the O vacancy state as an efficient generation–recombination center and to the N doping, which contributed to band gap narrowing by mixing with the O 2p VB, respectively. Narrow band gap semiconductors have been used as sensitizers to improve the optical absorption properties of TiO_2 nanomaterials in the visible light region by various groups. Hoyer et al. [51] reported the sensitization of a nanocrystalline TiO_2 matrix by small PbS nanoparticles (<2.5 nm), and they found that the photogenerated excess electrons could be directly injected from the PbS to the TiO_2, resulting in strong photoconductance in the visible region. Fitzmaurice et al. [52] found that excitation of the sensitizer AgI on TiO_2 nanoparticles resulted in a stabilization of electron–hole pairs with a lifetime well beyond 100 μs and in electron migration from AgI to TiO_2. Vogel et al. [51] studied the sensitization of nanoporous TiO_2 by CdS, PbS, Ag_2S, Sb_2S_3, and Bi_2S_3 and found that the relative positions of the energetic levels at the interface between the quantum size particles and TiO_2 could be optimized for efficient charge separation using the size quantization effect and that the photostability of the electrodes could be significantly enhanced by surface modification of the TiO_2 nanoparticles with CdS nanoparticles. Besides metal- or nonmetal-doped composite and narrow band gap semiconductors surface sensitization, organic dyes also have been widely employed as sensitizers for TiO_2 nanomaterial to improve its optical properties, such as DSSCs as was introduced in the solar cell [5].

These works do succeed in broadening the range of visible light photoresponse. According to the formula $E_g = hv$ (h is Planck's constant, v is light frequency), unfortunately, the photo potential energy excited by visible light is lower than that excited by UV light, thus generating holes with lower oxidation potential and weaker oxidative ability, which cannot degrade the organic compound completely or produce H_2 or O_2 efficiently. In order to get holes possessing oxidation potential high enough to destroy the organic pollutants or the H–O bond completely, it is necessary to excite the photocatalyst with UV light. However, UV light occupies only about 4 % of the sunlight spectrum, whereas visible light composes 43 % of the solar spectrum [53]. So researchers have considered whether there were some substances which could convert visible light into UV light. Besides visible light, a large fraction of the NIR light (~50 %) in sunlight still remains untapped for photocatalysis. How to utilize the visible and NIR light efficiently is an obstacle to the photocatalysis. Fortunately, the UC materials can do the work and meet these needs. Combining the advantages of UCNPs and the excellent photocatalysis properties of TiO_2, UCNPs might serve as an inter-medium which, after absorbing NIR light, can transfer energy to TiO_2 to generate strong oxidative holes and reductive electrons [12].

11.3.3 Recent Progress in Upconversion-Assisted Photocatalysis

Qin et al. [54] have reported a NIR photocatalysis based on YF_3:Yb^{3+},Tm^{3+}/TiO_2 core/shell nanoparticles. These nanoparticles were synthesized by a modified hydrolysis method using PVP as a coupling agent. Uniform TiO_2 shells were formed on the UC NCs by adjusting the experimental conditions. The decomposition of methylene blue (MB) was carried out by these core/shell nanoparticles under a 980-nm laser and NIR irradiations (λ > 700 nm) of sunlight. The results indicated that the YF_3:Yb^{3+},Tm^{3+}/TiO_2 nanoparticles had photocatalytic activity under NIR irradiation. Moreover, the core/shell structure made it convenient to modify TiO_2 further in order to develop new photocatalysts sensitive to UV, visible, and NIR irradiation. This study inspires a thought of utilizing the NIR energy of sunlight. The same strategy has great potential in improving the utility rate of solar energy for photochemical and photoelectrical applications based on TiO_2 materials.

Shi and coworkers [55] synthesized a series of UC luminescent Er-doped $SrTiO_3$ (ABO_3-type) photocatalysts with different initial molar ratios of Sr/Ti by a facile polymerized complex method. Er^{3+} ions were gradually transferred from the A to the B site with increasing ratio of Sr/Ti. The light response range was effectively extended from the UV (for $SrTiO_3$) to the visible light (for erbium-doped $SrTiO_3$) region due to intraband transition absorption of Er^{3+}, and the high-energy excited states of Er^{3+} were populated by UC processes. Compared with those with A-site occupancy, Er^{3+} ions with B-site occupancy are located in distorted BO_6 octahedra with dipole moments and thus subject to much stronger local internal fields, which promotes energy transfer from the high-energy excited states of Er^{3+} to the host $SrTiO_3$ and facilitates the generation of charge carriers in the VB and CB of the host $SrTiO_3$. Consequently, erbium-doped $SrTiO_3$ with B-site occupancy instead of A-site occupancy is validated as being novel, stable, and efficient photocatalysts for visible-light-driven H_2 or O_2 evolution. In addition, UC luminescent holmium- and neodymium-doped $SrTiO_3$ are also active in visible-light-driven photocatalytic H_2 evolution. It is generally suggested that the site-selected doping of UC luminescent agents into host semiconductors is an effective strategy in the design of photocatalysts for visible-light-driven H_2 or O_2 evolution.

More recently, Qi and Zhong's group [56] has synthesized a new nanocomposite consisting of UC material (YF_3:Yb^{3+},Tm^{3+}), TiO_2 (P25), and graphene (GR) which showed excellent sunlight activity. During the facile hydrothermal method, the reduction of GR oxide and loading of YF_3:Yb^{3+},Tm^{3+} and P25 were achieved simultaneously, and the functionalities of each part were integrated together.

The as-prepared ternary UC–P25–GR nanocomposite photocatalyst exhibited great adsorptivity of dyes: a significantly extended light absorption range, efficient charge separation properties, and superior durability. Indeed, the photocatalytic activity of this novel ternary nanocomposite under sunlight was improved compared to those of P25–GR nanocomposites and bare P25. Overall, this work could provide new insights into the fabrication of ternary composites as high-performance photocatalysts and facilitate their application in environmental protection issues.

In addition, there are many similar reports on the visible or NIR photocatalysis with the assistant of the UCNPs. In 2012, Obregón and Colón [57] reported the synthesis of Er^{3+}-doped TiO_2 by a surfactant-free hydrothermal method, and the presence of Er^{3+} does not affect the structural and morphological features of the TiO_2 significantly. The obtained Er^{3+}–TiO_2 composite exhibited good photoactivity under sun-like excitation for the degradation of phenol. Interestingly, the authors found the Er^{3+} cation responsible for the enhanced photocatalytic activity, which participates in different mechanisms under UV and NIR excitation (Fig. 11.3b, c). On the one hand, under UV excitation, the clear improvement of the photoactivity would be related to the electron scavenger role of Er^{3+}. Within this step, Er^{3+} could be reduced to Er^{2+} by trapping the photogenerated electron. This Er^{2+} species would participate in the O_2 reduction reaction. By this way, the presence of Er^{3+} would induce an effective charge carrier separation. On the other hand, upon NIR irradiation, the presence of Er^{3+} cations would promote an UC process pumping photons in the UV range into the TiO_2 structure. It has been widely reported that Er^{3+} shows UV PL emission between 390 and 410 nm after excitation at 980 nm. This UC process involves a sequential three-photon absorption ($^4I_{15/2}$–$^4I_{11/2}$, $^4I_{11/2}$–$^4F_{7/2}$, and $^4S_{3/2}$–$^2G_{7/2}$). Then, by a multi-phonon relaxation, the $^2G_{7/2}$ excited state decays to $^2G_{11/2}$ and $^2H_{9/2}$ lower states. The PL emission in the UV/violet range is then produced by the $^2G_{11/2}$–$^4I_{15/2}$ transition, giving small emissions at 390 and 406 nm. Thus, it can be assumed that the improvement of the photoefficiency might be related to the increasing number of available photons with the appropriate energy.

In 2011, Yan et al. [58] presented the fabrication of highly ordered crystallized mesoporous TiO_2 stabilizing at a temperature as high as 700 °C; by introducing rare earth elements in the synthetic process, they successfully achieved the NIR photocatalytic degradation of rhodamine B by the ordered mesoporous TiO_2 doped with Yb^{3+} and Tm^{3+}. The UV–Vis–NIR absorption spectrum of the TiO_2:Yb^{3+},Tm^{3+} sample exhibits two strong absorption bands (Fig. 11.3d). The absorption band at the UV region corresponds to the TiO_2 semiconductor band gap, which is located at about 400 nm. The relatively wide absorption range at the NIR region corresponds to the absorption of Yb^{3+}, which overlaps one of the maximum NIR energy distribution ranges of sunlight from 920 to 1,110 nm, and the relatively weak absorption bands at about 690 and 800 nm correspond to the absorption of Tm^{3+}. The UC luminescence spectrum of the TiO_2:Yb^{3+},Tm^{3+} upon excitation with 980-nm laser diode displays only one intense emission band at 700 nm, attributed to the radiative transitions from $^3F_3$ to $^3H_6$ of Tm^{3+}. Interestingly, the UV emission band of Tm^{3+} at about 360 nm is not observed, signifying energy transfer from Tm^{3+} to TiO_2. The heterogeneous photocatalytic degradation of rhodamine B (RhB) solution under 980-nm NIR irradiation is detected with a UV–Vis–NIR spectrophotometer after adsorption–desorption equilibrium. The characteristic absorption band of RhB located at 554 nm steadily decreases with increasing irradiation times for the RhB solution using ordered mesoporous TiO_2:Yb^{3+},Tm^{3+} as the catalyst. The photocatalytic activity of mesoporous pure TiO_2 photocatalyst was also measured under the same condition for comparison. Figure 11.3e displays the time profiles of C/C0 under NIR light irradiation, where C is the concentration of RhB at the irradiation time *t* and C0 is the initial concentration. After lighting on, the concentration of RhB decreased with irradiation time and the pseudo-first-order reaction was observed. After 2-h irradiation by 980 nm NIR light, 40 % of RhB was decomposed. As expected, the pure TiO_2 did not decompose RhB under the irradiation of the 980 nm light. Meanwhile, the worm-like mesoporous TiO_2:Yb^{3+},Tm^{3+} also exhibits the NIR photocatalytic activity. This result further confirms the energy transfer from Tm^{3+} to TiO_2. In this photocatalysis process, Yb^{3+} ions absorb the NIR light and excite Tm^{3+} ions by energy transfer and then electron transfer from an excited state of Tm^{3+} to TiO_2, followed by a hole generation by extraction of an electron from a VB of TiO_2 to Tm^{4+}.

A similar work was done by Feng et al. [59] They synthesized nanometer TiO_2 photocatalysts doped with Er^{3+}:$Y_3Al_5O_{12}$ (Er^{3+}:YAG) UC luminescence agent. Then, the visible light photocatalytic activities of as-synthesized samples were studied by photodegradation of MB. The results showed that the degradation rate in the presence of the doped sample was two times that of the undoped one. The results primarily showed that the oxidation mechanics of the photoproduction holes occurred and that the MB was degraded thoroughly in the presence of the doped samples. All the results proved that the doped TiO_2 had more excellent photocatalytic activities than the undoped one, and this method can be envisaged as a technology for treating dye wastewater using solar irradiation.

Beside the TiO_2, several other semiconductors have also been combined with UC materials as a novel photocatalysts.

As one of the simplest Aurivillius oxides, bismuth tungstate (Bi_2WO_6) has recently attracted a great deal of attention for its good photocatalytic performance in water splitting and organic contaminant decomposition under visible light irradiation. However, pure Bi_2WO_6 presents photoabsorption properties from the UV light region to visible light shorter than 450 nm, which occupy only a small fraction of the solar spectrum. In order to harvest more solar light to achieve efficient solar energy conversion, in 2010, Wang et al. [53] synthesized Er^{3+}-doped Bi_2WO_6 photocatalyst, in which Er^{3+} with UC properties could convert the visible light beyond the absorption edge of Bi_2WO_6 to UV light and then excite Bi_2WO_6 to generate electron–hole pairs. The photocatalytic activities of the photocatalysts were evaluated by the decolorization of rhodamine B (RhB) and the degradation of a typical colorless model pollutant, phenol, under simulated solar light irradiation ($\lambda > 290$ nm). The results showed that the photocatalytic activity of Er^{3+}-doped Bi_2WO_6 was much higher than the undoped one. In addition, Er^{3+}-doped Bi_2WO_6 exhibited photocatalytic activity under a 3 W LED with the wavelength beyond the absorption edge of Bi_2WO_6, further testifying the UC effect of Er^{3+} on the enhanced photocatalytic activity of Bi_2WO_6. The mechanism here was similar with the

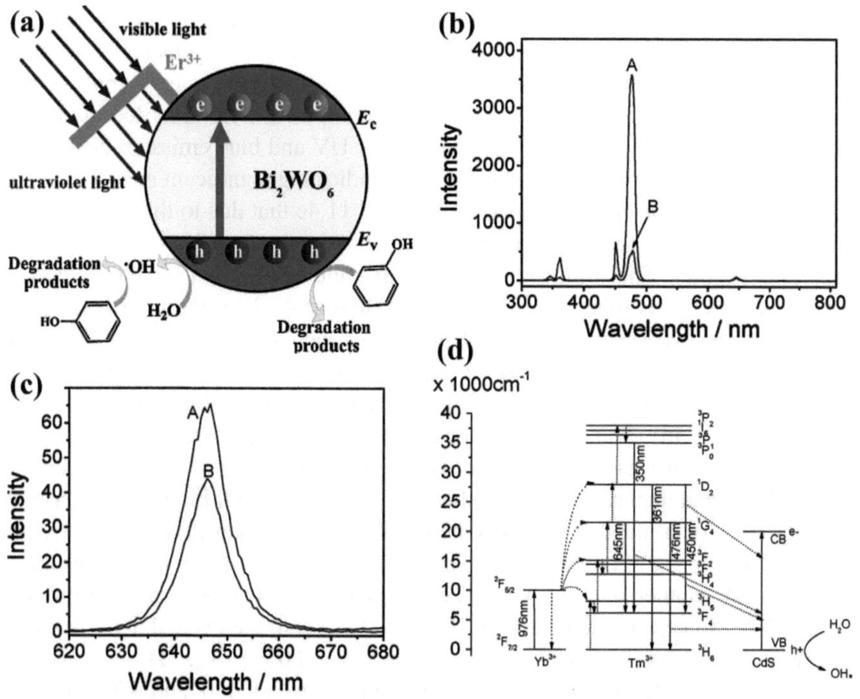

Fig. 11.4 **a** Photocatalysis mechanism of phenol by Er^{3+}-doped Bi_2WO_6 under visible light beyond the absorption edge of Bi_2WO_6. **b**, **c** Upconversion emission spectra of (*A*) $NaYF_4$:Yb,Tm and (*B*) $NaYF_4$:Yb,Tm/CdS. **c** is the magnified spectrum of (**b**). **d** Schematic illustration of energy transfer mechanism from $NaYF_4$:Yb,Tm to CdS upon 976-nm NIR irradiation. **a** Reprinted with the permission from Ref. [53]. Copyright 2010 Elsevier. **b–d** Reprinted with the permission from Ref. [60]. Copyright 2010 Elsevier

Er^{3+}-doped TiO_2 photocatalyst. Both the pure Bi_2WO_6 and Er^{3+}-doped Bi_2WO_6 samples showed photoabsorption from the UV light region to visible light shorter than 450 nm, whereas the wavelength of the LED used in this experiment was 465 nm, which was beyond the absorption edge of the photocatalyst. Even so, the Er^{3+}-doped Bi_2WO_6 sample exhibited excellent photocatalytic performance in the degradation of phenol. The photocatalysis mechanism of Er^{3+}-doped Bi_2WO_6 under visible light beyond the absorption edge of Bi_2WO_6 can be explained as follows: As shown in Fig. 11.4a, in the degradation process, Er^{3+} could be excited by visible light of 465 nm and transform the visible light into UV light, which can then activate Bi_2WO_6 to generate electron–hole pairs. The strongly oxidative holes excited by UV light then directly oxidized the organic pollutants or reacted with water molecules to form ·OH radicals, which was also a highly oxidative species and the main radical in the degradation of phenol, and they can degrade the contaminant indirectly.

In 2010, Yu et al. [60] reported the preparation of a novel NIR photocatalyst by combining a low-energy band gap semiconductor CdS with an UC material $NaYF_4$:

Yb,Tm. Particles of CdS and $NaYF_4$:Yb,Tm were pretreated with thioglycolic acid and mercaptoethanol. The reaction between these two organic molecules caused the formation of $NaYF_4$:Yb,Tm/CdS composite. To examine the energy transfer process, the UC emission spectra of the UC/CdS composite were recorded upon 976-nm NIR excitation (Fig. 11.4b). The intensity of UV and blue emission peaks were greatly reduced after the combination of CdS, indicating significant energy transfer. However, it could be clearly observed from Fig. 11.4c that due to the small overlap with the absorption spectrum of CdS, the red emission peak of $NaYF_4$:Yb,Tm only decreases slightly. The luminescence intensity ratio of I_{blue}/I_{red} for the composite was 11.9, which was much smaller than the value of 41.3 for the pure $NaYF_4$:Yb, Tm. Since the two spectra were recorded under identical conditions and the Tm content was the same, the decrease in I_{blue}/I_{red} strongly suggests an energy transfer from $NaYF_4$:Yb,Tm to CdS. Therefore, the mechanism can be explained as following (Fig. 11.4d): The absorption of pump photons populated the $^2F_{5/2}$ level in Yb^{3+}. A Tm^{3+} ion was excited to the $^3H_5$ level by the energy transferred from the excited Yb^{3+} and then relaxed non-radiatively to the $^3F_4$ level. Energy transfer from another Yb^{3+} ion to the Tm^{3+} caused the formation of a $^3F_2$ level. Subsequently, the Tm^{3+} relaxed to the $^3H_4$ level and then was excited to the $^1G_4$ level by absorbing energy from another excited Yb^{3+}. The sequential energy absorption from two excited Yb^{3+} ions promoted Tm^{3+} to $^1D_2$ and $^3P_2$. Then, the excited Tm^{3+} ions fell to lower energy levels. $^1D_2 \rightarrow {}^3F_4$, $^1G_4 \rightarrow {}^3H_6$, $^3P0 \rightarrow {}^3F_4$, $^1D_2 \rightarrow {}^3H_6$, and $^1G_4 \rightarrow {}^3F_4$ transitions produced the blue emissions at 450 and 470 nm, UV emissions at 350 and 361 nm, and red emission at 645 nm, respectively. Here, the $^1G_4$ level was the predominant excited state in this material as suggested by the strongest blue emission. For CdS, the energy gap from VB to CB is about 2.5 eV which was lower than the emitted blue and UV photon energy. It is known that semiconductors can be excited by the photons with an energy equivalent to or higher than the band gap energy, which causes the formation of photoelectrons and holes.

11.3.4 Conclusion and Future Opportunity for Upconversion-Assisted Photocatalysis

The relatively young field of UC-modified photocatalysts presents exciting possibilities in achieving enhancements of photocatalysis efficiency. For continued advancement of the field, however, it is necessary to recognize the unique challenges that exist in using UCNPs for the purposes described in this part.

Photocatalysis has a similar root-like solar cell, so the drawbacks of UC-assisted photocatalysis are the same as those in the solar cell, such as the poor efficiency, low absorption coefficient, and limited absorption wavelength of the UC process. Relative details can be found in Sect. 11.2.4. Besides the above-mentioned contents, there are some specific drawbacks for the UC-assisted photocatalysis:

1. As was discussed in the "obstacles and recent solutions for photocatalysis" portion, the photo potential energy excited by visible light is lower than that excited by UV light, thus generating holes with lower oxidation potential and weaker oxidative ability, which cannot degrade the organic compound completely or produce H_2 or O_2 efficiently. In order to get holes possessing oxidation potential high enough to destroy the organic pollutants or the H–O bond completely, it is necessary to excite the photocatalyst with UV light. However, UV light occupies only about 4 % of the sunlight spectrum, whereas visible light composes 43 % of the solar spectrum. To deal with this problem, researchers have developed visible–UV UC materials to enhance the photocatalysis efficiency; however, visible excited UC process is not as efficient as the NIR excited one.
2. The energy level of lanthanide ions is not well matched with the semiconductor, and the crystal lattice constant of UC materials is also not well matched with the semiconductor. As one can see in the above-mentioned papers, there are two main syntheses to combine UC materials and semiconductors (or some organic dye): firstly, doping lanthanide ions into the semiconductor crystal; as was mentioned before, the doping concentration will not be very high due to the lattice mismatch between the UC materials and the semiconductor crystal. Thus, the low doping concentration will result in low absorption of light and low efficiency. Secondly, the UC materials are combined with the semiconductor crystal using core/shell structure; this idea is good because the core/shell structure can make sure the largest contact surface area between UC materials and the semiconductor crystal. However, this core/shell nanocomposite is suffering phase splitting due to the lattice mismatch. Moreover, it is not stable after several turns of usage because this structure is vulnerable and easy to be broken.

Aiming at above drawbacks, we propose some items as an outlook to the UC-assisted photocatalysis:

1. The optimization of optical properties of UC materials such as efficiency, absorption wavelength tunability, and absorption cross section is an urgent task as described in solar cell part.
2. It is necessary to find more efficient visible–UV UC process.
3. In order to combine the UC materials with semiconductor with a smaller crystal lattice mismatch, optimization of UC host material or the core/shell structure is also very important. Recently, scientists have used SiO_2 as a media layer between UC core and noble metal layer for better performance in local field enhancement (LFE) effect. The idea of a media layer is thinkable in photocatalytic nanocomposite design. In order to combine the UC materials with semiconductor, the media layer should be optical transparent, thickness tunable, and easy to be modified.

In conclusion, recent research efforts on applying UC materials to photocatalysis have resulted in notable advances. At this stage, successful proof-of-concept have been reported on all fronts. These efforts can proceed even as new fundamental advancements in UC materials continue to arise, in addition to more advanced photocatalysts. The capacity to combine all these tools, as needed, represents a significant achievement in the ability to manipulate and convert light energy for sustainability purposes. Further application to other environmental technologies is also a real possibility with potential to benefit the developing world, wherein sunlight remains the most prevalent source of energy.

11.4 UCNPs for Security Application

As one of the photoluminescent materials, the unique optical properties of UCNPs provide opportunity in security application [2]. The majority of photoluminescent materials currently employed in security applications are based on downconversion luminescence, for which short-wavelength excitation produces long-wavelength emission. Many people, for example, have the shared experience of observing an airport security officer use a UV penlight to authenticate our driver's license. The authenticity of the license can be identified by confirming the correct patterns on it. Yet, counterfeiters are increasingly successful in mimicking security features based on downconversion photoluminescent materials. In response, UCNPs, for which long-wavelength excitation produces short-wavelength emission, have been suggested as the basis for the next generation of luminescent inks for security applications [61, 62].

There are several intrinsic advantages to the use of UCNPs as opposed to standard downconversion photoluminescent materials. UCNPs and appropriate NIR excitation sources (or "readers") are more difficult to duplicate than standard fluorescent dyes and UV sources. UC patterns can be written on highly fluorescent surfaces (e.g., papers and textiles with optical brighteners), since the NIR reader will not excite standard downconversion luminescence. Also, UCNPs can be formulated to produce the correct color rendering only under specific excitation power densities, such that a predetermined read condition becomes a security feature itself.

In 2009, Prasad et al. [63] reported a method for fabricating predefined photopatterns of UCNPs using a chemical amplification reaction for direct writing of films with multilayer color-coded patterning for security applications. To photopattern the nanocrystals film, they synthesized rare earth ion (Er^{3+}/Yb^{3+} or Tm^{3+}/Yb^{3+}) codoped sodium yttrium fluoride (α-$NaYF_4$) nanophosphors and functionalized the nanocrystals surfaces by incorporation of a photopatternable ligand such as *t*-butoxycarbonyl (*t*-BOC). The surface modification allowed photopatterning of the nanophosphor solid-state film. Furthermore, UC nanophosphors showed a nearly quadratic dependence of the UC PL intensity on the excitation light power, and tailoring the PL wavelength was possible by changing the lanthanide ions. They had demonstrated the capability of anchoring nanophosphors at desirable locations by a photolithography technique. The photopatterned films exhibited fixed nanophosphor structures clearly identifiable by strong UC PL under NIR illumination which was useful for a number of applications in security. Actually, the patterned UCNPs played a role as an ink in this work; after being patterned on the film, the UC signals were only detected upon 980-nm irradiation. Afterward, other researchers proposed the idea of UC ink formally.

In 2012, Luu et al. [64] developed a composite ink that was compatible with two separate printing systems (direct-write printing and aerosol jet printing), for producing highly resolved printed features which are invisible under ambient light, but produce luminescent images visible by the naked eye upon 980-nm excitation. The synthesis, printing, and imaging processes are as following:

1. Synthesis of β-$NaYF_4$:3%Er, 17%Yb UCNPs using thermal decomposition method (details can be seen in Chap. 2).
2. Ink dispersion: A 2 wt% dispersion of the UCNPs sample in toluene was prepared by stirring vigorously for 2 h, followed by sonication for 10 min. Poly (methyl methacrylate) beads (PMMA) were added at 1 wt% to the UCNPs dispersion along with enough methyl benzoate to bring the final composition of the solution to 90:10 v/v toluene/methyl benzoate. The resulting mixture was then stirred for 3 h to achieve complete dissolution of the polymer.
3. Printing process: Direct-write printing was performed with a Sono-Tek Exacta Coat three-axis robot with an Impact EDGE printhead system. Aerosol jet printing was performed with an Optomec-maskless mesoscale material deposition system (M^3D).
   Direct-write printing with Sono-Tek UC inks: The ink was deposited over a stainless steel stencil onto substrates of various characteristics. The laser-cut stainless steel screen contained shapes and letters ranging from 1 to 5 mm in size. The substrates tested were Kapton^R (high bond paper) and glass slides. The deposited material adhered well to the paper substrates, due to the absorption of the ink by the paper fibers. Deposition onto Kapton^R and glass slides was successful using the same printing parameters. Deposited features on the glass slides had slightly sharper edges compared to those on Kapton^R, because of the glass's higher surface energy. The higher surface energy allowed the ink to spread easily on the surface.
   M^3D uses the same printing processes as with Sono-Tek but operates under different mechanics. Compared to Sono-Tek, M^3D uses a programmable platen on which the substrate is placed. The platen travels in two dimensions to obtain the desired print pattern, while the printhead remains stationary. Also different from Sono-Tek, the ink is atomized prior to reaching the system's printhead. The ink can be atomized either through an ultrasonic bath or via pneumatic atomization. The atomized ink then travels through a polymer tube toward the printhead. The ink is deposited through a selected nozzle orifice ranging from 150 to 250 μm in diameter. Ink is focused to a circular shape using a sheath gas and has the capability of printing linewidths ranging from 25 to 250 μm. M^3D directly prints line-based elements; areas of features are filled in with either a spiral or serpentine array of lines. This allows M^3D printing to produce various features, shapes, and passes using Auto-CAD software. Materials within the viscosity range of 0.7–30 cP can be printed on M^3D using the ultrasonic

atomizer. Higher viscosities require the pneumatic atomizing system. Line-widths and film thickness are controlled by changing the flow parameters, the speed of the platen, and the tip size of the deposition head. The printing parameters for producing optimal lines were developed with PMMA solutions without nanocrystals.

As with Sono-Tek deposition, M^3D was used to deposit the composite UC ink on high bond paper, Kapton^R, and glass substrates. Using the ultrasonic atomizer at room temperature, sheath gas infusion set at 125 cm^3, the atomizer set at 40 cm^3, and the atomizer power set at 40 V, 45-μm-wide lines were successfully achieved with a 150-μm tip. These parameters produced lines with no overspray. The deposited lines produced films with an average thickness of 1.5 μm per layer. Line thickness can be decreased by increasing the speed of the platen, although this leads to the negative side effect of increasing distortion at the corners of the print pattern. When deposited with M^3D, the UC films are not as uniform in thickness as compared to Sono-Tek deposition, but the nanoparticles were dispersed homogeneously throughout the film.

4. UC images of printed features: UC luminescence images of printed features for "naked-eye" viewing were obtained using excitation from a 980-nm diode laser with a fiber coupling. Each substrate (paper, glass, Kapton^R) was evaluated with a range of excitation power densities (1–5 W/cm^2). Photographic images were obtained with using a 1–3 W/cm^2 range of excitation power density. The ink was deposited onto substrates of various surface energies through both direct-write and screen printing.

A crucial criterion for security printing is its invisibility under standard lighting conditions. The security features deposited from both printing systems were invisible under standard lighting conditions. This paper sample demonstrated the invisible film deposited by M^3D. Features became faintly visible only when relatively large amounts of material were deposited, either through multiple-layer deposition or increased infusion rates. The visibility of the features was a result of the faint yellow color of the UCNPs sample. This faint yellow color could be reduced or removed using more stringent washing procedures.

It is determined that, of the two printing methods, M^3D printing produces more sharply defined features and leads to less spreading of material. This increased definition is a result of the printing mechanics of the M^3D system. The key factor is the ability of M^3D to atomize the ink prior to deposition in combination with using a 150 μm tip. In comparison, Sono-Tek uses an atomizing nozzle with an orifice diameter of 760 μm. The printing is limited by the laser-cut manufacturing of the screen and the ability of the spray pattern to wet the substrate through the openings. UC images of printed features were recognizable down to 1 mm in size. Also, when using screen printing, there is a considerable waste of ink, because the majority of the ink is deposited on the masked part of the screen. With M^3D, however, essentially all of the deposited material contributes to the desired print feature. The printing time required using the Sono-Tek system, however, is much shorter than that for M^3D. To print identical arrays of shapes and letters as presented here, screen printing with the Sono-Tek system deposited a single layer ~9 times faster than M^3D [64].

Kellar et al. [65] also presented a work on security printing using UCNP ink. The details were similar with Luu's paper, and his highlight is that his UCNP ink contains both Yb^{3+}/Er^{3+} and Yb^{3+}/Tm^{3+} doped-$NaYF_4$ nanocrystals. That means his UCNP ink can achieve higher levels of security by printing both green and blue upconverting inks, based on combinations of Yb^{3+}/Er^{3+} and Yb^{3+}/Tm^{3+}, respectively, in a single QR code. This idea inspired other researchers to design and synthesize novel UC-based anti-counterfeit materials, because people were realizing that the multi-color property of UCNPs is a huge advantage among the reported anti-counterfeit materials. Later in 2014, the same group demonstrated a RGB (red, green, and blue) additive color printing system that produced highly resolved predefined patterns that were invisible under ambient lighting, but viewable as luminescent multi-color images under NIR excitation. Patterns were generated by independent deposition of three primary color (red, green, and blue) upconverting inks using an aerosol jet printer. The primary color inks were printed as isolated and overlapping features to produce images that simultaneously emit red, green, blue, cyan, magenta, yellow, and white UC luminescence. The dependence of the chromaticity of certain secondary colors (cyan and magenta) and white on NIR excitation power density could be exploited as an additional authentication feature. The UCNPs used in this work were β-$NaYF_4$:10%Er^{3+},2%Tm^{3+} (red), β-$NaYF_4$:25%Yb^{3+},0.3%Tm^{3+} (blue), and β-$NaYF_4$:17%Yb^{3+},3%Er^{3+} (green); the synthesis method was all the same except for the lanthanide dopants used. Green, blue, and red inks were made by adding 0.5 wt% green, 5 wt% blue, and 6 wt% red UCNCs, respectively, and 1 wt% poly(-methyl methacrylate) (PMMA) to a solvent of 90:10 v/v toluene–methyl benzoate. The mixture was vigorously stirred for approximately 24 h. The resulting solution was further mixed with a vortex mixer and then sonicated for 10–20 min to achieve dispersion of the nanocrystals. Different amount of UCNPs was used due to the difference of absorption and quantum yield among the three NCs. The primary color inks were printed as isolated and overlapping features to produce images that simultaneously emit red, green, blue, cyan, magenta, yellow, and white UC luminescence. The dependence of the chromaticity of certain secondary colors (cyan and magenta) and white on NIR excitation power density could be exploited as an additional authentication feature. The development of an RGB UC printing system paved the way for an entirely new arena in security printing [62].

During the past few years, nanoxerography has emerged as an innovative technique for directed assembly of colloids on surfaces. It relies on charge patterns written on electrets to trap colloidal particles from their suspensions, through electrostatic interactions. Sequential charge writing on electret thin films may be performed by different means such as an electron beam, focused ion beam, or polarized atomic force microscopy (AFM) tip, and parallel charge writing can be done by electrical nanoimprint lithography or electrical microcontact printing. Among the sequential charge writing techniques, AFM is especially advantageous, offering the ability to write both positive and negative charges in a single run with an enormous flexibility over the written pattern geometry. In 2013, Ressier et al. [66] presented the formation of 3D close-packed assemblies of upconverting $NaYF_4$ colloidal nanocrystals on surfaces by AFM nanoxerography. The surface potential of the charge patterns, the nanocrystals concentration, the polarizability of the nanocrystals, and the polarity of the dispersing solvent were identified as the key parameters controlling the assembly of $NaYF_4$ nanocrystals into micropatterns of the desired 3D architecture. This insight allowed the authors to fabricate micrometer-sized quick response codes encoded in terms of UC luminescence intensity or color. Topographically hidden messages could also be readily incorporated within these microtags. This work demonstrated that AFM nanoxerography has enormous potential for generating high-security anti-counterfeiting microtags.

The use of traditional UCNPs as anti-counterfeiting microtags utilizes the advantages of UCNPs, such as NIR irradiation, invisibility in ambient light, high stability, a lack in background, and simplicity of synthesis. Yet, the convenient synthesis methods also make the counterfeiters successful in mimicking security feature-based UCNP inks. To further improve the security level of UCNP inks, researchers pay more attention on the modification of the structure of UCNPs. Through synthesizing the UCNPs with unique nanostructures, novel optical properties can be obtained; this progress on novel UC inks extremely enhances the security level of UCNP-based anti-counterfeiting microtags.

In 2014, Liu et al. [67] reported the synthesis of novel luminescent crystals based on hexagonal-phase $NaYF_4$ UC microrods. The synthetic procedure involved an epitaxial end-on growth of UCNPs comprising different lanthanide activators onto the $NaYF_4$ microrods. This bottom-up method readily afforded multi-color-banded crystals in gram quantity by varying the composition of the activators. Importantly, the end-on growth method using one-dimensional microrods as the template enabled facile multi-color tuning in a single crystal, which is inaccessible in conventional UCNPs.

The lanthanide-doped $NaYF_4$ microrods were prepared by a well-established hydrothermal method. Emissions from individual rods doped with different sets of lanthanides could be identified due to the strong UC luminescence of the hexagonal-phase microrods (Fig. 11.5a–c). Scanning electron microscopic imaging revealed the high uniformity of the as-synthesized microrods ($\sim$1.9 μm in length and $\sim$140 nm in diameter) (Fig. 11.5d). It was important to note that the length of the rods could be tuned from 0.3 to 2 μm using the gadolinium doping approach. Furthermore, they found that upon addition of α-$NaYF_4$ nanocrystals as the precursors, successive end-on growth of the β-$NaYF_4$ microrods could be achieved, as evidenced by the optical microscopic images (Fig. 11.5e). This could be attributed to favorable epitaxial growth along the long (001) axis of the crystal. TEM imaging showed two dark-colored tips of the segmented rods, attributable to the high contrast of heavily Yb^{3+}-doped $NaYF_4$ (Fig. 11.5f). High-resolution TEM reveals the single-crystalline nature of the rod with a d-spacing of 0.30 nm, corresponding to the (110) plane of β-$NaYF_4$.

To confirm the tip growth onto the parent microrod, they intentionally doped Yb^{3+} with large concentration disparities into the β-$NaYF_4$:Yb/Er microrod (5 mol% of

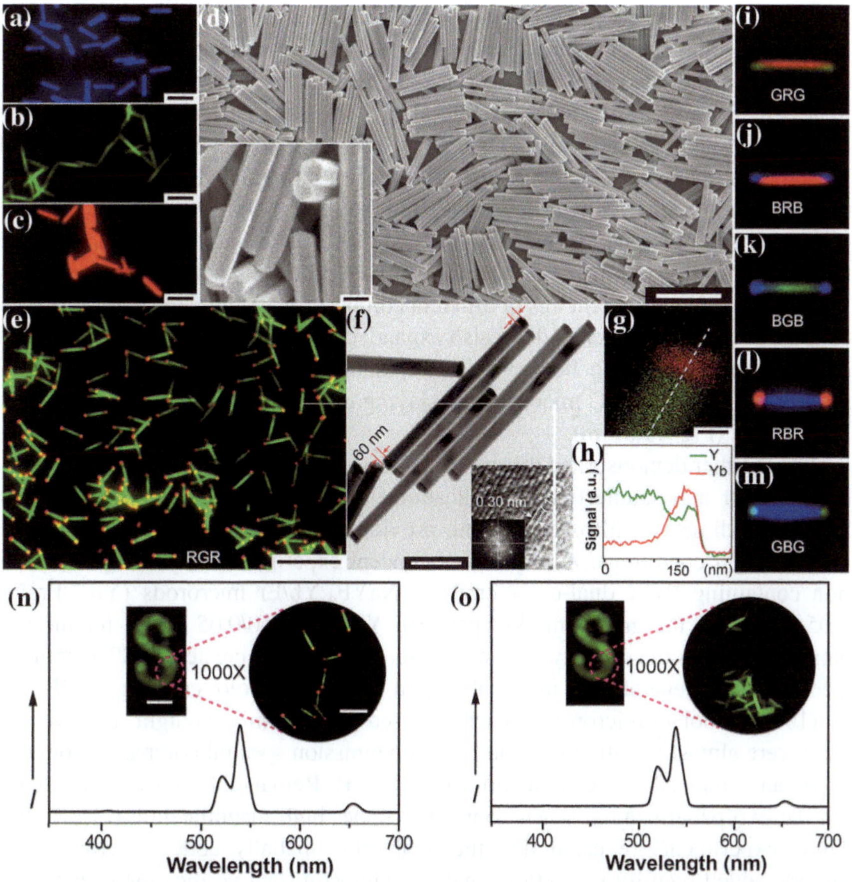

Fig. 11.5 **a–c** Optical micrographs of the parent $NaYF_4$ upconversion microrods doped with Yb^{3+}/Tm^{3+} (20/0.2 mol%), Yb^{3+}/Er^{3+} (5/0.05 mol%), and Yb^{3+}/Er^{3+} (50/0.05 mol%), respectively. Scale bars in **c–e** are 2 μm. **d** A typical SEM image of the $NaYF_4$ microrods. Scale bar is 2 μm. Inset shows the hexagonal prism morphology of the rods. Scale bar is 200 nm. **e** Optical micrographs of dual-color (RGR)-emitting $NaYF_4$ microrods after the end-on epitaxial growth with $NaYF_4$:Yb/Er precursors. Scale bar is 5 μm. Note that the red-emitting tips are heavily doped with 50 mol% of Yb^{3+}, while the green-emitting segment is doped with 5 mol% of Yb^{3+}. **f** TEM image of the tip-modified microrods (inset: HRTEM image and corresponding Fourier transform diffractogram of the segment joint). Scale bar is 500 nm. **g, h** Elemental mapping and corresponding EELS line scan conducted on a single tip-modified microrod. Scale bar in **g** is 50 nm. **i–m** Optical micrographs showing five additional sets of dual-color-banded upconversion microrods obtained by varying the composition of the dopants. Note that the appearance of a tinge color at the tip junction is due to the chromatic aberration and limited resolution of the microscope. **n** Luminescence image of a stamped letter S generated with dual-color microrods (shown in 1,000 × magnified image) as the ink. The corresponding upconversion emission spectrum is also shown. Scale bars are 3 μm (*right*) and 3 mm (*left*). **o** Luminescence image of a control sample generated with single-color microrods as the ink. Note that the emission spectra (or bulk color appearance) of these two patterns are almost identical. However, their differences in spatial distribution are clearly distinguishable under high magnification. **a–o** Reprinted with the permission from Ref. [67]. Copyright 2014 American Chemical Society

Yb^{3+}) and the α-$NaYF_4$:Yb/Er nanocrystals precursor (50 mol% of Yb^{3+}), respectively. Elemental mapping by STEM was performed on the resulting hybrid rods. The difference in the elemental distribution of Yb^{3+} and Y^{3+} over the rod tip clearly indicated the presence of a junction (Fig. 11.5g). The line scan and spot scan analyses revealed the Yb abundance (40 %) in the new tip, which was in good agreement with the Yb content in α-$NaYF_4$ precursors (Fig. 11.5h). Importantly, the use of α-$NaYF_4$ nanocrystals precursors rather than lanthanide ionic solutions was critical for high-yield synthesis of the segmented microrods, as the direct use of ionic precursors leads to phase separation dominant in the crystal growth process. It is worth noting that the introduction of luminescent tips of different composition not only enhanced the color contrast of the parent microrod but also expanded the scope of multi-color UC at the single crystal level. Using the step-by-step epitaxial growth method, they synthesized a series of dual-color UC microrods displaying different combinations of three primary colors (Fig. 11.5i–m).

They further demonstrated the use of the as-synthesized multi-color microrods as barcodes for anti-counterfeiting application. Note that these microrods were dispersible in dimethyl sulfoxide solvent, providing a transparent ink solution under ambient light conditions. As a proof-of-concept experiment, they stamped a solution containing RGR dual-color emitting $NaYF_4$:Yb/Er microrods (Yb^{3+}/Er^{3+}:5/0.05 mol% for the green-emitting part and Yb^{3+}/Er^{3+}:50/0.05 mol% for the red-emitting ends) as the security ink onto a paper substrate to create an "S" pattern. As a control, they generated an identical pattern using green color emitting $NaYF_4$:Yb/Er (10/0.05 mol%) microrods. When exposed to 980-nm laser light, the two patterns were almost indistinguishable based on emission spectral comparison or color appearance under low magnification (Fig. 11.5n). Remarkably, they could clearly tell the two patterns apart when zoomed in under high magnification (Fig. 11.5o). These experiments revealed that the dual-color spatially coded microrods can provide added security protection, enabling almost instant verification of patterned features.

In conclusion, UCNPs have been well exploited as anti-counterfeiting microtags. Compared with traditional anti-counterfeiting, UCNP-based ink benefits from multi-color properties, UC processes, invisibility in ambient light, and high stability. However, it is necessary to recognize the unique challenges that exist in using UCNPs as anti-counterfeiting microtags. Here, we provide several suggestions for the future of UCNP-based anti-counterfeiting microtags.

1. It is very difficult to obtain pure red/blue/green light by UCNPs due to its specific PL mechanism. Moreover, the ratio of different light depends on the doping ratio of lanthanide ions which can be hardly precise in one reactor. That means this ratio is not stable even the whole process is operated by one person. This error can be ignored in some bio-application, solar cell, or photocatalysis, but it is intolerable in security application which is very strict to the chromatic aberration. Besides the challenge in chromatic aberration, the ratio of different light is sensitive to the irradiation power density, which can also result in the chromatic aberration. To deal with these problems, measurement after synthesis

is very important; the final products should be identified by the fluorescence spectrum, rather than the doping ratio in the raw materials. What's more, a calibration curve can be established for these UC spectrums, so the UC inks can be more convenient.

2. Another advantage of UCNPs is the diversity of its morphology, such as nanoparticle, nanoplate, nanorod, nanowire, nanobelt, and nanoellipsoid. This property can also be exploited in anti-counterfeiting microtags. However, regular light microscopes generally have the best spatial resolution of ~200 nm constrained by the optical diffraction limit. To resolve two emission features (or spots), the two emitting objects need to be separated by a distance larger than the diffraction limit. Otherwise, the microscope would not have the necessary resolution power to distinguish them. Up to now, there is no report about morphology-based anti-counterfeiting because the size must be at least 200 nm. A few types of morphology can also be used in anti-counterfeiting, such as the nanorod and the nanowire because they can be easily synthesized to a size larger than ~1 μm, which is large enough to be seen under optical microscope.

11.5 UCNPs for Optical Storage, Remote-Control Photoswitching and Other Applications

Because there are some restrictions for UC bulk materials in the applications of lighting devices, researchers have obtained composites incorporating fluorides NCs in some matrices such as transparent glass ceramics, thin films, and polymers in order to develop fluorides as bulk materials. Wang et al. [68] obtained a transparent glass ceramic material embedding a series of UCNPs, such as $NaYF_4:Er^{3+}$ (tunable red-green emission), $YF_3:Yb^{3+},Er^{3+},Tm^{3+}$ (white light emission) [69], $LiYbF_4:Tb^{3+}$ (UV UC emission of Tb^{3+} and blue emission of Yb^{3+}) [70], $TbF_3:Yb^{3+}$ (cooperative energy transfer UC and quantum cutting downconversion) [71], and $REF_3:Ln^{3+}$ [72, 73]. These nanocomposites have not only comparative low phonon energies ascribed to the precipitated fluorides but also desirable transparency, high chemical, and mechanical stability related to the silica-based glass matrix. The key to the achievement of efficient luminescence in these glass ceramics lies in the partition of the optically active RE ions into the fluoride nanocrystals. Thin-film materials containing UCNPs are often produced by sol-gel techniques. Van Veggel et al. [74] first reported a simple method to produce UC white light from sol-gel-derived SiO_2 thin film made with three combinations of $La_{0.45}Yb_{0.5}Er_{0.05}F_3$, $La_{0.75}Yb_{0.2}Tm_{0.05}F_3$, and $Yb_{0.75}La_{0.2}Eu_{0.05}F_3$ nanocrystals. Then, they extended this method to obtain highly luminescent sol-gel-derived silica and alumina films with LaF_3 nanocrystals doped with lanthanide ions (Er^{3+}, Nd^{3+}, and Ho^{3+}) [75]. Again, they reported the generation of white light from sol-gel-derived SiO_2 and ZrO_2 thin film made with new combinations of $LaF_3:Ln^{3+}$ nanocrystals [76]. This method combined the advantage of the improved luminescent properties of Ln^{3+}-doped fluoride nanocrystals, and the simplicity of making thin films using the sol-gel method. More recently, Yan et al. [77] fabricated a novel composite film constructed from both β-$NaYF_4$:Yb^{3+},Er^{3+} ordered UC inorganic nanopatterns and photochromic diarylethene organic species by facile "bottom-up" assembly methods. This composite film is capable of rewritable optical storage with a nondestructive readout. The incorporation of UCNPs in polymer matrices to obtain transparent polymer composites is an available choice to achieve bulk composite materials. Among the broad variety of available polymers, PMMA has been one of the most widely studied in the last decades due to its outstanding and promising mechanical and chemophysical properties. Recently, several groups reported the synthesis of UCNPs–PMMA bulk transparent nanocomposites via in situ polymerization route without the need of post-deposition heat treatment. Lin et al. [78] and van Veggel et al. [79] applied thermal polymerization to prepare oleic acid-stabilized β-$NaYF_4$:Yb^{3+},Er^{3+}/PMMA monoliths, respectively. Lin et al. [80] also prepared β-$NaYF_4$:Yb^{3+},Er^{3+}(Tm^{3+})–PMMA bulk nanocomposites via an in situ radical photopolymerization method. Photopolymerization has advantages over thermal polymerization, including rapid polymerization rate, low polymerization temperature, low energy consumption, and easy process control. The ability to synthesize transparent UC polymer nanocomposites can facilitate the manufacture of a three-dimensional volumetric luminescence display [81].

References

1. Mader, H.S., Kele, P., Saleh, S.M., Wolfbeis, O.S.: Upconverting luminescent nanoparticles for use in bioconjugation and bioimaging. Curr. Opin. Chem. Biol. **14**, 582–596 (2010)
2. Haase, M., Schafer, H.: Upconverting nanoparticles. Angew. Chem. Int. Ed. **50**, 5808–5829 (2011)
3. Wang, F., Liu, X.G.: Recent advances in the chemistry of lanthanide-doped upconversion nanocrystals. Chem. Soc. Rev. **38**, 976–989 (2009)
4. Halme, J., Vahermaa, P., Miettunen, K., Lund, P.: Device physics of dye solar cells. Adv. Mater. **22**, E210–E234 (2010)
5. Hagfeldt, A., Boschloo, G., Sun, L.C., Kloo, L., Pettersson, H.: Dye-sensitized solar cells. Chem. Rev. **110**, 6595–6663 (2010)
6. Hoppe, H., Sariciftci, N.S.: Organic solar cells: an overview. J. Mater. Res. **19**, 1924–1945 (2004)
7. Kamat, P.V.: Quantum dot solar cells. Semiconductor nanocrystals as light harvesters. J. Phys. Chem. Cells **112**, 18737–18753 (2008)
8. Goetzberger, A., Luther, J., Willeke, G.: Solar cells: past, present, future. Sol. Energ. Mater. Sol. Cells **74**, 1–11 (2002)
9. Chapin, D.M., Fuller, C.S., Pearson, G.L.: A new silicon p-n junction photocell for converting solar radiation into electrical power. J. Appl. Phys. **25**, 676–677 (1954)
10. Goetzberger, A., Hebling, C., Schock, H.W.: Photovoltaic materials, history, status and outlook. Mater. Sci. Eng. R **40**, 1–46 (2003)
11. Wang, H.Q., Batentschuk, M., Osvet, A., Pinna, L., Brabec, C.J.: Rare-earth ion doped up-conversion materials for photovoltaic applications. Adv. Mater. **23**, 2675–2680 (2011)
12. Li, X.M., Zhang, F., Zhao, D.Y.: Highly efficient lanthanide upconverting nanomaterials: progresses and challenges. Nano Today **8**, 643–676 (2013)

13. Richards, B.S.: Luminescent layers for enhanced silicon solar cell performance: down-conversion. Sol. Energy Mater. Sol. Cells **90**, 1189–1207 (2006)
14. Trupke, T., Green, M.A., Wurfel, P.: Improving solar cell efficiencies by down-conversion of high-energy photons. J. Appl. Phys. **92**, 1668–1674 (2002)
15. Klampaftis, E., Ross, D., McIntosh, K.R., Richards, B.S.: Enhancing the performance of solar cells via luminescent down-shifting of the incident spectrum: a review. Sol. Energy Mater. Sol. Cells **93**, 1182–1194 (2009)
16. Wegh, R.T., Donker, H., Oskam, K.D., Meijerink, A.: Visible quantum cutting in $LiGdF_4$: Eu^{3+} through downconversion. Science **283**, 663–666 (1999)
17. Struempel, C., McCann, M., Beaucarne, G., Arkhipov, V., Slaoui, A., Svrcek, V., del Canizo, C., Tobias, I.: Modifying the solar spectrum to enhance silicon solar cell efficiency–an overview of available materials. Sol. Energy Mater. Sol. Cells **91**, 238–249 (2007)
18. Shalav, A., Richards, B.S., Green, M.A.: Luminescent layers for enhanced silicon solar cell performance: up-conversion. Sol. Energy Mater. Sol. Cells **91**, 829–842 (2007)
19. Imenes, A.G., Mills, D.R.: Spectral beam splitting technology for increased conversion efficiency in solar concentrating systems: a review. Sol. Energy Mater. Sol. Cells **84**, 19–69 (2004)
20. Wolf, M.: Limitations and possibilities for improvement of photovoltaic solar energy converters .1. considerations for earths surface operation. Proceedings of the Institute of Radio Engineers **48**, 1246–1263 (1960)
21. Barnham, K.W.J., Duggan, G.: A new approach to high-efficiency multi-band-gap solar-cells. J. Appl. Phys. **67**, 3490–3493 (1990)
22. Keevers, M.J., Green, M.A.: Efficiency improvements of silicon solar-cells by the impurity photovoltaic effect. J. Appl. Phys. **75**, 4022–4031 (1994)
23. Shan, G.B., Demopoulos, G.P.: Near-infrared sunlight harvesting in dye-sensitized solar cells via the insertion of an upconverter-TiO_2 nanocomposite layer. Adv. Mater. **22**, 4373–4377 (2010)
24. Su, L.T., Karuturi, S.K., Luo, J.S., Liu, L.J., Liu, X.F., Guo, J., Sum, T.C., Deng, R.R., Fan, H.J., Liu, X.G., Tok, A.I.Y.: Photon upconversion in hetero-nanostructured photoanodes for enhanced near-infrared light harvesting. Adv. Mater. **25**, 1603–1607 (2013)
25. Gibart, P., Auzel, F., Guillaume, J.C., Zahraman, K.: Below band-gap IR response of substrate-free GaAs solar cells using two-photon up-conversion. Jpn. J Appl. Phys. Part 1-Regul. Pap. Short Notes Rev. Pap. **35**, 4401–4402 (1996)
26. Richards, B.S., Shalav, A.: Enhancing the near-infrared spectral response of silicon optoelectronic devices via up-conversion. IEEE. T. Electron Dev. **54**, 2679–2684 (2007)
27. Shalav, A., Richards, B.S., Trupke, T., Kramer, K.W., Gudel, H.U.: Application of $NaYF_4$: Er^{3+} up-converting phosphors for enhanced near-infrared silicon solar cell response. Appl. Phys. Lett. **86**, 013505 (2005)
28. Chen, D.Q., Lei, L., Yang, A.P., Wang, Z.X., Wang, Y.S.: Ultra-broadband near-infrared excitable upconversion core/shell nanocrystals. Chem. Commun. **48**, 5898–5900 (2012)
29. Li, Z.Q., Li, X.D., Liu, Q.Q., Chen, X.H., Sun, Z., Liu, C., Ye, X.J., Huang, S.M.: Core/shell structured $NaYF_4$:Yb^{3+}/Er^{3+}/Gd^{+3} nanorods with Au nanoparticles or shells for flexible amorphous silicon solar cells. Nanotechnology **23**, 025402 (2012)
30. Trupke, T., Shalav, A., Richards, B.S., Wurfel, P., Green, M.A.: Efficiency enhancement of solar cells by luminescent up-conversion of sunlight. Sol. Energy Mater. Sol. Cells **90**, 3327–3338 (2006)
31. de Wild, J., Rath, J.K., Meijerink, A., van Sark, W., Schropp, R.E.I.: Enhanced near-infrared response of a-Si:H solar cells with beta-$NaYF_4$:Yb^{3+}(18 %), Er^{3+}(2 %) upconversion phosphors. Sol. Energy Mater. Sol. Cells **94**, 2395–2398 (2010)
32. de Wild, J., Meijerink, A., Rath, J.K., van Sark, W., Schropp, R.E.I.: Towards upconversion for amorphous silicon solar cells. Sol. Energy Mater. Sol. Cells **94**, 1919–1922 (2010)
33. Li, Y., Wang, G.F., Pan, K., Jiang, B.J., Tian, C.G., Zhou, W., Fu, H.G.: NaYF4:Er^{3+}/Yb^{3+}-graphene composites: preparation, upconversion luminescence, and application in dye-sensitized solar cells. J. Mater. Chem. **22**, 20381–20386 (2012)

34. Fujishima, A., Honda, K.: Electrochemical photolysis of water at a semiconductor electrode. Nature **238**, 37–38 (1972)
35. Augustynski, J.: The role of the surface intermediates in the photoelectrochemical behavior of anatase and rutile TiO_2. Electrochim. Acta **38**, 43–46 (1993)
36. Fang, W.Q., Gong, X.-Q., Yang, H.G.: On the unusual properties of anatase TiO_2 exposed by highly reactive facets. J. Phys. Chem. L. **2**, 725–734 (2011)
37. Bickley, R.I., Gonzalezcarreno, T., Lees, J.S., Palmisano, L., Tilley, R.J.D.: A structural investigation of titanium-dioxide photocatalysts. J. Solid State Chem. **92**, 178–190 (1991)
38. Reeves, P., Ohlhausen, R., Sloan, D., Pamplin, K., Scoggins, T., Clark, C., Hutchinson, B., Green, D.: Photocatalytic destruction of organic-dyes in aqueous TiO_2 suspensions using concentrated simulated and natural solar-energy. Sol. Energy **48**, 413–420 (1992)
39. Bacsa, R.R., Kiwi, J.: Effect of rutile phase on the photocatalytic properties of nanocrystalline titania during the degradation of p-coumaric acid. Appl. Catal. B-environ **16**, 19–29 (1998)
40. Yamazaki, S., Matsunaga, S., Hori, K.: Photocatalytic degradation of trichloroethylene in water using TiO(2) pellets. Water Res. **35**, 1022–1028 (2001)
41. Chen, H.H., Nanayakkara, C.E., Grassian, V.H.: Titanium dioxide photocatalysis in atmospheric chemistry. Chem. Rev. **112**, 5919–5948 (2012)
42. Linsebigler, A.L., Lu, G.Q., Yates, J.T.: Photocatalysis on TiO_2 surfaces—principles, mechanisms and selected results. Chem. Rev. **95**, 735–758 (1995)
43. Shiraiwa, M., Sosedova, Y., Rouviere, A., Yang, H., Zhang, Y.Y., Abbatt, J.P.D., Ammann, M., Poschl, U.: The role of long-lived reactive oxygen intermediates in the reaction of ozone with aerosol particles. Nat. Chem. **3**, 291–295 (2011)
44. Fujishima, A., Zhang, X.T., Tryk, D.A.: TiO_2 photocatalysis and related surface phenomena. Surf. Sci. Rep. **63**, 515–582 (2008)
45. Hoffmann, M.R., Martin, S.T., Choi, W.Y., Bahnemann, D.W.: Environmental applications of semiconductor photocatalysis. Chem. Rev. **95**, 69–96 (1995)
46. Chen, X., Mao, S.S.: Titanium dioxide nanomaterials: synthesis, properties, modifications, and applications. Chem. Rev. **107**, 2891–2959 (2007)
47. Mills, A., LeHunte, S.: An overview of semiconductor photocatalysis. J. Photochem. Photobiol., A **108**, 1–35 (1997)
48. Li, W., Wang, Y., Lin, H., Shah, S.I., Huang, C.P., Doren, D.J., Rykov, S.A., Chen, J.G., Barteau, M.A.: Band gap tailoring of Nd^{3+}-doped TiO2 nanoparticles. Appl. Phys. Lett. **83**, 4143–4145 (2003)
49. Morikawa, T., Asahi, R., Ohwaki, T., Aoki, K., Taga, Y.: Band-gap narrowing of titanium dioxide by nitrogen doping. Jpn. J Appl. Phys. Part 2-Lett. **40**, L561–L563 (2001)
50. Nakano, Y., Morikawa, T., Ohwaki, T., Taga, Y.: Deep-level optical spectroscopy investigation of N-doped TiO_2 films. Appl. Phys. Lett. **86**, 132104 (2005)
51. Vogel, R., Hoyer, P., Weller, H.: Quantum-sized PBS, CDS, AG_2S, SB_2S3, and BI_2S_3 particles as sensitizers for various nanoporous wide-bandgap semiconductors. J. Phys. Chem. **98**, 3183–3188 (1994)
52. Fitzmaurice, D., Frei, H., Rabani, J.: Time-resolved optical study on the charge-carrier dynamics in a TiO_2/AGI sandwich colloid. J. Phys. Chem. **99**, 9176–9181 (1995)
53. Zhang, Z.J., Wang, W.Z., Yin, W.Z., Shang, M., Wang, L., Sun, S.M.: Inducing photocatalysis by visible light beyond the absorption edge: effect of upconversion agent on the photocatalytic activity of Bi_2WO_6. Appl. Catal. B-environ **101**, 68–73 (2010)
54. Qin, W.P., Zhang, D.S., Zhao, D., Wang, L.L., Zheng, K.Z.: Near-infrared photocatalysis based on YF3:Yb^{3+}, Tm^{3+}/TiO_2 core/shell nanoparticles. Chem. Commun. **46**, 2304–2306 (2010)
55. Shi, J.W., Ye, J.H., Ma, L.J., Ouyang, S.X., Jing, D.W., Guo, L.J.: Site-selected doping of upconversion luminescent Er^{3+} into $SrTiO_3$ for visible-light-driven photocatalytic H–2 or O–2 evolution. Chem. Eur. J. **18**, 7543–7551 (2012)
56. Ren, L., Qi, X., Liu, Y.D., Huang, Z.Y., Wei, X.L., Li, J., Yang, L.W., Zhong, J.X.: Upconversion-P25-graphene composite as an advanced sunlight driven photocatalytic hybrid material. J. Mater. Chem. **22**, 11765–11771 (2012)

57. Obregon, S., Colon, G.: Evidence of upconversion luminescence contribution to the improved photoactivity of erbium doped TiO_2 systems. Chem. Commun. **48**, 7865–7867 (2012)
58. Li, Z.X., Shi, F.B., Zhang, T., Wu, H.S., Sun, L.D., Yan, C.H.: Ytterbium stabilized ordered mesoporous titania for near-infrared photocatalysis. Chem. Commun. **47**, 8109–8111 (2011)
59. Feng, G.F., Liu, S.W., Xiu, Z.L., Zhang, Y., Yu, J.X., Chen, Y.G., Wang, P., Yu, X.J.: Visible light photocatalytic activities of TiO_2 nanocrystals doped with upconversion luminescence agent. J. Phys. Chem. C **112**, 13692–13699 (2008)
60. Li, C.H., Wang, F., Zhu, J.A., Yu, J.C.: $NaYF_4$ Yb, Tm/CdS composite as a novel near-infrared-driven photocatalyst. Appl. Catal. B-environ **100**, 433–439 (2010)
61. Suyver, J.F., Aebischer, A., Biner, D., Gerner, P., Grimm, J., Heer, S., Kramer, K.W., Reinhard, C., Gudel, H.U.: Novel materials doped with trivalent lanthanides and transition metal ions showing near-infrared to visible photon upconversion. Opt. Mater. **27**, 1111–1130 (2005)
62. Meruga, J.M., Baride, A., Cross, W., Kellar, J.J., May, P.S.: Red-green-blue printing using luminescence-upconversion inks. J Mater. Chem. C **2**, 2221–2227 (2014)
63. Kim, W.J., Nyk, M., Prasad, P.N.: Color-coded multilayer photopatterned microstructures using lanthanide (III) ion co-doped $NaYF_4$ nanoparticles with upconversion luminescence for possible applications in security. Nanotechnology **20**, 185301 (2009)
64. Blumenthal, T., Meruga, J., May, P.S., Kellar, J., Cross, W., Ankireddy, K., Vunnam, S., Luu, Q.N.: Patterned direct-write and screen-printing of NIR-to-visible upconverting inks for security applications. Nanotechnology **23**, 185305 (2012)
65. Meruga, J.M., Cross, W.M., May, P.S., Luu, Q., Crawford, G.A., Kellar, J.J.: Security printing of covert quick response codes using upconverting nanoparticle inks. Nanotechnology **23**, 395201 (2012)
66. Sangeetha, N.M., Moutet, P., Lagarde, D., Sallen, G., Urbaszek, B., Marie, X., Viau, G., Ressier, L.: 3D assembly of upconverting $NaYF_4$ nanocrystals by AFM nanoxerography: creation of anti-counterfeiting microtags. Nanoscale **5**, 9587–9592 (2013)
67. Zhang, Y.H., Zhang, L.X., Deng, R.R., Tian, J., Zong, Y., Jin, D.Y., Liu, X.G.: Multicolor Barcoding in a single upconversion crystal. J. Am. Chem. Soc. **136**, 4893–4896 (2014)
68. Liu, F., Ma, E., Chen, D.Q., Yu, Y.L., Wang, Y.S.: Tunable red-green upconversion luminescence in novel transparent glass ceramics containing Er: $NaYF_4$ nanocrystals. J. Phys. Chem. B **110**, 20843–20846 (2006)
69. Chen, D.Q., Wang, Y.S., Zheng, K.L., Guo, T.L., Yu, Y.L., Huang, P.: Bright upconversion white light emission in transparent glass ceramic embedding Tm(3+)/Er(3+)/Yb(3+):beta-YF (3) nanocrystals. Appl. Phys. Lett. **91**, 251903 (2007)
70. Chen, D.Q., Yu, Y.L., Huang, P., Weng, F.Y., Lin, H., Wang, Y.S.: Optical spectroscopy of Eu^{3+} and Tb^{3+} doped glass ceramics containing $LiYbF_4$ nanocrystals. Appl. Phys. Lett. **94**, 041909 (2009)
71. Chen, D.Q., Yu, Y.L., Wang, Y.S., Huang, P., Weng, F.Y.: Cooperative energy transfer up-conversion and quantum cutting down-conversion in Yb^{3+}:TbF_3 nanocrystals embedded glass ceramics. J. Phys. Chem. C **113**, 6406–6410 (2009)
72. Chen, D.Q., Yu, Y.L., Huang, P., Wang, Y.S.: Nanocrystallization of lanthanide trifluoride in an aluminosilicate glass matrix: dimorphism and rare earth partition. CrystEngComm **11**, 1686–1690 (2009)
73. Chen, D.Q., Yu, Y.L., Huang, P., Lin, H., Shan, Z.F., Wang, Y.S.: Color-tunable luminescence of Eu^{3+} in LaF_3 embedded nanocomposite for light emitting diode. Acta Mater. **58**, 3035–3041 (2010)
74. Sivakumar, R., van Veggel, F., Raudsepp, M.: Bright white light through up-conversion of a single NIR source from sol-gel-derived thin film made with Ln(3+)-doped LaF_3 nanoparticles. J. Am. Chem. Soc. **127**, 12464–12465 (2005)
75. Sudarsan, V., Sivakumar, S., van Veggel, F., Raudsepp, M.: General and convenient method for making highly luminescent sol-gel derived silica and alumina films by using LaF3 nanoparticles doped with lanthanide ions (Er^{3+}, Nd^{3+}, and Ho^{3+}). Chem. Mater. **17**, 4736–4742 (2005)

76. Sivakumar, S., Boyer, J.C., Bovero, E., van Veggel, F.: Up-conversion of 980 nm light into white light from sol-gel derived thin film made with new combinations of LaF3:Ln(3+) nanoparticles. J. Mater. Chem. **19**, 2392–2399 (2009)
77. Zhang, C., Zhou, H.P., Liao, L.Y., Feng, W., Sun, W., Li, Z.X., Xu, C.H., Fang, C.J., Sun, L.D., Zhang, Y.W., Yan, C.H.: Luminescence modulation of ordered upconversion nanopatterns by a photochromic diarylethene: rewritable optical storage with nondestructive readout. Adv. Mater. **22**, 633–637 (2010)
78. Lin, C.K., Berry, M.T., Anderson, R., Smith, S., May, P.S.: Highly luminescent NIR-to-visible upconversion thin films and monoliths requiring no high-temperature treatment. Chem. Mater. **21**, 3406–3413 (2009)
79. Boyer, J.C., Johnson, N.J.J., van Veggel, F.: Upconverting lanthanide-doped $NaYF_4$-PMMA polymer composites prepared by in situ polymerization. Chem. Mater. **21**, 2010–2012 (2009)
80. Chai, R.T., Lian, H.Z., Hou, Z.Y., Zhang, C.M., Peng, C., Lin, J.: Preparation and characterization of upconversion luminescent $NaYF_4$:Yb^{3+}, Er^{3+}(Tm^{3+})/PMMA bulk transparent nanocomposites through in situ photopolymerization. J. Phys. Chem. C **114**, 610–616 (2010)
81. Downing, E., Hesselink, L., Ralston, J., Macfarlane, R.: A three-color, solid-state, three-dimensional display. Science **273**, 1185–1189 (1996)

Chapter 12
Conclusions and Perspectives

Lanthanide ion-doped upconversion luminescent nanomaterials are emerging as one of the most important members in the luminescence materials family. It has been well developed and pushed forward with unprecedented speed toward diverse applications in recent years, especially biological applications. The upconversion luminescence-based bio-probes inherit the unique optical properties of lanthanide ions, such as long-lived luminescence, large Stokes or anti-Stokes shifts, narrow emission bands, high resistance to photo-bleaching, and low toxicity. Most importantly, compared with the conventional biological labels, upconversion nanoparticles exhibit virtually zero auto-fluorescence background, less excitation light scattering, and deep tissue penetration because of excitation being in the NIR region that is within the optical transparency window. By taking advantage of the distinct lanthanide-based upconversion luminescence, a series of efficient analytical methods, such as heterogeneous time-resolved upconversion luminescence assays, homogeneous FRET-based upconversion luminescence assays, bio-imaging, PDT, targeted drug delivery, and optical–magnetic multimodal bio-sensing techniques, have been established based on these novel upconversion nanoparticles in the past decade.

In this book, we summarized the development of the lanthanide-doped upconversion nanoparticles, including the upconversion mechanism; excitation and emission energy levels guided design of nanostructure; the techniques to fabricate the nanostructure with optimum energy level structure; the fabrication of upconversion nanoparticle-based nanocomposites to extend the working range of upconversion nanoparticles by introducing the additional functional component; surface modification and bio-conjugation of upconversion nanoparticles; upconversion nanoparticles for biomedical imaging; and the bio-applications (bioassay, bio-sensing, thermal sensing, light-activated therapy, etc.) of the upconversion nanoparticles.

Xiaomin Li and Fan Zhang contributed together to this chapter.

F. Zhang, *Photon Upconversion Nanomaterials*,
Nanostructure Science and Technology, DOI 10.1007/978-3-662-45597-5_12

Despite the upconversion nanoparticles have attracted a great deal of attention in a wide range of applications due to their unique optical and chemical properties, practical applications of upconversion nanoparticles set demanding requirements on their optical properties, size, surface chemistry, as well as their ability to incorporate with "alien species." New advancements in diverse aspects are needed to make rapid progress in this field to fulfill requirements of the practical applications. So, we would like to highlight some future directions that can produce pronounced effects in the field.

The synthesis of high-quality upconversion nanoparticles or upconversion nanocomposites is the base of the applications. However, the brightness of obtained upconversion nanoparticles is still unsatisfactory, in particular, under low excitation density of 10^{-2}–10^{-1} W/cm^2. Manipulation of local environment around the lanthanide dopants or the spatial distribution of the dopants by delicately designing and optimizing the structure of the upconversion nanoparticles may be the promising direction to enhance the quantum yield. Throughout the development of upconversion nanoparticles, in order to obtain abundant high-quality optical properties (such as high efficiency, emission/excitation tunability, multi-functionality, and so on), the structures of the nanoparticles are becoming more and more advanced and ingenious. Thus, the structures, especially the core@multi-shell structures, are becoming more and more complicated, indicating that the traditional synthesis methods are hard to fulfill the practical needs. So the second anticipation is the development of the new synthesis approach which can fulfill the requirements for the fabrication of complicated multi-shell structured upconversion nanoparticles. Developing new or modified synthetic strategies to prepare sub-10 nm upconversion nanoparticles is also of crucial importance in the future. Furthermore, such an ultrasmall size of upconversion nanoparticles is always associated with a significant decrease of the luminescent efficiency due to the extremely high "surface-to-volume" ratio. Thus, nanochemistry dealing with ultrasmall-size-related quenching mechanisms also needs our attention. Last but not least, currently, most of the equipments used to character the optical properties of upconversion nanoparticles are based on the homemade instruments. Standard measurement facilities and protocols are highly required to allow the quantitative comparison of optical properties.

For the fabrication of the upconversion nanoparticle-based nanocomposites, concerted efforts are still required for success: (1) The reported strategies to synthesis the upconversion nanoparticle-based nanocomposites are often accompanied by a reduction of luminescent efficiency due to unwanted quenching between different domains of the nanocomposites. So, in the future research, attentions must be paid to the synthesis of high-efficiency upconversion luminescence multi-functional nanocomposites. (2) Despite the fact that some successful results were obtained, the development of multi-functional nanoprobes for multimodal imaging is still in its early stages. It is still a major challenge to design a multi-functional nanoprobe for simultaneous bio-imaging covering optical imaging, MRI, CT, and other kinds of monitoring. In addition, active targeted multimodal imaging and drug delivery system still need to be developed to realize the disease detection and therapy simultaneously.

Upconversion nanoparticles have been perceived as technological marvels with characteristics that could greatly improve biological small-animal imaging and therapy. In the near future, there are still a number of research areas that are particularly promising, but will require concerted effort for success. For instance, some key issues such as biocompatibility, toxicity, and the long-term stability of nanoprobes need to be tackled for in vivo application. Although previous studies have shown that functionalized upconversion nanoparticles with biocompatible coatings (e.g., PEG) appear to be safe to cells at certain concentration ranges and not obviously toxic to the treated animals over months, the effects that this class of nanomaterials would have on animals over an even longer time period, how they interact with immune systems, and whether they may interfere with the reproductive system and affect the next generation remain essentially unknown. Much more systematic investigations are still demanded before those nanomaterials can finally be used in clinics. Under the guidance of upconversion luminescence bio-imaging, the combined use of light-activated therapy (e.g., PDT and PTT) produced in upconversion nanoparticles is a promising direction to enable more effective therapeutic effect in vitro and in vivo. Moreover, the integration of light-activated therapy with light-regulated drug delivery in upconversion nanoparticles platform can become a new direction for more efficient imaging-guided therapy.

Index

F. Zhang, *Photon Upconversion Nanomaterials*,
Nanostructure Science and Technology, DOI 10.1007/978-3-662-45597-5

U

X

Z

MIX
Papier aus verantwortungsvollen Quellen
Paper from responsible sources
FSC® C105338

If you have any concerns about our products,
you can contact us on
ProductSafety@springernature.com

In case Publisher is established outside the EU,
the EU authorized representative is:
Springer Nature Customer Service Center GmbH
Europaplatz 3, 69115 Heidelberg, Germany

Printed by Libri Plureos GmbH
in Hamburg, Germany